Sustainable Fibres and Textiles

The Textile Institute Book Series

Incorporated by Royal Charter in 1925, The Textile Institute was established as the professional body for the textile industry to provide support to businesses, practitioners and academics involved with textiles and to provide routes to professional qualifications through which Institute Members can demonstrate their professional competence. The Institute's aim is to encourage learning, recognise achievement, reward excellence and disseminate information about the textiles, clothing and footwear industries and the associated science, design and technology; it has a global reach with individual and corporate members in over 80 countries.

The Textile Institute Book Series supersedes the former 'Woodhead Publishing Series in Textiles', and represents a collaboration between The Textile Institute and Elsevier aimed at ensuring that Institute Members and the textile industry continue to have access to high calibre titles on textile science and technology.

Books published in The Textile Institute Book Series are offered on the Elsevier web site at: www.elsevier.com/books-and-journals and are available to Textile Institute Members at a substantial discount. Textile Institute books still in print are also available directly from the Institute's web site at: www.textileinstitute.org

To place an order, or if you are interested in writing a book for this series, please contact Matthew Deans, Senior Publisher: m.deans@elsevier.com

Recently Published and Upcoming Titles in The Textile Institute Book Series:

Handbook of Technical Textiles, Volume 1, 2nd Edition, A. Richard Horrocks and Subhash C. Anand, 9781782424581
Handbook of Technical Textiles, Volume 2, 2nd Edition, A. Richard Horrocks and Subhash C. Anand, 9781782424659
Geotextiles, Robert Koerner, 9780081002216
Advances in Braiding Technology, Yordan Kyosev, 9780081009260
Antimicrobial Textiles, Gang Sun, 9780081005767
Active Coatings for Smart Textiles, Jinlian Hu, 9780081002636
Advances in Women's Intimate Apparel Technology, Winnie Yu, 9781782423690
Smart Textiles and Their Applications, Vladan Koncar, 9780081005743
Advances in Technical Nonwovens, George Kellie, 9780081005750
Activated Carbon Fiber and Textiles, Jonathan Chen, 9780081006603
Performance Testing of Textiles, Lijing Wang, 9780081005705
Colour Design, Janet Best, 9780081012703
Forensic Textile Science, Debra Carr, 9780081018729
Principles of Textile Finishing, Asim Kumar Roy Choudhury, 9780081006467
High-Performance Apparel, John McLoughlin and Tasneem Sabir, 9780081009048

The Textile Institute Book Series

Sustainable Fibres and Textiles

Edited by

Subramanian Senthilkannan Muthu

The Textile Institute

ELSEVIER

WP
WOODHEAD PUBLISHING
An imprint of Elsevier

Woodhead Publishing is an imprint of Elsevier
The Officers' Mess Business Centre, Royston Road, Duxford, CB22 4QH, United Kingdom
50 Hampshire Street, 5th Floor, Cambridge, MA 02139, United States
The Boulevard, Langford Lane, Kidlington, OX5 1GB, United Kingdom

Library of Congress Cataloging-in-Publication Data
A catalog record for this book is available from the Library of Congress

British Library Cataloguing-in-Publication Data
A catalogue record for this book is available from the British Library

ISBN: 978-0-08-102041-8 (print)
ISBN: 978-0-08-102042-5 (online)

For information on all Woodhead Publishing publications
visit our website at https://www.elsevier.com/books-and-journals

Publisher: Matthew Deans
Acquisition Editor: David Jackson
Editorial Project Manager: Andrae Akeh
Production Project Manager: Debasish Ghosh
Cover Designer: Victoria Pearson

Typeset by TNQ Books and Journals

This book is dedicated to:

The lotus feet of my beloved Lord Pazhaniandavar.

My beloved late father.

My beloved mother.

My beloved wife Karpagam and daughters Anu and Karthika.

My beloved brother.

Last but not least,
To everyone working in the global textile supply chain
who make it sustainable.

Contents

List of contributors

K. Amutha Bharathiar University, Coimbatore, India

Asim K. Roy Choudhury KPS Institute of Polytechnic, Belmuri, West Bengal, India; Ex-professor, Government College of Engineering and Textile Technology, Serampore, West Bengal, India

Sanjoy Debnath ICAR–National Institute of Research on Jute and Allied Fibre Technology, Kolkata, India

Seiko Jose ICAR–Central Sheep and Wool Research Institute, Avikanagar, India

T. Karthik PSG College of Technology, Coimbatore, India

Ajay Kumar ICAR–Central Sheep and Wool Research Institute, Avikanagar, India

Shanmugasundaram O. Lakshmanan K.S. Rangasamy College of Technology, Tiruchengode, India

Faqeer Mohammad Jamia Millia Islamia (A Central University), New Delhi, India

Lesile L. Prince ICAR–Central Sheep and Wool Research Institute, Avikanagar, India

Shanthi Radhakrishnan Kumaraguru College of Technology, Coimbatore, India

Guruprasad Raghavendran ICAR-Central Institute for Research on Cotton Technology, Mumbai, India

R. Rathinamoorthy PSG College of Technology, Coimbatore, India

Anbalagan Saravanan SSN College of Engineering, Chennai, India

Ponnusamy Senthil Kumar SSN College of Engineering, Chennai, India

Mohd Shabbir Jamia Millia Islamia (A Central University), New Delhi, India

Shahid-ul-Islam Jamia Millia Islamia (A Central University), New Delhi, India

Subburaj Suganya SSN College of Engineering, Chennai, India

Preface

The word 'sustainability' and its importance has been well-defined and is very popular now. I would say most industries practise sustainability in their daily business operations, and it has become a routine part of today's business in any industrial sector. *Sustainability in Textiles* is not a new subject these days. The textile industry is certainly up to speed in terms of making the entire supply chain sustainable. There are a lot of sustainable practices for producing conventional and common fibres and textile materials and there are a lot of innovative fibres and production methods and technologies available to make textile products sustainable. Apart from the point of view of industrial production, there are many industrial initiatives and collaborative efforts currently in the industry to make the textile industry a sustainable one. Addressing the entire spectrum of activities and efforts in one platform to bring out as a complete guide is the key objective of this book.

This book is structured in two parts with a total of 13 highly informative chapters. Part 1 is dedicated to sections dealing with fibres. Eight chapters are devoted to the sustainable production of various textile fibres. Topics covered in this first part of book are: Introduction to Sustainable Fibres and Textiles, Sustainable Cotton Production, Sustainable Production of Bast Fibres, Sustainable Wool Production, Ecological Dyeing of Wool With Biomordants, Sustainable Silk Production, Sustainable Production of Regenerated Cellulosic Fibres and Sustainable Production of Synthetic Fibres. Part 2 deals with sustainable textile and apparel production. Five chapters examine this subject. These five chapters discuss advanced topics pertaining to sustainable textile production: Low Water Consumption Technologies for Textile Production, Sustainable Chemical Technologies for Textile Production, Sustainable Wastewater Treatments in the Textile Sector, Sustainable Chemical Management and Zero Discharge and Sustainable Production and Application of Natural Fibre-Based Nonwovens.

I have no doubt that for all who seek to enhance their knowledge on this subject of sustainable fibres and textiles, this book will inevitably be in their library.

I take this wonderful opportunity to thank all of the contributors of this book for their earnest efforts and hard work to bring out the book. I am very sure that the readers of this book will benefit.

Dr Subramanian Senthilkannan Muthu

Introduction to sustainable fibres and textiles

1

Ponnusamy Senthil Kumar, Subburaj Suganya
SSN College of Engineering, Chennai, India

1.1 Introduction

In the 15th century, European countries traded commodities and high-quality goods by sea, a method of transport which required no taxes for the goods. The British started the East India Company, which was followed by Dutch trading 2 years later. They bought Indian goods, especially textiles. Spices were the primarily favoured in exchange for gold and silver (World Acrylic Fiber, 2013). The immense response from Europe to Indian textiles made them fashionable, trendy and traditional. Calico, pyjamas, gingham, dungarees, chintz, and khaki attained the peak of popularity in Eastern and European countries. Many textile products are named based on where they are manufactured. For instance Kashmir shawls and Kanjipuram silk exhibit the ancient art of weaving. Second only to agriculture, textile is the largest industry in India; 17% of India's gross domestic product is in textiles, and whole exports are as one of the largest employers (CAPMAS, 2014). However, the textile sector is divided into two broad categories: organized and disorganized. The presence of modern machineries and updated technology makes it a grand success in large-scale production, which is called organized. Small-scale producers using conventional methods such as handlooms, handicrafts and sericulture with traditional tools are called disorganized.

Although the garment industry has become the backbone of many nations, there are no proper guidelines maintained by manufacturers. It poses a human threat, such as the tragedy of the Ranipet Tannery disaster that occurred in Tamil Nadu, India, in which 10 workers drowned in a toxic pollutant tank. The tank in which the people drowned contained chromium and other hazardous materials to mummify animal skins. In addition, the garment building collapse in Savar became a critical issue in Bangladesh as well as the rest of the world as a result of the death of thousands of labourers. From this incident, the media and activists raised questions about the safety and sustainability of the textile industry and its employees.

Many garment industries ignore minimum safety guidelines regarding the raw materials, machinery and technologies adopted for rapid production, and have no uniform labour policy. A sufficient investment does not include the proper route of waste discharge or recycling unit. As a result, hazardous chemicals are released in the open environment, which directly or indirectly affects ecological species as well as humans. The slow release and high toxicity of the chemicals make natural resources such as soil infertile. The result is reflected in the next generation, which is born with illness, and in the depletion of natural sources to consume and with which to

Sustainable Fibres and Textiles. http://dx.doi.org/10.1016/B978-0-08-102041-8.00001-9

survive. Manufacturers should obey government policies and avoid illegal steps in waste management.

The textile processing industry consumes large amounts of water along with chemicals for various purposes. It is recommended to reduce the generation of waste and its life-threatening impacts on the surroundings by modifying improper production procedures, reducing packaging, altering raw materials, enhancing handling, considering energy efficiency, using less water consumption with a proper discharging unit, and not ignoring the use of toxic metals in industry (CAPMAS, 2014; Rock and Angel, 2007). Few garment industries make preventive plans to monitor pollution levels, optimizing energy by selecting suitable equipment and methodology. Textile wastewater is released in the form of solids, liquids and gas, which is a serious issue for researchers working with social environmental concerns.

Organic fibre may be a promising tool to resolve all of these important issues associated with the garment, apparel, tannery and textile industries, and which needs to be endorsed by the public. Plants and animals have been used globally to produce natural fibres for textiles since ancient and medieval times. Later, because of the lack of available natural resources, man-made or synthetic fibres were used. To overcome such a demand, experts suggest organic farming. The cultivation of cotton, silk, Tencel, ramie, organic linen, milk silk, corn fibre, bamboo fibres, alpaca, soy silk, pineapple and banana for fibres and hemp are advised for their highly drought-tolerant ability. In addition, synthetic fibres such as polyester, nylon, viscose, acrylic and polypropylene recycled polyester, jute, black diamond fibre, polylactic acid fibre, Lycra, Lyocell, organic silk and organic wool, etc., can be recycled through waste-to-wear technology (Rock and Angel, 2007).

1.2 Principles of sustainability

Meeting human needs without overwhelming nature or society is called sustainability. To improve this, organic textile must cover the cultivation of raw material, mass production, manufacturing, processing, packaging, labelling and distribution of organic textiles globally. The final product would be in the form of fibres, yarns, fabrics and garments including home textile products. The garment industry is sustainable with basic requirements framed by the Global Organic Textile Standard (Lo et al., 2012). It also investigates the quality of organic fibres according to high-level environmental and social criteria. It includes organic certification, labelling and licensing that ensure the organic status of textiles from harvesting raw materials through environmentally socially responsible manufacturing units to provide credible assurance to the consumer.

1.2.1 Raw materials

Raw materials of textiles can be derived from natural and synthetic fibres. Seed hairs such as cotton, stem (bast) fibres such as flax and hemp, leaf fibres such as sisal, and husk fibres coconut are derived from plant fibres. Wool, hair and secretions such as silk are derived from animal fibres. Abaca is a source of rope that is used instead of glass fibres in automobiles to save energy. Alpaca wool is a high-end luxury fabric with a yield of 5000 tons per year. Coir is a short fibre extracted from the outer shell of coconuts; it is used in ropes, mattresses, brushes, geotextiles and automobile seats.

Angora wool is derived from angora rabbits; it is silky white, fine and soft and is used in high-quality knitwear. Camel hair is found in Mongolia (Fresner, 1998). Bactrian camels are the source of this soft fibre. Baby camel hair is in high demand. Cotton is a world-famous natural fibre made of pure cellulose. It is still the undisputed king of the global garment industry. Flax is called one of the world's strongest fibres once it is harvested, spun and woven. Cashmere is famous for its soft touch and great insulation properties. Hemp opens the door for cottonization with high quality.

Silk is called the 'queen of fabrics'. Such a royal fibre has been used since ancient times. Mohair is a white, fine and silky fibre noted for its softness, brightness and receptiveness to rich dyes. Jute is a farmer-friendly fibre which provides a livelihood to millions of small farmers because of its strong threads used in sackcloth. Ramie is one of the strongest natural fibres, similar to flax. It is white with silky lustre absorbency and the density of silk lustre. Sisal is strongly believed to replace glass fibres that are too coarse for textiles used in composite materials, cars and furniture. Wool is the world's premier textile fibre which produces a limited supply but unique traits.

Over the past half century, synthetic or man-made fibres have begun to displace natural fibres in clothing, household furnishings, industries and agriculture. Man-made fibres are given names such as acrylic, nylon, polyester and polypropylene. The success of synthetic fibres is due to their low cost and inexpensive treatment methodology. Commonly used synthetic fibres are mass produced from petrochemicals to uniform strengths, lengths and colours which are easily customized to specific applications.

The emerging 'green' economy is based on energy efficiency, renewable feedstocks in polymer products and industrial processes for recycling materials that reduce carbon emissions. Carbon fibres are made from polyacrylonitrile, about 90% (http://en.wikipedia.org/wiki/AirDye) of which is obtained from precursors and the remaining percentage of which is used for rayon or petroleum pitch. These materials are specifications from long strings of molecules and bound together this exact variation of precursor from one to another. The process by which polyester is manufactured from the main ingredient of ethylene is called polymerization. Polyester is frequently used for paper; the general term is for a high molecule with an ester bond structure, but it refers to polyethylene terephthalate in the case of fibre. It is characterized by high strength, superior heat and water resistance. Spandex is a lightweight, synthetic fibre made of a long chain called polyurethane that is useful for stretchable clothing such as sportswear.

1.2.2 Factors affecting the textile industries

The fundamental strength of this industry flows from its strong production base of a wide range of fibres/yarns from natural fibres such as cotton, jute, silk and wool to synthetic/man-made fibres such as polyester, viscose, nylon and acrylic. With escalating demand for textile products, textile mills and their wastewater have been increasing proportionally, causing a major problem of pollution in the world. Many chemicals used in the textile industry cause environmental and health problems. Among the many chemicals in textile wastewater, dyes are considered important pollutants. Worldwide environmental problems associated with the textile industry are typically those associated with water pollution caused by the discharge of untreated effluent and those because of the use of toxic chemicals especially during processing (Fresner, 1998). The effluent is of critical environmental concern because it drastically decreases oxygen concentration as a result of the presence of hydrosulphides and blocks the passage of light through the water body, which is detrimental to the water ecosystem. Textile effluent is the cause of a significant amount of environmental degradation and human illnesses. About 40% of globally used colourants contain organically bound chlorine, a known carcinogen (Fresner, 1998). Chemicals evaporate into the air we breathe or are absorbed through our skin; they show up as allergic reactions and may cause harm to children even before birth. Owing to this chemical pollution, the normal functioning of cells is disturbed, which in turn may alter the physiology and biochemical mechanisms of animals, resulting in impairment of important functions such as respiration, osmoregulation, reproduction and even mortality. Heavy metals, which are present in textile industry effluent, are not biodegradable; hence they accumulate in primary organs in the body and over time begin to fester, leading to various symptoms of diseases. Thus untreated or incompletely treated textile effluent can be harmful to both aquatic and terrestrial life by adversely affecting the natural ecosystem and causing long-term health effects (Fresner, 1998; http://en.wikipedia.org/wiki/AirDye).

The common practice of low process efficiency results in substantially wasted resources and severe damage to the environment. The main environmental problem causes by textile industry is typically associated with water body pollution through the discharge of untreated effluents. Air emission, notably volatile organic compounds, excessive noise or odour as well as work space safety, are other equal important environmental issues influenced by the garment industry.

Various types of textile supply chains and customer interactions with the industry significantly affect company strategies regarding the design, product development, manufacture and marketing of textiles and textile products. Trends in textile markets have major implications for textile products and processes. Because of ever-increasing environmental awareness and the possible impact of environmental regulations on the whole textile supply chain, the industry is facing many challenges in the future. The textile manufacturing process involves the high consumption of resources such as water and fuel and a variety of chemicals in a long process that generates a significant amount of waste. Obsolescence in technology and machinery threatens the textile industries. In terms of poor infrastructure, labour skills, structural anomalies, high power consumption, poor productivity, economic norms and social obligations have been driving the garment industry to the point of poor productivity.

The textile and clothing trade in the global market is changing quickly with the scaling up of textiles in diverse areas. Asian countries including India have a dominant role in international trade in the global market. China has the major share in the textile and clothing trade in the international market. Both Bangladesh and Hong Kong have a significant share (Texmin). However, India is still on the back seat. It is reported that Asian counties export most textile and apparel to Europe, North America, the United States, etc. At the World Textile Summit, which was held for the first time in connection with the International Textile Machinery Association 2011 trade fair in Barcelona, the agenda was designed to offer a global perspective on the opportunities and challenges likely to face the textile industry in the years ahead. The importance of cooperation across the supply chain to drive sustainability and innovation, sustainable programmes incorporating the reduction of energy and water consumption, the growing influence of technical textiles, China's increasing challenges and growing opportunities for India were some of the key points of discussions at this summit (http://jetro.go.jp/j-messe/tradefair_en/ITMA2011_31426).

Fibre creators made suggestions to overcome these issues. They recommended that the transfer of knowledge from academic institutions to industry could be effective to raise productivity and reduce the environmental burden. Academic research establishments should be prioritized to make better use of knowledge and ideas generated from low class researchers, scientists and students to facilitate the right environment for interaction with textile and clothing companies. It may transform scientific facts in innovative experimentation which extend to applications on a large scale. The garment industry is experiencing challenges as a result of several structural changes made by the global textile industry. To avoid this, developing countries may become a partner of the World Trade Organization, to access its products easily all over the world. The Traditional Quota System (TQS) was introduced in 2004; it provides opportunities to countries that are restricted (World Trade Organization, 2001). The recession in 2008–09 had a negative effect on the global economy similar to that which occurred in 1930. Undergraduate and postgraduate textile education could be redesigned to enhance the nature of materials and production processes, to fulfil the future needs of trends in fashion and suit the needs of scientists, engineers and technicians in both industry and academia (http://jetro.go.jp/j-messe/tradefair_en/ITMA2011_31426; World Trade Organization, 2001).

1.2.3 Impact on environment

1. The pesticides used in farmland can harm wildlife as it grows, contaminating other products which get into food.
2. Chemicals such as sodium hypochlorite, sodium chlorite, sodium hydrosulphite, sulphuric acid and other optical brightening agents that are used to bleach and colour textiles can damage the environment and people's health and result in slow poisoning, health disorders and death.
3. Decomposing old clothes that are thrown away take up precious space in landfill sites and release methane, a harmful greenhouse gas which is a significant contributor to global warming.
4. Most textile machinery causes noise, sound and air pollution. In addition the use of dyes and chemicals in fabric and other components of clothing and shoes can leach into the soil, contaminating both surface and ground water.

5. Overuse of natural resources such as plants and water depletes or disturbs the ecological balance. Working conditions in the textile and clothing industry are substandard with no recycling units.
6. Exploitation of animals often goes hand in hand with intensive farming practices that damage the environment as a whole.

1.2.4 Trends in fibre production

Biomimicry is the use of natural designs to synthesize man-made products artificially. The stain-resistant properties and antimicrobial properties of crab and lobster shells are mimicked by researchers to help the textile industry discover a way to add strength to cloth. On the basis of performance and sustainable attributes, existing fibres are developed as solution-dyed nylon. Solution dying, a process of locking colour into fibre itself, is widely used in carpets, upholstery and apparel (Ozturk, 2005). It retains colour after repeated laundering and is free from deterioration.

Rapidly renewable, biodegradable, recyclable wool that can be produced organically is the route to making sustainable fibre production. Wool traceability standards and animal welfare standards must be followed for its production. Highly durable, with inherent flame-resistant properties, natural water repellency and oil-repellency as a perfluorinated finish, wool's performance attributes are laudable without added chemistry.

Cutting-edge research contributes high performance to customers and producers based on their notions of self-cleaning fabrics. However, products from the carpet, apparel and garment industries require regular care and maintenance, preferably with biodegradable detergents and nontoxic cleaners. These industries must remember traditional technologies such as naturally water- and flame-resistant wool, even as they pursue and purchase innovative new materials (Ozturk, 2005; http://www.fiber2fashion.com/industry-article/22/2142/airdye-technology coloringtextiles-without-the-use-of-water1.asp).

1.3 Sustainable fibre production

1.3.1 History of fibres

Since the dawn of civilization, textiles have been a fundamental part of human life. The history of fibre began with silk in the 27th century BC. Jute and coir were also cultivated since antiquity. The oldest wool carpet was found in Siberia, dating from 500 BC, and the oldest wool textile, found in Denmark, dated from 1500 BC (Texprocil). These reports prove that the existence of reliable sources of fabric have changed greatly. Since then, their function has been the main part of the textile industry.

Most natural fibres are used in clothing and containers and to insulate, soften and decorate our living spaces. Increasingly, however, traditional textiles are being used for industrial purposes as well as in components of composite materials, in medical implants, and in geo- and agro-textiles. A total of 35 million tons of natural fibres is harvested by farmers using a variety of plants and animals. This amount includes sheep, rabbits, goats, camels and alpacas, and fibres from cotton bolls, abaca and sisal leaves and coconut husks, and the stalks of jute, hemp, flax and ramie plants. Those fibres are made into yarn, followed by spinning, weaving, knitting, and then into fabric (US EPA, 1996).

Relentless competition between natural and synthetic fibres defines the current global economic impact. It hinders the livelihoods of millions of people who depend on natural fibre production and processing. The International Year of Natural Fibres aimed to raise global awareness about the importance of natural fibres not only for producers and the industry but also for consumers and about the environment. Natural fibres provide natural ventilation, which is why cotton cloth is mostly comfortable in summer and sweat suits used for weight reduction are 100% synthetic. Wool garments act as insulators against both cold and heat. Coconut fibres used in mattresses have natural resistance to fungus and mites. Hemp fibre has antibacterial properties, and studies show that linen is the most hygienic textile for hospital bed sheets (World Trade Organization, 2003).

1.3.2 Organic fibres

Health-conscious consumers have traditionally favoured natural fibres (primarily cotton) over synthetic clothing; cotton is one of the most environmentally friendly crops grown. Consumers now have their choice of an increasing selection of clothing, bedding and other products made from organic cotton, wool, linen, hemp and flax. The Organic Trade Association acknowledges that the growing sector of apparel is making producers and customers environmentally conscious about products made by organic fibre (Sami, 2013). The natural properties of wool make additives unnecessary; hence there is no harmful off-gassing. The slowly growing organic wool industry has steadily been gaining attention over synthetic fibres owing to its nonchemical bedding material requirements. Sustainable agriculture is appreciated by small, local, independent growers of organic wool.

The National Centre for Appropriate Technology's Appropriate Technology Transfer in Rural Areas program found that 30,000 pounds of organic wool was harvested from approximately 2300 sheep raised organically in the United States and Canada during the 2001 season. It encourages the growing of organic fibre and sheep production. Hemp is a burgeoning natural fibre which can be grown without herbicides or pesticides. It has been used for centuries in household linens and work clothes. This fibre is four to six times stronger than cotton. Landscaping workers are suggested to wear hemp fibre because of its natural hypoallergenic properties. Organic cotton in apparel is a quickly growing market for hygiene products, home furnishings and more; it occupies 67% of growth among other fibres (https://www.wto.org/english/tratop_e/tariffs_e/tariff_data_e.htm).

1.3.3 Processing of organic fibres

Processing of sustainable organic fibre starts with fibre extraction and yarn production followed by bleaching, dyeing, softening, printing and drying. Organic or green decorticated fibre production depends on skilled labourers, an integrated infrastructure and the sustainability of the fibre. It leads to the market price for each fibre: for example, Anjengo yarns may yield around Rs. 25/kg, whereas Vycome is quoted at Rs. 17.50/kg. Export prices for finished coir products such as hand-loomed mats and matting, rugs and carpets range between Rs. 70 and 80/kg (Sengupta and Singh, 2013).

The Indian state of Kerala encourages local producers to work for 750 cooperatives of the handloom industry. Yet this industry is under increasing economic and technical pressure because of inefficiencies within the system, competition from power loom cloth, a lack of quality and shrinking markets. Public sector support is being introduced to enhanced investment from the commercial sector with changes to mechanized coir extraction and the motorized spinning of yarn, in an effort to accommodate the social changes required. Greater productivity can be raised within the industry upon which the community depends. Alternatively, additional economic activities have to be introduced to augment or replace the main source of income. Hydrogen peroxide recipes have been developed to bleach natural fibres with substantial environmental benefits. A pretreatment step that uses considerably less chlorine-free bleaching powder is cost-effective and can be combined with cold (peroxide and chlorine) bleaching processes.

Dyeing is done conventionally and or is mechanized to improve uniformity and consistency of shades. The conventional method is broadly applicable for small-scale and medium-scale industry whereas mechanized dyeing is for large-scale industry. The total current scenario of the Indian market for dyestuffs in the textile industry is estimated at 50 tons per year. The cost of dyes is estimated to range at Rs. 6–10/kg. Semibulk-scale trials of spraying vegetable oil emulsions (e.g., castor oil) onto the fibres (0.2% w/w) have been tried successfully over chemical processes, with considerable improvements noted for yarn quality (Sengupta and Singh, 2013; Briga-Sá et al., 2013).

Reactive dyes are well known for fastness, but owing to their high cost the industry cannot use them widely to cover surfaces. Printing is an alternate route to reactive dyestuffs (e.g., Procion Brilliant Red M8B, urea and sodium bicarbonate); it is used regularly and costs Rs. 1.05 per mat (0.34 m^2). Drying takes place after printing to remove water from the fibre. The energy required to remove 25% weight water (requiring at least 2.5×10^6 J/kg) adds approximately. This scenario may be changed in the synthetic and optical fibre industry. Recyclable fibre employs paper as a virgin raw material because it is easily recycled. Paper is derived from recyclable fibre and virgin fibre based on the population in each country. Recycled fibre is combined with virgin fibre, which use natural resources efficiently and supports sustainable development. Cigarette paper, wallpaper, books and hygienic tissue paper are some nonrecyclable papers (Briga-Sá et al., 2013).

Synthetic fibres are processed by polymerization, drying, melt spinning and drawing the fibre, followed by winding. Several methods are available to manufacture polymers based on the final form. Te synthetic fibre may be filament, staple, tow or fibrefill. Dimethyl terephthalate is treated with ethylene glycol in the presence of catalyst at 150–210°C. The resulting material is reacted with terephthalic acid by raising the temperature to 280°C. The resultant polyester is clear and molten, and forms long ribbons. It is allowed to cool until the ribbons become brittle. The material is then melted at 260–270°C to form a syrup-like solution and is placed in spinneret which determines the size of yarn. Chemicals are added to this solution to become flame retardant, antistatic, or easier to dye. Elongation to 5 m total length is possible after this. Strength, tenacity, and resilience of the fibre are improved when the filaments dry. The fibres become solid and strong instead of brittle. The fibre is finally wound on large bobbins to be woven into material.

1.3.4 Technology upgrading

Biotechnology is expected to have an increasingly important role in the textile industry worldwide. This report outlines the current developments and highlights in areas where biotechnology is applicable. Biotechnology holds a promise in the production of modified fibres with high yield. Cotton is vulnerable to many insects that affect yield. These insects are managed with large amounts of pesticides. Biotechnology focuses on genetically engineering insect, disease and herbicide resistance in cotton plants because they have low tolerance levels. The modification of quality and properties (e.g., length and strength) of fibres leads to the development of high-performance cotton. For example, transgenic intensely coloured cotton is developed (e.g., blues and vivid reds) with the selection tool which one day could replace the need for bleaches and dyes (Sarwar and Ali, 2013).

To enhance special traits in animals, biotechnology is expected to have a large impact on animal breeding for sustainable fibre production. In vitro fertilization and embryo transfer, diagnostics, genetically engineered vaccines and therapeutic drugs are practiced, as is genetic modification of sheep to resist attack from blowfly larvae by engineering a sheep that secretes an insect repellent from its hair follicles and 'biological wool shearing'. An artificial epidermal growth factor, when injected into sheep, interrupts hair growth; using this latest technology, the broken wool fibre and fleece are pulled off.

Fermentation helps in the formation of biopolymers. Such a novel fibre is developed from the bacterial storage compound polyhydroxybutyrate (PHB) by Zeneca Bioproducts, named 'Biopol'. This high-molecular weight linear polyester has good thermoplastic properties (melting point of 180°C) and can be melt-spun into fibres. Biocompatibility and biodegradability makes PHB fibres ideally suited for surgical use; sutures made from PHB are slowly degraded by the body's enzymes. Zeneca Seeds is experimenting with a genetically engineered variety of rapeseed which can synthesize Biopol. Use of Biopol in conventional plastic applications such as shampoo bottles is considered costly, so ultimately Biopol might be produced by plants (Sarwar and Ali, 2013; Nordas, 2004).

Other biopolymers such as polysaccharides chitin, alginate, dextran and hyaluronic acid are applied in wound-healing applications. Unitika is a Japanese company which produces fibre made from chitin from seafood (shellfish) waste, used for wound dressings. Chitin and chitosan fibre is used to produce inexpensive wound dressings and other novel materials. These materials have wound healing–accelerating properties better than synthetic fibres. 'Sorbsan' is polysaccharide-derived fibre extracted from brown seaweed. *Leuconostoc mesenteroides* is a bacterial strain used in the fermentation of sucrose to manufacture Dextron. In addition, hyaluronic acid, a polydisaccharide of D-glucuronic acid and *N*-acetyl glucosamine found in the connective tissue matrices of vertebrates, is also present in the capsules of some bacteria.

Cellulose is an extracellular polysaccharide produced in two ways. One is by in vitro cultivation of plant cells, especially various strains of *Gossypium* for cotton fibres; the other is by a number of different bacteria in the form of ribbon-like microfibrils (Organisation for Economic Co-operation and Development, 2004). These can be used to produce moulded materials of relatively high strength. Without climatic or geographic limitations, and free of contamination from pests, desirable traits can be introduced. New genetic manipulation techniques focus on genetically modified proteins as hosts to synthesis fibres by fermentation. In the way that the fibres are grown, they are eco-friendly and improve sustainability by peroxide bleaching and green dyeing. They can withstand colourizing agents, dyes which induce mutagenic, carcinogenic and allergic properties. Innovative technologies are integrated to attract global trend by three-dimensional printing technology for intuitive designs, less financial risk, less effort to make and less production material needed. The processing and supply chain also ensure the well-being of the planet by reducing CO_2 release into the environment. Technology upgrading optimizes the use of raw materials by requiring less concentrated feed. For instance, cellulose fabric is a wood-based fibre processed with ionic liquid. It helps melt cellulose fibre into pulp followed by spinning into yarn as the final product. To make the product viscous, carbon disulphide and sodium hydroxide are used, which is harmful to the environment. Hence in the manufacture of cellulosic pulp, waste can be converted into cellulosic fibres and ionic liquid is completely recycled by evaporation. Yet, if most manufacturing units are not installed with recyclable units, such a system is not as useful.

Biotransformation has a particular edge over traditional chemical methods; it is defined as the biocatalytic transformation of one chemical into another, for which enzymes are used with great specificity and stereoselectivity. It can be operated at mild conditions of pH and temperature, which make it successful. It cannot be applied to the production of bulk chemicals using oil-based processes. Cellulosic enzymes are employed in biopolishing, the removal of fuzz from the surface of cellulosic fibres, which eliminates pilling and makes fabrics smoother and cleaner looking. An attempt to use protease enzymes is in current demand for wool. Pectolytic enzymes (Organisation for Economic Co-operation and Development, 2004) are synthesized from various microorganisms to degrade pectin in the middle lamella of these plant fibres.

Another breakthrough in medical textiles is nano-bleed stop technology. Here, electrospun nanofibre is coated on a conventional dressing to assist in controlling fluid drainage, to enhance barrier properties and moisture management. It also intensifies cellular compatibility, anchoring the characteristics of cells and proliferation, repairing tissues and supporting the function of organs. By combining technology and fashion, and plant or mineral-derived dyes, waste-free products may become popular for the organic clothing industry. Available technologies have a significant role in transforming business into healthy environments by phthalate-free printing, less consumption of water and energy, less exposure to hazardous materials and high process efficiency. Life-cycle assessments add value to agricultural performance systems, which may replace carbon with cotton 50% in weight into durable fibres.

Sufficiently available technologies play significant role in transforming business into healthy surrounding by phthalate free printing, less consumption of water and energy, less exposure to hazardous materials and high process efficiency. Life cycle assessment adds value to agricultural performance system which may replace cotton by carbon 50% in weight into durable fibres.

1.4 Sustainability of textile industries

1.4.1 Fibre characteristics

Rayon, modal and Lyocell are familiar cellulosic fibres made of cellulose, a component of all plants. The cellulose is often derived from wood pulp, which has an average cellulose content of 40% and sometimes bamboo. Hence these fibres are often referred to (and sometimes marketed as) 'environmentally friendly'. Rayon/viscose, the first-generation cellulose fibre, has been produced since 1938. 'Artificial silk' called 'rayon', has a highly lustrous quality, giving it a bright shine. Modal is the second generation and is known for its softness. Lyocell is third-generation technology. Its advantages include the environmental friendliness of its processing combined with its softness, drape and antibacterial properties. All three are referred to generically as 'regenerated cellulosic fibres' owing to their manufacturing process, which is similar to that of natural cellulosic fibres such as cotton, flax (linen), hemp and jute, more than those of petroleum-based synthetic fibres such as nylon or polyester (http://ec.europa.eu/trade/issues/sectoral/industry/textile/memo100605_en.htm). They may be called 'semisynthetic fibres' to mimic the combination of the natural raw cellulosic material and the chemical manufacturing process that breaks down the cellulose so it can be 'regenerated' into a fibre from the original pulp.

Rayon is a versatile fibre with the comfort properties of natural fibres; it is easily dyed in a wide range of colours. Rayon is breathable and does not insulate body heat, which makes it ideal for use in clothing worn in hot and humid climates. Rayon is more moisture absorbent than cotton and does not build up static electricity. It is comfortable, soft to the skin and has moderate dry strength and abrasion resistance. Like other cellulosic fibres, it is not resilient, which means that it will wrinkle. Rayon withstands ironing temperatures slightly less than those of cotton.

There are many different processes for manufacturing rayon; they vary among the chemicals used and their subsequent impact on the environment. Basically, the production of rayon (which also applies to modal and Lyocell) chemically converts purified cellulose into a soluble compound. The solution is passed through a spinneret (similar to the holes in a showerhead) to form soft filaments that are then converted or 'regenerated' into almost pure cellulose. The purified cellulose is converted to xanthate, which is dissolved in dilute caustic soda and the cellulose regenerated from the product as it emerges from the spinneret.

Like rayon, natural and synthetic fabrics have different strength and stretch characteristics created by adjusting the drawing process applied in spinning. Low wet strength is a typical unstable result; the fabric may stretch or shrink when wet. Formaldehyde-free finishes are sometimes applied to make fabrics washable and limit shrinkage to 3%.

Fabrics are made with high absorption and moisture-holding properties for disposable diapers, hygiene and incontinence pads and medical supplies. Many fibres are made strengthen showing versatility and ability to blend easily with many fibres. Modal is a 'high–wet modulus' (Brito et al., 2008; Fiji Times, 2006) rayon, second-generation technology which has virtually high wet strength and extra softness, which makes it especially useful for body contact clothing such as lingerie and undergarments. Modal is wear resistant and can be machine washed and tumble dried without shrinking or getting pulled out of shape, and it is lustrous. Like modal, other fibres are about 50% more water-absorbent per unit volume than cotton. Fibres are made to resist shrinkage, fading and greying and to be colour-fast. They are often blended with cotton, wool or synthetic fibres, and take and retain dyes well.

Lyocell, the third-generation rayon, absorbs excess liquid (perspiration) and quickly releases it into the atmosphere. It also acts as a protective shell to regulate body temperature and maintain water balance. At the same time, Lyocell's moisture management does not give bacteria a chance to grow. Moisture is directly absorbed from the skin and transported to the inside of the fibre, rather than on the surface where bacteria could grow. It results in odour-free, multiple wear and saves on water because of the need to wash it less (Fiji Times, 2006). Energy storage and this antibacterial property are inherent in the fibre without chemical additives that are used on synthetics and many cotton products.

By contrast, synthetic fibres do not absorb moisture. Nanofibrils are the key to resolve it with hydrophilic characteristics and to optimize absorption of moisture with excellent cooling properties. Manipulating or controlling the fibrils, the very fine hairs found on the outer fibres, produces a wide variety of fabrications from rugged denim to suede like surfaces, and to a clean, smooth silky touch.

1.4.2 Fibre quality

The cell wall and lumen help evaluate the maturity index of any fibre which depends on the thickness of the cell wall. A fibre is to be considered mature when the cell wall of the moisture-swollen fibre represents 50%–80% of the round cross-section; immature is represented by 30%–45%; and dead is less than 25%. Modern technology introduced a maturity tester at the laboratory scale. Dry matter is measured when forage is harvested by machine; the amount above the leftover stubble can be dried. Because 70%–90% of herbage is water, determining the amount of dry matter is usually referred to per acre (World Trade Organization, 2005). If the forage is consumed by grazing animals, there are other ways to test for quality. Forage quality can be determined by organoleptic observation, chemical composition and feed trial evaluations. Organoleptic observations means using the sensory organs (eyes, nose, taste, ears and touch) to evaluate leafiness, vivid colour, little foreign material, an appropriate stage of maturity, and no mould.

Touch can also be used to check for the right consistency, and silage cannot be too brittle and slimy. The right time for harvesting also makes the forage more palatable and digestible. Plant maturity keeps on balancing crop growth and yield. Livestock should be digestible when it attains maturity. Harvesting at the best times wisely uses the regrowth potential of grasses and legumes. At different maturity stages of fibre production, the moisture content is analysed. It also interprets the presence of proteins, carbohydrates, fats, nonorganic components and humidity.

Mineral ash content is evaluated to remove organic material. The Kjeldahl method is implemented to find nitrogen concentration. The content Ca, Mg, K, and Na is determined using the dry-ashing procedure followed by combustion, which indicates the mineral content. Neutral detergent fibre (NDF) is determined when a sample is extracted with a neutral detergent solution. The NDF score can predict dry matter intake because a high NDF score means the animal feels full longer, because certain components are taking longer to be digested. Often deficiencies or toxicities are determined by observation or symptoms (Neil Kearney, 2003).

1.4.3 Extent of fiber life span

Accumulation of dirt and dust from poor or careless industrial practices and the entry of insect-infested items into the collection may affect the life span of fibres. Infestation causes fibre cracks as a result of condensation by high humidity. The presence of biological growth such moulds or fungi, insects and rodents attack organic materials when both temperature and humidity are uncontrolled. Environmental conditions cannot alter or halt the proliferation of mould spores, and lead to the digestion of materials before they begin to grow. This results in loss of strength and the deterioration of materials. The reason for this is that a thin, stagnant pocket of moist air is formed which favours mould growth. The basic nutrition paradigm improves the longevity of fibres' life span. Immortality of fibre can indeed be promoted in the apparel industry (Barclay and Buckley, 2000).

1.4.4 Recyclable eco-friendly fabrics

Organic production does not involve the use of chemicals and other artificial methods. Certain environmental concerns make them free from hazardous chemicals, and are now expanded to address a broad range technical quality and human toxicity criteria and minimum social standards. The Global Organic Textile Standard is recognized as the world's leading processing standard for textiles made from organic fibres (Foreign Investment Advisory Services, 2004). It defines high-level environmental criteria along the entire organic textiles supply chain and requires compliance with social criteria as well.

In Kenya, some textile manufacturing companies bottle biogas made from waste generated from slaughterhouses, which they plan to sell to consumers at about half the cost of petroleum gas. The facilities also convert waste material into a fertilizer for Maasai herdsmen in the hopes of creating faster-growing grass for their grazing cattle. In this way, a carbon-intensive process takes place. In Zurich, researchers at the Swiss Federal Institute of Technology developed yarn comparable to Merino wool from slaughterhouse waste as the base for the material, specifically collagen from skin, bone and tendons. It is able to compete with petroleum-based synthetic fibre, the fibre's creators said. Producing synthetic fibre alone is an energy-intensive process.

Turning slaughterhouse waste into biofuel is an innovative idea developed by scientists at the University of Louisiana at Lafayette. They convert alligator fat into biofuel using a continuous reactor instead of a batch reactor, to speed the process. The main advantage of this is that it does not require a catalyst, and so it leads to less waste. The researchers are looking into whether other animal fats, such as chicken, pork, and beef, can be converted into biofuel using the same method (http://modernfarmer.com/2015/08/recycle-slaughterhouse-waste/; European Commission, 2005).

1.5 Future trends

1.5.1 Intuitive designs

Electrospun nanofibre technology has revolutionized the medical industry. However, nanofibre cannot be used on its own, so conventional dressings are coated with electrospun nanofibre to harvest their full potential. Electrospun nanofibre dressings enhance moisture management and offer unmatchable barrier properties in addition to assisting with the control of fluid drainage. A polymer solution is electrospun into very fine fibres which are then collected on a grounded electrode. The basic electrospinning unit contains a needle nozzle, a high-voltage power supply, a container for spinning fluid and an electrode collector. Electrospun nanofibre (Mirjalili, 2016) technology is also widely used in tissue engineering to repair, maintain, replace or enhance the function of a particular organ. Electrospun nanofibre mats are used for burn wounds because they exhibit good cellular compatibility and support cell attachment and proliferation. Whereas electrospun nanofibre technology is applicable for regular medical care, developments promise to further smooth postoperative patient care.

Digital textile printing is not only a highly creative form of textile printing, it also provides greater design flexibility compared with screen printing. Designers prefer digital printing for small prints because it is more cost-efficient and eco-friendly, with less physical inventory levels, which thus leaves a smaller carbon footprint. Digital printing involves directly printing onto fabrics with printers; it saves almost 95% of water, and energy use is reduced by 75%, leading to minimum textile waste.

1.5.2 Adopted technology

Traceability and transparency are guaranteed throughout the entire production and supply chain of brand labels. An organization defines the waste policy and ensures that it is appropriate to the nature and scale of the organization's use and consumption. It reduces fibre waste by using continuous flow operations instead of batch operations, reducing waste generation by raising awareness, introducing the importance of waste management to operators and training them in waste reduction practices.

Important policies include the reuse of nonhazardous solid wastes such as papers and plastics, the recovery of steam condensates and reuse for heating purposes and recycling of waste fibre by liquefying them and reinjecting them back into the process, to ensure the reduction of waste generation (http://www.millerwastemills.com/about/reclaimation/).

1.5.3 Modernization

Improved optical properties through fibres hold promise in the future. Silica glass with high fluoride content is the most promising for optical fibres, with attenuation loss even lower than today's highly efficient fibres. Fibres experimented upon and drawn from glass contained 50%–60% zirconium fluoride (ZrF_4), which showed losses in the range of 0.005–0.008 dB/km, whereas earlier fibres often had losses of 0.2 dB/km.

The improvement in the process caused researchers to produce more optical fibres from refined materials. The most sophisticated manufacturing processes use high-energy lasers to melt the preforms for the fibre draw. Fibres can be drawn from a preform at the rate of 10–20 m (32.8–65.6 ft) per second, and single-mode fibres from 2 to 25 km (1.2–15.5 miles) in length can be drawn from one preform. A fibre manufacturing company reported creating fibres of 160 km (99 miles) on a small budget, and the frequency is currently retooling as often as every 18 months. It takes research to the next dimension, and yet still greater innovations lie ahead. These advances will be seen in the growing use of optical fibres in computer networks and in increasing demand for the technology in burgeoning international markets.

It is classified as single-mode fibre about 9 μm in diameter and double-mode fibre at about 62.5 μm. Single-mode fibre transmits six times the wavelength of light whereas in double-mode fibre transmission is about 50 times the wavelength of light. The signal can be sent by a smaller core diameter, which minimizes the dispersion of light pulses with the increased distance. On other hand, a large core diameter, multimode fibre experiences more dispersion than single mode. A graded index core in which layers of light have a lower index of refraction is able to minimize dispersion but complicates the manufacturing process. The pure silica tube is mounted on a lathe equipped with a special heat torch to form soot. As the gasses flow inside the tube, they react to the heat by forming solid submicron particles, in the vicinity of the heat zone called 'soot'. It is deposited on the inner wall of the tube. As the burner traverses over the deposited soot, the heat transforms these solid white particles into pure, transparent glass in a process called vitrification. The deposited material by vitrification will form on the core region of optical fibre (http://www.arcelect.com/fibercable.htm; David et al., 1998).

1.5.4 Advanced machineries

1.5.4.1 Air dye technology

Air dye technology adds colour to textiles without using water. The process of making textiles can require several dozen gallons of water for each pound of clothing. The air dye process employs air instead of water to help dyes penetrate, a process that uses no water and requires less energy than traditional methods of dyeing. This technology works only on synthetic materials, not on natural fibre products. Advantages of this technology are that it is reliable and does not pollute water in the colour application process. No hazardous waste is emitted while replacing water by air and no water is wasted. The management of air dye technology greatly reduces energy requirements, thereby lowering costs and satisfying the strictest standards of global responsibility. There is no need to engage boilers, screen printing machines, drying ovens, or cleaning and scouring chemicals, which eliminates major sources of pollution (http://www.debscorp.com/eng/airdye-technology/). Eliminating water in the colour application step simplifies the process, creating revolutionary possibilities of new industry and employment in unfarmable, arid regions of the world.

1.5.4.2 Herbal textile

Herbal textiles are an alternate way to dye entirely with herbal extractions. Vegetable dyes are employed without using chemicals to bring medicinal value to apparel. Even bleaching of cloth does not adopt a chemical process; it is done naturally by exposing the cloth to sunlight. Because it is environmentally friendly, it does not pollute the environment by contaminating water resources in areas close to processing units. Those herbs provide many shades of red, yellow, brown, orange, green, etc. Herbal textiles can bring health benefits and fight diseases such as hypertension, heart ailments, asthma and diabetes. The healing effects of herbal textiles is proven by indigo; cuscus crass against asthma; turmeric, which can cure pain and is beneficial for enhancing skin qualities; and sandalwood, which fights stress with its soothing effects. Some other herbal dyes are catechu, pomegranate, tamarind, madder, castor oil, sweet basil, lime, wild turmeric, henna, curry leaf tree, aloe and certain herbal fruits (http://textile-learner.blogspot.in/2013/02/herbal-textile-manufacturing-process-of.html).

Sources of further information and advice

Progress being made in biotechnology and the level of investment by governments and private sectors has had enormous commercial implications for many sectors of the industry in the years ahead. Biotechnology has developed new products such as artificial protein, enzyme, fibres, starch and dyes, opened up new markets, speeded up production and helped to clean up the environment. The textile industries of developed countries must not lose out on these exciting developments. According to the Department of Trade and Industry, many companies in developing countries, and even in the United Kingdom, are potentially missing business opportunities because they are unaware or reticent of biotechnology. Biotechnology and its applications in the textile industry was identified as a key sector in which opportunities are available. Few companies launched ‘Biotechnology Means Business’ with experts who could help them deploy biotechnology to give a competitive edge to their business and win new markets.

In this chapter, technologies available for the enormous growth of textile industry were discussed. Biotechnology and nanotechnology are considered potential applications of conventionally produced textile materials on the nanoscale. Fermentation, downstream processing, three-dimensional printing technology and electrospun nanofibres account for at least 70% of production costs in the garment industry. There is a need for improved filtration and separation techniques to avoid the discharge of toxic chemicals into the environment.

References

Barclay, S., Buckley, C., 2000. Waste Minimisation Guide for the Textile Industry: A Step towards Cleaner Production, vol. II. Water Research Commission of South Africa, Durban, South Africa. University of Natal.

Briga-Sá, A., Nascimento, D., Teixeira, N., et al., 2013. Textile waste as an alternative thermal insulation building material solution. Constr. Build. Mater. 8, 155–160.

Brito, M.P.D., Carbone, V., Blanquart, C.M., 2008. Towards a sustainable fashion retail supply chain in Europe: organization and performance. Int. J. Prod. Econ. 114, 534–553.

CAPMAS, 2014. Egypt in Figures Report. Central Agency for Public Mobilization and Statistics.

David, J.W., Jeff, T., Dennis, W., et al., 1998. Fabrication and characterization of single mode electro-optic polymer optical fiber. OSA Publ. 23 (23), 1826–1828.

European Commission, 2005. EU-China Textile Agreement, Memo, 12 June 2005.

Fiji Times (online), 8 September 2006. Australia to Relax Fiji Garment Import Fees. www.fijitimes.com.

Foreign Investment Advisory Services, December 2004. Cambodian Garment Sector: Buyers' Survey Results. World Bank, Washington, DC.

Fresner, J., 1998. Starting continuous improvement with a cleaner production assessment in an Austrian textile mill. J. Clean Prod. 6, 85–91.

Guidelines for Implementation of Catalytic Development Programme (CDP) in Project Mode during XII Plan Period (2012-17). http://texmin.nic.in/schemes.

Lo, C.K.Y., Yeung, A.C.L., Cheng, T.C.E., 2012. The impact of environmental management systems on financial performance in fashion and textiles industries. Int. J. Prod. Econ. 135 (2), 561–567.

Mirjalili, M., 2016. Review for application of electro spinning and electrospun nanofibers technology in textile industry. J. Nanostructure Chem. 6, 207–213.

Neil Kearney (General Secretary, International Textiles, Garment and Leather Workers' Federation), 2003. Trade in Textiles and Clothing after 2005. http:/www.ers.usda.gov.

Nordas, H.K., 2004. The Global Textile and Clothing Industry and Clothing. Discussion Paper No. 5. WTO, Geneva.

Organisation for Economic Co-operation and Development (OECD), 2004. Structural Adjustment in Textiles and Clothing in the Post-ATC Trading Environment. TD/TC/WP 2004/4/FINAL, Paris.

Ozturk, H.K., 2005. Energy usage and cost in textile industry: a case study for Turkey. Energy 30, 2424–2446.

Profile of the Indian Cotton Textile Industry, Cotton Textile Export Promotion Council. http:/www.texprocil.com.

K. Rajendran Nair: India'sHandloom Sector @ www.pib.nic.in. Best Management Practices for Pollution Prevention in the Textile Industry, US: Environmental Protection Agency 1996.

Rock, M.T., Angel, D.P., 2007. Grow first, industrial transformation in East Asia. Environment 49, 10–19.

Sami, A.M., 5 March 2013. Technology of Eco Friendly Textile Processing-A Route to Sustainability.

Sarwar, M.I., Ali, M.A., 5 March 2013. Sustainable and Environmental Friendly Fibers in Textile Fashion-A Study of Organic Cotton and Bamboo Fibers Report No. 2010.9.14.

Sengupta, S., Singh, B.R., 5 March 2013. Natural Green Dyes for the Textile Industry Technical Report No. 57. University of Massachusetts Lowell. 2003.

World Acrylic Fiber, 2013. Industry Report: Trend in Demand & Supply. Centerac Technologies Limited, India.

World Trade Organization (WTO), 23 November 2001. Accession of the People's Republic of China – Decision of 10 November, WT/L/432, Annex 5B. Geneva.

World Trade Organization (WTO), 2003. Anti-dumping Actions in the Areas of Textiles and Clothing, Proposal for a Specific Short-Term Dispensation in Favour of Developing Members Following Full Integration of the Sector into GATT 1994 from January 2005.

World Trade Organization (WTO), 28 June 2005. Options for Least-developed Countries to Improve Their Competitiveness in the Textiles and Clothing Business, WT/COMTD/LDC/W/37. WTO, Geneva.

Part One

Sustainable fibre production

Sustainable cotton production

2

Shanthi Radhakrishnan
Kumaraguru College of Technology, Coimbatore, India

2.1 Introduction

The quality of any product starts with the quality of the raw materials. Cotton is one such material that controls the quality of the end product and also proves to be a performance fibre. The cultivation of cotton has long been marked by the use of unsustainable methods such as mono-cropping, extensive irrigation and the use of synthetic fertilizers and pesticides, but there is a change in the trend towards sustainability. Many developmental programmes to promote sustainability in cotton cultivation have been undertaken to protect the livelihood of the farmers and their habitat. In recent years there has been keen interest in implementing production risk management focused on safeguarding natural resources and the environment. Regulatory measures have come into existence on both the domestic and global levels and these regulations have made large retailers realize that the overall footprints of their products need to be reduced to strengthen the confidence of customers who purchase their products. As a result of these measures, a number of production standards and systems have evolved with the objective of nurturing sustainable farming and products.

2.1.1 *Importance of cotton*

The leading cash crop in the United States is cotton; the use of \$5.3 billion worth of supplies and services is essential for the production of 1 year's crop at the farm level. According to the National Cotton Council of America, this crop occupies the first place as a value-added crop, because annual business revenues exceed \$120 billion in the United States (National Cotton Council of America 2016). More than 100 countries and about 350 million people are engaged in cotton farming and industrial manufacture, and the world production of cotton is estimated at 25.5 million tons of cotton seed grown from 34.8 million hectares. The finest cotton in terms of quality is from Australia and Egypt whereas the highest cotton production is bagged by the United States and Israel (NETAFIM, 2016; International Trade Center, 2016a,b). In 2014, the International Cotton Advisory Committee reported that the major producers of cotton were China (6,700,000 tons), India (6,371,000 tons), the United States (2,811,000 tons), Pakistan (2,076,000 tons), Brazil (1,644,000 tons) and Uzbekistan (920,000 tons) (Food and Agriculture Organization of the United Nations, 2015).

About 40 species in the genus *Gossypium*, family Malvaceae are included under the name cotton plant. According to the United Nations (UN) Conference on Trade and Development report in 2005, Upland cotton (*Gossypium hirsutum*), which is found in Central America, South Florida and the Caribbean, New World species (*Gossypium*

Sustainable Fibres and Textiles. http://dx.doi.org/10.1016/B978-0-08-102041-8.00002-0

Table 2.1 Uses of the cotton plant (National Cotton Council of America, 2016)

Sl. no.	Description	Use
1.	Cotton fibre/lint	Spinning to yarn, weaving of fabric, development of nonwoven
2.	Linters: short fuzz on seed	Raw material as cellulose for making plastics, explosives, paper products, as batting for padding mattresses and cushions for furniture
3.	Cotton seed: crushed to form three components (cotton seed oil, meal and hull)	Cotton seed oil (18%–24% oil and 20%–40% protein) used for shortening, cooking oil and salad dressing; also used as high-protein concentrate in food products Meal and hull may be used (separately or combination) as feed for livestock, poultry and fish feed and as fertilizer
4.	Stalk and leaves of cotton plant	Used as soil enrichers and plowed into soil

barbadense), which is popular in South America, Tree cotton (G. arboreum), which is grown in South Asia, and Old World/Asiatic cottons (*Gossypium herbaceum*), which is found in South Africa are some of the commercially grown cotton species (UNCTAD, 2016). The quality of the fibre depends on various factors such as the variety, environment and crop management practices. The world cotton supply may be divided into six categories based on quality: extra-fine, fine, high-medium, medium, coarse count and waste/padding. Combed yarn is produced from extra-fine/fine/high-medium cottons obtained from *G. barbadense* with a staple length of 32 mm, a micronnaire value below 4 and strength of up to 40 g/tex. Carded ring-spun yarn uses Old World cotton in the medium category with a staple length less than 25 mm with a micronnaire value above 6. Coarse count cotton is used to produce open end yarns (Hustvedt and Dickson, 2009). Many industries pay a higher price for cotton with better fibre characteristics to obtain better efficiency with state-of-the-art high-speed machinery.

The usefulness of the cotton plant is explained in Table 2.1.

2.1.2 Consumers' emotional preference for cotton

Currently the fashion industry and the textile market consider organic cotton to be the highest order of sustainability. The global production of organic cotton is 0.2% of global cotton production (Blackburn, 2009) and water issues and labour are not completely tackled by the standards and regulatory measures. The niche market for organic cotton is small and prices are not yet low enough for average people to include it in their wardrobe. At this juncture organic cotton can be considered a pillar in the organic movement and sustainability. Cultivators have to move towards biotechnology because it is safe and can be used during adverse conditions under which the environment faces the impacts of pollution.

Cotton is the most preferred fibre for consumers and organic cotton seems to have a special weight. In a study the psychological profiling of the market segment that uses organic cotton content in purchasing decisions revealed that consumers preferred organic cotton because it indirectly improved the health and well-being of individuals or the family. Although manufacturers do not make claims regarding health, the results showed that consumers believed it had an impact on the aura of the person concerned. The study also highlighted that most consumers preferred organic farming because it had environmental benefits and they also supported marketing messages that promote organic cotton. Furthermore, consumers who used organic cotton and nonusers in this segment had positive attitudes towards organic and sustainable agriculture and preferred to buy locally made products (Hustvedt and Dickson, 2009).

Consumers' choice of organic cotton is considered to be a healthy because it is the cotton is free of chemicals and good for our environment. Feminine products, clothing, blankets and infant clothing are worn close to the body and we can be sure that there are no harmful chemicals and toxic poisons being absorbed into the body when it is made of organic cotton (Ellison, 2010). The purchase of organic cotton also supports farmers, manufacturers and retailers who are earnestly trying to make a difference to our society and planet.

2.1.3 Promotion of cotton

The global market share of cotton (40.6%) is approximately similar to that of polyester (42%) (International Trade Center, 2016a,b); the market share of cotton is on the rise as a result of the price advantage over polyester. Polyester is a significant competitor of cotton owing to the immense variations that can be incorporated into the fibre to produce new products which have qualities similar to cotton. Despite these facts cotton is a favourite fibre of consumers, which facilitates the improvement of cotton's market share. Promotional activities with regard to cotton fibre and preference are being carried out in producing, trading and consuming countries. Cotton organizations in the United States are the most active, followed by Australia, Egypt, India and South Africa; Germany, Poland and the United Kingdom undertake activities through the funding and support of organizations such as the Bremen Cotton Exchange and the Gdynia Cotton Association. Thc promotion of cotton has also been undertaken on a state level and private sector–focused programmes.

The promotion of cotton is done through two routes. Onc is competition between cotton from two producing countries in which one country takes up promotional activities against another country. The second type is to promote all cotton against chemical and synthetic fibres. From the consumer's point of view, the cotton brand is identified as the country of origin and certain representative logos and icons certify that the cotton used abides with or fulfils certain specified standards. In some cases, logos also identify whether the product is made from 100% cotton or has a blend of other fibres. One promotional activity carried out at the retailer and consumer level is 'Demand Enhancement', which creates a positive image about the product and can result in a measurable impact. This technique is most suited to products that have limited ways of distinguishing themselves among competitors. Export-focused cotton promotion is done between cotton-producing countries and textile-manufacturing countries whereas domestic promotion is linked to in-house demand. The world's largest producer of cotton, textiles and apparels is China, yet it imports cotton from other countries.

2.1.4 Integration of economy and ecology: from niche market to mass market

Today all manufactured products have the responsibility of showcasing the ecological background to the consumer, apart from the usefulness of the design and functionality. The manufacture, use and disposal of textiles are a burden to the environment; the imbalance in the manufacture is another problem: that is, the beginning and end of the textile chain are in the hands of developing countries whereas the added value sections are with the industrialized countries. Commitment of the entire textile chain to ecology and the environment requires optimization of all processes and procedures along the entire textile chain. Furthermore, the success of any product or technology depends on the bonding of economy and ecology and the competitive position of the business organization. An organization catering to the mass market will limit itself to ecological problems whose solutions will not greatly affect the cost structure whereas an organization serving niche markets will tend to tackle a great number of ecological problems because it will gain great leverage in marketing to environmentally conscious customers. The market and the product involved have a great role in influencing cost and differentiation. Currently eco-textiles are for niche markets where consumers pay a high price for eco-friendly products, but the real strength to be achieved is when eco-textiles cater to the mass market. This will become an achievable target when the research benefits of innovative solutions and clean technologies reach all sections of the manufacturing world, both developing and industrialized nations, and eco-textiles are manufactured on a large scale at competitive prices (Hummel, 1997).

2.1.5 Case studies

Since 2500 BC, naturally coloured organic cotton has been produced in Andean South America. Pakucho in Southern Peru produces the best superior certified long-staple white cotton with a yield double the national average, named Organic Tangius cotton. Peru is a major producer of high-quality coloured and white organic cotton for to many reasons. For one thing, Peru, the Andean region of South America, is the origin of the cotton species *G. barbadense*. Cotton can be planted in different regions of Peru throughout the year. All cotton is handpicked and any variety of long-staple organic cotton is available for organic farming. Traditional and indigenous farming methodologies have been adopted for farming. Four different cotton varieties in three natural colours (white, brown and green) with natural colour in the lumen area are cultivated representing the cotton species *G. barbadense* and *G. hirsutum* (Vreeland, 1997). Pakucho has vertical integration because it uses the cotton to produce ring-spun yarn for both knitted and woven textiles and is certified by many international organizations. This region has set a successful traditional model for other farmers in different parts of the globe.

In Egypt, organic cotton is cultivated as a joint effort among farmers, scientists and manufacturers with the final product designed by designers from Europe and Egypt. Organic cotton farms in the southern delta area and the north of Egypt use multicropping cotton cultivated in Egypt. It has good fibre elasticity and quality that surpasses conventional varieties (Vreeland, 1997). The cotton is spun, knitted and woven without chemicals by mechanical and thermic means in the SEKEM textile factory in Egypt. Under the brand name 'Cotton People Organic', children and teenager's clothes are produced along with a German sales partner. Three shops in Cairo and a retail chain in Cairo and Alexandria successfully sell Cotton People Organic textiles.

Organic cotton has become the livelihood for many people in countries around the globe. When worked in union with technology, traditional methods can produce good results and create a niche market for highly valued products that are carefully grown in a sustainable manner. In turn, these methods also safeguard the environment and the habitat of the water bodies where agriculture is predominant. Conventional methods of cotton cultivation have disregarded these concerns and have created various sustainability issues including deterioration of the environment and health hazards for all living beings. An illustrative quotation on sustainability states,

I do not want to protect the environment
I want to create a world where the environment does not need protecting
Durr (2016)

Cotton has been cultivated for many years but no care has been taken as to whether the cultivation procedures or techniques were sustainable. All of the resources available to humanity have been used lavishly and wasted without concern for the environment. The only aim was to increase the production of cotton and work on economics for profit rather than efficiency. Conventional farming techniques blended with the intensive use of chemical fertilizers and pesticides brought many hazardous consequences to living beings and their habitat and environment. Hence a review of key sustainability issues in cotton farming is essential to understand its relationship to sustainability.

2.2 Key sustainability issues in cotton farming

Agriculture is by far the biggest user of water. For this reason, the water that a person 'eats' everyday contained in food products is much larger than the volume of water a person drinks. Of all fresh water withdrawn for human use, industrial and household uses account for 20% and 10%, respectively, whereas agriculture consumes around 70% and much more in some locations. Fresh water constitutes less than 2.5% of the total volume of water on the earth; two-thirds of this water is inaccessible owing to its location in glaciers and ice caps. A report also states that only 0.77% of this water is found in aquifers, pores of soil, lakes, swamps, rivers, plant life and atmosphere, which is less than 10,665,000 km^3 (Gleick, 1993). The alarming rate at which fresh water is being withdrawn and the pollution levels of water show that a freshwater crisis will emerge in the years to come. According to the UN World Water Development Report, the average supply of water per person will drop by a third in the next 2 decades. Findings show that cotton is an important cause of the destruction of freshwater systems on a regional and global scale (WWF, 1999).

Three crops (rice, wheat and cotton) are considered to be the major users of fresh water. It has been estimated that in Pakistan about 31% of the fresh water used for agriculture is groundwater, and in China extensive use of groundwater for irrigation has brought about a fall in the water table levels. Some of the impacts of cotton cultivation on freshwater ecosystems and biodiversity are the result of runoff from cotton cultivated fields, contamination in drainage, pesticide use, withdrawal of water for irrigation, extensive irrigation and dam construction and land reclamation for cultivation of cotton, as shown in Table 2.2.

Table 2.2 Impacts of cotton cultivation on freshwater ecosystems and freshwater biodiversity (WWF, 1999)

Sl. no.	Mechanism	Pollutant/change	Impact	Cases
1.	Runoff from fields	Fertilizer, pesticides and sediments	Eutrophication and wildlife contamination	
2.	Drainage	• Saline drainage water • Pesticide/fertilizer-contaminated drainage water	• Salinization of fresh water • Pollution of fresh water	China Egypt Uzbekistan
3.	Use of pesticides	• Insecticides, fungicides, herbicides and defoliants • Spray drift (aerial application) • Leakage of equipment	• Wildlife contamination • Adjacent wetlands, groundwater and surface water contamination • Groundwater and surface water contamination	
4.	Withdrawal of water for irrigation	Use of groundwater and surface water	• Change in level of water table/depletion of groundwater • Degradation of water bodies and wetlands	New South Wales Australia Aral Sea Yellow River Valley
5.	Extensive irrigation	Water logging	• Rise in water tables • Salinization of soil surface	Australia Indus River Valley Uzbekistan Pakistan
6.	Construction of dams	Regulated water flow	• Habitat destruction • Change in water table • Change in water flow	
7.	Land reclamation	Change in vegetation	• Habitat destruction	

2.2.1 Runoff from fields

Water is considered to be in excess when the surface of agricultural land is saturated to the fullest capacity and cannot absorb water that comes its way. This water flows to adjacent water bodies and is termed 'runoff'. Improper water management and irrigation inexpertise are some reasons for runoff, which not only moves excess water but contains residues of pesticide, fertilizers and salts that have been administered to the cotton crop (Beven, 2004; Mcmahon, 1992; Jordanien et al., 2011).

2.2.2 Use of insecticides

Contamination as runoff from fields pollutes water bodies, resulting in fish mortality. Concentration and content, length of exposure, temperature, water pH, characteristics of fish species, physiological condition and prior exposure are some criteria that influence fish mortality. Insecticides such as organochlorine and organophosphate are considered more toxic than herbicides. Of the two, organochlorine is more toxic and persistent; it promotes bioaccumulation and results in the magnification of pesticides in the aquatic food chains because of intimate contact between aquatic organisms and water. Fig. 2.1 shows the impact of pesticide application on surface water and groundwater contamination.

The characteristics of selected pesticides are given in Table 2.3. From the table it can be understood that the lethal concentration that affects fish is very low in the case of organochlorine compared with organophosphates. Therefore a low-level concentration of organochlorine in runoff from fields can affect aquatic life. Furthermore, these low concentration levels can produce a high relative environmental hazard up to 14.2 compared with organophosphate, which shows a moderate relative environmental hazard up to 9.8 (Willis, 1982). Bio-magnification is a sorption process which brings to equilibrium the concentration of pesticides in the water and in the tissues of aquatic animals; it is in the same range as the coefficients of soil organic matter and water.

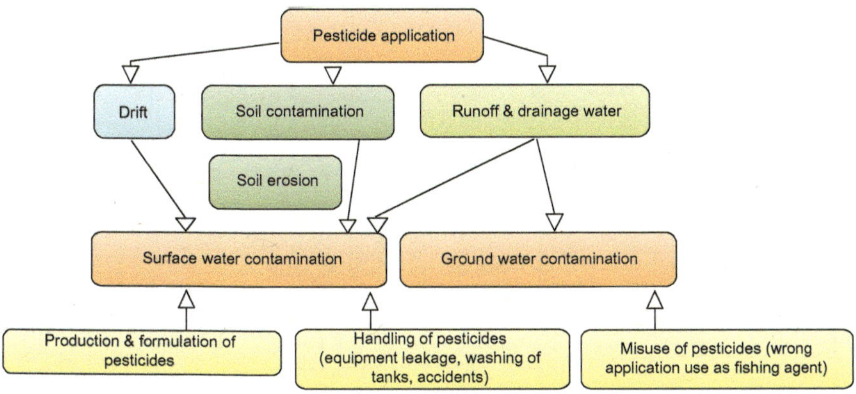

Figure 2.1 Impact of insecticides on freshwater ecosystems (Willis, 1982).

Table 2.3 **Comparison of properties of organochlorine and organophosphate pesticides (Willis, 1982)**

Sl. no.	Pesticide	Water solubility (mg/L)[a]	Predominant transport mode[b]	Toxicity, fish LC (mg/L)[c]	Relative environmental hazard[d]
1. Organochlorine					
	Aldrin	0.01–0.2	S	0.028 (G,M); 0.003 (B)	14.2
	Dichlorodiphenyltrichloroethane	0.001–0.04	S	0.027 (G); 0.032 (M)	14.2
	Dieldrin	0.1–0.25	S	0.16 (M); 0.037 (G)	14.0
	Endosulfan	–	S	0.001 (B)	12.4
	Endrin	0.1–0.25	S	0.001 (M); 0.002 (G)	14.3
	Lindane	7.3–10	S	0.062 (M)	11.6
	Methoxychlor	0.1–0.25	S	0.056 (G)	11.8
	Toxaphene	0.4–3	S	0.006 (G); 0.008 (M)	13.8
2. Organophosphate					
	Azinphos methyl	33	S	0.093 (M)	9.1
	Disulfoton	15–66	S	0.040 (B)	9.3
	Malathion	145	W	0.019 (B); 12.5 (M)	7.1
	Parathion	24	S	0.047 (B),2.0 (M)	8.0, 9.2
	Phorate	50–85	SW	0.0055 (B)	9.8
	Carbaryl	40–99	SW	1.0 (B); 12.0 (M)	6.6
	Carbofuran	415–700	W	0.21 (B)	8.9
	Trichlorfon	–	W	0.16 (B)	–

[a]Solubility at 20–25°C.
[b]As runoff, water [W] and sediment [S].
[c]LC, lethal concentration that affects 50% of the test sample (*B*, bluegill or trout; *G*, goldfish; *M*, fathead minnow).
[d]Assessment done in terms of (toxicity to rats; toxicity to fish; persistence; bioaccumulation) the sum of the value given to the four parameters (1 = low; 4 = highly severe). This value may range from 4 to 16.

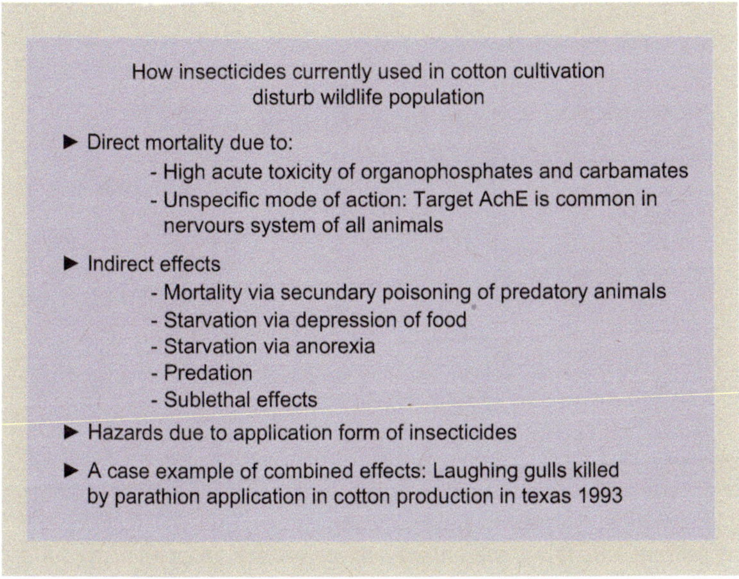

Figure 2.2 Impact of insecticides on wildlife population (WWF, 1999).

The impact of pesticides is twofold: the effects on macroorganisms and microorganisms in the water body, as shown in Fig. 2.2. Pesticides cause direct harm to aquatic macroorganisms, including death, changes in genetics or physiology or loss of strength; alterations in behavioural change that either eliminate or reduce the population of necessary organisms; indirect effects such as lesser food and oxygen; and alterations in the environment that is required for the growth and well-being of the organisms. In the case of microorganisms and phytoplankton, there is a direct impact on growth and certain important metabolic activities such as photosynthesis, and indirect effects which cause the deterioration of segments of the population. Damage to fish and wildlife is irregular and the impact becomes serious when the entire population, major sections or the whole ecosystem becomes affected. Aquatic and animal species have a high reproductive capacity, and hence the effect is temporary but when the turnover is low, the chronic effects become extremely important (Roger et al., 1994; National Wild Life Federation, 2015). Reports also show that when these negative effects are often repeated, the microbial community becomes permanently affected and is prone to harm.

Dichlorodiphenyltrichloroethane (DDT), organochlorine, is considered to be the most 'pernicious' of global pollutants. This insecticide was banned in 1972 but its impact on birds continues. At the end of World War II, by the 1960s, the concentration of DDT was found to be very high in the mud of Long Island, New York marshes which were sprayed with DDT as part of a mosquito control programme. Upon measuring the gulls, mergansers and cormorants which fed on these marshes, the presence of DDT was found to be 1000 times higher than the concentration in the soil of the marshes. Mortality among birds was common as DDT poisons the

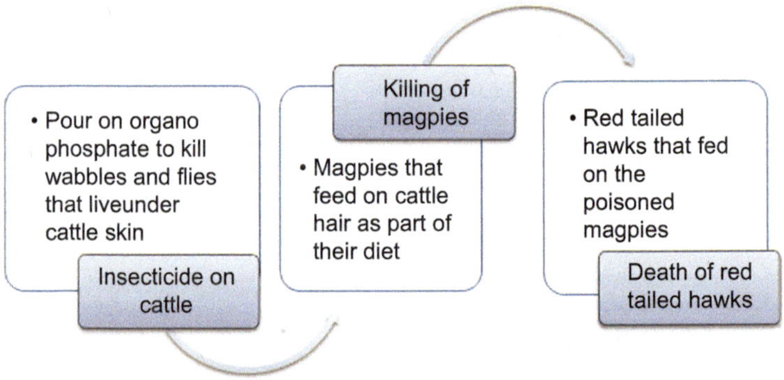

Figure 2.3 Indirect effects: secondary poisoning by insecticides (Henny et al., 1985).

nervous system, and local disappearance of birds was recorded (Woodwell, 1963). Analysis of organochlorine in nonmigratory birds such as the great horned owls of the Virginia Wildlife Center (Wildlife Center of Virginia, 2016; Cox, 1991) indicated the presence of DDT, which may have resulted from the ingestion of contaminated birds. Aerial forestry applications also showed an abundant reduction in singing males in terms of the number present and in the number of species present. About 30,000–40,000 birds were found to be dead after the application of organophosphates, because even small doses are lethal (Cox, 1991). Around 16–20 mg/kg body weight of organophosphate is considered to be acutely toxic and around 50 mg is considered most toxic.

Secondary poisoning is an indirect effect of insecticide, as shown in Fig. 2.3. Starvation is another effect caused by insecticides; the chemicals destroy the food source or cause anorexia, resulting in the death of birds such as water fowls and grackles (Ornes and Raloff, 2014; Rattner et al., 1982; Grue et al., 1982; Farage, 1989). Furthermore, the impact of insecticide includes a reduction in speed in birds and animals, and they fall prey to other animals. Apart from these, sublethal effects have been noticed in many bird species. DDT was found to feminize the development of embryos, resulting in skewed sex ratios and a reduction in breeding males of the gull population in California. Other cases were noticed such as the stoppage of egg production in quails, disruption of parental care to nestlings and the reduction of brain cholinesterase levels and enlarged livers (Fry and Toone, 1981). Sometimes the effects are humongous when all of these impacts work in combination, as in the case of a breeding colony of laughing gulls near Corpus Christi, Texas. The insecticide Parathion was applied to cotton fields to eradicate the bollworm. Over 100 adult birds were found dead near the gulls' nest, which was 3 miles from the cotton fields, because they ate insects from the cotton fields. A large number were also found dead near water bodies adjacent to the cotton fields. The analysis revealed poisoned insect parts in the birds' intestinal tract, and brain acetylcholinesterase activity was about half efficiency compared with that of healthy birds. Subsequent examination also showed the presence of poisoned insects in the dead chicks and many chicks died of starvation and neglect (White et al., 1979).

2.2.3 Drainage and leaching

A major problem in cotton cultivation is saline soil or saline irrigation. It reduces the availability of soil water to crops and is similar to water deficiency. Cotton is resistant to salinity and can exclude sodium ions. This may lead to a situation in which problems are understood only when they accumulate. It has been reported that more than half the irrigated area in 24 major irrigation countries is affected by salinity, which is highly predominant in the desert and the Mediterranean climate in the irrigated cotton belt that stretches from Spain to Central Asia and in America. Salinity problems have been found in Egypt, Pakistan, India, Uzbekistan (50% of irrigated area) and Northwest China.

Drainage systems are planned to avoid water logging and salinity of soil. To prevent salinity or ensure the salinity of the soil is reduced, water is allowed to flow downward through the roots of cotton plants through additional irrigation. The resultant water logging is remedied by providing adequate drainage to remove excess water, which in turn leads to the risk of a rising water table. Leaching of the additional irrigation water also reduces the field water use efficiency (WUE). The next problem associated with this system is the disposal of the saline drainage water which has to be returned to the water bodies, causing environmental pollution. Discharge of the saline drainage water is to the ocean, as in Pakistan, or to evaporation basins, as in Australia; this leads to wastage of land and recycling of water to irrigate the salt-tolerant crops.

Overirrigation practices adopted to reduce the salinity of the soil and consequential leaching cause a rise in water tables, which leads to secondary salinization. To leach out the salts in the profile caused by the rising water tables, it is essential to apply excess irrigation in relation to the ET requirement. The other side of the situation is shallow water tables caused by overirrigation practices and leaching for salinity control. Efficient drainage systems are the only priority to solve these problems. Many remedial measures are being undertaken by different countries to control this situation. In Pakistan, although 10% of irrigated land is affected by shallow water tables, over the 3 decades or so, 6 million hectares have been reclaimed. A series of Salinity Control and Reclamation Projects in Pakistan has installed 20,000 tube wells and vertical drains and excavated 10,000 km of drains. In Egypt, with the assistance of the World Bank, a national drainage scheme has been implemented. Drainage facilities are important in China along the Yellow River and the Yangtze River Valley for both salinity control and flooding because of heavy rainfall in the region. Drainage systems are generally adequate in China, Egypt and Pakistan but they are poor in Uzbekistan and Mexico.

2.2.4 Withdrawal of water for irrigation

Cotton production at the field level results in three types of effects on water use: evaporation of infiltrated rain water (green water use) for the growth of cotton, withdrawal of ground/surface water (blue water use) for irrigation and pollution of water caused by leaching of fertilizers and insecticides (FAO, 2002; Wani et al., 2009; Falkenmark, 2000, 2003; Falkenmark and Rockstrom, 2013). The amount of water used is measured in terms of volumes used per year. The quality of water is estimated in terms of

Table 2.4 **Water footprint of cotton products (Cottonconnect, 2014)**

		Water in litres			
Sl. no.	**Type**	**Blue water**	**Green water**	**Grey water**	**Total water**
1.	A pair of jeans	4900	4450	1500	10,800
2.	A cotton T-shirt	1230	1110	380	2720
3.	A pair of diapers	370	330	110	810
4.	A cotton swab	1.6	1.5	0.5	3.6

the quantity of emitted chemicals after use into the volume dilution that is required to calculate the pollution level. When cotton cultivation is at the field level, there is a depletion of resources (green water, blue water and fertilizers and pesticides) from the environment; in return polluted water is sent back, which results in the pollution of resources. Grey water is water from bathrooms, kitchens and others which can easily be recycled. The water footprint is the amount of water present in the product, not in terms of content in the product but rather the amount of water used to manufacture the product. The water footprint of cotton products is given in Table 2.4. The stages of production are carried out in different parts of the world to manufacture a finished product. For instance, Malaysia exports cotton clothes to European markets with the combined ventures of importing raw cotton from China and uses India and Pakistan to process and manufacture the product for export. Hence tracing the origin of a product is essential for analysing the impact of a product on the environment.

It has been reported that more than half of global cotton fields are irrigated and a major volume of water has been withdrawn for irrigation. A major portion of the water requirement of India and Pakistan is met by supplementary irrigation, and in Pakistan about 31% of all irrigation water is drawn from groundwater. The province of Xinjiang, China is entirely irrigated and intensive freshwater use has led to a decline in the water tables. Most cotton production depends on furrow irrigation, and sprinkler and drip irrigation methods are scarce. The estimate is that 0.7% of the land in the world is irrigated by these methods (Sandra et al., 1996).

2.2.5 *Contamination and degradation of water bodies*

Cotton cultivation is usually associated with the use of chemicals, and the runoff water contains nutrients, salts and pesticides. Excess irrigation compared with ET requirements and leaching of water to maintain salinity create favourable conditions for groundwater pollution. Excess irrigation and leaching in cotton cultivation call for the withdrawal of a huge volume of water from water bodies, creating the need for replenishment; when effective replenishment measures are not performed, there will be a shortage of water in the downstream area, resulting in the degradation of water bodies and the environment. Restrictions in water supplies call for the effective use of water for agriculture and crop management. This is a major problem in Central Asia, the Aral Sea and Uzbekistan.

2.2.6 Extensive irrigation

Cotton crops cultivated by extensive irrigation have the tendency to produce increased runoff into the groundwater because of greater percolation, which leads to rising groundwater tables. This culminates in the establishment of shallow water tables, as seen in the lower Macquarie Valley of New South Wales, Australia. Irrigation affects deep percolation with an increase ranging from 17 to 202 mm/year, which could be equated to leaching rates ranging from 3% to 25%. The potential rise in groundwater varies from 37 to 524 mm/year. It was also noticed that the largest increase in deep percolation rates were present when the water table was closest to the soil's surface. This study suggests that recharge resulting from irrigated agriculture is related to the development of shallow water tables (Willis and Black, 1996; OECD, 1998). This condition also leads to a dry climate and the salinization of soils. This condition is inevitable in regions where evapotranspiration (ET) exceeds rainfall and the amount of freshwater used for irrigation, e.g., Pakistan (15% of irrigated area), Uzbekistan (50% of irrigated area) and Brazil's reports of salinity (Gillham, 1995; OECD, 2010).

2.2.7 Construction of dams

The construction of artificial storage systems such as damns interrupts the natural water flow and adjacent ecosystems. This disruption of inhabitants depends on the location and size of the dam. A dam is considered to be large when the dam wall is higher than 15 m and has a crest length over 500 m with a reservoir capacity greater than 1,000,000 m^3. Such a dam is usually built to generate hydroelectricity, water for irrigation and domestic and industrial uses, and to control flooding. It has been estimated that five countries (the United States, India, China, Spain and Japan) together have more than 5000 dams higher than 30 m, which constitute 60% of the world's total dams of that size. Europe, Asia and North America have 88% of the world's dams; Africa has only two dams. Rogun Dam and Nurek Dam in Tajikistan are among large dams with a height × length of 335 × 660 m and 300 × 704 m, respectively (OCED, 1998). Whatever the statistics, dams generally have considerable impacts on the environment in the upstream and downstream areas as well as at the regional and global levels.

The environmental consequences of large dams may be directly related to the properties of the rivers and riparian or streamside environments. Fish migrations may be blocked by the dam wall and the breeding habitats may be separated from the rearing habitats. Sediments which are essential to maintain the processes and downstream habitats are blocked, which affects the maintenance of productive deltas, barrier islands, fertile flood plains and coastal wetlands. When the river is deprived of the sediment load, it tries to recapture it by eroding downstream river beds and banks, which can extend to tens or hundreds of kilometres below the dam.

In the upstream area the natural flow of the river is transformed into an artificial slack water reservoir, which alters the temperature, chemical composition and dissolved oxygen levels, making it unsuitable for aquatic plants and animals that thrive in this ecosystem. The life dependent on the river is conditioned to the timing and quantities of the flow of the river. Altered water flows are similar to dewatering the river and affect the aquatic and riparian life, unravelling the ecological web of the river system. Extinction of fish and aquatic species, disappearance of birds in flood plains, huge losses in forest, farmlands and wetlands and erosion of coastal deltas are some of the innumerable impacts of dams.

River bed deepening will lower water tables and water availability will be reduced. In 2002, low flows below the dam killed thousands of salmon in the Klamath Basin in North America. Nonnative and invasive species that find settlement in reservoirs further destabilize the natural communities of plants and animals that are found locally in abundance along the river. Studies have also identified another problem associated with dams and reservoirs. Greenhouse gas emissions from reservoirs are on par with the aviation industry (International Rivers, 2016; Rosenberg et al., 2016). It has also been shown that sediment and nutrient flow shifted biological processes of the Congo River far into the Atlantic Ocean, and the river also serves as a reservoir for greenhouse gases.

2.2.8 Land reclamation

The increase in arable land has been linked with increased agricultural activity and the change from natural ecosystems and landscapes. Cotton cultivation results in saline soils owing to water logging and leaching. Salt-affected soils are generally classified as sodic soils and saline soils based on their characteristics and genetic profile. Soils that contain excess water-soluble salts (sodium bicarbonate/carbonate or sodium silicate) and sufficient interchangeable sodium tend to interfere with the growth of plants and are called alkali or sodic soils. Saline soils, on the other hand, contain neutral salts (chlorides and sulphates of sodium, calcium or magnesium) in quantities that affect the growth of crops. The characteristics of sodic and saline soils are given in Table 2.5.

Table 2.5 Characteristics of alkali and saline soils (Silvertooth, 2001; Indrajit, 2016)

Sl. no.	Type of salt-affected soil	pH (1:2)	Electrical conductivity (dS/m)	Sodium (%)	Salts in excess
1.	Alkali/sodic soil	>8.5	Variable	>15	Carbonate and bicarbonate of sodium salts
2.	Saline soil	<8.5	>4	<15	Chloride and sulphate of calcium, magnesium and sodium salts
3.	Saline-sodic soil	<8.5	>4	<15	Chloride, sulphate, carbonate and bicarbonate of calcium, magnesium and sodium salts

It was reported in studies by the Central Soil Salinity Research Institute in the Sharda Sahayak Irrigation project (Uttar Pradesh), the Indira Gandhi Irrigation Project (Rajasthan), the Kakrapar Right Bank Canal (Gujarat) and the Western Yamuna Canal and Bhakra Canal Systems (Haryana) that there was a loss of yield in crops in salt-stressed soils compared with normal soils. The cotton yield in normal soil was 1.63, in salt-affected soils it was 0.61 (a reduction of 63%) and in water-logged soil it was 0.37 (reduction of 77%) (Tripathi, 2009). This clearly indicates the necessity of land reclamation and the management of problematic soils.

The most preferable areas for agriculture and irrigation schemes are the flood plains and wetlands owing to their flat shape and fertile soil. The monocultural cultivation of cotton and excessive drainage of the soil make it an unsuitable habitat for the original plants and animals. Furthermore, the remaining natural habitats are small and fragmented and cannot sustain continued existence of the natural ecosystem. This affects the terrestrial ecosystems and wetlands as well as the freshwater ecosystems in rivers or lakes. Deforestation, overgrazing and land reclamation practices, together with the growing population, have reduced the protective cover area to 20% of the entire area in the Loess Plateau of China's Yellow River Basin (FAO, 2011). Although the area for cotton cultivation has been stable since 1930, there is a need for land reclamation because of the change in overused farmlands to new cultivable areas. This may be attributed to inappropriate irrigation practices, water logging and agricultural mismanagement. According to a National Council of Educational Research and Training publication, in India, alkalinity and salinity have affected 8 million hectares of land and another 7 million hectares have lost fertility by water logging; the excessive use of pesticides has caused the accumulation of toxic components to the soil (NCERT, 2015–2016). Reports also stated that damage by salinity of the irrigated area in top cotton-producing countries is between 12% and 36% (WWF, 1999; Shortle and Griffin, 2016).

Issues regarding conventional cotton cultivation have led to the massive use of resources, causing various concerns in many parts of the globe. This calls for various rehabilitative and protective measures that will help grow healthy sustainable crops to develop products and preserve the environment for future generations. Sustainable crop management helps to foresee all of the forthcoming difficulties in cotton farming and tries to resolve it with the resources on hand.

2.3 Sustainable crop management

After World War II, agriculture saw vast growth because of new technologies, mechanization, the use of chemicals and the specialization of government policies promoting productivity. Technological improvements and reduced labour demands increased productivity but reduced raw materials and resources in many countries across the world. However these farming practices had another side: top soil depletion, a decline in family farms, the contamination of groundwater, increasing costs of production, paltry living and working conditions for farm workers and the breakup of economic and social conditions in rural communities. Over the past 2 decades, the focus has been on promoting agricultural practices that alleviate these agricultural and social problems. Sustainable crop production and management are becoming widely accepted because they address many problems and opportunities for all stakeholders in the entire food system.

Sustainable agriculture envisages three goals: environmental health, economic profitability and social and economic equity. In this context it is necessary to maintain and enhance land and natural resources and take into account the conditions of agricultural labourers, the needs and requirements of rural people, consumer safety and health for the present as well as the future. Sustainable crop production is the ecological and ethical way to raise crops by adhering to practices that do not harm the environment, provide fair treatment to workers and support rural and local communities. In the United States, industrial crop production existed and agricultural produce was grown as a commodity to serve as raw materials for industrial goods rather than direct human consumption. Industrial crop production relied on mono-cropping, intensive application of fertilizers and pesticides, genetically engineered seeds, intensive irrigation and heavily mechanized farming methods, which led to an increase in corn and soy production. Animals and consumers were fed with these agricultural products either in their natural form or as 'junk foods', which brought widespread problems such as heart disease, obesity and type 2 diabetes (Grace Communication Foundation, 2016).

Sustainable crop management focuses on agricultural techniques that are sustained over a long period without damage to the environment, or with only minimal damage. The next section provides a view of the principles which help in sustainable crop production and manufacture.

2.3.1 Importance of traditional methods of agriculture

Traditional agriculture safeguards agro-biodiversity, genetic diversity and local ecological knowledge. Despite the industrialization of agriculture, many farmers continue to work with small-scale diverse agro-systems. In Latin America, 75 million traditional farmers have farms with an average size of 1.8 hectares and contribute to 41% of agricultural output for domestic consumption (Altieri, 2004). The distribution and extent of traditional agriculture are given in Table 2.6.

Furthermore, knowledge of traditional agro-systems and associated plant diversity results from the union of natural and social systems leading to diverse agricultural landscapes for multiple use and local food self-sufficiency. Ethnobotany is folk knowledge about agriculture, land characteristics and environmental conditions; it analyses the best traditional methods which were most adaptive and passed on from generation to generation. Most traditional agriculture is place or location specific and may not be suitable for other places. Small farmers generally tend to mimic nature by designing a crop system which simulates the wild species of the particular habitat. This coexistence with natural surroundings and biotic intricacy result in successful cropping systems and high productivity which promote both productive and protective roles.

Table 2.6 Traditional agriculture statistics in the developing world (Altieri, 1999)

Sl. no.	Region	Area	Contribution
1.	Latin America	38% of total agricultural land (60.5 million hectares)	41% of domestic consumption
2.	Brazil	30% of total agricultural land (4.8 million farms)	50% of land for food crops
3.	Cuba	1.5 million hectares	10% of all food crops
4.	Africa	100–150 million hectares	80%–95% of food crops
5.	Asia	27.8 million hectares	Supports 250 million people
6.	Global estimate for developing world	50–100 million hectares	30%–50% of basic food crops

2.3.2 Multicropping

The selection of crops for cultivation is an art and requires careful planning in agriculture. Rapid industrialization in farming has brought about mono-cropping in place of multicropping. An agricultural method of growing multiple crops on one piece of land during the same or successive growing seasons is known as multicropping. Practices such as changing crops from season to season (crop rotation) or planting two or more different crops in close proximity to neighbors, or as cover crops (intercropping), helps guard plants from insect and pest infestation, fixes nitrogen, provides shade, improves soil health and increases water retention. The cotton plant has slow initial growth and when there is a wide space between rows of plants, short-duration crops such as green gram, black gram or soybean can be grown to increase soil fertility and maximize monetary returns without affecting the growth of the cotton plant (Dury et al., 2011; Sankaranarayanan et al., 2011). Similar results were noted in multitier cropping system studies in which cotton was grown with radish, cluster bean and beetroot as one crop system and cotton, radish, beetroot and coriander as another. The study revealed that the root volume and leaf area of the base crop (cotton) and the intercrops (vegetables) were significantly influenced by the multitier system when estimated at 45 days after sowing, as shown in Fig. 2.4 (Sankaranarayanan et al., 2011).

Multispecies or multicropping systems provide environmental services apart from agricultural products (Fig. 2.5). In addition to controlling pests and insects and increasing productivity, the greater diversity of plants in the field hosts a number of microbial organisms, insects, birds and mammals above and below the ground, leading to soil enrichment and fertility – conservation of biodiversity; different cropping plants have different root systems which bring up nutrients from the deeper soil layers to increase the use of nutrients and prevent nutrient leaching from the soil layers – nutrient recycling; multispecies agriculture also increase soil cover in the top soil, preventing runoff on the soil surface, thereby improving the watershed scale and water quality of rivers

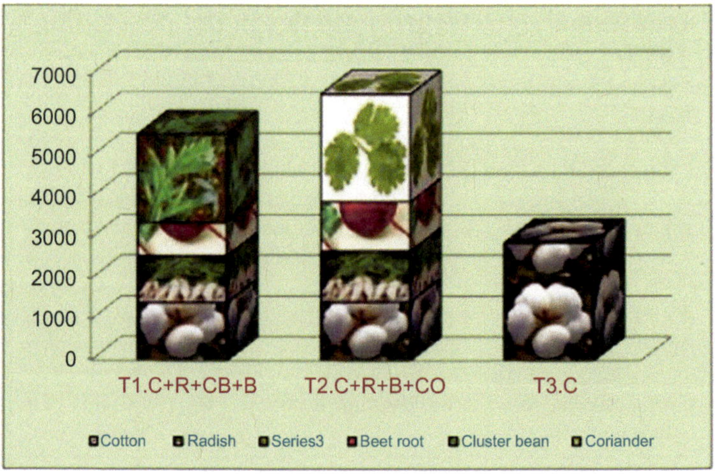

Figure 2.4 Leaf area at 45 days after sowing of multitier cropping system (Sankaranarayanan et al., 2011).

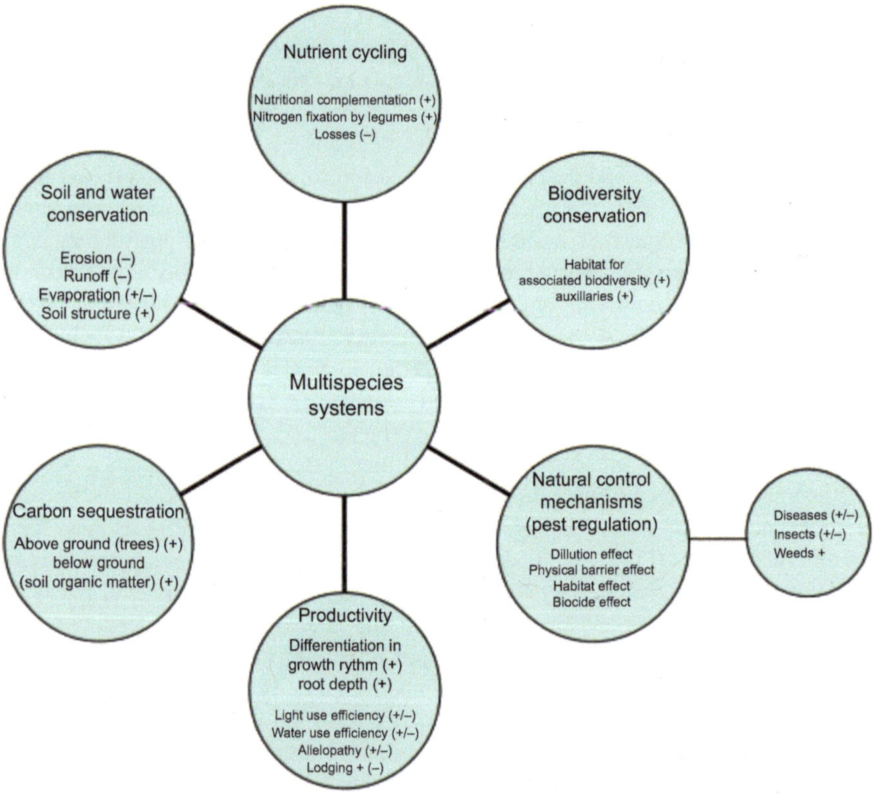

Figure 2.5 Benefits of multispecies systems (Malezieux et al., 2009).

and reducing the intensity of floods – soil conservation and water quality; multispecies systems sequester carbon over the crop stands, thereby enhancing the carbon content of the soil (Malezieux et al., 2009; Altieri, 2002; Swift et al., 2004; Perfecto et al., 1996).

2.3.3 Minimal/no pesticide use

Crop protection is essential to survive competition from weeds, animals, pests, pathogens and viruses. An estimate for losses despite crop protection practices during the period 1996–98 across 17 regions for a few crops (wheat, rice, maize, barley, potatoes, soybeans, sugar beets and cotton) varied from 50% to 80%, with 40% for cotton. It was also noted that weeds occupied the highest potential loss (32%) followed by animal pests (18%), pathogens (15%) and viruses (7%–1%) (Oerke and Dehne, 2004). The use of chemicals as pesticides in the form of herbicides, insecticides, microbicides and fungicides (Fig. 2.6) has increased agricultural production but there are many side effects to their use. These consequences include damage to agricultural lands, fisheries and flora and fauna; unintentional destruction of the beneficial killers of pests, which increases the virulence of many agricultural pests, e.g., brown planthopper (*Nilaparvata ingens*) and the rice gall midge (*Orseolia oryzea*); and increased mortality and morbidity of humans as a result of exposure to pesticides and consumption of food grown in the environment (Tisdell & Wilson, 2001; Fitt, 2000). Much discussion is being held in developed and developing countries regarding integrated pest management (IPM) and safe crop protection to prevent negative effects on the environment. Experience in ecosystems show that therapeutic interventions provide relief only in the short term and are soon neutralized by changes within the system. Pest management strategies should take place over the long term in nature, at multitrophic levels with built-in preventive systems and therapeutic tactics as backups to these natural regulators (Tisdell and Wilson, 2001). Suggested alternatives to pesticide use are IPM, intercropping and companion planting, mulching, ground cover and manual control and release of beneficial insects and organisms.

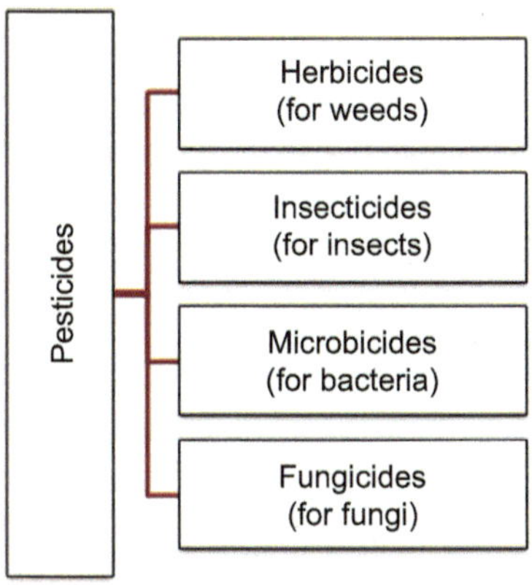

Figure 2.6 Types of pesticides.

2.3.4 Soil health focus

Soil management is important for maintaining and enhancing the structure and fertility of the soil. Studies show that plants grown in a sustainable environment are rich in vital micronutrients and macronutrients owing to good soil health, which is a result of organic growing methods and sustainability practices. Soil structure includes the characteristics of soil particles and the soil pores which determine air, water and nutrient movement, root growth and space for soil organisms. Microorganisms and organic matter have a major role in soil health and fertility. Microorganisms convert plant material to humus, which becomes a binding agent for stable soil structure (BCI, 2013).

Two extremes are prevalent across the globe: overuse of fertilizers in the European Union has threatened the sustainability of 70% of nature whereas the underuse of fertilizers in sub-Saharan Africa exports soil nutrients with the crop without replenishment, leading to soil degradation (FAO, 2016; BCI, 2013). The fertilizers nitrogen and phosphorus are a boon to farmers and a disadvantage to the environment. The use of mineral fertilizers per hectare is highest in Asia and Europe and results in environmental problems such as soil and water acidification, contamination of surface and groundwater contamination and increased greenhouse gas emissions. In China, the uptake efficiency of nitrogen is less than 20%–28%; the rest is lost into the environment (FAO, 2016). Some sustainable practices to promote soil health are organic fertilizer use, crop rotation, intercropping and mulching, less or no tilling of soil and a reduction of heavy machinery use.

2.3.5 Choosing sustainable seed and plant varieties

Heirloom crops are those that are have been grown in the past and generally are not used for industrial crop production. They are chosen for their taste and nutritional value and are grown in a particular environment that makes them more resistant to the local pests and best suited to the local climate. Hybrid varieties are available for large-scale industrial production in which the seeds are sterile and new seed stock is usually controlled by agricultural business agents. Genetically engineered crops are used to control pests or allow the greater application of herbicides, or to suit consumer demands, e.g., genetically modified apple varieties are grown which do not become brown when they are cut. Sustainable agriculture does not promote gene modification because of the potential impact it would have on the environment, their contribution to the health of the consumer and the large amounts of chemicals that would be used for their growth.

Genetically engineered cotton produces toxic proteins from the soil bacteria *Bacillus thuringiensis* that kills the pink bollworm (*Pectinophora gossypiella*) that bores into the balls of cotton. The tissues of these plants contain Bt toxins, and when the worms or insects eat the plant they die; such pests cannot be controlled by Bt sprays because they are embedded in the plant (Ronald, 2011). Bt toxin sprays are another form of biotechnology-based pesticides which are target specific and do not cause harm to beneficial insects, wildlife and people. These Bt toxins have a long history of testing and assessment. Only when it is confirmed as nontoxic to human safety do the Environmental Protection Agency and the Food and Drug Administration departments allow such a spray for commercial use. In Arizona, the IPM for Bt cotton is effective; it reduced the use of insecticide by 70% and saved more than $200 million in 1996–2008 (Naranjo and Ellsworth, 2009).

2.3.6 Water conservation and sustainable irrigation

Water conservation techniques are important to safeguard the quality of soil and environment. Overirrigation increases the salinity of the soil and results in declining yield. Low-volume irrigation, rain water catchment and planting of drought-resistant crops are some measures adopted for sustainable water conservation in cultivation. Studies have highlighted that water conservation techniques include increasing the output per unit of ET (WUE), reducing losses of usable water into sinks, reducing water pollution (sediments, salinity, nutrients and other agrochemicals) and using water for high-value needs (Howell, 2001). The greatest challenge for the agricultural sector is to produce more food with less water by increasing crop water productivity (CWP). CWP increases when the same production exists with less water resources or when there is higher crop production with the same water resources. A study conducted referring to CWP as the marketable crop yield over actual ET estimated it using the following equation:

$$CWP = \frac{Y_{act}}{ET_{act}} \quad (kg/m^3)$$

where Y_{act} = the actual marketable crop yield (kg/ha); and ET_{act} = actual crop water consumption by ET (m^3/ha).

Through improved land and water management practices and research in agronomy, e.g., plant breeding, the CWP score increased compared with previous studies. Grismer reported in 2002 that in Arizona and California, CWP values for irrigated cotton increased considerably compared with the report presented by Doorenbos and Kassam in 1979 (Grismer, 2002; Doorenbos et al., 1988).

Apart from all of these measures, sustainable crop production has reached a greater dimension. There are many challenges to be addressed and the scarcity of fresh water, which is finite, has to go hand in hand with the increasing global demand for agriculture, industry and consumption by humans. The challenge of meeting water demands in the coming decade will require global efforts towards two goals: (1) improving freshwater ecosystems to perform their ecological functions by ensuring the receipt of the quantity, quality and timing of flow; and (2) increasing water productivity by getting twice the service, satisfaction and benefit from each unit of fresh water obtained from water bodies.

In South Africa's postapartheid democratic government there is a strong environmental policy regarding water management. 'The quantity, quality and reliability of water required to maintain the ecological functions on which humans depend should be reserved so that the human use of water does not individually or cumulatively compromise the long term sustainability of aquatic and associated eco systems' (Postel, 2000). The Murray-Darling river basin states of Australia agreed to allocate 25% of the river's natural flow to maintain ecological health. The Central Valley Project in California has been working to restore the habitat and ecological health of the Sacramento–San Jaoquin river and restore salmon and anadromous fish natural production; the US Fish and Wildlife Service has created the Endangered Species Act to recreate the natural habitat of endangered chubs, suckers and squawfish in the Flaming Gorge Dam on the Green River in Utah. All of these efforts have been performed to restore the natural flow of rivers and natural habitats that depend on rivers.

Irrigation accounts for about two-third of global water extraction. Studies have shown that drip irrigation has the potential to double crop yield per unit of water for 20%–50% of crops cultivated and reduce water use by 30%–60% (Sivanappan, 1994; INCID, 2016). Similarly, dams across rivers are being taken into consideration to restore the natural flow of rivers and safeguard the environment, e.g., Edward Dam on Maine's Kennebec River and the Elwha and Glines Canyon dams in Washington State, United States (Collier et al., 2000; Poff et al., 1997). Apart from these efforts, there are numerous other sustainable crop production methods. They include raising aquatic animals in a symbiotic environment with plants (aquaphonics), raising trees and shrubs along with crops for mutual benefit (agroforestry), combining several agricultural principles such as agroforestry, intercropping, mulching, rain water catchment, etc. (permaculture) and other methods such as rooftop farms, small backyard plots and vacant spaces (urban agriculture).

There are many excellent available techniques for crop management, but the backbone of all of these is the selection of cotton varieties for farming, and the key to this is biotechnological advances that have come into existence. All of the traits that are required for healthy growth and development along with good-quality fibre and yield can be monitored and engineered into the genes of the plant to create efficient hybrids. A lot of research is being carried out to generate hybrids necessary for the future. A look at this aspect of cotton farming with specially engineered varieties is given next.

2.4 Sustainability and selective breeding

The population of the world is estimated to rise from 6.7 to 9 million by 2050 and the world agricultural production needs to rise by 50% by 2030 to cater to the increased demand (Hamilton, 2009). The natural resource of 'land' is limited and there is little undeveloped land that is cultivable because the major portion is lost to urbanization, salinization, desertification and environmental degradation. Similar prospects are predicted for the limited resource of 'water', which is low and insufficient to meet current demands. Reports state that fresh water available per person has decreased by four times over the past 60 years and about 70% of water available has been used for agriculture (Carol, 2016; Dennis et al., 2008). Many rivers do not reach the sea, 50% of the world's wetlands have disappeared and because of water mining, the water tables in many parts of the world such as India, China, Mexico and North Africa are declining by 1 m/year (Yu et al., 2012). Predictions state that climate change will melt glaciers, causing low-level agricultural land to submerge and river systems to have shorter and concentrated seasonal flows, resulting in floods, whereas crop production will decline at temperatures over 30° and as a result of the increase in pests and diseases (Jiang et al., 1998; Yu et al., 2012; Small et al., 1999; Smith and Cothren, 1999).

This situation calls for advances in agriculture to produce crops that are resistant to high temperature, low water availability, rising salinity and changing pathogens and insect threats. Sustainability in agriculture can be achieved by effective land and water policy, IPM, reductions in harmful inputs and the development of new-generation, stress-tolerant agricultural crops (Ming et al., 1998; Annon., 2016; Azimov and Akhmedov, 2013) Austrian monk Gregor Mendel demonstrated that plant characteristics were inherited and not acquired from the surrounding environment. He also highlighted that crossing two plants with different traits would create a plant with improved traits or different characteristics compared with the parental plants. In the 1940s, DNA was identified as genetic material, and the 1960s and 1970s saw a green revolution when crop yield was high compared with previous years owing to advances in biotechnology. In the 1980s and 1990s, the commercial adoption of agricultural biotechnology led to the introduction of specific genes into the same crop or different species. In the early 2000s, the plant genome was sequenced and many plant genome sequences were documented in the GenBank (Hamilton, 2009).

2.4.1 Role of biotechnology for new creations

The involvement of living organisms (biotechnology) in cotton cultivation and growth phases has been beneficial. Biopesticides include biochemicals derived from microorganisms and natural sources, microbial pesticides, and involve modification of the plant genes to express the encoding insecticidal toxins, e.g., control of white fly on cotton by neem products (Mazid et al., 2011). Bollcure is a commercialized pesticide developed by TERI under the patronage of the Department of Bio Technology. This biopesticide affects the growth of the bollworm larvae, reducing it by 88%, as a result of which the larva does not reach the mature stage to damage cotton crops (The Energy Resources Institute TERI, 2016). Genetically modified (GM) cotton is another application of biotechnology in which the GM cotton is engineered with traits that resist herbicides and/or produce toxins that kill insects, e.g. *B. thuringiensis*. GM cotton is currently grown on 25 million hectares around the world in India, China, Pakistan, the United States and other countries such as South Africa, Brazil, Argentina, Egypt and Australia (CBAN, 2013). GM crops can be grown through conventional breeding or mutation breeding methods. Variation breeding or mutagenic breeding is the use of chemicals or radiation to generate mutants with desirable traits to be bred with other cultivars to produce mutagenic plants or seeds.

Recombinant DNA technology is a branch of genetic engineering that removes a known DNA sequence and introduces it into another organism or vector, thereby altering the genotype of the recipient. It involves the study of the arrangement, expression and regulation of genes, the modification of genes to form a changed protein product and the alteration of gene expression to enhance or suppress a particular characteristic followed by making multiple copies of the nucleic acid segment and artificially introducing it into another organism to produce a transgenic with desirable or modified traits (Rao, 2006). This technology is best presented in the study patented in 1991. The tissues obtained from hypocotyl pieces of immature cotton plants were inoculated with a culture of *Agrobacterium tumefaciens* to obtain a regenerable somatic embryo; the bacteria from the infected tissues was removed and simultaneously tissues resistant to a selectable marker were selected and amplified for transformed clonal growth of regenerable tissues to obtain whole plant development (Umbeck, 1991). This process is useful for general genetic engineering of cotton plants and lines.

Selective breeding was known as ‘artificial selection’ by Charles Darwin, a biotech process by which new organisms and plants are developed with desirable characteristics or traits. Cabbage, broccoli, cauliflower, Brussels sprouts and kale have been derived from the same plant *Brassica oleracea*, the wild cabbage. Similarly corn or maize has been developed from a grass called teosinte and physical dissimilarities exist between teosinte and corn by only five genes (Carol, 2016). This concept has the advantage of producing hybrids with desirable qualities but it also has the disadvantage of producing undesirable traits owing to recessive genes found in the parental plants.

2.4.2 Methods of selective breeding

Plants with the phenotype of interest are picked out. A combination of good genes results in desirable traits; this becomes the phenotype of interest for propagation. The plant is allowed to grow by either self-fertilization or cross-fertilization with other plants of similar phenotype of interest. In the F1 generation, the plant expressing the required phenotype of interest is used as the parent of the next generation. The desirable traits are passed down to other generations continuously. This method of selective breeding was carried out unintentionally in the past and has resulted in many domesticated crops available today.

In the second method, two different plant species are crossed to obtain desirable traits of both species. Cross-pollination of genetically compatible trees has been used since the Neolithic period and has resulted in diversified hybrid varieties; e.g., peaches, plums, apricots, cherries and almonds belong to the same family of trees and are compatible. Different hybrids such as pluots, plumcots, apriums, cherrums, pluerries, nectacotums and peacotums emerged from cross-pollination of this family of trees.

In the third method, two members of the same species that possess dominant traits for particular genes are crossed, e.g., long life and quick metabolism of one plant crossed with another possessing fast growth and high yield. The underlying principle is that some offspring produced by the crossing of the dominant genes will have all of the desirable characteristics. Such an offspring obtained from two dissimilar parents is termed a ‘hybrid’ and has most of the desirable traits that were taken into consideration. This selection of plants with desirable traits is carefully studied. When required, breeders take up different combinations to produce plants with much more desirable characteristics than the previous generation. The heterozygous offspring with an allele for each characteristic will have one dominant gene and one recessive gene. Professional breeders document these

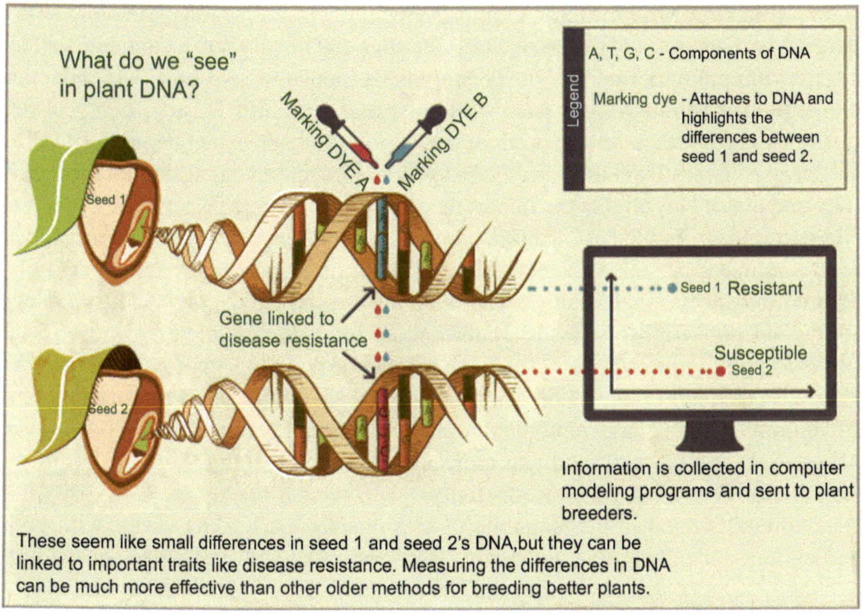

Figure 2.7 Diagrammatic representation of marker-assisted selective breeding technique (Anon., 2016).

combinations in the gene bank for these qualities and perform different combinations of crossing to derive various heterozygous offspring, e.g., AAbb with AAbb will produce an offspring which can be combined with aaBB to produce a heterozygous offspring with all of the desirable traits that are passed on from generation to generation (Anon., 2016).

In selective breeding, breeders use 'marker-assisted breeding' in which scientists examine the DNA responsible for disease resistance, drought tolerance, yield, etc., and markers are used to test all of the plants available for breeding. Accurate selection of the seeds is done to produce plants with desirable traits (Monsanto Company, 2015; Dennis et al., 2008). The plants in which the genetic material (DNA) has been altered do not naturally occur by mating or natural recombination; they are termed GM crops. The process is given in Fig. 2.7.

2.4.3 Research studies on selective breeding of cotton

Allotetraploid Upland cotton (*G. hirsutum*) and Sea Island or Egyptian cotton (*G. barbadense*) vary in quality and quantity of the epidermal fibres on the cotton seed. The chromosomal locations and subgenomic distributions of the quantitative trait loci (QTLs) were determined by a detailed restriction fragment length polymorphism map in a cross between these QTLs influencing the fibre quality and yield located in the 'D' subgenome, derived from the ancestor that provides fibres which do not possess spinnable qualities. Superior fibre yield and the quality of 'D' subgenome QTLs may be partly attributed to the domestication and breeding of tetraploid cottons compared with the parallel improvement of 'A' subgenome diploid cottons. Merging two genomes of different characteristics and history into a common nucleus provides unique opportunities for phenotypic response to selection. The reduction in quantitative variation associated with polyploid formation is partly compensated for and can be a reason for the prominence of polyploids in extant angiosperms. The discovery of genes from a noncultivated genome for increased agricultural productivity and quality highlights the importance of preserving and investigating exotic or wild germplasm. In the case mentioned previously, *G. hirsutum* and *G. barbadense* were combined, but this can be carried out with other nondomesticated AD polyploid cotton species including *Gossypium tomentosum*, *Gossypium mustelinum* and *Gossypium darwinii*, to provide opportunities for valuable alleles to arise. Such synthetic polyploids created by humans have contributed to improving cotton (Yu et al., 2012; Small et al., 1999; Smith and Cothren, 1999). High-density mapping and alignment of the large genome of cotton with small genomes, e.g., *Arabidopsis* (Ming et al., 1998), may facilitate molecular cloning of cotton fibre QTLs for the evolution of fibre quality and productivity.

Cotton cultivation in the Republic of Uzbekistan is a leading sector with great implications for the national economy. Many varieties with fast ripening and high yield are grown in this region but the use of chemicals to control pests and insects produced plants with high toxicity as well as embryotoxicity, carcinogenicity and mutagenicity. Hence a new gene and a selective breeding technique introducing a new cotton variety was invented which had high resistance to insects and extreme ecological conditions as well as economic traits. Hybrids obtained by cross-breeding of allotetraploid midstaple varieties Korotkostebelniy 1 and Akademiya Nauk 14 with synthetic amphidiploid introgressive line IL-296, which was obtained using wild diploid species *Gossypium trilobum Skovsted*, was studied for resistance to *Verticillium* wilt in the first and second filial hybrids. The parameters studied were the stem height, number of sympodial branches, length of the vegetation period, and fibre yield and overall susceptibility to *Verticillium* wilt. All of these qualities showed incomplete dominance in reciprocal first filial hybrids obtained by cross-breeding of the parental cotton species. Traits in reciprocal second filial hybrids were the main stem height, number of sympodial branches and raw cotton weight per boll and susceptibility to *Verticillium* wilt were inherited by type of incomplete dominance. Taking the example of fibre yield and susceptibility to *Verticillium* wilt (Table 2.7), four reciprocal hybrids, F, were in the intermediate position whereas in the second filial hybrid combinations, F, the values of two combinations were dominant over parental varieties. A similar pattern was observed in reciprocal second filial hybrids. This study showed that cross-breeding of synthetic amphidiploid introgressive line IL-296 with cultivated allotetraploid cotton varieties KS-1 and AN-14 inherited a type of incomplete dominance with clear demarcation towards the side of the best initial variety (Azimov and Akhmedov, 2013). Reciprocal first and second filial hybrids showed deviation towards the side of the wilt-resistant line IL-296, which indicated the potential use of this line in gene and selective breeding research to build wilt-resistant cotton varieties.

Table 2.7 **Fibre yield and susceptibility to *Verticillium* wilt for first and second filial hybrids compared with parental varieties (Azimov and Akhmedov, 2013)**

Position of variety	Variety used	Fibre yield (%)	Susceptibility to *Verticillium* wilt (%)
Initial parental varieties	KS-1	37.2±1.36	39.6
	AN-14	35.3±1.86	50.2
	IL-296	32.3±1.17	2.7
First filial hybrids	IL-296×KS-1	35.8±1.27	10.2
	KS-1×IL-296	35.7±1.73	9.6
	IL-296×AN-14	33.5±1.64	13.9
	AN-14×IL-296	34.7±1.67	9.8
Second filial hybrids	IL-296×KS-1	36.1±1.63	16.8
	KS-1×IL-296	36.4±1.35	18.9
	IL-296×AN-14	36.2±1.65	19.4
	AN-14×IL-296	35.8±1.73	21.4

2.4.4 Sustainable cotton varieties

The first discovery of *B. thuringiensis* was by a Japanese biologist, Ishiwata Shigetane, in 1901, followed by scientists in Germany. Since the 1920s spores and crystalline insecticidal proteins produced by *B. thuringiensis* have been used as liquid sprays to ward off insects such as Lepidoptera (moths and butterflies), Diptera (flies and mosquitos), Coleoptera (beetles), Hymenoptera (wasps, bees, ants and sawflies) and nematodes. Upon sporulation, *B. thuringiensis* forms protein-based insecticidal toxin crystals encoded by cryogenes located in the plasmids. When insects ingest these crystals, the alkaline digestive tract converts these crystals into soluble form and combines it with protease enzymes to liberate the toxin from the crystal into the insect gut, paralysing the digestive tract. The insect stops eating and starves to death. Similarly other Bt insecticidal proteins have been discovered, including vegetative insecticide proteins and parasporin (Dean, 1984; Hilder and Boulter, 1999; Ferre and Rie, 2002; Rao et al., 1998; Brar et al., 2006; Oppert, 1999).

In 2009, scientists observed that the pink bollworm that affects the cotton plant had become resistant to first-generation Bt cotton which had one Bt gene, *Cry1Ac* (Monsanto Company, 2009). This was noticed in Gujarat, India, Australia, China, Spain and the United States. A second-generation Bt cotton with multiple Bt proteins was introduced. In China, mirids completely destroyed Bt cotton cultivation and in India mealy bugs and aphids were problems for small farmers, but on the whole the use of pesticides reduced the infestation (Bagla, 2010; Tabashnik et al., 2008; Lu et al., 2010; Wang et al., 2009; Zhao et al., 2011; Goswami, 2007; Sarin, 2007; Stone, 2011; Krishna and Qaim, 2012).

Roundup Ready Flex Cotton has a modified gene technology that tolerates applications of glyphosphates. The main action of glyphosphates is to interrupt the production of proteins essential for the plant's survival. Glyphosphate acts through a specific biochemical process called the 'shikimate pathway' which hinders the conversion of photosynthesis sugar into the EPSPS (5-enolpyruvyishikimate-3-phosphate synthase), which is essential for the amino acids phenylalanine, tryptophan and tyrosine; these amino acids are responsible for the vital biochemical reactions of the plant (Monsanto Company, 2016a,b; CFS, 2007; SourceWatch, 2016). Soon these proteins are not available and growth-inhibiting processes occur, leading to senescence. The modified form of EPSPS enzyme, the CP4 strain, allows the production of amino acids and is resistant to herbicides, as shown in Fig. 2.8.

Transgenic cotton resistant to lepidopteran pests was first commercialized in Australia and the United States in 1996–97. Bollgard II contains two *B. thuringiensis* genes, Cry1Ac and Cry2Ac, which control certain lepidopteran larvae when they feed on it. Two Bt genes have been incorporated to broaden the spectrum of worms that can be curtailed and increase the time span of resistance buildup against one to two genes or delay resistance buildup. Cry1Ac levels are expressed as 1–3 ppm and Cry2Ac as 7–19 ppm (ICAC Recorder, 2003; Kurmanath, 2015). Furthermore, insects with tolerance to one protein will generally be removed by susceptibility to the other protein (Monsanto Company, 2016a,b). This technology reduces the number of insecticide sprays that ought to be used, saving money and effort and enhancing the growth of the beneficial insect population.

Weed control in cotton is important to increase the yield and quality of cotton. Cotton is sensitive to weeds in the early stages, which results in unacceptable stands and reduced yield. BXN cotton is a cotton variety genetically engineered with the BXN gene to resist the herbicide Buctril4EC. When applied to non-BXN cotton plants, this herbicide results in instantaneous death. There are two varieties, BXN 47 and BXN 49B. Credit facilities and cash awards were given to growers who purchased this variety of cotton (Edmisten and York, 1995; Robinson, 2002).

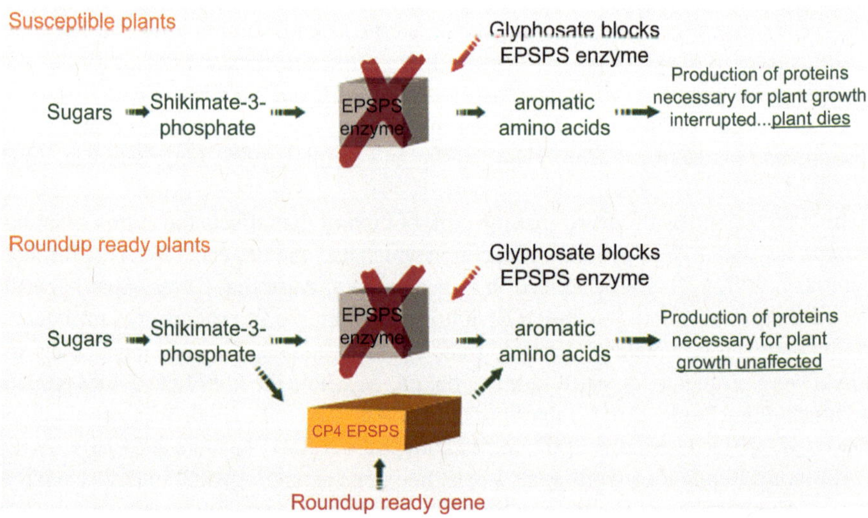

Figure 2.8 Mechanism of Roundup Ready Flex cotton (Monsanto Company, 2016a,b).

Sustainability does not stop with the cultivation of cotton; it continues to all stages of production and moves across the supply chain. Any product which is manufactured is promoted along the lines of sustainability, and consumers look for credentials in the product which certify that the product is eco-friendly or made with eco-friendly raw materials and processes. The selection of raw materials and techniques for processing are important for securing a label or certification for eco-friendliness or sustainability. Sustainable cotton processing can bring manufacturers the most sought-after certification to market their products and give consumers a reason to buy the products.

2.5 Sustainable cotton processing

Sustainability in cotton processing is the goal for any government, buyer, retailer, processor, manufacturer and intermediaries of the supply chain as well as the consumer. Sustainability will not be fulfilled without the words 'organic cotton' or 'eco-friendly processing' and without an impact on the environment, economics and society, which are the three pillars of sustainability.

2.5.1 Sustainable cotton fibres

2.5.1.1 Organic cotton

Organic cotton has been defined in many ways. Cotton grown without using synthetic chemicals either as pesticides or as growth promoters, thereby supporting biodiversity and biological cycles, may be termed organic cotton. In the United States, cotton obtained from plantations that meet the requirements of the National Organic Programme (NOP), US Department of Agriculture (USDA) is considered 'organic'. The worldwide annual production of organic cotton was 26 million tons in 2014–15 and was grown in 35 countries. About 90% of all the cotton worldwide is of the *G. hirsutum* species (Organic Cotton, 2016a,b; Organic Cotton Plus, 2016). According to the Textile Exchange, 2010 Global Market Report on Sustainable Textiles, the global sales of organic cotton apparel and home textile products was $ 5.16 billion; it showed an increase of 20% compared with the 2009 market (Organic Trade Association, 2012). The benefits of organic cotton are given in Fig. 2.9.

2.5.1.2 Naturally coloured cotton

Peruvian Pima and Tangüis cotton are naturally coloured in a range including oranges, browns and greens. The fibres have short staple length but are advantageous because dyeing is not necessary. Scientists can work on coloured cotton fibres and produce long coloured cotton fibres with the help of biotechnology. Most coloured cotton varieties are organically grown and needs only minimal maintenance with little water, pesticides and fertilizers. These fibres have inbuilt fire retardant qualities and the seeds also have antibiotic properties.

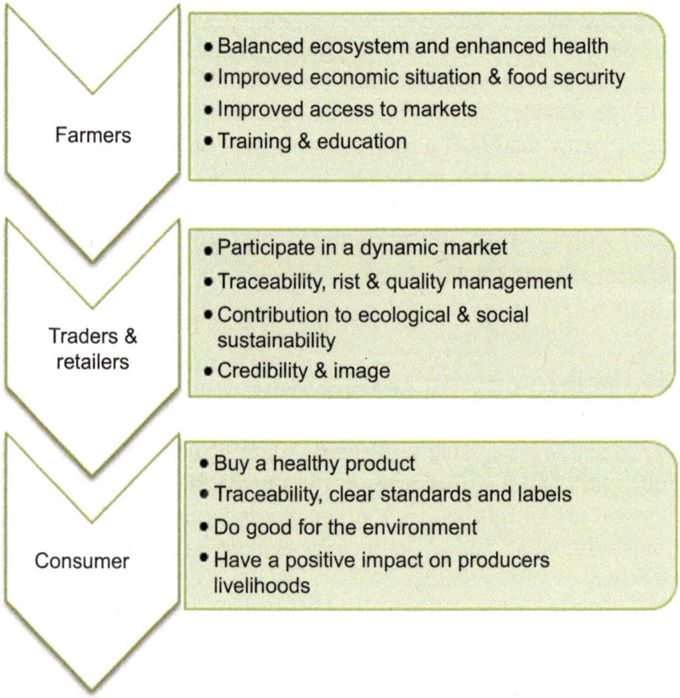

Figure 2.9 Benefits of organic cotton (Organic Cotton, 2016a,b).

2.5.1.3 Fair trade cotton

Cotton that fulfils the requirements of the Fair Trade Standard for seed cotton and is harvested with the seed and fibre attached is certified with the Fair Trade mark. This certification mark ensures that farmers obtain a fair and stable market price for their products without price exploitation, because they are on the last rung of the supply chain ladder. The Fair Trade Organization sets a basic price at the farm level on the basis of the actual cost of the sustainable produce. The Fair Trade Organization aims to improve the livelihood of small and marginalized farmers in developing countries and to widen benefits for the community. Although criteria for Fair Trade cotton differ from those of organic cotton, farmers implement the integrated crop management system and biological pest control systems and efficient water management for easy transition from conventional to organic farming. However Fair Trade does not promote GM seeds (SFF, 2016).

2.5.1.4 Bt cotton

The US and Australia cotton industries have promoted research in the field of biotechnology to reduce the use of insecticides which cause harm to the environment and habitats. Gene modifications are being done to develop GM cotton which is resistant to insects. Estimates report that there is around a 70% reduction in the use of insecticides and fertilizers, according to the Sustainable Cotton Project in California, through biological IPM techniques and through biological agricultural systems in cotton.

2.5.2 Sustainability in conversion of cotton to finished products

Working with nature is the principle of organic farming. Sustainable practices are followed in cotton cultivation in terms of soil, weed control, pest control and harvest mechanisms. Ginning of cotton removes fibres from the seed and the cotton seed oil is used to make many products. The cotton seed residues of conventional cotton are used as animal feed whereas the organic cotton seed remnants are used to make food products.

The manufacturing process includes spinning, weaving, knitting, dyeing and finishing. In the manufacture of organic clothing, petroleum-based scouring agents, formaldehyde, antiwrinkling agents, chlorine bleaches and other inauthentic materials are not used. Natural spinning oils that biograde easily are used for spinning; potato starch is used for starch or to size the yarns; coloured cotton, low-impact dyes, natural dyes and coloured clays are used for dyeing or colouring; and natural, mineral or vegetable inks and binders are used for printing, to eliminate toxic substances in the organic clothing. The whole process is monitored and alternatives are used for conventional chemicals and finishes (Menezes, 2016).

Pretreatment is a process required for all cotton textiles before any other value-added process such as dyeing, printing and finishing. Enzymes have replaced conventional pretreatment stages including desizing, scouring and bleaching. An approach was employed in 2004 in which all of the pretreatment processes were performed in a single bath with various enzymes such as enzymatic desizing, bioscouring and hydrogen peroxide bleaching after optimizing the different enzymes for application (Ibrahim et al., 2004). The fabrics were checked for residual size, weight loss, wettability, yellowness and whiteness index and dyeability with reactive dyes. Optimal conditions for all of the various enzymes to be used in a single bath were determined and the properties proved to match the accepted quality limit standards. Enzyme processes have also been enhanced using ultrasonic assistance in pretreatment and dyeing (Yachmenev et al., 2002; Eren and Erismis, 2013; Agrawal et al., 2007; Calafell et al., 2005).

In cellulosic reactive dyeing, a low material liquor ratio is used, from 1:8 to 1:10. After dyeing the hydrolysate has to be removed, which requires more washing and rinsing. The equipment employed for dyeing uses air to propel the fabric through the dye machine. This allows the dyeing of fabric to take place near the wet pickup of the material. This may require capital investment, but in the long run it produce benefits for energy and water savings. Reactive dyes are available as bifunctional or heterofunctional dyes which contain two or more reactive sites that help easy application and fixation. Low-salt dyes form the bridge between the gap of two dye molecules, providing easy exhaustion and high colour value as a result of two chromophores in one dye molecule.

Cationization of cotton is a sustainable process which helps in good dye uptake and 100% dye use. Cationized cotton is beneficial because it uses less of time, water

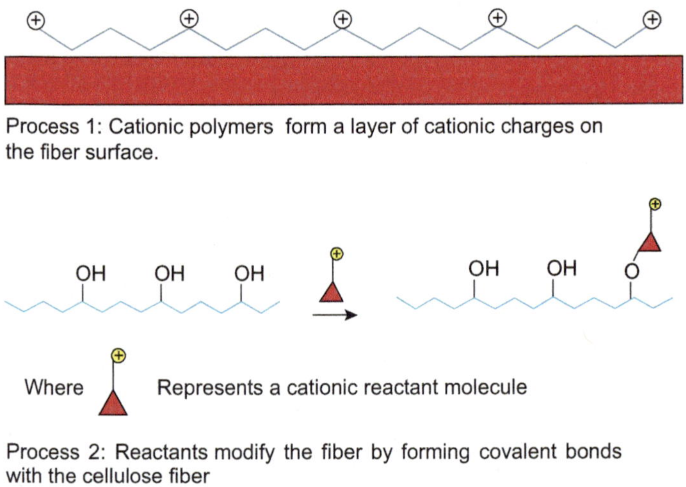

Figure 2.10 Cationic pretreatment processes (Hauser, 2000).

and energy and provides other benefits: it eliminates salt, gives shade reproducibility, requires fewer rinses and has a low material liquor ratio and cycle time (Farrell, 2012). The most commonly used cationic agents are choline chloride-4 and *N*-(3-chloro-2-hydroxypropyl) trimethyl- ammonium chloride-9. Fundamentally two types of cationic pretreatments are given to cotton, as shown in Fig. 2.10. In one process the cationic charges form a layer on the fibre surface; in the other process the reactants modify the fibre surface by forming covalent bonds with the cellulosic fibre. The cationic charges attract anionic dye particles, which fix themselves tightly to the pretreated material compared with the one which is not cationized (Hauser, 2000; Ristic and Ristic, 2012).

Finishing is an area that uses water, energy and chemicals. Enzymes are replace harsh chemicals and reduce the impacts on the environment and habitats. Denim abrasion and colour change with enzymes consumes less water and energy. Clariant Business Textiles uses a patented Foam EcoCare technology to apply dyes and finishes to fabric without immersing them into water; this saves water and increases the fabric strength. Central Textiles, Hong Kong, a leader in denim processing, uses vibrating membrane technology to separate indigo dye from the waste stream to reuse dye and water (Anon., 2013). Digital printing on cotton uses computers for printing; it demonstrates good efficiency and sustainability. The recycling and reuse of materials also adds to sustainability. For example, ultrafiltration is used to recover sizing agents such as polyvinyl alcohol and carboxymethyl cellulose, and polyacrylate and alkali can be recovered from spent lye after mercerization by a caustic soda recovery process (Zietlow, 2013).

Sustainability of the cotton supply chain can be well addressed by guiding and partnering all personnel in the supply chain. The enhancement of facilities and knowledge and empowerment can create more awareness and enrichment, helping to raise the livelihoods of all involved. Efficiency of operations can be estimated by making alliances with organizations such as Field to Market: The Alliance for Sustainable Agriculture during planning and decision making. Following responsible labour and International Labour Organization practices, clean technology and healthy environment will serve to address the social aspects of sustainability.

Looking into the importance of cotton, the key issues associated with cotton cultivation, methods of sustainable crop development and management followed by environmentally friendly methods of processing can result in products that will fulfil the mandatory requirements for sustainability that have become essential. However many steps need to be taken for future improvement and development along the path of sustainability. Voluntary sustainable initiatives and the road map for sustainable cotton production are appraised here to raise consciousness about sustainability.

2.6 Future trends and conclusion

The diversity of the farming system in different geographic areas, challenges owing to the specific nature of sustainability and the collaborative association of different components of sustainability pose many problems to stakeholders who raise and process sustainable cotton. All stakeholders need to understand potential threats that are relevant to the sustainability of a particular farming system and think of methods to ensure the measurement and benchmark for sustainable production and the improvement of livelihoods. Policies and governance structures need to be holistic in their approach and work in a cross-sectional or interdisciplinary manner across the three pillars of sustainability. This process will preserve and enhance the quality of natural resources without compromising the quality and quantity of agricultural production. All industries are expected to use natural resources in a sustainable manner and the results of this attempt need to be revealed to consumers as sustainability credentials. Sustainability monitoring and reporting are essential, and a set of sustainability indicators need to be followed to evaluate the system adopted towards establishing the goal.

A rise in voluntary sustainability initiatives (VSIs) has occurred over the past decade. Initiatives, standards or methodologies that enhance and measure sustainability outcomes in agricultural production systems are termed VSIs. These initiatives not only consider agricultural systems but also stages of any value chain. These initiatives and endorsements will help consumers to identify eco-friendly and environmentally preferable products. Large retailers are focused on showcasing the social and environmental integrity of their products whereas public sector developmental programmes encourage sustainable agricultural practices and protecting farmers and the environment. Many countries across the world have started looking into sustainability issues and developing resource management systems aimed at the responsible use and conservation of natural resources.

Some important VSIs that are generic to agriculture but have a potential impact on cotton, as well as those specific to cotton, are given in Fig. 2.11.

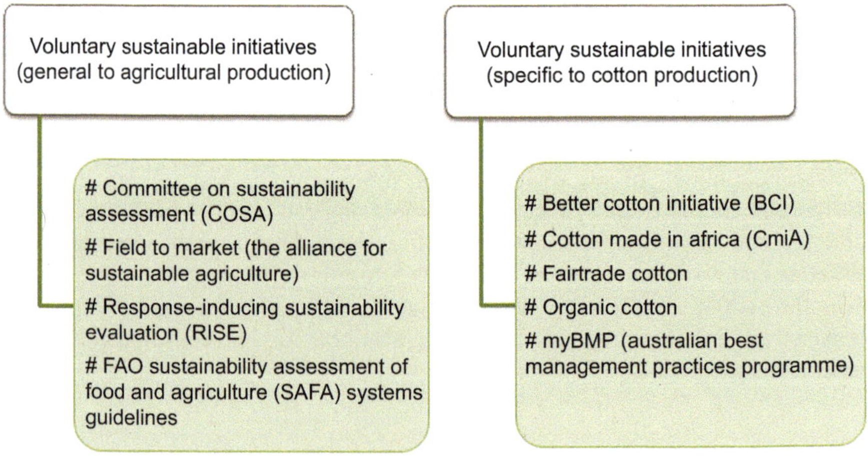

Figure 2.11 Voluntary sustainable initiatives (FAO, 2015).

2.6.1 Voluntary sustainable initiatives pertaining to cotton production

2.6.1.1 Better Cotton Initiative

In 2005, the World Wildlife Fund (WWF) invited experts from different parts of the globe to find sustainable solutions for farmers and the environment and the future of the that sector. The Better Cotton Initiative (BCI) was first started; it was supported by retail giants such as Adidas, Gap Inc., H&M, ICCO, IFAP, IFC, IKEA, Organic Exchange, Oxfam, PAN UK and WWF. In 2009, BCI published the Better Cotton Standard System based on better cotton production principles and criteria. The six principles included crop protection, water, soil, habitat, fibre quality and decent work. A set of criteria were used to evaluate the particular principle. The geographical focus was on four regions: Brazil, India, Pakistan and West and Central Asia, and the first harvest of better cotton was in 2010–11. An online tracking system, 'The Better Cotton Tracer', was launched in 2013 to facilitate BCI members to record all transactions of Better Cotton through the supply chain. By 2020, BCI intends for Better Cotton to be recognized as a sustainable commodity and for 5 million BC farmers to produce around 8.2 million metric tons of Better Cotton, accounting for 30% of global cotton production (BCI, 2016a).

2.6.1.2 Better Cotton Initiative experiences from the field

In India, a major concerns was domination by commission agents: farmers lost around 8–10 kg per quintal owing to faulty weighing by traders. Through funds provided by the project, weighing scales were purchased and the problem was resolved. Mahadev Kumar from Hoovinakala had sold 13 quintals of cotton and lost around 130 kg, when BCI project members intervened and the remaining 2 quintals was sold through MYKAPS, a nongovernmental organization (BCI, 2016b). An example of sharing experiences with farmers was evident on March 6, 2012, when Gandhabhai Chauhan narrated the benefits of drip irrigation and savings through IPM to a group of 820 farmers from 14 villages in Ambujanagar, India; group discussions were followed by awards for the best learning group, best facilitator and best extension volunteer. This encouraged interaction and transparency among participants.

In Pakistan, BCI farmers prepared botanical pesticides with 500 L neem seed extract and 1200 L bitter melon extract to fight pests such as thrips, whitefly, mites, jassid and army bollworm. Biopesticides were given without cost to neighboring farmers for a pilot study. Later around 78,000 L bitter melon extract and other plant extracts were used during the entire season on approximately 28,000 acres of cotton crop (BCI, 2016c). Jose Brasil, a small farmer from Catuti, Brazil, had worked with cotton since age 8. After interacting with BCI he learned to work with protective equipment and safeguard himself while applying pesticides which otherwise would cover him while spraying cotton plants. He also understood the problems of soil erosion and subsequently paid attention to planting seeds along soil contour lines to retain moisture and save saplings from being washed away by rain (BCI, 2016c).

2.6.1.3 Better cotton initiative experiences from the supply chain

Mr Jayesh Patel, of Patel Cotton Industries, one of the largest ginners in Gujarat, India, with a legacy of 60 years in the sustainable textile industry, and Mr Marazban F. Velati, of Welspun, Gujarat, India, the largest integrated towel manufacturer in Asia, shared their experience in procuring of Better Cotton from ginners registered under the BCI traceability system and farmers from Better Cotton farms. They found an improvement in yarn quality owing to a reduction in foreign fibre and access to new markets and buyers because of the use of Better Cotton, and they predicted a huge scope for the cultivation of sustainable cotton in the untapped remote parts of India and a positive impact on the environment (BCI, 2016d).

Mr Marco Baenninger, Vice President and Head of Sustainability, Paul Reinhart AG, Winterthur, Switzerland, a leading worldwide cotton merchant since 1788, and Mr Nurul Islam, Chairman and Founder, Norman Group, one of the largest textile industrial groups in Bangladesh, with 28 sister concern factories, stated that using Better Cotton has given them credentials to use eco-friendly labels on their products. This has made sourcing easier because of the tracking system and has provided a platform to discuss ideas and opinions across the supply chain (BCI, 2016d).

BCI is striving hard to provide a platform for the discussion and management of significant global issues associated with cotton cultivation. The Production Principles and Criteria focus on six issues that were identified at the consultation phase of BCI on a global level. BCI does not cover all potential problems and solutions, but it provides a start to identify and address global issues in cotton cultivation and to create awareness about impacts on the economy, environment and society (BCI, 2013). Comprehensive solutions may emerge based on experiences and sharing among different parts of the globe.

2.6.1.4 Cotton made in Africa initiative

Sub-Saharan Africa is the fifth largest global cotton exporter, with 3.4 million small farmers and 20 million people directly or indirectly involved in the cotton business (Cotton Made in Africa CMAI, 2016). Despite this advantage, farmers are unable to use this potential fully to improve their living conditions because of challenges such as fluctuating prices, poor infrastructure and low productivity. In 2005, the Cotton Made in Africa Initiative (CMAI) took up this problem with the aim of improving the living of small holding farmers in Africa sustainably, according to the principle of helping people to help themselves through trade. Agricultural training in efficient and environmentally friendly cultivation methods is provided by CMAI experts and the organization also has an international alliance with textile companies across the globe who buy cotton cultivated in Africa with a licensing fee to use the seal. This fee is used for developmental activities and is reinvested in project regions of Africa.

The sustainability goals are three 'P's: *people* (social equality for African farmers with small holding), *planet* (environmental protection) and *profit* (higher yields and income for cotton farmers). Two components, CmiA and the CmiA-Organic, contribute to the CmiA Standards with a focus on economic, environmental and social criteria which reflect the sustainability features for cotton growing and processing from field to gin. The standards have exclusion criteria and use the traffic light system. The implementation status is awarded a red, yellow or green rating in which green indicates sustainable development. The farmers are not expected to follow the sustainability criteria immediately, but they must provide an improvement plan and show progress in sustainable production.

The impacts speak volumes about the work and effort contributed by CmiA. As of March 2016, more than 670,000 small holder farmers with over 5.6 million people were involved in the cotton business. Fair contracts and prompt payment to farmers saw an average improvement of income by 30%, which in turn improved the standard of living (Cotton Made in Africa CMAI, 2016). Because sustainable agricultural practices are followed, such as exclusive rain-fed agriculture and no artificial irrigation, threshold spraying of bioinsecticides, building of compost pits and harvest by hand, the life-cycle assessment of CmiA cotton results in 40% fewer greenhouse gas emissions compared with conventional cotton and saves more than 2100 L water/kg for cotton fibre compared with the global average. Apart from this, improvement in educational infrastructure, a better drinking water supply and strengthening of the rights of women are some of the social aspects addressed by CmiA.

2.6.1.5 Fair Trade cotton

Fair Trade Labelling Organizations International (FLO) was established in 1997 in Bonn, Germany to unite the national Fair Trade organizations under one roof and harmonize worldwide standards and certification. The Fair Trade Standards include standards for small producer organizations, hired labour, contract production, trader standards, climate standards and textile standards based on the underlying principles of democracy and transparency, Fair Trade minimum price, capacity building, opportunities to adapt, labour conditions, environment and a reduction in end buyer emissions (Fair Trade International FAI, 2016a,b). A prohibited material list and hazardous material list were published. The Fair Trade Standards address the imbalance in trading relationships, unstable markets and injustices in conventional trade; a product with the Fair Trade mark shows that all of these standards have been met by producers and traders.

Developing countries of the world house around 90% of small-scale farmers; FLO focuses its efforts on them. According to the 2015 Fair Trade Monitoring and Impact Report, 22 Fair Trade producer organizations for seed cotton represent 54,700 farmers, 65% of whom hold organic certification. Of the 22 organizations, 13 are small producer organizations in West Africa and the remaining nine are contract production organizations in India that sell cotton to intermediary companies which support them to form independent cooperatives. About 41% of the Fair Trade Premium is invested in developmental activities (education, health care and infrastructure), and 34% on services to farmers as training and tools (Fair Trade International FAI, 2016a,b). The Fair Trade Premium is paid apart from the selling price of cotton; it has proved to be an additional income for farmers that makes it possible for them to apply for loans before the harvest. These standards also empower farmers to protect the environment as an integral part of farm management. The Fair Trade Cotton Programme, 2014, creates a synergy between textile business partners and cotton cooperatives to improve the livelihood of farmers.

2.6.1.6 Organic Trade Association

Cotton termed 'organic' must be grown according to NOP standards without using persistent and toxic pesticides and fertilizers or genetic engineering. USDA third-party certification is essential to claim the organic label. Because this certification pertains to the compliance standards only for raw materials, Global Organic Textile Standards (GOTS) are used for postharvest manufacturing technology and the conversion of material to a finished product. Dyes, auxiliary agents and finishes need to be chosen carefully to receive this label and certification. GOTS claim for high level of sustainability in environmental and social criteria along the entire organic product supply chain. The Organic Trade Association, Soil Association (United Kingdom), International Association Natural Textile Industry (Germany) and Japan Organic Cotton Association all are under the umbrella of GOTS (Organic Trade Association OTA, 2016a,b). Organic operations should demonstrate the protection of natural resources, the conservation of biodiversity and the use of approved crops. Livestock and processing methods, synthetic pesticides and fertilizers, sewage sludge, ionizing radiation and GM organisms are prohibited in organic production technologies. Three labels are used in product labelling ('100% organic', 'organic' and 'made with organic ingredients'). These must be certified by a USDA-accredited certifying agent. A survey conducted in 2013–14 reported a 14% increase (from 15,973 in 2013 to 18,234 in 2014) in the number of acres planted with organic cotton, which represented the largest number of US acres dedicated to organic cotton since 1995. The prediction in upcoming years is an increase in organic cotton–planted land, to 19,818 acres (Organic Trade Association OTA, 2016a,b).

2.6.1.7 myBMP, an Australian best management practices programme

myBMP began in 1997; it is a voluntary farm and environmental management system which has built-in self-assessment mechanisms, tools and audit processes that ensure Australian cotton is produced according to best practice. This programme allows the host to understand the industry's requirement and social license, thereby ensuring the industry for best management practice. The programme is structured into modules with key areas specified under each module; each key area has standards, checklist items and additional information and resources for better understanding.

Voluntary initiatives general to agricultural production such as the Committee on Sustainability Assessment, Field to Market: The Alliance for Sustainable Agriculture, Response-Inducing Sustainability Evaluation and Sustainability Assessment of Food and Agriculture Systems all work towards sustainable agricultural production with a focus on economics, environment and social improvement. Each organization has developed a mechanism or pathway to achieve this goal by taking all concerned stakeholders under its surveillance, evaluating and certifying enrolled participants based on their performance and the attainment of sustainability.

2.6.2 Road map to sustainable cotton production

Sustainable cotton processing is vital because it serves as a means of livelihood for many. The diversity of geographical locations calls for many sustainable adaptive measures to suit local environmental needs. Factors such as climate, natural resources, water and chemical inputs and outputs, capital and farm efficiency and local pests vary in their aspects and require modified approaches. No single solution is readily available to attain sustainability because it is a moving target and cannot be fixed indefinitely. However some effort has been made here to provide indicators or tools for people involved in cotton growth, processing, manufacturing and retailing to work towards sustainable goals. Some of the inputs which are grouped under the heading of technology, agriculture and production, including the selection of the cotton variety to be planted, the environment in which the cotton plan grows, the technologies to be followed before harvest for a sustainable crop and the technological methods to be followed after the cotton harvest, as well as the social side of the processes, are mentioned here. Sustainability will not be achieved without human intervention across the supply chain. It calls for individual and collective contributions from all quarters. They include the roles of the farmer, manufacturer, retailer, consumer and government in terms of domestic and global approaches, as shown in Fig. 2.12.

Plant biotechnology is rapidly evolving and all stakeholders must be updated regarding technologies such as transgenic technologies, molecular breeding and marker-assisted selection, mutagenesis and tilling, and cotton genome sequencing, which produce hybrids with superior qualities and high yield. To increase yields, high-density planting is preferred. This calls for plants that are compact and erect and can withstand biotic and abiotic stresses in the environment; plant architecture has to be refined with the help of biotechnology to favour land use efficiency and the mechanization of agricultural operations. Cotton faces many problems during the growth stage, such as climate change, erratic rainfall, prolonged natural conditions, saline soil conditions and other stresses. Genes are responsible for drought resistance (P5CS and DREB), thermal resistance and cold stress (Rubisco activase and superoxide dismutase) and enhance adaptability to a wide range of environments

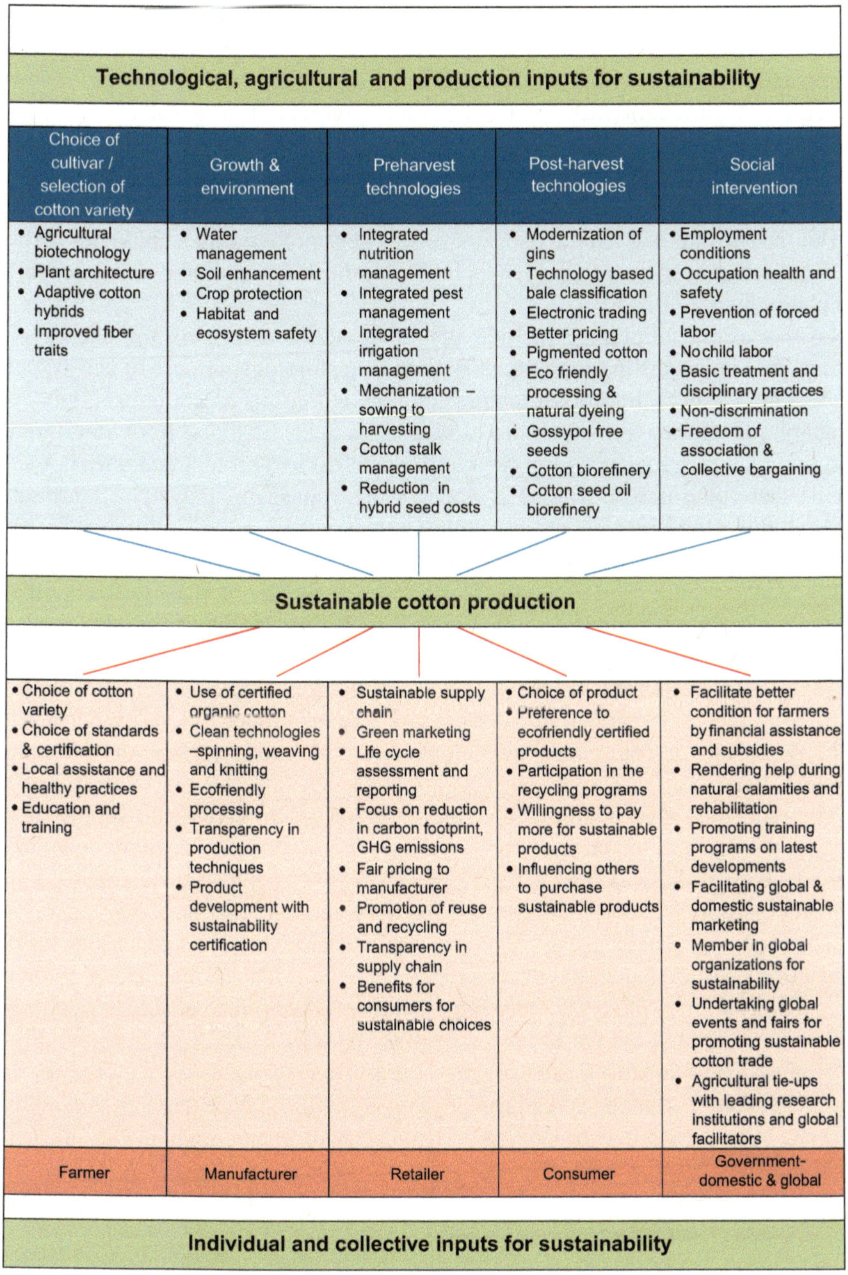

Figure 2.12 Road map to sustainable cotton production.

(Saha, 2011). Bt cotton has also saved plants from damage caused by insects such as bollworm and leaf caterpillar, sap-sucking insects, viral, bacterial and fungal diseases, and nematodes. Many hybrids produce high yield but the strength is lower than for conventional cotton; hence the right combination of genes based on fibre qualities needs to be selected to address this problem. The environment for cotton plants to grow requires best management practices with regard to water management, soil enhancement and crop protection management without hindering the welfare of the surrounding habitat and ecosystem. Thus preharvest technologies include well-defined management practices which help in the growth and development of the cotton plant.

Postharvest technologies include the modernization of gins for energy efficiency, pricing according to the quality of their technology, the use of eco-friendly processing, clean technologies, natural dyeing and coloured cotton to attain zero discharge. Gossypol-free seeds are produced by gene-silencing RNA interference technology, which makes cotton seed a source of oil and protein that does not need not be detoxified before use, as is done in the Vanaspathi Industry (Sunilkumar et al., 2006). Cotton stalks are available after harvest, e.g., 30 million metric tons in India. This can be used as a resource for converting biomass to biofuels (ethanol and butanol) or bulk food products (sugar and vinegar) by biomass destruction and enzymatic saccharification technology. Selective separation technologies are used to recover minor components and using selective enzymes to convert cotton oil to useful oil for stakeholders (FCCI, 2012). Sustainability cannot be achieved without addressing the social side, and International Labour Organization standards need to be practised for community development and to enhance farm and plantation workers' lives.

The road map also points to the role of stakeholders across the supply chain, because the demand for products will rise only if sustainability is addressed and all individuals work individually as well as collectively to achieve sustainability in the context of sustainable cotton processing and development. The problems of consumers' lack of willingness to pay higher prices for sustainable products and insufficient efforts by producers have to change. These aspects will become economical when a mass effort is made and only eco-friendly products are available to customers. Governments around the world have started to focus on features of sustainability in all products. Solutions to these problems need to be discussed on a global level in conferences and forums to spread research findings from laboratories to commercial markets, to transform the framework and mindset of people and society about the environment.

> *'About one thousand farmers die every day from pesticide poisoning or suicide debt which is more than in all the wars going on across the world put together' says Katharine Hamnett, the British designer and activist for organic cotton when she expressed her anguish that the media promotion for organic cotton was not sufficient. Her quote for the day ...*
> *Have you COTTONED on yet?*
> *I choose to support farmers and protect environment. I choose ORGANIC COTTON*
> *Katharine Hamnett, The Organic Cotton Initiative (Chua, 2015).*

References

Agrawal, P.B., Nierstrasz, V.A., Klug-Santner, B.G., Gübitz, G.M., Lenting, H.B.M., Warmoeskerken, M.M.C.G., 2007. Wax removal for accelerated cotton scouring with alkaline pectinase. Biotechnol. J. 2, 306–315.

Altieri, M.A., 1999. Applying agroecology to enhance the productivity of peasant farming systems in Latin America. Environ. Dev. Sustain. 1, 197–217.

Altieri, M.A., 2002. Agroecology: the science of natural resource management for poor farmers in marginal environments. Agric. Ecosyst. Environ. 1–24. http://agroecology.pbworks.com/f/NRMfinal.pdf.

Altieri, M.A., 2004. Linking ecologists and traditional farmers in the search for sustainable agriculture. Front. Ecol. Environ. 2, 35–42.

Anon., 2016. Methods of Selective Breeding. https://sites.google.com/site/asoftsb/home/methods-of-selective-breeding.

Anon., 2013. A New Era of Innovation for Cotton Manufacturing. https://www.forumforthefuture.org/greenfutures/articles/new-era-innovation-cotton-manufacturing.

Azimov, A.A., Akhmedov, K., 2013. Inheritance of morpho-biological and economic traits in breed of reciprocal cotton hybrids. IJCPS 4, 97–101.

Bagla, P., 2010. Hardy cotton-munching pests are latest blow to GM crops. Science 327, 1439.

BCI, 2016a. BCI History. http://bettercotton.org/about-bci/bci-history/.

BCI, 2016c. Botanical Pesticides: An Organic Solution for Cotton Pests. http://bettercotton.org/wp-content/uploads/2014/01/13_botanical.jpg.

BCI, 2013. Better Cotton Production Principles & Criteria Explained. http://bettercotton.org/wp-content/uploads/2014/01/Better-Cotton-Production-Principles-and-Criteria-Explained_Final-2013_eng_ext.pdf.

BCI, 2016b. A Definite Measure of Progress. http://bettercotton.org/wp-content/uploads/2014/01/3_measure_of_progress.jpg.

BCI, 2016d. Stories from the Supply Chain. http://bettercotton.org/about-better-cotton/stories-supply-chain-2/.

Beven, K., 2004. Robert E. Horton's perpetual model of infiltration processes. Hydrol. Process. 18, 3447–3460. http://dx.doi.org/10.1002/hyp.5740.

Blackburn, R.S. (Ed.), 2009. Sustainable Textiles Life Cycle & Environmental Impact, vol. 1. CRC Press, Florida.

Brar, S.K., Verma, M., Tyagi, R.D., Valero, J.R., 2006. Recent advances in downstream processing and formulations of *Bacillus thuringiensis* based biopesticides. Process. Biochem. 41, 323–342.

Calafell, M., Klug-Santner, B., Guebitz, G., Garriga, P., 2005. Dyeing behaviour of cotton fabric bioscoured with pectate lyase and poly galacturonase. Color. Technol. 121, 291–297.

Carol, M., 2016. What Is Selective Breeding? http://www.treehugger.com/natural-sciences/what-selective-breeding.html.

CBAN, 2013. Genetically Modified Cotton, CBAN Factsheet. http://www.cban.ca/Resources/Topics/GE-Crops-and-Foods-On-the-Market/Cotton/Genetically-Modified-Cotton-CBAN-Factsheet.

CFS, 2007. Cotton Concentration Report an Assessment of Monsanto's Proposed Acquisition of Delta and Pine Land. http://www.centerforfoodsafety.org/files/cfs-cta_monsanto-dpl_merger_report_public_release_-_final__2_.pdf.

Chua, J.M., 2015. Quotes of the Day. http://www.ecouterre.com/katharine-hamnett-i-want-to-sell-as-much-organic-cotton-as-possible/.

Collier, M., Webb, R.H., Schmidt, J.C., 2000. Dams and Rivers: A Primer on the Downstream Effects of Dams. https://pubs.er.usgs.gov/publication/cir1126.

Cotton Made in Africa CMAI, 2016. Initiative. http://www.cottonmadeinafrica.org/en/about-us/the-initiative.

Cottonconnect, 2014. More Crop per Drop-Water Report on the Cotton Industry. http://www.cottonconnect.org/media/26055/cottonconnect_more_crop_per_drop.pd.

Cox, C., 1991. Pesticides and Birds: From DDT to Today's Poisons. http://eap.mcgill.ca/MagRack/JPR/JPR_14.htm.

Dean, D.H., 1984. Biochemical genetics of the bacterial insect-control agent *Bacillus thuringiensis*: basic principles and prospects for genetic engineering. Biotechnol. Genet. Eng. Rev. 2, 341–363.

Dennis, S.E., Ellis, J., Green, A., Llewellyn, D., Morell, M., Tabe, L., Peacock, W.J., 2008. Genetic contributions to agricultural sustainability. Philos. Trans. R. Soc. Lond. B Biol. Sci. http://rstb.royalsocietypublishing.org/content/363/1491/591.

Doorenbos, J., Kassam, A.H., Bentvelsen, C.L.M., Branscheid, V., Plusje, J.M.G.A., Smith, M., Uittenbogaard, G.O., Van Der Wal, H.K., 1988. Yield Response to Water. http://203.200.22.249:8080/jspui/bitstream/2014/11987/1/Yield_response_to_water.pdf.

Durr, D., 2016. Famous Quotes about Sustainability. http://quotesgram.com/img/famous-quotes-about-sustainability/137350/.

Dury, J., Garcia, F., Bergez, J.E., 2011. Models to support cropping plan and crop rotation decisions. Agron. Sust. Dev. http://www.cicr.org.in/pdf/multitier.pdf.

Edmisten, K., York, A., 1995. Buctril 4ec Herbicide Registered for Bxn Cotton. http://www.cotton.ncsu.edu/ccn/1997/ccn_95_5a.html.

Ellison, M., 2010. Why You Should Buy Organic Cotton. http://www.naturalnews.com/029904_organic_cotton.html.

Eren, H.A., Erismis, B., 2013. Ultrasound-assisted bioscouring of cotton. Color. Technol. 129, 360–366.

Fair Trade International FAI, 2016a. Cotton. http://www.fairtrade.net/products/cotton.html.

Fair Trade International FAI, 2016b. Our Standards. http://www.fairtrade.net/standards/our-standards.html.

Falkenmark, M., Rockstrom, J., 2013. Balancing Water for Humans and Nature: The New Approach in Ecohydrology. Routledge, UK. http://dx.doi.org/10.4324/9781849770521.

Falkenmark, M., 2000. Competing freshwater and ecological services in the river basin perspective. Water Int. 25, 172–177.

Falkenmark, M., 2003. Freshwater as shared between society and ecosystems: from divided approaches to integrated challenges. Science 2037–2049.

FAO, 2002. Land–Water Linkages in Rural Watersheds. ftp://ftp.fao.org/agl/aglw/docs/lw9e.pdf.

FAO, 2011. The State of the World's Land and Water Resources for Food and Agriculture Managing Systems at Risk. Earthscan, New York, USA.

FAO, 2015. Measuring Sustainabillity in Cotton Farming Systems–Towards a Guidance Framework. Food and Agriculture Organization of the United Nations.

FAO, 2016. Soil Health. http://www.fao.org/ag/save-and-Grow/en/3/index.html.

Farage, E.M., 1989. Enzyme and behavioral changes in young chicks as a result of carbaryl treatment. J. Toxicol. Environ. Health. 26, 119–131.

Farrell, M.J., 2012. Sustainable Cotton Dyeing. https://repository.lib.ncsu.edu/bitstream/handle/1840.16/7452/etd.pdf?sequence=1&isAllowed=y.

FCCI, 2012. Cotton 2020: Roadmap for Sustainable Production. http://www.tierraseedscience.com/pdf/cotton-2020-roadmap-india.pdf.

Ferre, J., Ric, J.V., 2002. Biochemistry and genetics of insect resistance *Tobacillus thuringiensis*. Annu. Rev. Entomol. 47, 501–533.
Fitt, G.P., 2000. An Australian approach to IPM in cotton: integrating new technologies to minimise insecticide dependence. Crop Prot. 19, 793–800.
Food and Agriculture Organization of the United Nations, 2015. Measuring Sustainability in Cotton Farming Systems: Towards a Guidance Framework. https://www.icac.org/getattachment/Home-International-Cotton-Advisory-Committee-ICAC/measuring-sustainability-cotton-farming-full-english.pdf.
Fry, D.M., Toone, C.K., 1981. DDT-induced feminization of gull embryos. Science 213, 922–924.
Gillham, F.E.M., 1995. Cotton Production Prospects for the Next Decade. World Bank Publications, Washington DC, USA.
Gleick, P.H., 1993. Water in Crises: A Guide to the World's Fresh Water Resources. https://www.2000litergesellschaft.ch/fileadmin/user_upload/Referenzen_und_Literatur/GLEICK_1993_Water_in_Crisis.pdf.
Goswami, B., 2007. Making a Meal of Bt Cotton. http://www.countercurrents.org/goswami120907.htm.
Grace Communication Foundation, 2016. Industrial Crop Production. http://www.sustainabletable.org/804/industrial-crop-production.
Grismer, M.E., 2002. Regional cotton lint yield, ET_c and water value in Arizona and California. Agric. Water Manage. 54, 227–242.
Grue, C.E., Powell, G.V.N., McChesney, M.J., 1982. Care of nestlings by wild female starlings exposed to an organophosphate pesticide. J. Appl. Ecol. 19, 327–335.
Hamilton, R., 2009. Agriculture's Sustainable Future: Breeding Better Crops. https://www.scientificamerican.com/article/agricultures-sustainable-future/.
Hauser, P., 2000. Cationic Pretreatments of Cotton. http://www.textileworld.com/textile-world/textile-news/2000/10/cationic-pretreatments-of-cotton/.
Henny, C.J., Blus, L.J., Kolbe, E.J., Fitzner, R.E., 1985. Organophosphate Insecticide (Famphur) Tropically Applied to Cattle Kills Magpies and Hawks. http://fresc.usgs.gov/products/papers/194_henny.pdf.
Hilder, V.A., Boulter, D., 1999. Review genetic engineering of crop plants for insect resistance: a critical review. Crop Pro. 18, 177–191.
Howell, T.A., 2001. Enhancing water use efficiency in irrigated agriculture. Agron. J. 93, 1–10.
Hummel, J., 1997. Integrating Economy and Ecology – Ecotextiles on the Way from Niche Market to a Mass Market. http://collections.infocollections.org/ukedu/en/d/Jgq971e/2.8.html.
Hustvedt, G., Dickson, M.A., 2009. Consumer likelihood of purchasing organic cotton apparel: influence of attitudes and self-identity. JFMM 13 (1), 49–65. http://dx.doi.org/10.1108/13612020910939879.
Ibrahim, N.A., Hossamy, M.E., Morsy, M.S., Eid, B.M., 2004. Development of new eco-friendly options for cotton wet processing. J. Appl. Polym. Sci. 1825–1836.
ICAC Recorder, 2003. Bollgard II: A New Generation of Bt Genes Commercialized. https://www.icac.org/cotton_info/tis/biotech/documents/recorderdocs/march_03.pdf.
INCID, 2016. Irrigation & Drainage. http://www.cwc.gov.in/main/INCID/welcome.html.
Indrajit, 2016. Saline and Alkaline Soil: Nature. Characteristics and Reclamation.
International Rivers, 2016. Environmental Impacts of Dams. https://www.internationalrivers.org/environmental-impacts-of-dams.
International Trade Center, 2016a. Marketing the Importance of Cotton Promotion. http://www.cottonguide.org/cotton-guide/cotton-marketing-the-importance-of-cotton-promotion/.

International Trade Center, 2016b. The Importance of Cotton in World Trade. http://www.cottonguide.org/cotton-guide/the-world-cotton-market/the-importance-of-cotton-in-world-trade/.

Jiang, C.X., Wright, R.J., El-Zik, K.M., Paterson, A.H., 1998. Evolution Polyploid formation created unique avenues for response to selection in *Gossypium* (cotton) quantitative variation cotton fiber quality crop evolution inter genomic interactions comparative quantitative trait locus analysis. Proc. Natl. Acad. Sci. USA 95, 4419–4424.

Jordanien, P., Aufleger, M., Mett, M., 2011. Handshake across the Jordan: Water & Understanding. Books on Demand, Norderstedt.

Krishna, V., Qaim, M., 2012. Bt cotton and sustainability of pesticide reduction in India. Agric. Sys. 107, 47–55.

Kurmanath, K.V., 2015. Wily pink bollworm survives Monsanto's Bollgard-II. Hindu Bus. Line. http://www.thehindubusinessline.com/economy/agri-business/wily-pink-bollworm-survives-monsantos-bollgardii/article7814810.ece.

Lu, Y., Wu, K., Jiang, Y., Xia, B., Li, P., Feng, H., Guo, Y., 2010. Mirid bug outbreaks in multiple crops correlated with wide-scale adoption of Bt cotton in China. 328, 1151–1154.

Malezieux, E., Dupraz, C., Ozier, H., Tourdonnet, S.D., 2009. Mixing plant species in cropping systems: concepts, tools and models. Agron. Sust. Dev. https://www.researchgate.net/profile/Christian_Dupraz/publication/41701453_Mixing_Plant_Species_in_Cropping_Systems_Concepts_Tools_and_Models_A_Review/links/0912f50b730a762e76000000.pdf.

Mazid, S., Rajkhowa, R.C., Kalita, J.C., 2011. A review on the use of bio pesticides in insect pest management. 1, 169–178.

Mcmahon, T.A. (Ed.), 1992. Global Runoff: Continental Comparison of Annual Flows & Peak Discharges. Catena Verlag, Germany.

Menezes, E., 2016. Organic Cotton Processing. http://www.fibre2fashion.com/industry-article/3373/organic-cotton-processing?page=1.

Ming, R., Liu, S.C., Lin, Y.R., Silva, J.D., Wilson, W., Braga, D., Deynze, A.V., Wenslaff, T.F., Wu, K.K., Moore, P.H., Burnquist, W., Sorrells, M.E., Irvine, J.E., Paterson, A.H., 1998. Detailed alignment of saccharum and sorghum chromosomes: comparative organization of closely related diploid and polyploidgenomes. Genetics 150, 1663–1682.

Monsanto Company, 2009. Pink Bollworm Resistance to GM Cotton in India. http://www.monsanto.com/newsviews/pages/india-pink-bollworm.aspx.

Monsanto Company, 2015. Modern Breeding Techniques. http://www.monsanto.com/improvingagriculture/pages/modern-breeding-techniques.aspx.

Monsanto Company, 2016a. Roundup Ready Flex Cotton Technical Manual. http://www.monsanto.com/global/au/products/documents/roundup-ready-flex-technical-manual.pdf.

Monsanto Company, 2016b. Roundup Ready Flex Cotton. http://www.monsanto.com/products/pages/genuity-roundup-ready-flex-cotton.aspx.

Naranjo, S.E., Ellsworth, P.C., 2009. Fifty years of the integrated control concept: moving the model and implementation forward in Arizona. Pest Manag. Sci. 65, 1267–1286.

National Cotton Council of America [US], 2016. The Story of Cotton. https://www.cotton.org/pubs/cottoncounts/story/upload/The-Story-of-Cotton-73k-PDF.pdf.

National Wild Life Federation, 2015. Five Years & Counting Gulf Wildlife in the Aftermath of the Deepwater Horizon Disaster. http://www.nwf.org/~/media/PDFs/water/2015/Gulf-Wildlife-In-the-Aftermath-of-the-Deepwater-Horizon-Disaster_Five-Years-and-Counting.pdf.

NCERT, 2015–2016. Land Resources and Agriculture. http://ncert.nic.in/ncerts/l/legy205.pdf.

NETAFIM, 2016. Cotton. http://www.netafim.co.za/offering/irrigation/agriculture/field/cotton/?id=77.
OECD, 2010. Sustainable Management of Water Resources in Agriculture. http://www.oecd.org/tad/sustainable-agriculture/49040929.pdf.
OECD, 1998. Sustainable Management of Water in Agriculture: Issues and Policies – the Athens Workshop. OECD Publishing, Paris.
Oerke, E.C., Dehne, H.W., 2004. Safeguarding production—losses in major crops and the role of crop protection. Crop Prot. 23, 275–285.
Oppert, B., 1999. Protease interactions with *Bacillus thuringiensis* insecticidal toxins. Arch. Int. Physiol. 42, 1–12.
Organic Cotton Plus, 2016. Sustainability Learning Center: Organic Cotton 101. http://organic-cottonplus.com/pages/learning-center.
Organic Cotton, 2016a. World Cotton Production. http://www.organiccotton.org/oc/Cotton-general/World-market/World-cotton-production.php.
Organic Cotton, 2016b. Benefits of Organic Cotton. http://www.organiccotton.org/oc/Organic-cotton/Benefits-of-organic-cotton/Benefits-of-oc.php.
Organic Trade Association OTA, 2016a. Fibers and Textiles. https://www.ota.com/advocacy/fiber-and-textiles.
Organic Trade Association OTA, 2016b. Organic Standards. https://www.ota.com/advocacy/organic-standards.
Organic Trade Association, 2012. Organic Cotton Facts. https://ota.com/sites/default/files/indexed_files/Organic-Cotton-Facts.pdf.
Ornes, S., Raloff, J., 2014. Bug Killer Linked to Decline in Birds. https://www.sciencenewsfor-students.org/article/bug-killer-linked-decline-birds.
Perfecto, I., Rice, R A., Greenberg, R., Voort, M.E.V.D., 1996. Shade coffee: a disappearing refuge for biodiversity. Biosci. Rep. 46, 598–608.
Poff, N.L., Allan, J.D., Bain, M.B., Stromberg, J.C., 1997. The natural flow regime: a paradigm for river conservation and restoration. BioScience 47, 769–783.
Postel, S.L., 2000. Entering an era of water scarcity: the challenges ahead. Ecol. Appl. 10, 941–948.
Rao, S.P.N., 2006. Genetic Engineering/Recombinant DNA Technology. http://microrao.com/micronotes/genetic_engineering.pdf.
Rao, M.B., Tanksale, A.M., Ghatge, M.S., Deshpande, V.V., 1998. Molecular and biotechnological aspects of microbial proteases. Microbiol. Mol. Biol. Rev. 62, 3597–6351.
Rattner, B.A., Sileo, L., Scanes, C.G., 1982. Hormonal response and tolerance to cold of female quail following parathion ingestion. Pest Biochem. Physiol. 18, 132–138. 10.1016/0048-3575(82)90096-7.
Ristic, N., Ristic, I., 2012. Cationic modification of cotton fabrics and reactive dyeing characteristics. JEFF 7, 113–121.
Robinson, E., 2002. Program Makes BXN More Affordable. http://www.southeastfarmpress.com/program-makes-bxn-more-affordable.
Roger, P.A., Simpson, I., Oficial, R., Ardales, S., Jimenez, R., 1994. Effects of pesticides on soil and water microflora and mesofauna in wetland rice fields: a summary of current knowledge and extrapolation to temperature environments. Aust. J. Exp. Agric. 34, 1057–1068.
Ronald, P., 2011. Plant genetics, sustainable agriculture and global food security. Genetics 188, 11–20.
Rosenberg, D.M., McCully, P., Pringle, C.M., 2016. Environmental Effects of Hydrological Alterations. https://www.internationalrivers.org/environmental-effects-of-hydrological-alterations.

Saha, S., 2011. Role of Biotechnology in Sustainable Development of Cotton. https://www.icac.org/wp-content/uploads/2012/08/e_tis_2011.pdf.

Sandra, L.P., Gretchen, C.D., Paul, R.E., 1996. Human appropriation of renewable fresh water. Science 271, 785–788.

Sankaranarayanan, K., Nalayini, P., Sabesh, M., Rajendran, K., Nachane, R.P., Gopalakrishnan, N. (Eds.), 2011. Multi-Tier Cropping System for Profitability & Stability in Bt Cotton Production. CICR, Coimbatore. https://www.cabdirect.org/cabdirect/abstract/20093120021.

Sarin, J., 2007. Bug Makes Meal of Punjab Cotton, Whither Bt Magic? http://twocircles.net/2007sep02/bug_makes_meal_punjab_cotton_whither_bt_magic.html.

Small, R.L., Ryburn, J.A., Wendel, J.F., 1999. Low levels of nucleotide diversity at homoeologous Adh loci in allotetraploid cotton (*Gossypium* L.). Mol. Biol. Evol. 16, 491–501.

Shortle, J.S., Griffin, R.C., 2016. Irrigated Agriculture and the Environment, the Management of Water Resources Series. Edgar Edward Publishing, UK.

Silvertooth, J.C., 2001. Saline and Sodic Soil Identification and Management for Cotton. https://cals.arizona.edu/crop/cotton/soilmgt/saline_sodic_soil.html.

Sivanappan, R.K., 1994. Prospects of micro-irrigation in India. Irrig. Drain. Syst. 8, 49–58.

Smith, C.W., Cothren, J.T. (Eds.), 1999. Cotton: Origin, History, Technology, and Production. John Wiley & Sons, New York, USA.

SourceWatch, 2016. Roundup Ready Cotton. http://www.sourcewatch.org/index.php/Roundup_Ready_Cotton.

Stone, G.D., 2011. Field versus farm in Warangal: Bt cotton, higher yields, and larger questions. World Dev. 39, 387–398.

Sunilkumar, G., Campbell, L.M., Puckhaber, L., Stipanovic, R.D., Rathore, K.S., 2006. Engineering cottonseed for use in human nutrition by tissue-specific reduction of toxic gossypol. Proc. Natl. Acad. Sci. USA 103, 18054–18059.

Sustainable Fibres & Fabrics SFF, 2016. Cotton Processing. http://sff.arts.ac.uk/Fibre%20Processing/cottonprocessing.html.

Swift, M.J., Izac, A.M.N., Noordwijk, M.V., 2004. Biodiversity and ecosystem services in agricultural landscapes—are we asking the right questions? Agric. Ecosyst. Environ. 104, 113–134.

Tabashnik, B.E., Gassmann, A.J., Crowder, D.W., Carriére, Y., 2008. Insect resistance to Bt crops: evidence versus theory. Nat. Biotechnol. 26, 199–202.

The Energy Resources Institute TERI, 2016. Bollcure – New Bio Pesticide for Cotton Bollworm. http://www.teriin.org/technology/bollcure.php.

Tisdell, C.A., Wilson, C., 2001. Why farmers continue to use pesticides despite environmental, health and sustainability costs. Ecol. Econ. https://www.researchgate.net/profile/Clement_Tisdell/publication/4840284_Why_Farmers_Continue_to_Use_Pesticides_Despite_Environmental_Health_and_Sustainability_Costs/links/0fcfd50cfc08db1cd7000000.pdf.

Tripathi, R.S., 2009. Alkali Land Reclamation: A Boom for Development. Mitthal Publications, New Delhi, India.

Umbeck, P.F., 1991. Genetic Engineering of Cotton Plants and Lines. https://docs.google.com/viewer?url=patentimages.storage.googleapis.com/pdfs/US5004863.pdf.

UNCTAD, 2016. Cotton. http://unctad.org/en/PublicationsLibrary/INFOCOMM_cp05_Cotton_fr.pdf.

Vreeland, J.M., 1997. The Case of Peru: The Native Cotton Project. http://collections.infocollections.org/ukedu/en/d/Jgq971e/2.4.html.

Wang, Z., Lin, H., Huang, J.K., Hu, R., Rozelle, S., Pray, C., 2009. Bt cotton in China: are secondary insect infestations offsetting the benefits in farmer fields? ASC 8, 83–90.

Wani, S.P., Rockstrom, J., Owes, T.Y., 2009. Rainfed Agriculture: Unlocking the Potential. CABI, Oxforshire, UK.

White, D.H., King, K.A., Mitchell, C.A., Hill, E.F., Lamont, T.G., 1979. Parathion causes secondary poisoning in a laughing gull breeding colony. Bull. Environ. Contam. Toxicol. 23, 281–284.

Wildlife Center of Virginia, 2016. Wildlife Center of Virginia's Position on the Lead Toxicity in Raptors. http://wildlifecenter.org/sites/default/files/WCV-Position-on-Lead4.pdf.

Willis, G.H., 1982. Pesticides in Agricultural Runoff and Their Effects on Downstream Water Quality. http://onlinelibrary.wiley.com/doi/10.1002/etc.5620010402/epdf.

Willis, T.M., Black, A.S., 1996. Irrigation increases groundwater recharge in the Macquarie Valley. Aust. J. Soil. Res. 34, 837–847.

Woodwell, G.M., 1963. Toxic Substances and Ecological Cycles. http://www.cfr.washington.edu/classes.esrm.456/Woodwell.1967.pdf.

WWF, 1999. The Impact of Cotton on Fresh Water Resources and Eco Systems. http://d2ouvy59p0dg6k.cloudfront.net/downloads/impact_long.pdf.

Yachmenev, V.G., Bertoniere, N.R., Blanchard, E.J., 2002. Intensification of the bio-processing of cotton textiles by combined enzyme/ultrasound treatment. J.Chem. Technol. Biotechnol. 77, 559–567.

Yu, L.Z., Kohel, R.J., Fang, D.D., Cho, J., Deynze, A.V., Ulloa, M., Hoffman, M.S., Pepper, A.S., Stelley, D.M., Jenkins, J.N., Saha, S., Kumpatia, S.P., Shah, M.R., Hugie, W.V., Percy, R.G., Wendel, J., 2012. A high-density simple sequence repeat and single nucleotide polymorphism genetic map of the tetraploid cotton genome. G3 (Bethesda) 2, 43–58.

Zhao, J.H., Ho, P., Azadi, H., 2011. Benefits of Bt cotton counterbalanced by secondary pests? Perceptions of ecological change in China. Environ. Monit. Assess. 173, 985–994.

Zietlow, B., 2013. Best Available Techniques in the Textile Sector Brigitte Zietlow. http://www.igep.in/live/hrdpmp/hrdpmaster/igep/content/e48745/e49028/e58164/e58169/BAT_Textile_Vortrag_GPCB_.pdf.

Sustainable production of bast fibres

3

Sanjoy Debnath
ICAR–National Institute of Research on Jute and Allied Fibre Technology, Kolkata, India

3.1 Introduction

Food, shelter and clothing are the three basic needs of human civilization. Of the three basics, in the case of clothing, textiles have a pivotal role. As far as the textile materials are concerned, in antiquity natural fibrous materials from plant and animal sources were the main sources. With the development of science and technology, synthetic fibres developed in the mid 19th century. These synthetic fibres could be produced according to needs, and thus are also known as engineered fibres. On the other hand, natural fibres mostly depend on nature, which is mostly controlled by the environment, climate, weather, soil, etc., and by the methods of extraction as far as human intervention is concerned. In the early 20th century, there was more awareness of the use of natural fibres they became popular. Natural fibres are susceptible to natural degradation by biological means such as microorganisms, fungi and pests in the soil, and with time. Synthetic fibres are mostly unaffected by microorganisms and living organisms, although they are degraded to some extent by ultraviolet rays present in sunlight, which takes several decades. Further during its life span, synthetic/plastic fibres became a nuisance and thus forms synthetic-debris causes hindrance to natural development. Each day, the minimal use of synthetic fibres is being practiced in developed and developing countries, considering the future life of the earth (Jan, 2016). Hence popularization of the natural fibres has become essential (Memon, 2016). Composite is another promising area in which natural fibres are used as a reinforcing material to replace glass fibre used conventionally (Nechwatal et al., 2003).

Natural fibres are mainly classified into two different categories based on their sources: (1) plant fibre and (2) animal fibre. Plant fibres are further classified into various categories based on their plant sources: leaf fibres from leaves, bast fibre from plant stems, seed fibre from the seeds of plants, fruit fibre from plant fruits, etc. On the other hand, animal fibres are categorized as hair fibre, insect secretion fibre, etc., from different sources of the animal kingdom. Animal fibres are mostly protein-based and plant fibres are cellulosic/lingo-cellulosic in nature. Apart from protein and cellulose as major components of animal and plant fibres, respectively, these fibres contain many other components. Because all of these natural fibres are from biological sources, renewability or sustainability is ensured as long as eco-friendly processes were been adopted during cultivation, processing and production. Many natural fibre-based products have become important in the fashion industry (Debnath, 2016, 2017) in terms of revenue generation. On the contrary, developments of environ-friendly

Sustainable Fibres and Textiles. http://dx.doi.org/10.1016/B978-0-08-102041-8.00003-2

product and process are in full swing, considering process and recyclability are developing un-parallel. Jute, an important bast fibre, is covered in another chapter in detail. Hence, considering these factors, this chapter emphasizes the sustainable production of some important bast fibres other than jute that are used in the textile industry. Similar to jute, most natural fibres that absorb moisture well (Debnath et al., 2006; Debnath and Madhusoothanan, 2010; Debnath et al., 2010) have extensive potential in the area of agriculture and different industrial applications as far as absorbent material is concerned. As we know, textiles are not restricted to the area of clothing (a basic need), but extend to technical and industrial textiles (Debnath, 2014), which have an important role in modern life.

3.2 Production of some important bast fibres: status of fibre extraction, mechanical processing (spinning and weaving) of bast fibres, chemical processing, product development, product diversification and added value

When we talk about the popular bast fibres, the list starts with flax, jute, nettle, sunn hemp, hemp, ramie, banana, etc. Among these fibres, commercial marketing of flax was globally established mainly in Europe and on part of the Asian and America continents; jute was mainly from Asia and part of Africa; ramie was mainly from Asia; nettle was from Asia, Europe, America, etc. Other fibres were explored locally and the popularity and demand of other bast fibres has increased in different developed and underdeveloped countries. Jute is a fibre known for more than a century for industrial applications such as its use in sacks and packaging, geotextiles and carpet backing (Debnath et al., 2009); flax fibre also has great potential in the fashion arena. However, over the past few decades, there have been lots of times when jute has been used in fashion products in different wearable and unwearable ways. The demand for conventional products has increasingly declined owing to the availability of low-cost synthetic materials (Debnath, 2017). The demand for exports has also been reduced because of stringent norms caused by the presence of band residual chemical traces. Hence many industries are now concentrating on development jute and other natural fibres blended into diversified products from fashion to packaging, and even in agricultural applications (Indu and Senthilkumar, 2016). This chapter discusses most of these fibres: their processes of extraction, the conversion of fibre to yarn and fabric, and finally product diversification for added value and economic growth.

3.2.1 Flax

Flax fibre comes under the category of bast fibre (Fig. 3.1) and is extracted from the plant of the linseed/flax plant (*Linum usitatissimum* L.). This plant is popular for two reason: one is flax fibre and another is linseed oil (used for industrial applications) extracted from the seed of the plant (Basu and Dutta, 2014). According to historical

Figure 3.1 Flax cultivation (Anonymous, 2016a).

records, the earliest example of preserved linen appears to be a needle-netted linen headpiece from Nahal Hemar Cave in Israel 8500 years ago, and Swiss Lake Dwellers used a native flax to make cloth 5000–6000 years ago. It was also found that ancient Egyptians used clothing, bed linen, and shrouds for mummies and for ships' sails prepared mainly from linen in those days. The earliest Egyptian linen cloth dates from the Old Kingdom, but flax appears to be have been grown for linen about 5000 years ago in the Early Dynastic period (Debnath, 2016). Sharma et al. (1999) described and compared different retting methods of flax fibre, including water, dew and enzyme-retted, on physical chemical and thermal properties of flax fibre. As far as fashion industry is concerned, flax and its blend with other fibres (natural/man-made) is popular, and hence flax-based fashion will be discussed.

In today's scenario, when we talk about flax fibre it is well-known in the textile industry as a linen fashion as well as formal apparel for men and women. There are many leading manufacturers of these linen-based products in the world. Different blended linen products (Basu and Dutta, 2014; Basu et al., 2006; Basu and Roy, 2008) are linen-cotton (warp cotton, weft linen; and warp cotton, weft cotton-linen blended yarn), linen-based-cotton (warp cotton, weft linen-jute or linen-ramie or linen-pineapple or linen-silk waste, etc.), linen-synthetic (warp polyester/cotton blended or polyester/ viscose blended, weft linen or linen-jute or linen-cotton, etc.), and so forth. Apart from these is a huge variety from which to develop many other blended materials from linen. Elite men and women currently prefer linen-based apparel mainly for comfort in addition to its aesthetic look. Linen fibre material can absorb moisture quickly from body perspiration, which provides coolness and comfort in humid conditions when such clothes are worn. As far as the properties of the fibre are concerned, linen fibre material swells and improves in strength under wet conditions. This is why linen/flax spinning is normally preferred to be done under wet rather than dry conditions. As far as the international market is concerned, a lot of pure/blended linen products are available, such as ramie-linen tunic tops for women for casual wear, linen blend trousers (50% viscose, 35% cotton, and 15% linen; white: 65% viscose, 25% cotton, and

10% linen) by Joanna Hope, linen mix trousers and shorts (55% linen and 45% cotton) by Southbay, a single-breasted linen mix blazer (55% linen and 45% cotton; lining: polyester) by Williams and Brown, lightweight fabric that will keep you cool but stylish in warm weather (54% cotton and 46% linen) by Black Level Jacamo, linen mix three-quarters pants (55% linen and 45% cotton) by Southbay, a ladies fashion linen bow decor straw braid summer sun hat (100% linen), a linen summer women's wide-brim church wedding/sea beach sun hat (100% linen) by Kentucky Derby, a ladies pleated crisscross fashion sexy linen dress, a women's white linen dress (100% linen), a female slim short jacket blazer/linen blazer/ladies coat, ladies thong linen underwear/briefs (85% linen, 12% nylon, and 3% spandex), women's linen shorts (85% linen and 15% nylon), and so on. All of this shows that a good fashion market exists internationally (Debnath, 2016). Apart from these apparel applications, flax fibre has wide potential in the area of industry in the form of nonwovens (Maity et al., 2014). Optimum blending of flax and polyester fibre to make needle-punched nonwoven fabric can replace currently used 100% synthetic nonwovens in geotextile applications (Rawal and Anandjiwala, 2007). In these geotextiles there is another area where the quantum added value of this flax fibre is possible. Research by Alimuzzaman et al. (2014) disclosed that the biodegradability of nonwoven flax fibre reinforced polylactic acid biocomposites. Because flax is a strong fibre it can be used in composites (Xue et al., 2011) as flax nonwoven mats through a vacuum-assisted resin transfer molding method. Many researchers are working on using flax fibre as a reinforcing material in the form of nonwoven mats (Fortea-Verdejo et al., 2016) and as a natural/synthetic resin to develop biodegradable composites (Nechwatal et al., 2003; Alimuzzaman et al., 2014), for sustainable composites in the near future.

3.2.2 Nettle

Nettle is a common herbaceous plant which encompasses around 30 to 45 species. This bast fibrous material is part of the Urticaceae family such as ramie (Asian nettle, *Boehmeria nivea*) and belongs to the genus *Urtica* (Fig. 3.2). The stinging nettle (*Urtica dioica*) is the most prominent species in Europe and in Himalayan ranges of Asia (Debnath, 2015). Since early 77 BC, nettle fibre has been used to make hand-spun yarn and to prepare hand-spun twine using a flywheel length of 30 cm (Stewart, 1994; Vogl and Hartl, 2003). Huang (2005) suggested that the best fibre properties can be obtained after several processes. He recommended that after harvesting the green bundled plants be retted for 2 weeks in a pond. After this the bast portion can easily be separated from the woody core of the nettle plant. These retted fibres are unsuitable for spinning. There are 15 stages involved in the biological and chemical processes to clean nettle fibre efficiently from green harvested nettle fibres to processing the plant successfully in spinning machinery for yarn preparation. Overall the process produces well-separated nettle fibre with good spinning performance, but a huge amount of chemicals and steps are involved in making nettle fibre ready for spinning. In a study by Dreyer et al. (2002), green nettle plants were retted followed by drying and then fibres were separated by a mechanical extractor. These mechanically extracted nettle fibres were treated separately with three different treatments: chemical, enzymatic

Figure 3.2 Nettle plant (Anonymous, 2016b).

and steam-exploded processes. In a comparison of these three treatments, the authors concluded that the enzymatic process of separating coarse nettle fibre has good potential for producing fine and strong nettle fibre. A similar amount of fibre separation can be obtained from chemical treatment in which alkali is used as the major chemical. The authors concluded that considering the environmental impact and other related aspects, the enzymatic process is best among these three methods of separating nettle fibres whereas chemical treatment leads to environmental pollution and steam explosions requiring a huge amount of heat energy. Considering these aspects, Bacci et al. (2010) experimented with different methods of retting and fibre extraction. Their studies confirmed that both controlled microbiological retting with the combined use of anaerobic and aerobic bacteria and vat enzymatic retting, especially when a solution of Pectinex Ultra SP-L with ethylene diamine tetraacetic acid was used, improved fibre quality.

As far as the products of nettles are concerned, Dogan et al. (2008) reported that the stems and fibres of stinging nettle are used to prepare traditional handicrafts in several Balkan countries. The nettle fibrous material in Bulgaria locally known as *kopriva* is used for the sustainable manufacture of cloth, sacks, cord, and nets. In Romania, nettle is known as *urzica* and it substitutes for cotton in fishing net and is used to make paper. It is known as *kopriva* in Serbia, where nettle fibre is considered one of the major textile fibres and is used in the spinning industry to produce textile products. Overall, there is a wide range of possible handicraft products (doormats, flower vases, wall hangings, door chains, carpets, handbags, table mats, beach umbrellas, lampshade, etc.) made of nettle, from fibre, yarn, fabric or combination of these. All of these products have a huge profit margin owing to the high cost to benefit ratio (Debnath, 2015). Most of the handicraft products fall under fashion items. In her findings, Dunsmore (2006) described different handicraft products made of nettle fibres and handspun yarn from Nepal. That study also discussed sustainable rural livelihood from the cultivation of nettle to handicraft product development out of nettle. This making of handicrafts from nettle fibrous material for rural hill people has created an alternative source of income during slower economic period of agricultural activities. Economic development in the nettle processing community of Nepal has been created through proper marketing strategies and exporting fashion textiles and handicraft products to Europe and America (Dunsmore, 1998). Bacci et al. (2010) also said that for sustainable handicraft products from nettle it is essential to achieve enzymatic retting to obtain the best quality fibre. Deokota and Chhetri (2009) reported in their research that hand-loom articles and handicrafts move side by side to promote nettle-based products in Nepal. These include coarse hand-woven cloth, sacks, bags, fishnets and *namlo* (head straps to carry loads), which are sold in the local market or in some cases are bartered for food or other necessary items in rural communities. Anonymous (1998) demonstrated various sustainable fashion products made of nettle and its blends, such as hats, jackets, room decorations and various handicraft products. Many such products are available in international markets (Debnath, 2015).

3.2.3 Ramie

Similar to nettle, ramie is another bast fibre extracted from the bark of the ramie plant (Fig. 3.3). Ramie (*Boehmeria nivea*) is a bast fibre extracted from the stems of the ramie plant. China is a pioneer in ramie fibre production. In China it is known as China grass and different fashion products such as women's dress material, shirts, suits and handicraft products are popular there. However, ramie-based technologies and fashion products have not spread to a greater extent in other part of the world. The important concern about ramie is the removal of gum content from the extracted fibres, which contain as much as 30% gum per weight of fibre. If gum is present the fibre is not spinnable and the removal of gum is essential.

Paul and Bhattacharyya (1979) showed several mixed bacterial cultures isolated from different sources, of which four were efficient methods for removing the gum of ramie

Figure 3.3 Ramie plant (Anonymous, 2016c).

fibre to a considerable extent with no adverse effects on the fibre cellulose. By using the best culture, a simple and inexpensive microbial process was developed for degumming ramie fibre. They also developed a combined microbial and chemical process. They concluded that this combined process is simpler and more economical than the chemical process itself. With these combinations, smaller amounts of chemicals were consumed and fewer processing difficulties were encountered (Paul and Bhattacharyya, 1979). Pandey (2007) critically reviewed research work and developments in physical and related properties of ramie, a long vegetable bast fibre. The properties of the fibre at different stages of plant growth were discussed including the aspects of topography, optical microscopy, electron microscopy, regaining of moisture, density, strength, tenacity, fineness and mechanical properties, as well as thermal, infrared, Fourier transform infrared and X-ray properties. He also emphasized different developments in the area of technologies concerning ramie fibre processing. The economical process of gum extraction from ramie fibre was developed and the extracted natural gum was used as a natural resin to develop jute particleboard. With this composite technology, the whole process of ramie gum used to develop a green particleboard/biocomposite has been demonstrated (Anonymous, 2002a). Thakur et al. (1999) described detailed findings on the chemical composition of ramie fibre. They found that complete removal of the gum from the fibre is similar to that of pure cellulose, about 94%–95% α-cellulose. As far as the fibre is concerned, ramie is a strong, lustrous, soft and fine bast fibre obtained from inner bark of the plant *B. nivea (L) Gaud.* Research shows the processing technology of cotton-ramie blends (Anonymous, 2002b, Anonymous, 2003) on a short-staple spinning system (cotton spinning). Since both fibres have their origins in natural plants, it is expected that the products will be environmentally green. *Lasingphee*, fancy ramie-cotton woven towels, fancy designed fabrics, etc., are some novel fashion products developed from cotton-ramie blended yarns. Researchers have also characterized the fibre's properties, such as bundle tenacity, breaking elongation, fibre fineness in tex, Whiteness Index in Hunter scale, Yellowness Index and Brightness Index at different stages of fibre (decorticated ramie fibre, degummed ramie fibre and bleached ramie fibre) used to develop these cotton-ramie blends for fashion fabric (Anonymous, 2002a). Furthermore, Ahmed et al. (2004) concluded that a reduction in gum content from 20%–30% to 2%–3% using alkali treatment and subsequent bleaching of ramie fibre in blended cotton ramie yarn spun in the cotton spinning system resulted in a deterioration of the yarn's strength.

It was been found in the literature (Anonymous, 2015) that Ramie fibre is also known as China grass, grass linen, rhea and grass cloth. It is said that ramie fibres are one of the oldest natural vegetable fibres, used for thousands of years for fabric and clothing, including ancient Egyptian mummy wraps and shrouds. The fibre is white, lustrous, and fine like silk. However it is somewhat stiff and brittle, which is great for coarser products such as twine, rope, wallpaper and nets. Ramie is often used as a substitute for cotton in the fashion industry (Debnath, 2016). When spun wet it produces a high-luster soft yarn. Dry spinning results in a harsh, hairy yarn. The versatility of ramie fibre allows it to be made into fine yarn for garments ranging from dresses and suits to sportswear and jeans. Fabrics made of 100% long and fine ramie fibres are lightweight and silky. The fibres are uneven, which gives ramie fabric an appearance similar to linen. To produce fabrics with various improved characteristics, ramie is often blended with other fibres such as cotton. By doing this, creating fabrics from fine linens to coarse canvas gives ramie fibres almost unlimited potential. By blending with wool, the fabric is lighter and shrinking is minimized. Cotton blends result in increased strength, colour and luster whereas rayon blends result in a higher wet strength. It is also commonly used as a substitute for flax/linen and can be blended with silk fibres. Ramie/China grass fibres are commonly used in blends for sewing threads, fashion sweaters, fancy clothing and linens. Because the fashion industry is becoming increasingly eco-conscious, the popularity of ramie will continue to rise. Interest in ramie is being rekindled and is appearing more often in the fibre content of clothing and textiles. Because of the trend toward natural fibres, expect to see ramie become even more popular. Ramie fabric is used for apparel including suits, skirts, jackets, dresses, shirts, blouses, pants and handkerchiefs. It is also commonly used in home fashion articles such as draperies, upholstery, linens and thread (Anonymous, 2015). Industrial uses of ramie include parachute fabrics, fire hoses, and canvas. High-quality paper goods such as bank notes and cigarette papers are produced from the short fibres. Research demonstrates the processing technology of cotton-ramie blends (Anonymous, 2002a,b) on short-staple spinning system (cotton spinning). Because both of the fibres are of natural plant origin, the final products are environmentally friendly. Different fashion items such as *Lasingphee*, fancy ramie-cotton woven towels, fancy designed fabrics, etc., are some of the novel fashion products developed from cotton-ramie blended yarns. However less effort has been documented regarding the sustainability of the product. The main problem associated with the ramie fibre is its gum content. Ramie fibre usually contains as much as 30% gum. Two processes are normally used to extract the gum from the fibre: one is chemical (alkaline treatment) and the other is a microbial/enzymatic process. Although the process is more sustainable in the case of the microbial process, it is costly and time-consuming. China is popular for the development of eco-friendly, sustainable, ramie-based textiles for fashion products. Different international companies (Joanna Hope linen-blend trousers, for example) market ramie-linen *kurti* for women for casual wear. Apart from conventional fashion textiles, when ramie fibre is used in industrial applications, it is suitable as an electrical insulation material when used in dry application areas because of its good dielectric properties (Sao et al., 1990). To obtain such good dielectric properties, ramie fibre is treated with an 18% concentration alkali (NaOH) solution.

3.2.4 Banana

The banana plant contains good-quality textile-grade fibres popularly known as banana fibre. This fibre is another unexplored natural fibre used for the fashion and technical textile industries for sustainable product development. These fibres are extracted from the pseudostem of the banana plant. Suitable banana fibre extractors can be used to extract the fibres from the pseudostems of the banana plant. The fibres are bleached and blended with jute fibre for diverse value-added fashionable textiles (Debnath and Das, 2012). There are also different fashion products from banana-based textiles. Research contributed by Sinha (1974a,b) discloses a suitable process to spin in banana fibre in a jute spinning system. They made lot of effort for blending banana-based fibres and further different products development thereof. This work also elaborated on the use of white jute, tossa jute, and kenaf, which were blended with banana fibre at different proportions to develop different sustainable products. Anonymous (2012a) discussed conventional hydrogen peroxide bleaching of banana-based fibrous material, which has been used to bleach banana fibre and further dyed it for added value. Jute-banana blended yarns have been made with different blend ratios (100:0, 75:25, 50:50, 25:75 and 0:100) and the properties of the banana-blended yarns have been evaluated. Conventional jute spinning machinery has been used to spin these banana-based yarns. The authors concluded that because of coarseness and brittleness of the banana fibre, 100% banana fibre in spinning shows inferior results compared with banana-blended yarns. Furthermore, Anonymous (2012b) disclosed that bleached and dyed jute-banana fibre blended yarns are used to develop ornamental fibre using a jacquard attachment in hand looms. The decorated fabrics are used for fashion jackets and garment development. Hence there is enormous potential to design and develop green banana fibre textiles (Basu et al., 2006; Basu and Roy, 2008).

3.2.5 Sunn hemp

Among different bast fibres, sunn hemp (Fig. 3.4) has an important role in maintaining the agricultural soil nutrient. Sunn hemp is an important legume crop and fixes nitrogen in the soil, apart from its multiple uses in textiles as commented on by Chaudhury et al. (2015).

Researchers have developed various value-added textiles from sunn hemp fibres (Anonymous, 2005). They established that the fibre procured from Nagpur, India contained 60% more good spinnable fibre than fibres obtained from other parts of India. The raw fibre had an acceptable level of strength and fineness for subsequent textile

Figure 3.4 Sunn hemp cultivation (Anonymous, 2016d).

processing. A normally used textile wet processing technology can be adopted for the chemical processing of sunn hemp fibre. Chemical treatment resulted in a slight loss of strength but it was sufficiently strong to spin good-quality yarn. After bleaching treatment, acceptable whiteness can be achieved and different colour shades can be applied. Miniature jute spinning as well as commercial jute spinning systems can be adopted to spin sunn hemp yarn. The physical properties of spun yarn are compatible with conventional jute yarn of a similar linear density. Sunn hemp–cotton union fabrics were developed for different applications in which the cotton and sunn hemp yarns were used in warp and weft directions, respectively. This union fabric can absorb dye uptake satisfactorily with the proper application of a reactive dye. Finished products from sunn hemp–based fibres have an attractive look and feel (Anonymous, 2005).

3.2.6 Hemp

Hemp, the cannabis plant (Fig. 3.5), is well known for its use as a psychotropic drug. Apart from its seed, it is particularly as used for fishing bait and its fibres are used for textile applications.

Processed hemp fibre can be used to manufacture a fabric ideal for summer clothing with good ventilation and moisture absorption as well as antibacterial and anti-ultraviolet properties (Paolo and Gianpietro, 2004). Hemp fibres are rigid and rough to handle. Hence, clothing made with the untreated fibres appears to be wrinkled and feels prickly. As a result of the high levels of friction and tension generated during spinning and weaving, parts of the short hemp fibre stand out of the surface of yarns (hairiness) and fabric made of these hairy yarns feels hairy. Unlike flax, ramie and pineapple fibres, hemp is also popularly known in the luxury textile industry. Hence, further detailed coverage has not been made. Apart from these textiles in the area of clothing, Chen et al. (2007) made a comparative study of composites made of hemp

Figure 3.5 Hemp cultivation (Anonymous, 2016e).

fibre. They found that composites made of hemp fibre are comparable to composites made of other natural fibres. Pietak et al. (2007) treated hemp fibre using steam, alkaline treatment and combined steam–alkaline and enzymatic treatments; in addition, acetylation was used as a final treatment to the combined steam–alkaline treated samples to decrease the polarity of the fibre. They selected NaOH as the alkaline treatment. After chemical treatment (combinative steam–alkaline and enzymatic treatments, plus acetylation) the wettability of the fibre improved, resulting in high strength of the final composite material.

3.2.7 Other unpopular bast fibres

Nature gifted us with a wide range of plants grown in various agroclimatic zones that have the potential for fibres to be extracted from their stem. Because many of those plants are not highly explored or are not popularly cultivated specifically for the purposes fibres, many of those unpopular fibres are still unknown. There are a few researchers who explored these unknown sources of plant fibres from the plant kingdom. In their research, Ghali et al. (2014) extracted alfa fibres from the plant *Stipa tenacissima*, or esparto grass (alfa is the Arab name for esparto grass), which grows in the dry regions of North Africa. These fibres are used extensively in the production of pulp for the paper industry. It has been recorded that this plant is highly available in Tunisia and that to make it a successful alternative crop it must be incorporated into value-added products such as nonwovens for domestic textiles materials. Ghali et al. (2014) further concluded that this alfa fibre can be used to making nonwovens for diverse applications. In the area of composites, Bessadok et al. (2007) used alfa fibre; they proved that this fibre has tremendous potential to reinforce composite materials. The main problem with this natural fibre being used as a reinforcement is that it is hydrophilic. Therefore, the ageing of composite materials can be pronounced because of the diffusion of water molecules leading to swelling. Moreover, adhesion between natural fibres and the polymer matrix is insufficient, as reported in their study. In this work, various chemical surface treatments (acetylation, acrylic acid, styrene and maleic anhydride) were performed on alfa fibre. Treatments reduced the overall water uptake of alfa fibres. However, in particular, styrene treatment significantly increased the moisture resistance of these fibres (Debnath, 2016).

Khimp fibre (*Leptadenia pyrotechnica*) is another grass category of bast fibre that has not been much explored but has good potential in the textile industry. Kundu et al. (2005) extracted khimp fibre from the khimp plant and analysed the fibre's properties. Jamal et al. (2001) highlighted the importance of khimp fibre for the development of indigenous technologies in the pulp and paper industries. This khimp fibre contains high α-cellulose with a favourable length/breadth ratio of the ultimate cell, which makes it preferable for use in blending with cotton or polyester fibres to produce value-added products. Jamal et al. (2001) also examined the innovation process in selected firms engaged in paper making in India. The current firm-level analysis focuses on incremental innovations as a process to achieve indigenous technological capability, especially to use imported technologies more effectively through adaptation and improvement. Their study also confirms that 'learning by doing' is a good indicator for these firms in building the capability to indigenize and improve upon sustainable technology. The in-house R & D units at a firm level provide strong support to achieve indigenous technological capability as reported by them.

3.3 Sustainability of bast fibre in future

Apart from cotton, silk and wool as major natural fibres, many other unexplored plant fibres such as jute, banana, flax and ramie are competent to survive sustainably in coming days. When we discuss the sustainable production of these fibres, most of these bast fibre plants do not need special attention and care during the production of the fibre (jute, nettle, banana, flax, sunn hemp, ramie, etc.). For this reason, the production of many these fibres, either on a commercial scale or locally, has being ongoing for several centuries despite much progress in science and technology. Main advantages of these natural fibres are that they are from plant sources and production to fibre extraction is sustainable (Debnath, 2015). With advances in fibre technology, many synthetic/artificial fibres have been developed and used for the past 60 decades. However, over the past 15–20 years, special attention has been given to using smaller quantities of synthetic fibre and more emphasis has been on the use of natural fibres in different areas of fibre application. With this production of natural fibres, special attention and application of these fibres in different textile fields including fashion also increased. Because of the scarcity of agricultural land and urbanization in limited areas, a higher amount of fibre is being produced (for example, 85–90 lakh bales/year raw jute were produced in 1990–95 and production increased to 115–120 lakh bales/year in 2010–13). However, with this improved productivity more importance is being placed on the sustainability of the production system.

As far as the flax industry is concerned, there are different methods of retting and the involvement of human intervention exists. For the sustainable production of this flax fibre, one needs to concentrate not only on fibre retting but also on the further processing. With the fibre extraction and processing, the final product from of this fibre-based material becomes costly and hence product diversification is required to sustain the industry. This diversified product will hold the niche market which will fetch a high cost–benefit ratio. Apart from traditional flax products which have a known market, innovative product design and newer products may be attractive for the sustainable development of the flax industry.

Products from nettle fibre are costly. Currently, many of the producers associated with the nettle fibre industry face challenges from a sustainable point of view. The nettle extraction process is tedious. In addition to this, products made from this fibre are not known in many parts of the globe. Furthermore, the producers are restricted mostly to local markets. Local market demands fulfil the requirement and the profits are not high. However, in the current scenario, more emphasis needs to be given to online marketing of nettle-based products in developed countries, which may lead to high profits for producers to sustain this industry. Suitable chemicals must be used during fibre processing, as well as other types of chemical processing including natural dyeing; together, they will make the product environmentally friendly, which may sustain the nettle industry.

Ramie is a fibre which requires serious attention as far as a suitable degumming process is concerned. Although there are a lot of studies on enzymatic and microbial degumming of ramie, the most economic is chemical degumming. Much more work is required to sustain this industry because the eco-friendly (enzymatic/microbial) process is still costly. On the other hand, the products are mostly marketed in producing countries. There is a huge demand for ramie-based products in European countries, America, Japan, etc. Ramie-based products must certify green processing and reach the niche market with a wide product range and diversified products, which may sustain the ramie fibre-based industries.

Other bast fibre products such as sunn hemp, banana and hemp still use many chemicals which are not acceptable as far as green chemicals are concerned. Hence more importance must be given to designing a product made with green chemical processing. More emphasis must be given to enzymatic/microbial treatment added to herbal/natural dyeing to sustain the bast fibre industry. Products made from these bast fibres must be made widely available so that more markets can be attended.

In cottage and small-scale industries, fashion products are being produced directly from raw plant fibres without much use of hazardous chemicals and machinery consuming high amounts of electrical energy. Some of the fancy items are also being produced in decentralized sectors. However, large industries including composite textile mills are producing fashion products in huge quantities where the consumption of power is high because of automation. Many common products can be made from unexplored plant fibres, such as sacks, cloth for packaging, agricultural application, and so on. However, there will be a huge added value if we switch to the production of diversified fashion and lifestyle items from conventional products. Many of the industries processing natural fibres worldwide are now concentrating on added value in product design. For instance, jewellery majewelryjewelrye from natural fibres require a much smaller amount (a few grams) of fibre and the added value is many fold because the products are sold at a high price. Thus we have to search for some alternative uses for which there a high cost to benefit ratio exists. Composite is another alternative which is gaining in popularity in automobile industries. This may lead to greener technology and sustainable development. Hence, the product diversification from conventional products is the order of the day for the sustainable development of natural bast fibres.

In many cases these natural plant fibres require tedious processes including human drudgery to extract the fibres from the plant component (Basu and Dutta, 2014). In addition, different toxic chemicals and higher energy-consuming processes are being used to convert those fibrous materials into yarn, and toxic gases as well as pollutants released during the processing of these fibres are alarming as far as greenhouse gases are concerned. The time has come to understand and implement environmentally friendly processes starting from agricultural practices to the final products, so that the products is bio-friendly and disposal will not be an issue. Hence, more efforts are needed and in-depth research is necessary to eliminate toxic chemical processing (Basu and Dutta, 2014). It is also essential to use energy economically to make the overall process sustainable. For instance, most of the bast fibre industry uses petroleum-based oil (batching oil) for ease of processing, which may be replaced by vegetable oil; more in-depth research is needed on this topic (Debnath, 2016, 2017).

Because most the bast fibres are affected by microbial attack, eco-friendly (preferably herbal) treatment may enhance the lifespan of the end product. In this area, more attention is required. Natural plant fibres have various advantages, as reported by Anonymous (2015). For instance, let us take an example of ramie fibre, which is highly absorbent; has the natural ability to resist stain; has a lustrous appearance; is strong and durable, is naturally resistant to bacteria, moulds and mildew; has low elasticity so it does not easily shrink; withstands higher water temperatures; has good dimensional stability; is resistant to light, rot and insect attack; dyes easily; its fibre can be bleached; and it absorbs heat and releases moisture, which making it comfortable to wear in warm weather, etc. The clothing and textile industry recognizes ramie as a premium product because it is one of the strongest natural fibres. On the other hand, the positive aspects of strength can be better used for high-modulus composite manufacturing, which can minimize the cost of cars in the automobile industry. Overall, to make bast fibre sustainable, it is important to move toward the application of greener technologies as well as a search for newer diversified applications for a better profit margin.

3.4 Conclusions and suggestions for further research for sustainable development of bast fibres

Natural fibres are the ultimate resource for future survival. The use of synthetic fibre should be minimized since the petroleum reserve is limited, on the contrary, natural fibres are annually renewable and many of the cases does not require additional care for growing, henceforth the use of natural fibres should be boosted. Moreover, these bast fibres produce a huge amount of oxygen and release a comparatively lesser amount of carbon dioxide during their lives in addition to providing soil nutrients. Therefore natural fibres can minimize greenhouse gases and keep the earth green. They make the overall process sustainable. An eco-friendly process has to be adopted, but more profitable product diversification is also necessary, from fashion textiles to industrial textiles and composites.

To sustain the bast fibre industry, it must concentrate on the sustainable quality of the product, awareness of the global market (design and colour range, product range, etc.); supply-chain management, consistent supply of the product to the market, especially in global markets; good logistics and packaging; clustering of the bast fibre industry (for example, clustering of the ramie industry, the nettle industry, etc.); and online marketing (attached to global online marketing agencies such as Amazon, Alibaba, etc.) to sell products faster and reach the most customers. Finally, product diversification is needed to make the bast fibre industry sustainable.

References

Ahmed, M., Chattopadhyay, S.K., Chaphekar, A.K., Gaikwad, R.S., Dey, S.K., 2004. Characteristics of degummed ramie fibre and its cotton blended yarn. Indian J. Fibre Text. Res. 29 (3), 362–365.

Alimuzzaman, S., Gong, R.H., Akonda, M., 2014. Biodegradability of nonwoven flax fiber reinforced polylactic acid biocomposites. Polym. Compos. 35, 2094–2102. http://dx.doi.org/10.1002/pc.22871.

Anonymous, 1998. http://www.vads.ac.uk/learning/csc/collingwood/essay. html.

Anonymous, 2002a. Annual Report 2001-2002. National Institute of Research on Jute & Allied Fibre Technology, Kolkata, India. pp. 34–35 and pp. 45–47.

Anonymous, 2002b. Annual Report 2001-2002. National Institute of Research on Jute & Allied Fibre Technology, Kolkata, India, pp. 58–60.

Anonymous, 2003. Annual Report 2002-2003. National Institute of Research on Jute & Allied Fibre Technology, Kolkata, India, pp. 76–77.

Anonymous, 2005. Annual Report 2004-2005. National Institute of Research on Jute & Allied Fibre Technology, Kolkata, India, pp. 69–70.

Anonymous, 2012a. Annual Report 2011-2012. National Institute of Research on Jute & Allied Fibre Technology, Kolkata, India, pp. 76–78.

Anonymous, 2012b. Annual Report 2011-2012. National Institute of Research on Jute & Allied Fibre Technology, Kolkata, India, pp. 15–17.

Anonymous, 2015. http://textilefashionstudy.com/top-sisal-producers-country-of-the-world-hard-fiber-production/.

Anonymous, 2016a. http://www.urbanfarm.org/blog/2016/01/14/flax-on-flax off/.

Anonymous, 2016b. https://bioweb.uwlax.edu/bio203/2011/homolka_kail/adaptation.htm.

Anonymous, 2016c. http://davesgarden.com/guides/pf/showimage/338016/#b.

Anonymous, 2016d. https://drapjt.wordpress.com/2014/08/18/sunn-hemp/.

Anonymous, 2016e. http://www.hempwickbeeline.com/illegalhemp/.

Bacci, L., Lonardo, S.D., Albanese, L., Mastromei, G., Perito, B., 2010. Effect of different extraction methods on fiber quality of nettle (*Urtica dioica* L.). Text. Res. J. 81 (8), 827–837.

Basu, G., Datta, M., 2014. Potentiality of Indian Flax. National Institute of Research on Jute and Allied Fibre Technology, Kolkata, India, pp. 1–104.

Basu, G., Roy, A.N., Bhattacharya, G.K., 2006. Processing of Jute in Blends with Natural and Man-Made Fibres. In: Bhattacharyya, S.K. (Ed.). National Institute of Research on Jute and Allied Fibre Technology, Kolkata, India, pp. 1–54.

Basu, G., Roy, A.N., 2008. Blending of jute with different natural fibres. J. Nat. Fibers 4 (4), 13–29. http://dx.doi.org/10.1080/15440470801893323.

Bessadok, A., Marais, S., Gouanvéa, F., Colassea, L., Zimmerlinc, I., Roudeslib, S., Métayera, M., 2007. Effect of chemical treatments of Alfa (*Stipa tenacissima*) fibres on water-sorption properties. Compos. Sci. Technol. 67 (3–4), 685–697.

Chaudhury, J., Singh, D.P., Hazra, S.K., 2015. Sunnhemp. http://assamagribusiness.nic.in/sunn-hemp.pdf.

Chen, Y., Sun, L., Negulescu, I., Wu, Q., Henderson, G., 2007. Comparative study of hemp fiber for nonwoven composites. J. Ind. Hemp 12 (1), 27–45. http://dx.doi.org/10.1300/J237v12n01_04.

Debnath, S., 2014. Jute-based sustainable agrotextiles, their properties and case studies. In: Muthu, S.S. (Ed.), Roadmap to Sustainable Textiles and Clothing. Springer Publisher, pp. 327–355. ISBN:978-981-287-064-3 (Print) 978-981-287-065-0 (Online).

Debnath, S., 2015. Chapter 3. Great potential of stinging nettle for sustainable textile and fashion. In: Gardetti, M.A., Muthu, S.S. (Eds.), Handbook of Sustainable Luxury Textiles and Fashion, Environmental Footprints and Eco-design of Products and Processes. Springer Science+Business Media, Singapore, pp. 43–57. http://dx.doi.org/10.1007/978-981-287-633-1_3.

Debnath, S., 2016. Chapter 1. Unexplored vegetable fibre in green fashion. In: Muthu, S.S., Gardetti, M.A. (Eds.), Green Fashion, Part of the Series Environmental Footprints and Eco-design of Products and Processes. Springer Science+Business Media, Singapore, pp. 1–19. http://dx.doi.org/10.1007/978-981-10-0245-8_1. ISBN:978-981-10-0244-1 (Print) 978-981-10-0245-8 (Online).

Debnath, S., 2017. Chapter 7. Sustainability in jute-based industries. In: Muthu, S.S. (Ed.), Sustainability in Textile Industry, Part of the Series Textile Science and Clothing Technology. Springer Nature Singapore Pte Ltd, Singapore. ISBN: 978-981-10-2638-6, pp. 139–147.

Debnath, S., Das, R., 2012. Leaflet on Extraction and Value Addition of Banana Fibre. National Institute of Research on Jute and Alled Fibre Technology, Kolkata, India, pp. 1–4.

Debnath, S., Ganguly, P.K., De, S.S., Nag, D., 2010. Control of soil moisture and temperature by light weight jute fabrics. J. Inst. Eng. 90TX (2), 16–19.

Debnath, S., Madhusoothanan, M., 2010. Water absorbency of jute-polypropylene blended needle-punched nonwoven. J. Ind. Text. 39 (3), 215–231. http://dx.doi.org/10.1177/1528083709347121.

Debnath, S., Nag, D., De, S.S., Ganguly, P.K., Ghosh, S.K., 2006. Studies on mechanical and hydraulic properties of JGT for geo-technical applications. J. Inst. Eng. TX86 (2), 46–49.

Debnath, S., Roy, A.N., Basu, G., Chattopadhyay, S.N., 2009. NIRJAFT's technologies for rural development, Section-V, entitled Natural fibres and geotextile applications. In: Shukla, J.P. (Ed.), New Technologies for Rural Development Having Potential of Commercialisation. Allied Publishers Pvt. Ltd, India. ISBN: 978-81-8424-442-7, pp. 136–142.

Dogan, Y., Nedelcheva, A.M., Petkovic, D.O., Padure, I.M., 2008. Plants used in traditional handicrafts in several Balkan countries. Indian J. Tradit. Knowl. 7 (1), 157–161.

Deokota, R., Chhetri, R.B., 2009. Traditional knowledge on wild fiber processing of allo in Bhedetar of Sunsari district, Nepal. Kathmandu Univ. J. Sci. Eng. Technol. 5 (I), 136–142.

Dreyer, J., Mussig, J., Koschke, N., Ibenthal, W.D., Harig, H., 2002. Comparison of enzymatically separated hemp and nettle fibre to chemically separated and steam exploded hemp fibre. J. Ind. Hemp 7 (1), 43–59. http://dx.doi.org/10.1300/J237v07n01_05.

Dunsmore, J., 1998. Microenterprise development: traditional skills and the reduction of poverty in Highland Nepal. Himal. Res. Bull. 18 (2), 22–27.

Dunsmore, S., 2006. The Nettle in Nepal: Tradition and Innovation, second ed. The John Dunsmore Nepalese Textile Trust, Nepal.

Fortea-Verdejo, M., Lee, K.Y., Zimmermann, T., Bismarck, A., 2016. Upgrading flax nonwovens: nanocellulose as binder to produce rigid and robust flax fibre performs. Composites: A 83, 63–71.

Ghali, L., Halimi, M.T., Hassen, M.B., Sakli, F., 2014. Effect of blending ratio of fibers on the properties of nonwoven fabrics based of alfa fibers. Adv. Mater. Phys. Chem. 4, 116–125.

Huang, G., 2005. Nettle (*Urtica cannabina* L) fibre, properties and spinning practice. J. Text. Inst. 96 (1), 11–15. http://dx.doi.org/10.1533/joti.2004.0023.

Indu, G.K., Senthilkumar, P., 2016. Development of nonwovens from natural fibres for various applications. Int. J. Appl. Eng. Res. 11 (3), 1879–1882.

Jamal, T., Prakash, N.B., Khumbhar, P.P., 2001. Patterns of technological change—an innovation process to achieve indigenous technological capability in Indian paper industry. J. Sci. Ind. Res. 50 (8), 629–639.

Jan, E.G., 2016. Environmental benefits of natural fibre production and use. In: Proceedings of the Symposium on Natural Fibres. Van Dam Wageningen University, The Netherlands. ftp://ftp.fao.org/docrep/fao/011/i0709e/i0709e03.pdf.

Kundu, S.K., Mojumdar, P., Bhaduri, S.K., Das, B.K., 2005. Physical characteristics of khimp fibre. Indian J. Fibre Text. Res. 30 (2), 153–156.

Maity, S., Gon, D.P., Paul, P., 2014. A review of flax nonwovens: manufacturing, properties, and applications. J. Nat. Fibers 11 (4), 365–390. http://dx.doi.org/10.1080/15440478.2013.861781.

Memon, N.A., 2016. Nonwovens: Global Demand Is Expected to Reach 9.1 Million Tonnes in 2017. http://www.ptj.com.pk/Web-2014/12-2014/PDF-December-2014/Technical-Textiles-and-Nonwovens-DrNoor.pdf.

Nechwatal, A., Mieck, K.P., Reumann, T., 2003. Developments in the characterization of natural fibre properties and in the use of natural fibres for composites. Compos. Sci. Technol. 63, 1273–1279.

Pandey, S.N., 2007. Ramie fibre: Part II. Physical fibre properties. A critical appreciation of recent developments. Text. Prog. 39 (4), 189–268. http://dx.doi.org/10.1080/00405160701706049.

Paolo, R., Gianpietro, V., 2004. Hemp as a raw material for industrial applications. Euphytica 140, 1–6.

Paul, N.B., Bhattacharyya, S.K., 1979. The microbial degumming of raw ramie fibre. J. Text. Inst. 70 (12), 512–517. http://dx.doi.org/10.1080/00405007908658895.

Pietak, A., Korte, S., Tan, E., Downard, A., Staiger, M.P., 2007. Atomic force microscopy characterization of the surface wettability of natural fibres. Appl. Surf. Sci. 253, 3627–3635.

Rawal, A., Anandjiwala, R., 2007. Comparative study between needlepunched nonwoven geotextile structures made from flax and polyester fibres. Geotext. Geomembr. 25, 61–65.

Sao, K.P., Samantary, B.K., Bhattacherjee, S., 1990. Electrical property of ramie fibre. J. Mater. Sci. Lett. 9, 466–469.

Sharma, H.S.S., Faughey, G., Lyons, G., 1999. Comparison of physical, chemical, and thermal characteristics of water, dew and enzyme-retted flax fibers. J. Appl. Polym. Sci. 74, 139–143.

Sinha, M.K., 1974a. The use of banana-plant fibre as a substitute for jute. J. Text. Inst. 65, 27–33.

Sinha, M.K., 1974b. Rope making with banana-plant fibre. J. Text. Inst. 65, 612–615.

Stewart, H., 1994. Indian Fishing-Early Methods on the Northwest Coast. Douglas and Mcintyre Ltd, Canada, pp. 29–80. 10:0295958030.

Thakur, R., Sarkar, C.R., Sarmah, R., 1999. Chemical composition of some varieties of ramie and their fibre characteristics. Indian J. Fibre Text. Res. 24 (4), 276–278.

Vogl, C.R., Hartl, A., 2003. Production and processing of organically grown fiber nettle (*Urtica dioica* L.) and its potential use in the natural textile industry: a review. Am. J. Altern. Agric. 18 (03), 119–128.

Xue, D., Miao, M., Hu, H., 2011. Permeability anisotropy of flax nonwoven mats in vacuum-assisted resin transfer molding. J. Text. Inst. 102 (7), 612–620. http://dx.doi.org/10.1080/00405000.2010.504566.

Further reading

Nag, D., Debnath, S., Ganguly, P.K., Ghosh, S.K., 2010. Efficient management of soil moisture by jute geotextile for cultivation of horticultural crops in red lateritic zone. J. Inst. Eng. AG-91 (1), 21–24.

Shah, A., Gong, R.H., Akonda, M., 2014. Biodegradability of nonwoven flax fiber reinforced polylactic acid biocomposites. Polym. Compos. 35, 2094–2102. http://dx.doi.org/10.1002/pc.22871.

Sustainable wool production in India

4

Ajay Kumar, Lesile L. Prince, Seiko Jose
ICAR–Central Sheep and Wool Research Institute, Avikanagar, India

4.1 Introduction

Globally, wool contributes only 2%–3% to total fibre production and has the status of a luxury fibre. To meet the global fibre demand, over the past decade, fibre production across the world increased by 73% whereas man-made fibres grew by 104%. In contrast, global wool production dropped nearly half, from 2.5% in 2000 to 1.3% in 2014 (Wool Production and Supply, Trend, Drivers and Prospects, 2015). Wool has unique aesthetic properties and thus used for premium apparels, carpets upholstery, etc. Australia and New Zealand, the largest producers of wool in the world, have witnessed a significant drop in wool production over the past decade, mainly owing to a fall in the wool sheep population. Wool production during this period declined by 49%, from 2.28 million tons in 2000 to 1.13 million tons in 2014. China, the second largest wool producer, observed an increase in the first half but stagnated in the later half of the decade for wool production during the same period. India, although ranked third in sheep population after Australia and China, was rank sixth as a wool-producing country owing to a lower productivity rate of 0.9 kg/sheep/year compared with the world average of 2.4 kg/sheep/year. It was to reported increase marginally by 1.9%, from 47.6 million kg in 2000 to 48.5 million kg in 2015, accounting for 1.8% of total world wool production. The globally declining trend in fine wool production includes China and India. However, medium (24–30 μ) and coarse-grade wool (above 30 μ), which account for 48% of global wool production, showed an increasing trend because of their specific uses in furnishing fabrics, carpets, upholstery, blankets, etc. Indian wool is almost exclusively of larger micron and is used to manufacture carpets and rugs. Of the wool produced in the country, about 70% is carpet-grade wool, 10% is apparel grade, and 20% is coarser-grade wool for making *darries* (rugs), etc.(Annual Report Textile Ministry, 2006). On the plus side, wool is sustainable and renewable, and it has a low impact as a fibre resource. Sheep do not need massive inputs of artificial stuff; they can survive on grasses and pasture produced on uneven topology that are not useful for other forms of agricultural production. Wool is not a highly processed material. It is clipped from the sheep, scoured and spun/felted until it is ready for end use. It is reusable and recyclable and is biodegradable. It is warm and comfortable (resilient). It is used for apparels but also for insulating carpets and other furnishing materials for insulation in our homes, which can reduce the need for heating energy. The global impact of wool and woollens is provided in Table 4.1.

Sustainable Fibres and Textiles. http://dx.doi.org/10.1016/B978-0-08-102041-8.00004-4

Table 4.1 **Major importers and exporters of woollens**

Products	Major importers	Major exporters
Wool types	Italy, China, Germany, Turkey, South Korea	China, Uruguay, Argentina, Czech Republic, Germany
Wool worsted yarn	China, Italy, Hong Kong, Germany, Japan	China, Hong Kong, Germany, Italy, Poland
Wool woollen yarn	Belgium, Hong Kong, Australia, Denmark, United States	New Zealand, Belgium, China, Hong Kong, Lithuania
Wool knotted carpets	United States, Germany, Czech Republic, United Kingdom, Turkey	India, Iran, Afghanistan, Pakistan, Nepal
Wool woven carpets	United States, United Kingdom, United Arab Emirates, Germany, Russian Federation	China, India, United Kingdom, Belgium, United Arab Emirates
Wool tufted carpets	United States, United Kingdom, Germany, Australia, Canada	India, Belgium, New Zealand, China, Netherlands

Source: IWTO

4.2 Sheep and wool production trend in India

The long relationship between sheep and humans has led to the development of a large number of breeds. Different breeds of sheep have been developed to meet different environmental conditions and to satisfy the human needs for clothing and food. After sheep were domesticated, they were selected to produce larger quantities of wool that did not shed. Centuries of selection for wool qualities are reflected in today's extraordinary range of fleece types, textures and colours and categorized into major types, including fine wool, carpet wool and hairy mutton breeds. Carpet wool breeds are generally adapted to extreme environments, and the characteristics of the wool are part of that adaptation. Hair sheep breeds are adapted to hot, humid climates. They lack wool altogether and are covered instead by short hair and are mainly selected for mutton. Selection for wool type, flocking instinct and other economically important traits over the centuries has resulted in more than 200 distinct breeds of sheep occurring worldwide. The diversity of sheep breeds and production systems has allowed for the wide distribution of sheep around the world.

In the 19th century under British rule, the Indian farming system was deprived and oriented to cater to the market needs of western countries. No steps were taken by the British to empower animal husbandry or wool production in India; as a result, there was a downfall in the wool and carpet industry in the country. Somehow artisans were barely able to keep alive the centuries-old art of carpet weaving. After independence, the Indian government took drastic steps to increase food grain and livestock

Table 4.2 Sheep breeds reared in different agro-climatic regions of India

North western arid and semiarid region	Southern peninsular region	Eastern region	Northern temperate region
Chokla (CW)	Bellary (MCW)	Balangir (MCW)	Bhakarwal (CW)
Jaisalmeri (MCW)	Coimbatore (MCW)	Bonpala (MCW)	Changthangi (CW)
Jalauni (MCW)	Daccani (M)	Chottanagpuri (MCW)	Gaddi (CW)
Kheri (MCW)	Hassan (M)	Ganjam (MCW)	Gurez (CW)
Magra (CW)	Kanguri (M)	Garole (MP)	Karnah (AW)
Malpura (MCW)	Kilakarsal (M)	Tibetan (CW)	Kashmir Merino (AW)
Marwari (MCW)	Madras red (M)	Kendrapara (MP)	Poonchi (CW)
Muzaffarnagar (MCW)	Mandya (M)		Rampur Bushair (CW)
Nali (CW)	Mecheri (M)		
Patanwadi (CW)	Nellore (M)		
Pugal (MCW)	Nilgiri (AW)		
Sonadi (MCW)	Rammand white (M)		
Munjal (M)	Tiruchy black (M)		
	Vembur (M)		
	Chevaadu (M)		
	Katchaikatty (M)		

Letters in parentheses indicates the major uses of the breed: *AW*, apparel wool; *CW*, carpet wool; *M*, mutton; *MCW*, mutton and carpet wool; *MP*, mutton prolificacy.

production. Schemes were launched to empower artisans to revive the art of carpet weaving and the carpet industry in the country. Because of consistent efforts by the government and growing demand, Indian wool production increased from 27.5 million kg in 1951–1952 to 40 million kg by the 1990s. India has 42 registered sheep breeds, but not all are wool producing in a real sense. However, 13 breeds of the northwest arid and semiarid region contribute maximally to carpet-quality wool production in India. This is why Rajasthan ranks first in wool-producing states in India. Apart from these groups of sheep breeds, eight breeds of the northern temperate region in India produce fine and semicoarse wool. Breeds of the southern peninsular and eastern region hardly contribute wool; they are primarily mutton-producing sheep. Wool production in India divided into four regions on the basis of the agro-climatic conditions and the types of sheep available is shown in Table 4.2.

Fine wool production in the country is mainly concentrated in the northern temperate region in the state of Jammu andKashmir, Himachal Pradesh and Uttarakhand; about 20% of the total wool production is of a lesser staple length. This has restricted its use in the woollen industries and it is mostly used locally for manufacturing shawls. Growing tourism and affordable prices have popularized local brands of Kullu and

Kashmir shawls. Keeping in mind the topography and the demand for meat and wool, sheep rearing and wool are a sustainable source of income to local people. Wool production in this region is increasing at a rate of 3% per year.

The northwestern central arid and semiarid region is most important for the country because of its carpet wool breeds. This region has second largest population of sheep and is composed of the states of Punjab, Haryana, Rajasthan, Gujarat, the plains of Uttar Pradesh and Madhya Pradesh, with carpet wool-type sheep breeds including Chokla, Nali, Marwari, Magra, Jaiselmeri, Pugal, Malpura, Sonadi, Pattanwadi, Muzaffarnagar, Jalauni, Hissardale and Kheri. The produced wool is mostly suited for manufacturing carpets. Acharya (1984) reported that nearly 50% of the sheep population in this region produces two-thirds of carpet-grade wool. Rajasthan alone with 16 million sheep contributed 40% to Indian wool production (Nagashanker and Bohra, 1984). Livestock farming, mainly sheep production, depends extensively on pastures for nutrition in this region. Sparse vegetation and a low carrying capacity, i.e., adult cattle units per hectare (ha), and most of the year's drought conditions compel sheep breeders to migrate for their livelihood. It is full of hazards and hardships for farmers and sometimes causes a social crisis in migration areas, ending up in a loss of interest in sheep farming. This can result in a fall in sheep population. Thus, wool production shows a declining trend especially in Rajasthan from 19.67 to 12.27 million kg over the past decade. Since the 1980s sheep population increased 1.5 times, from 48.4 million to 71.6 million (18th Livestock Census, 2007). The major growth in the sheep population observed in the south peninsular region was a coarser wool breed mainly reared for meat production. This caused a drop per unit wool production from 0.9 kg/sheep/year to 0.7–0.8 kg/sheep/year. However, over the years, wool production increased at a marginal growth rate of 1%, from 42.0 million to 48.4 million kg. The eastern region of country has consistently been a marginal contributor to national wool production. The trend in Indian wool production over the past few years is shown in Fig. 4.1.

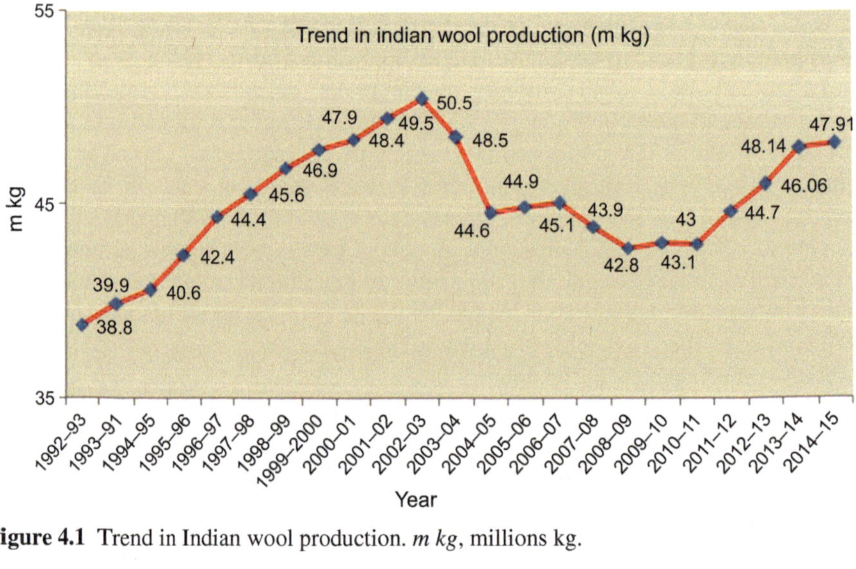

Figure 4.1 Trend in Indian wool production. *m kg*, millions kg.

4.3 Quality of Indian wool

Indian wool is almost exclusively of larger micron and is best suited for carpet manufacturing because of to its unique characteristics of better pile coverage, handle and springiness (Bhagle and Whitley, 1963). Burns et al. (1940) defined ideal wool for carpet as having 15% by count or 35% by weight heterotypical fibres with a mean fibre diameter at least of 30 μm and fibre length of 3–4 inches. On the basis of gross dimensional properties, the prerequisites for ideal carpet wool were redefined by Sule (1984). The most important fibre properties are fibre diameter and medullated fibre proportion, which were identified to be 30–40 μm and a diameter coefficient of variation percent less than 45%. The fibre composition should have pure fibres (40%–60%), heterotypical medullated fibres (20%–25%), hairy-type medullated fibres (30%–40%) and a proportion of kempy fibres less than 2%.

Indian wools are classified as medium to course or course-grade wool worldwide; most sheep breeds in India produce wool with a fibre diameter less than 30 μm with a varying proportion of medullated fibres, except for fine cross-breed wool, wool produced in the north temperate region and coarse Daccani wool. Lemon (1961), in his study on the distribution of sheep breeds and the wool produced, observed that wool mainly produced in the northwestern region of India has good fullness and handle, i.e., it is lofty owing to higher springiness and the fibres are well separated from each other and are most suitable for carpet manufacturing.

Despite having a large sheep population, we have to import carpet-grade wool to cater the large needs our carpet industry because of its huge potential for carpet exports. About 25–30 million kg of indigenous wool and nearly double the quantity of carpet-grade imported wool is used in Indian carpet industries to manufacture high-areal density, resilient and lustrous carpets which have a high demand in export markets.

The carpet industry is mainly established in the decentralized woollen sector. It is one of the prime industries and is entirely driven by demand. It is growing at an annual rate of 1%. The industry's needs are met by using all of the carpet-grade wool produced in the country, which is only one-third of the total consumption of 50–55 million kg of raw wool. To meet the growing demand, indigenous wool production increased from 38.8 million kg to 49.5 million kg, from 1992–1993 to 2001–2002. The maximum wool production was observed at 50.5 million kg in 2002–2003; afterward a fall was recorded owing to drought, when production declined to 44.6 million kg in 2004–2005. Over 1992–2002, the compounded annual growth rate (CAGR) was –0.4%. In 2008–2009 wool production further dropped and reached a minimum of 42.8 million kg. Afterward, Indian wool production increased with marginal growth of 1%. Current wool production in India is at 48.14 million kg and it is expected that it will produce 65 million kg of wool by 2050.

4.4 Wool marketing practices in India

The marketing of wool in country is controlled by the private sector. There are more than 60 marketing centres (wool *mandies*) in different states of the country where wool is sold in variable quantities. About 50% of the produced wool is brought to the

Table 4.3 **Style grading of Indian wool**

Wool grade	Mean fibre diameter (μm)	Staple length (cm)	Cleanliness	Colour	Vegetable matter content (%)
A	Below 35	8–10	Back washed and skirted	White and lustrous	Below 3
B	35–40	8–10	Back washed and skirted	Yellow and lustrous	Below 3
C	Above 40	5–8	Back washed and skirted	Yellow and dull	3–6
D	Above 40	5–8	Unwashed and unskirted	Yellow, choky and dull	Above 6

nearest wool *mandies* and sold through open auction. The sheep owners, who belong to a highly illiterate and economically weaker section of society and live in remote areas, still depend, by and large, on local brokers (middlemen) to sell their stock.

4.5 Sorting and grading of wool

The wool that arrives in *mandies* is manually sorted for coloured fibres, high burr lumps and coarser wool from the belly, legs and tail portions. Most of the carpet grade wool in India followed style grading for marketing in mandies as detailed below. (Table 4.3).

4.6 Price structure of wool that arrives in *mandies*

The wool arrives in *mandies* in variable quantities year-round. In the wool *mandies* of the northwestern region, e.g., Bikaner, Kekri and Bewar in Rajasthan, during the shearing seasons which coincide with three major festivals such as Holi (Feb. to Mar.), Rakhi (Jul. to Aug.) and Deepawali (Oct. to Nov.), the wool arrives in the largest quantities. In the rest of the country only two clips are taken in a year in Feb. through Mar. and Sep., and are accordingly known as '*Chetu*' and '*Sawanu*' clips. The qualities and types of wools received in wool markets of Rajasthan vary largely in their fleece characteristics, owing to wide divergence in the characteristics of the breeds, the places of their origin and intermixing/interbreeding during transhumance. There is overlapping of breeds in some areas and sheep belonging to more than one breed are found in certain areas. Wools arriving in wool *mandies* are therefore known more by the area or region than by their breeds. Prices for wools with similar quality do not vary much in the main wool *mandies* of Rajasthan. Prices fetched by different wools are given in Table 4.4.

Table 4.4 **Raw wool prices in wool *mandies* of Rajasthan (2011–2012)**

S. No.	Breed	Scouring yield (%)	Clip	Rate (Rs/kg) in *mandies*	
				Bikaner	Kekri
1.	Chokla	70	*Chetu*	92–95	–
			Asharhu	69–70	–
			Syalu	62–66	–
2.	Bikaneri Chokla	85	*Chetu*	150–155	–
			Asharhu	95–110	–
			Syalu	90–110	–
3.	Magra	85	*Chetu*	100–105	–
			Asharhu	80–82	–
			Syalu	82	–
4.	Nali	75	*Chetu*	82–85	–
			Asharhu	54–60	–
			Syalu	60–62	–
5.	Marwari Dhop (scoured fleeze)	70	*Chetu*	67–68	54
			Asharhu	55–58	52
			Syalu	56–57	50
6.	Marwari greasy	60	*Chetu*	32	34–36
			Asharhu	25–28	33
			Syalu	24–25	30
7.	Jaisalmeri	75	*Chetu*	85	–
			Asharhu	57–58	–
			Syalu	60–62	–
8.	Malpura	80	*Chetu*	–	84
			Asharhu	–	75–77
			Syalu	–	68–69

Source: www.woolboard.nic.in – Raw wool price.

Table 4.4 also shows that the *Asharhu* clip wools which are generally canary coloured fetch a smaller price than do the wools of the other two clips that have a white colour. The rates of yellow wools also vary depending on the intensity of the yellowness. The highest price is given to Bikaneri Chokhla (*Magra*) wool, which is considered the best carpet wool; it ranges between Rs.150 and 155/kg for the Mar. (*Chetu*) clip, whereas the yellow wool of the same quality was sold for only Rs. 90–110/kg.

Most wool produced in the northwestern region is marketed in three wool *mandies*: Bikaner, Kakrei and Beawer. Out of them Bikaner is the major wool trading centre. The wool *mandi* of the Bikaner is the largest wool *mandi* in Asia. About 60% of the wool produced in the state and wool from adjoining states, including Punjab, Haryana, Utter Pradesh, Madhya Pradesh and Gujarat, and carpet-grade imported wool from African and Middle Eastern countries arrive here. According to an estimate, more than 10 million kg of wool is sold in the Bikaner wool *mandi* annually.

There are more than 400 registered wool traders and 250 commission agents in Bikaner who fulfil more than 50% of the demand for raw carpet wool to the carpet yarn industries of Bikaner.

4.7 Import of wool to India

The Indian carpet industry is an export-oriented industry and the economic recession in western countries resulted in a fluctuation in exports of carpets from India. Despite this, the carpet industry in India grew with a CAGR of 1%. The domestic availability and quality of native wool is a matter of concern; about half of the wool produced is not up to the requirements of our carpet industry in view of its huge potential for carpet exports. To cater to the needs of the carpet industry, about two-thirds of the required quantity of 50–55 million kg of raw wool is imported. Indian consumption of raw wool is estimated to double, from 114.2 million kg in 2008–2009 to 206.8 million kg by 2019–2020.

For the period between 2009–2010 and 2014–2015, imports were estimated to grow at a CAGR of 11.7%, and for 2015–2016 to 2019–20, they are estimated to grow at a CAGR of 10.1% (Ministry of Textile, 2012). The wool that is imported is of two types; apparel and carpet grade. For apparel and knitwear-grade wool, India depends on Australia for its needs and currently imports about 20.9 million kg (2012–2013) of apparel-grade wool. Carpet-grade wool of various qualities was imported from more than 50 countries. The quantity of wool imported over the past decade is detailed in Table 4.5.

Table 4.5 **Annual imports of carpet-grade wool in India**

Year of import	Quantity (million kg)	Import value (Rs. crore)
1998–99	(53.968-Aus)	
1999–2000	36.771	208.83
2000–01	28.757	174.689
2001–02	46.572	264.87
2002–03	53.077	333.3
2003–04	68.471	418.58
2004–05	66.681	425.98
2005–06	72.326	436.156
2006–07	72.151	462.895
2007–08	74.458	520.966
2008–09	47.745	435.557
2009–10	62.971	564.663
2010–11	74.98	741.69
2011–12	57.5	909.785
2012–13	88.9	1808.02
2013–14	96.13	1952.66

Wool ranging from the coarsest type to medium fine New Zealand wools, cross-bred wools, lime wool and pulled wools are coming to India in large quantities. The major sources of imported carpet wool are from New Zealand and China. Although indigenous wools are superior to New Zealand wools, particularly with respect to pile coverage, resilience and bulk, they lack lustre. The import of wool in the country has had an adverse effect on the prices of indigenous wools. To meet the needs of wool for carpet, carpet manufacturers use a mix of imported carpet-grade wool and Indian wool to produce carpet yarns.

4.8 Diversity in Indian sheep breeds

According to the Food and Agricultural Organization World Watch List (1995), there are 57 breeds of sheep in India. This list includes both well recognized and lesser known breeds, along with some wild species. Although there are about 42 breeds of sheep, most of the sheep population (around 75%) does not belong to any defined breeds. India's share of the total number of breeds in the world is 6.41%. India is endowed with a biologically diverse sheep population composed of sheep producing fleeces with a fibre diameter from below 30 μm to totally hairy sheep. Some of the breeds from the adjoining areas have been identified as separate breeds although they may be similar to each other genetically. The phenotypic differences may be due to selection for adaptation to specific agro-climatic conditions. The unique characteristics of Indian sheep breed are described in Table 4.6.

Table 4.6 **Breeds with unique characteristics**

Breed	Unique characteristics
Magra	Lustrous wool, excellent for carpet manufacture
Changthangi	Alpine sheep of high altitude for fine wool
Chokla	Fine carpet-quality fleece
Garole	High fecundity; twins and triplets common. Survival under saline conditions
Mecheri, Madras red	High-quality skin and mutton
Mandya	Excellent meaty conformation, high quality and meat palatability
Patanwadi	Carpet wool and good milk producer
Nellore	Tallest sheep breed of India
Muzaffarnagar	Best mutton-producing breed
Marwari, Deccani Jaisalmeri	Hardy and capable of walking long distances during migration
Kendrapara	High fecundity; twins and triplets common

4.9 Apparel wool breeds

The apparel wool breeds are Hissardale, Nilgiri, Kashmir Merino and Bharat Merino. These are mostly cross-breeds of native and exotic fine wool/dual-purpose exotic breeds of sheep. The number and distribution of Hissardale and Nilgiri are limited. Kashmir Merino, although large in number, has no definite level of exotic fine wool inheritance and has involved almost all native breeds of Jammu Kashmir.

Hissardale: Hissardale evolved at the Government Livestock Farm, Hissar, by cross-breeding Australian Merino rams with Bikaneri (Magra) ewes and stabilizing the exotic inheritance at about 75%. There is a small flock of this breed at the Government Livestock Farm, Hissar. Earlier, rams of this breed were distributed in the hilly regions of Kullu, Kangra, etc.

Nilgiri: The Nilgiri breed evolved during the 19th century. It originated from a cross-bred base and contains an unknown level of inheritance of Coimbatore, the local hairy breed, Tasmanian Merino, Cheviot and Southdown. The animals are medium sized and white, except with brown patches on the face and body. The face line is convex, with a typical Roman nose. The ears are broad, flat and drooping. Males have horn buds and scurs; females are polled. Attempts have been made by Sheep Breeding Farm, Sandynallah, to improve these sheep through crossing with Rambouillet and Russian Merino, which resulted in substantial improvement in fleece weight and quality. Means of body weight at birth, 3, 6, 9, and 12 months of age were 2.74, 9.64, 14.06, 17.04 and 19.44 kg, respectively. Average 6- monthly wool yields were 438 g with a fibre diameter of 24.3 μm and staple length of 4.5 cm (Terminal Report, 1986).

Kashmir Merino: This breed originated from crosses of different Merino types with predominantly migratory native sheep breeds such as Gaddi, Bakharwal and Poonchi. The level of inheritance varies from very low to almost 100% Merino, although a level from 50% to 75% predominates. The animals are highly variable because of the involvement of a number of native breeds. The annual greasy fleece weight (kg) is 2.80 with an average fibre diameter of 20.4 μm.

Bharat Merino Sheep: Bharat Merino sheep developed by cross-breeding indigenous Chokla and Nali sheep with Rambouillet and Merino rams and were stabilized at 75% exotic inheritance. They have potential as an import substitute for exotic fine wool inheritance. Annual greasy wool production is 2.5 kg with a fibre diameter of 19–20 μm, medullation less than 1% and staple length in annual clip at Mannavanur of 7–8 cm. From ICAR-CSWRI, Southern Regional Research Centre, Mannavanur Bharat Merino sheep were distributed to the Kolar District of Karnataka and in the Sathyamangalam and Thalavadi areas of the Erode District in Tamil Nadu. The average body weight of Bharat Merino sheep at Kolar District was found to be 45 and 32 kg, and ranged from 38 to 70 kg and 28 to 40 kg for adult males and females, respectively (Shakyawar et al., 2007).

4.10 Carpet wool

Northwestern arid and semiarid regions comprising the states of Rajasthan, Haryana, Gujarat and Madhya Pradesh produce good-quality carpet wool. These regions produce almost two-thirds of the total wool produced in the country. Except for the wool produced from Malpura and Sonadi breeds of sheep, the wool produced is of good carpet quality. Wool produced in the northern temperate region is suitable for apparel and finer carpets. Carpet wool has been defined by various workers in different manners. Wool with average fibre fineness ranging from 30 to 40 μm containing hetero and hairy-type medullated fibre admixed with pure fibre with an adequate fibre length is generally termed carpet wool. Promising Indian breeds producing carpet-type wool are Chokla, Magra, Nali, Patanwadi, Marwari and Jaisalmeri. In general, wool with average fineness ranging from 30 to 50 μm with medullation ranging from 15% to 68% is considered to be carpet wool.

4.11 Superior carpet wool breeds

Chokla, Nali, Marwari, Magra, Jaisalmeri, Pugal, Patanwadi, Tibetan, Bonpala, Gaddi, Rampur Bushair, Poonchi, Karnah, Gurez and Changthangi are superior carpet wool breeds. Performance details of some important breeds are described in detail next.

Chokla: This breed is distributed in Churu, Jhunjhunu, Sikar and bordering areas of Bikaner, Jaipur and Nagaur districts of Rajasthan. The animals are light to medium-sized. The face, which is generally devoid of wool, is reddish brown or dark brown, and the colour may extend up to the middle of the neck. The skin is pink. The coat is dense and relatively fine, covering the entire body including the belly and the greater part of the legs. Chokla is a fine carpet wool Indian sheep and is reared basically for its wool quality and suitability for migration. Chokla grows the finest carpet wool of all Indian breeds and range in quality from 54 to 60s count. The wool produced by Chokla sheep is heterogeneous and is generally mixed with coarser fleece of other sheep before it is used as carpet wool. The average first 6-monthly greasy fleece yield (GFY) and adult annual GFY of Chokla sheep under farm conditions range between 0.9 and 1.0 kg and 2.2 and2.4 kg, respectively. Fibre diameter ranges between 28 and 32 μm with medullation between 20% and 35%. Staple length ranges from 4.8 to 6.2 cm.

Magra: Magra sheep are distributed in Bikaner, Nagaur, Jaisalmeri and Churu districts of Rajasthan. The animals are medium to large in size. A white face with light brown patches around the eyes is characteristic of this breed. The fleece is of medium carpet quality, extremely white and lustrous and not very dense. The average adult annual GFY is about 2.087 kg. Fibre diameter ranges between 32.4 and 36.1 μm with a staple length from 5.8 to 6.8 cm. Average fibre diameter, medullation and staple length are 34.40 μm, 32.40% and 6.30 cm (NWPSI Annual Report, 2015–16).

Marwari: Marwari sheep are distributed in Jodhpur, Jalore, Nagaur, Pali and Barmer districts extending up to Ajmer and Udaipur districts of Rajasthan and the Jeoria region of Gujarat. The animals are medium size with a black face. The fleece is white and not very dense. The adult annual GFY is about 1.526 kg. Average fibre diameter, medullation and staple length are 38.83 μm, 55.72% and 5.01 cm, respectively (NWPSI Annual Report, 2015–16).

Nali: The Nali sheep is found in Ganganagar, Churu and Jhunjhunu districts of Rajasthan, the southern part of Hissar and Rotak districts of Haryana. The animals are medium-sized. The face colour is light brown and the skin colour is pink. Both sexes are polled. The ears are large and leafy. The tail is short to medium and thin.

The fleece is white, coarse, dense and long-stapled. The forehead, belly and legs are covered with wool. The overall means for the first 6-monthly and adult annual GFY are 1.01 and 2.84 kg, respectively, and means of birth, 3, 6, 9 and 12 month's weight of lambs are 2.43, 10.74, 14.93, 17.13 and 19.64 kg, respectively, under farm conditions. Average fibre diameter, medullation and staple length are 29.89 μm, 41.14% and 6.79 cm, respectively (Terminal Report, 1986).

Jaisalmeri: The Jaisalmeri breed is found in Jaisalmer, Banner and Jodhpur districts of Rajasthan. Pure specimens are found in southwestern Jaisalmer, extending up to northwestern Barmer and southern and western Jodhpur. The animals are tall and well built with a black or dark brown face and colour extending up to the neck, and with a typical Roman nose. Long drooping ears, generally with a cartilaginous appendage, are seen. Both sexes are polled. The tail is medium to long. The fleece colour is white and the fleece is of medium carpet quality and not very dense.

Pugal: The Pugal area of Bikaner district is its home tract. It is distributed over the Bikaner and Jaisalmer districts of Rajasthan. The animals are fairly well built. The face is black with small light brown stripes on either side above the eyes; the lower jaw is typically light brown. The black colour may extend to the neck. The ears are short and tubular. Both sexes are polled. The tail is short to medium and thin. The fleece is white.

Patanwadi: The Patanwadi is also called Desi, Kutchi, Vadhiyari and Charotari. Three distinct strains have been described.

1. nonmigratory, red-faced type with small bodies and yielding relatively finer fleeces. These are typical Patanwadi and are located in northeastern Saurashtra
2. migratory type, with large bodies and long legs, typical Roman nose and long tubular ears, yielding coarse fleeces, found in western and northeastern Gujarat
3. meat type, with big bodies, low stature and coarse fleeces, observed in southeastern areas around Palitana. In general, the animals are medium to large with relatively long legs. They have a typical Roman nose with a brown face which may be tan in a few cases. The overall least square means for first 6-monthly and adult annual GFY is 566 and 1063 g, respectively, under farm conditions. Average fibre diameter and medullation are 29.11 μm and 38.41%, respectively (Terminal Report, 1986) at its native tract.

Gaddi: Gaddi sheep are distributed in Kistwar and Bhadarwah tehsils in Jammu province of Jammu and Kashmir, Ramnagar, Udampur and Kullu and Kangra valleys of Himachal Pradesh and Dehradun, Nainital, Tehri Garhwal and Chamoli districts of Uttar Pradesh. The animals are medium sized, usually white, although tan, brown and black and their mixtures are also seen. Males are horned; 10%–15% of females are also horned. The tail is small and thin. The overall least square means for first 6-monthly and adult 6-monthly GFY is 417 and 454 g, respectively, and least square means of birth, 3, 6, 9 and 12 month's weight of lambs are 2.22, 7.64, 11.01, 14.01 and 16.19 kg, respectively, under farm conditions. Average fibre diameter and medullation were 27.7 μm and 20.6%, respectively (Terminal Report, 1986).

Avikalin: The Avikalin strain evolved from a Rambouillet × Malpura half-bred base through interbreeding and selection for greasy fleece weight, producing about 1.50–1.71 kg annual greasy wool with a 28- to 33-μm diameter, 27%–35% medullation and staple length of about 5.00 cm in half-yearly clip. This breed is suitable as a dual-purpose sheep for carpet wool and mutton production.

4.12 Coarse wool breeds

Malpura, Sonadi, Kheri, Muzaffarnagar, Jalauni, Deccani. Bellary, Coimbatore, Chhotanagpuri, Balangiri, Ganjam, Bhakharwal and Shahabadi are classified as coarse carpet wool breeds. Performance details of some of the important breeds are described in detail next.

Malpura: Malpura sheep are found in Jaipur, Tonk, Sawaimadhopur and adjacent areas of Ajmer, Bhilwara and Bundi districts in Rajasthan. The animals are fairly well built, with long legs. The face is light brown. The fleece is white, extremely coarse and hairy. The belly and legs are devoid of wool. The overall least square means for the first 6-monthly and adult annual GFY is 752 and 954 g, respectively (MSSP Annual Report, 2015–16). Wool from Malpura sheep is coarse with a fibre diameter of 49.9 µm, 71% medullation and 6.02 cm staple length. Scanning electron microscopy images of coarse wool are shown in Fig. 4.2. It is suitable for felt production and other diverse uses such as making, floor covering, toys, wall hanging arts, etc.

Sonadi: Sonadi is found in Udaipur and Dungarpur districts and to some extent the Chittorgarh district of Rajasthan; the breed also extends to northern Gujarat. The animals are fairly well built, somewhat smaller than Malpura, with long legs. The fleece is white and extremely coarse and hairy. The belly and legs are devoid of wool. The means of the first 6-monthly and adult annual GFY are 279 and 612 g, respectively (MSSP Annual Report, 2015–16).

Muzaffarnagar: This breed is distributed in the Muzaffarnagar, Bulandshahr, Saharanpur, Meerut, Bijnor and Dehradun districts of Uttar Pradesh and parts of Delhi and Haryana. Pure specimens are found in the Muzaffarnagar district. The animals are medium to large in size. The fleece is white, coarse and open. The belly and legs are devoid of wool. The average adult annual GFY is 1.262 kg. The means of wool quality attributes such as fibre diameter, hetero fibres, hairy fibres, medullation and staple length are 38.39 ± 1.36 µm, 14.35 ± 1.37%, 46.89 ± 3.42%, 61.03 ± 4.33% and 5.09 ± 0.30 cm, respectively (NWPSI Annual Report, 2015–16).

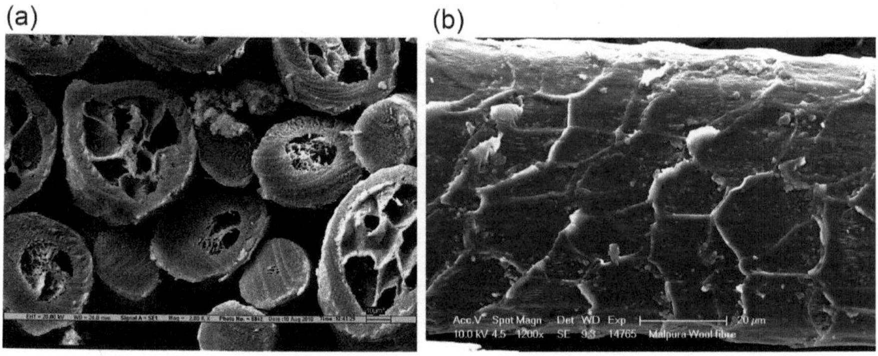

Figure 4.2 (a) Scanning electron microscopy images of a cross-section of coarse wool; (b) scale structure of coarse wool.

Deccani: This breed is spread over the greater part of the central peninsular region, comprising the semiarid areas of Maharashtra (especially the Pune Division), parts of Kumool, Mahbubnagar, Nalgonda, Nizamabad, Anantapur, Warangal and the entire districts of Madak and Hyderabad in Andhra Pradesh, and Bidar, Bijapur, Gulbarga and Raichur districts in Karnataka. The animals are medium-sized. The fleece is extremely coarse, hairy and open. The belly and legs are devoid of wool. The average means for lambs' first and second 6-monthly and adult annual GFY are 0.476, 0.421 and 0.951 kg, respectively (NWPSI Annual Report, 2015–16).

Coimbatore: These sheep are distributed in the Coimbatore district and the adjoining Dindigul district of Tamil Nadu. They are medium-sized animals, white with black or brown spots. The ears are medium-sized and directed outward and backward. The tail is small and thin. The fleece is white, coarse, hairy and open. The animals are maintained on grazing with no supplementation. The sheep are penned in harvested fields. The rams are both horned and polled whereas ewes are hornless. Average body weight at birth and 3, 6 and 12 months of age is 2.16, 7.50, 10.83 and 14.77 kg, respectively. Average 6-monthly greasy fleece weight is 365 g with an average fibre diameter of 41 μm and medullation percent of 58. The medullation in the coarse wool is shown in Fig. 4.3.

4.13 Production system and farming practices

Sheep production in India is largely an extensive system in which flocks move in search of food and water. Over the years a system has been developed to use available biomass optimally for growth and reproduction of the flocks. In the northern temperate region, sheep move to alpine pastures during summer months; in the northwestern region, flocks migrate from the scarcity zone to areas where stubble

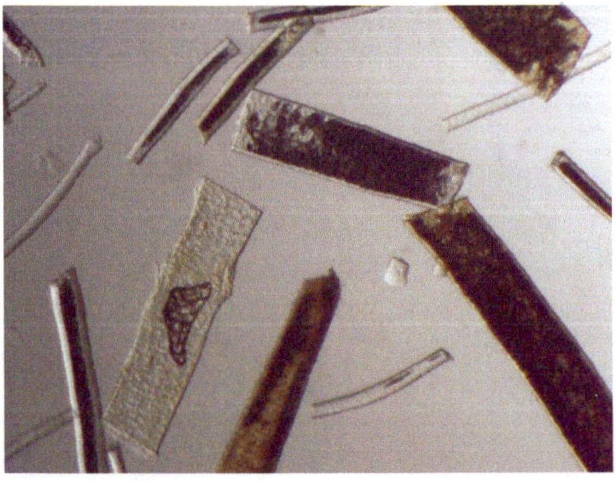

Figure 4.3 Medullation in coarse wool.

and other grazing resources are available during summer months. In Rajasthan about 70% of the sheep population goes on temporary/permanent migration. Temporary migration depends on the monsoon season. Introduction of temperate mutton breeds may not be favoured over improving well-adapted breeds through selection. However, the local shepherd is keen to take advantage of the heterosis by crossing indigenous sheep with rams from superior breeds of nearby areas. In the process the purity of breed has been diluted. One of the breeds that evolved in this process is Kheri in Rajasthan. Similarly, Munjal sheep in Hanumangarh, Sri Ganganagar of Rajasthan, Muktsar and Bhatinda of Punjab are being developed by local farmers.

The notable change observed in sheep farming practices in India during the past decade or so is related to the gradual reduction in income from wool from sheep compared with other sources, i.e., the sale of live animals, meat, manure, etc. Sheep farmers have also shifted emphasis in selecting rams for body weight rather than wool. The age at the marketing of lambs has also declined to about 8 months, from10–14 months in some breeding and declined to 4 months from 9 months in most breeds. Some type of supplementary feeding in cultivated fodders is also offered to lambs so that marketing weight is achieved at a younger age.

Management practices and systems of sheep production in the country are unique and diversified depending on geo-ecological situations and agricultural cropping systems in different parts of the country. In arid and semiarid regions, sheep are raised partially on migration to places far from their native habitats. The flocks follow well-established routes and remain away for 5–6 months in the year for want of grazing, and are brought home for the rest of the year. The flocks are grazed on crop stumps in harvested fields, forests and wastelands, and during lean periods on tree branches. Special attention and care are given to animals in advanced pregnancy and during the lactation period with supplementary concentrates and conserved fodder. Flocks are mainly penned in open fields but some flocks are kept in courtyards on a temporary basis. Some flock owners prefer to keep their animals in houses where they themselves sleep; of course in this case the flock size is very small. Breeding is mainly confined to seasons when grazing resources are adequate, although they breed year-round as males are kept with the flocks.

In the subtropical and temperate regions, most flocks start migrating in April, when they move to alpine pastures to graze in high mountains and start migrating down in October. During winter, animals are fed with stored dry fodder, tree leaves and concentrate. They are protected during winter with closed housing.

On organized and government farms, sheep are reared with semiintensive feeding. Animals are allowed limited grazing and browsing for 6–8 h/day and fed with supplementary concentrates and seasonal green fodders and dry hay. The intensive production management is out of reach of sheep farmers because of the costs but it is common with urban people who keep a few sheep or goats to meet their household needs for milk and a cash income through the sale of lambs and kids.

Raising sheep in India is primarily in the hands of people who lack economic and technical resources and follow traditional methods of management. This reflects the ignorance of the farmers and also a reluctance to accept progressive methods and technologies owing to inadequate awareness. There is a need to make newer technologies available to such sheep and goat farmers and also for methods to be updated so that such technologies are acceptable to traditional farmers. Therefore, intensification of efforts to provide technology in rural areas will be a positive step toward enhancing the productivity of sheep and goats.

4.14 Breeding strategies for sustainable wool production

4.14.1 *Cross-breeding to improve wool production*

Sporadic attempts to improve the productivity of sheep through cross-breeding were made even during the preindependence period. Most earlier efforts were aimed at improving sheep for fine wool production by introducing exotic fine wool inheritance or upgrading using indigenous improving wool or dual-purpose breeds on a limited scale. Although reasonable success was achieved in developing new breed or strains, because of the discontinuity of efforts and the population size involved, achievements could not be sustained for perceptible results.

4.14.2 *Cross-breeding among native breeds*

Among native wools, Chokla has the overall best performance for the functional characteristics of carpets, followed by Magra, Jaisalmeri, Malpura, Nali and Sonadi, in that order. Initially limited research efforts were made to achieve improvement in wool yield and quality among indigenous sheep breeds by upgrading low-producing local sheep with indigenous superior carpet wool breeds such as Chokla and Magra. Cross-breeds of south Indian mutton such as Bellary at Hosur (Tamil Nadu) with north Indian improver Magra breed resulted in an advantage in wool production and quality. Some improvements in local breeds of Madhya Pradesh and Uttar Pradesh were also observed when they were crossed with Chokla and Magra rams. Increases in greasy fleece weight and fibre length and decreases in fibre diameter and medullation percentage were reported (Chaudhry and Chaudhary, 1991). Bohra et al. (1993) identified a new breed of sheep: i.e., Kheri evolved under field conditions in migratory flocks by using Marwari/Jaisalmeri rams with local sheep.

4.15 Cross-breeding involving exotic breeds: for fine wool production

Attempts to cross-breed native sheep with exotic fine wool breeds to increase wool production and quality were made even during the preindependence period. Many cross-breeding experiments to improve fine wool were reported by Das and Rajgopalan (1956), Amble et al. (1967), Amble and Malhotra (1968), Raina et al. (1973), Acharya (1974), Chaudhry and Chaudhary (1991) and Singh and Singh (1998). In the early

1900s, superior indigenous sheep breeds were crossed with cape Merino, Southdown, Romney, Marsh, etc., in Maharashtra, Utter Pradesh, Mysore, Punjab and Tamil Nadu, and new strains such as Hissardale, Nilgiri and Kashmir Merino were developed. However, these programmes also experienced a lack of continuity. Results from earlier cross-breeding experiments were not conclusive because of the small number of rams used and the lack of availability of contemporary performance records. In general, however, the experiments indicated the superiority of cross-breeds over indigenous breeds with respect to body weight, wool weight and quality traits. Furthermore, survivability of the half-breeds was similar to that of native animals, but the higher crosses exhibited higher mortality. These results also indicated that the Rambouillet/Merino crosses thrived better than those from the British breeds.

With the establishment of the Central Sheep and Wool Research Institute (CSWRI) and the launch of the All India Coordinated Research Projects (AICRP) on sheep breeding, concerted efforts were made in this direction. The cross-breeds that evolved as well as their genetic composition are presented in Tables 4.7 and 4.8. Avivastra, a

Table 4.7 Sheep strains developed to improve wool production and quality

Genotype with utility/ location	Parent breeds		Level of exotic inheritance (%)
	Indigenous	Exotic	
Bharat merino, for fine wool CSWRI, Avikanagar	Chokla, Nali, Malpura, Jaiselmeri	Rambouillet, Russian Merino	75
Avivastra, for fine wool CSWRI, Avikanagar	Chokla, Nali	Rambouillet, Russian Merino	50
Nilgiri synthetic, for apparel wool Tamil Nadu Veterinary and Animal Sciences University, Sandynallah	Nilgiri	Rambouillet, Soviet Merino	62.5/75
Patanwadi synthetic, for carpet wool Gujarat Agricultural University, Dantiwada	Patanwadi	Rambouillet, Soviet Merino	50
Avikalin, for carpet wool CSWRI, Avikanagar	Malpura	Rambouillet	50
Kashmir merino, for fine wool Jammu and Kashmir state	Gaddi, Bhakarwal, Poonchi	Delaine Merino Rambouillet, Soviet Merino	50–75
Hissardale, for apparel wool Haryana state	Bikaneri	Australian Merino	75

Source: CSWRI, Central Sheep and Wool Research Institute.

Table 4.8 Fleece characteristics of newly developed strains in India

Breed	Greasy fleece weight (kg)	Average staple length (cm)	Average fibre diameter (μm)	Medullation (%)
Hissardale	1.46–2.72	5.52–6.15	21.53–24.45	0.0–0.55
Kashmir Merino	2.45–2.80	4.77–5.60	20.40–20.94	0.0
Nilgiri	0.20–1.40	7.01–8.70	21.57–27.40	8.40–18.0
Deccani Merino	1.50	5.75	22.1	1.6
Nilgiri synthetic	1.0–1.5	4.6–5.0	20.6–21.6	1.00–3.20
Patanwadi synthetic	1.2–1.7	4.6–5.2	19.3–23.2	9.0–20.8
Bharat Merino	2.61–2.8	2.88–9.40	17.61–20.24	1.00–1.08
Avivastra	1.90	3.50	21.6	7.60
Avikalin	1.72	4.26	25.2	24.9

Source: Singh, V.K., Singh, G., 1998. Breeding strategies for improving wool production, Paper Presented in National Seminar on "Golden Jubilee Seminar on Sheep, Goat and Rabbit Production and Utilization," from 24–26th April, 1998, Organized by Indian Society for Sheep and Goat Production and Utilization at Jaipur.

new fine wool strain, evolved through crossing Chokla, Nali and Jaisalmeri sheep with Rambouillet/Merino; they are capable of producing about 2.5 kg of greasy wool annually. They have a fibre diameter of 21–22 μm, a medullation of 15% and a staple length 4.5 cm. They are thus potential fine wool–producing animals meeting apparel-quality characteristics. Avivastra has been upgraded and merged with Bharat Merino, a fine wool strain with 75% inheritance of Rambouillet/Merino, because they do not differ significantly in growth and wool quality traits. The Bharat Merino strain of sheep was developed by CSWRI and maintained in the Kodai hills; they are capable of producing about 3.0 kg greasy wool annually with a fibre diameter of 20.0 μm, staple length of 7.5 cm and medullation of less than 1%. The Rambouillet and Merino were crossed with Nilgiri at the Sheep Breeding Research Station, in Sandynallah. They also have the potential to evolve into a fine wool breed at a 50%–62% level of exotic inheritance.

There was substantial improvement in fibre diameter (14.0%–35%) and medullation percentage (3.0%–37.0%) as well as gain in body weight when the superior carpet wool breeds such as Chokla and Nali were crossed with exotic fine wool breeds. However, there was little increase in wool yield when they were crossed with fine wool breeds. When crossed with exotic mutton/dual purpose breeds, the poor-quality carpet wool breeds resulted in an improvement in weaning weights and greasy fleece weights as well as quality traits except staple length. Except in some cases, the very coarse to hairy wool breeds produce medium-quality carpet wools at about 25–35 μm diameter and 30%–40% medullation even in the first cross (Chaudhry and Chaudhary, 1991; Singh and Singh, 1998).

Results indicate that almost all cross-breeds have shown improvement in wool quality but the staple length range between 3.5 and 5 cm at 6-monthly clip. This remains the limiting factor for which rigorous selection and/or appropriate shearing practices need to be developed to make the wool suitable for appropriate processing. Furthermore, results suggest that increasing exotic inheritance beyond 75% has little advantage. It requires a higher level of management and feeding input than for half-breeds and native sheep.

4.15.1 Carpet wool production

The cross-breeding of extremely coarse and hairy breeds such as Malpura with Rambouillet at CSWRI, Avikanagar, showed that half-breeds from such crosses yield more wool of excellent carpet quality than do native Malpura. The Avikalin strain evolved from this cross-bred base is able to produce about 2 kg greasy fleece annually with 25-μm fibre diameter, 25% medullation and about 4.5 cm staple length. The wool produced by Avikalin is superior to carpet wool produced by indigenous breeds. Improvement in wool yield and quality was also observed in Avimaans and Awassi × Malpura cross-breeds. Avikalin × Avimaans crosses are being upgraded with Avikalin as dual-purpose sheep (carpet wool and mutton production).

The performance of newly evolved strains through cross-breeding for carpet wool is given in Table 4.7. Cross-breeding experiments have given encouraging results and shown substantial improvement in wool quality approaching the targets of ideal carpet wool. The cross-breeding of hairy breeds in the southern region of India with mutton-type breeds such as Suffolk and Dorset also have the appearance of fleece cover. Similar trends are exhibited by synthetics evolved under AICRP unit of Andhra PradeshAgricultural University, Palamaner (Andhra Pradesh) using the Nellore and Mandya local breeds (Chaudhry and Chaudhary, 1991).

4.16 Sustainability and selective breeding

The focus of sheep development in the past was essentially on improving the quality and quantity of wool by using different types of exotic fine wool breeds. A number of strains were developed by cross-breeding native breeds with exotics (Task Force, 1996). Developed genotypes demonstrated their production potential in terms of body weight, wool quantity and quality under experimental farm management conditions, but these could not outdo the natives under field conditions, except for a few genotypes, owing to the lack of availability of necessary nutritional and climatic conditions.

A number of the cross-breeds so produced have not survived when run together with the natives, particularly when the advantage of special care and management at the experimental farms was withdrawn. The general situation is that under the prevailing realities of sheep husbandry, it may not be possible for cross-breeds within village conditions of management and survival to outdo indigenous well-adapted breed types on the basis of lifetime production. Therefore it is important to undertake an evaluation of indigenous breeds, starting with those that are numerically more important, and in keeping with local product needs, such as meat and wool.

4.17 Breeding strategies

Improvement of sheep through breeding strategies should involve decisions regarding objectives and approaches to bring about genetic improvement. This will depend on socioeconomic and ecological considerations, existing genetic resources, their productivity, the possibility of improvement through selection within a breed, grading with indigenous improver breeds or replacing it an indigenous breed with an existing breed if it is arises from a similar ecological region and thus will be adaptable, or cross-breeding to evolve a new breed combining the adaptation and hardiness of the local breed and the higher productivity of the exotic breed.

- Surveys, evaluation, conservation and productivity improvement should be attempted in native stock. Research work on breed improvement involving the farmers' flocks and characterization of genetic resources have been initiated. Genetic improvement of sheep for wool using both farm and field units should be strengthened.
- Important indigenous breeds of sheep should be properly characterized both phenotypically and genetically. Experiments should be undertaken to assess their population and production performance and to exploit their production potential under ideal feeding and health management conditions. Elite flocks of different native breeds should be developed to produce genetically superior rams. Open nucleus flocks should be established for genetic improvement.
- Development of fine wool production in arid and semiarid regions has little purpose. In view of current trends in the globalization of trade, the import of quality wool apparel will continue to fill the gap in the need for added value. Because it has not been possible to obtain fine wool with a desirable staple length from higher crosses in the arid and semiarid regions, apparel wool production may be intensified only in temperate areas such as the northern temperate hilly region and the Nilgiri and Kodai hills of the southern region. In these areas, three-quarter crosses of Rambouillet or Merino including Bharat Merino may be propagated and annual clips may be obtained to meet the requirements for apparel manufacture.
- Indigenous carpet wool, which has several desirable properties, is preferred by the hand-knotted carpet industry. Carpet wool production and quality should be concentrated in Rajasthan, Gujarat and parts of Haryana and Uttar Pradesh. Further improvement in terms of fineness and higher yield would be beneficial to the wool industry. This could be achieved by using improver breeds in appropriate areas.
- Carpet-type wool-producing breed such as Nali, Chokla, Patanwadi, Marwari, Magra, Pugal, Gaddi and Jaisalmeri should be further improved through selective breeding by distributing good-quality rams of the respective breeds. Wool quality traits of these breeds are closer to the requirements of the industry for carpet wool.
- Efforts should also be made to introduce lustre in indigenous carpet wool breeds to harvest maximum returns through the sale of quality carpets.
- Elite male germ plasm centres should be developed to provide good breeding rams. Such farms should evolve a system of open nucleus breeding in which elite germ plasm may be multiplied through embryo transfer technology of selected quality rams; they should also seek the participation of some commercial farms.
- Within-breed selection may be taken up systematically in recognized breeds in the country. Selection of any one of the following schemes may be adopted in descending order of preference: (1) selection and rearing of ram lambs in the farmers' flocks and their use in the flocks and for distribution to other flocks which are to be improved; (2) selection of ram

lambs in farmers' flocks and rearing in ram lamb rearing centres for distribution to the farmers' flocks; (3) establishment of open nucleus breeding flocks through animals objectively selected on the basis of trait(s) of interest and selection within the nucleus to produce genetically superior rams for use in the nucleus flocks and for distribution to farmers' flocks to be improved. Existing state sheep breeding farms may also be considered for conversion into open nucleusbreedingschemes. The productivity of each breed may be improved by bringing as large a number of farmers' flocks as possible under the improvement programme, which result in improvement through different breeding approaches and also involve provision of a totality of inputs and services related to genetic improvement, health cover, feed resources development, conservation and their strategic use, management and marketing of the produce.

- Under field conditions, two important characteristics to be considered for selection for improvement in carpet wool production and quality are: (1) the first 6-monthly greasy fleece weight, and (2) the medullation percentage, especially in breeds in which it is high and selection against it is required to improve wool quality.
- Grading up of nondescript sheep through continuous breeding with selected rams of desired descript breed(s) should be undertaken, depending on characteristics to be improved. As far as possible, the rams of a breed from tracts close in agro-ecology to the flock to be improved should be selected.
- The National Ram Production Programme, which has been implemented for a long time, may be suitably modified to take up production of genetically superior rams, preferably by selecting open nucleus breeding flocks. Currently rams supplied from state sheep breeding farms may help farmers improve the productivity of their flocks to some extent, but it is doubtful whether these rams are genetically superior to the rams being produced by farmers in their own flocks. Because the farm maintains a limited population and closed flocks, any selection being carried out is less effective. Moreover, production of rams through the current system will remain small compared with the farmers' needs for superior rams.

4.18 Processing strategies for sustainable wool production

4.18.1 Quality improvement of coarser wools

Sheep-rearing practices in the country are mainly for mutton production. In the northern temperate and northwestern regions, sheep are reared for the dual purposes of mutton and wool production. Coarse wools are mainly produced in a major part of the country from central India to the southern peninsular region. With the increasing domestic demand for mutton, there is a further increase in the annual production of coarser wools (mean fibre diameter >40 μm) to 40% of total wool production, e.g., 18–20 million kg. Because of its inherent fibre properties of low crimp, high bending rigidity and high proportion of medullated fibre, these wools cannot be made into yarn and have limited end use in traditional *Namda* making. This has resulted in a very low price for such wool in the market, Rs. 25–35/kg; it does not even meet the annual shearing cost for the animal. Wool shearing results in a loss for mutton sheep. On the other hand, medium coarse wool of dual-purpose sheep in the northwestern region is in high demand and fetches four to five time the price, up to Rs. 100–120/kg owing to

Figure 4.4 Slow-speed carding unit.

its fibre properties, which are ideally suited to quality carpet manufacturing. Because of this, there is an urgent need to improve the quality of coarser wools. The dehairing technique developed to separate coarse guard hairs with an average fibre diameter less than 50 μm from fine fibre (10–13 μm) from the Pashmina fleece is a promising method. The process is somewhat easier for Pashmina fleeces because of the large difference in fibre diameter as well as the fibre weights of the fine undercoat and guard hairs, whereas for coarser wool fleeces such as Malpura, Sonadi, Patanwadi and Bellary, which have an admixture of fibres ranging in diameter, and separation of the coarsest fibre from the fleece is not only difficult but uneconomical. The possibility of removing highly medullated fibre with a mean fibre diameter of less than 50 μm from coarser wool has been envisaged.

4.19 Process of removing coarser fibres

Removal of the coarsest fibres from coarse fleece is not possible manually. Commercially available machines to do this work that were developed in western countries are costly and require a large quantity of fibres for processing. The economical separation of fine fibre from coarse fleece is possible with indigenously developed slow-speed carding units (Fig. 4.4).

The mechanism of separating highly coarse fibres from the fleece admixture involves the combined action of centrifugal force and the gravitational pull of coarser and heavier fibres around the surface of the carding cylinder. Medium coarse fibres adhere to wire points and are transferred to the doffer roller. The fibre properties for coarse and improved Malpura wool are given in Table 4.9.

The fibre diameter of medium coarse Marwari wool used for carpet making is nearly 0.58 times that of coarser Malpura wool. The prespinning process to remove coarser and medullated fibre from the stock to improve the fibre quality of Malpura wool results in a fibre diameter and medullation of 28.5% and 8.5%, respectively. The improved Malpura wool has fibre properties similar to medium coarse wools and may be used in a blend for carpet yarn manufacturing. The process yields 60% medium

Table 4.9 Physical properties of fibres in fleece and improved Malpura wool

Fibre characteristics	Medium coarse Marwari wool	Coarse Malpura wool	Improved Malpura wool
Mean fibre diameter (μm)	36.93	52.43	38.00
Hetero type	32.17	16.66	33.33
Hairy type	8.74	56.00	28.66
Total medullated fibers (%)	40.91	72.66	61.99
Staple length (cm)	6.56	5.94	–

coarse fibres from the coarser fleece. This controls the country import bill for such medium coarse wools and also doubles the price wool growers may realize for their coarser produced wools. The by-product (35%–40%) of this process may be used for geo-textiles in agriculture or horticultural crops.

4.20 Geo-textile application of coarser wools

The slow degradation rate of wool is the result of its hardy form of protein; because of keratin and its capacity for high moisture retention it can be used in agricultural applications. An experiment has been conducted to employ coarse and unused wool in the form of geo-textiles to improve crop productivity, soil health and moisture retention. Coarser wool was been applied at three different levels of loose wool, i.e., 500, 1000 and 1500 grams per square metre (GSM) at a depth of 15 cm, and two crops of guar and pea were sown for the *kharif* and *rabi* seasons, respectively. Guar (variety RGC 1002) was sown at the rate of 40 kg/ha whereas the seed rate for peas was kept at 30 kg/ha. Controlled irrigation was uniformly given to all plots, according to needs, throughout the experiment. No significant improvement was observed in either plant growth or in crop yield for *kharif* season 1. However, the soil moisture content was observed to be highest by 12% compared with 8.5% for the control plot. The observed value was 42% higher over the control plot moisture retention. Decomposition of wool placed as a geo-textile occurred in the initial phase during *kharif* season 1, which thus rose to the highest moisture retention of 42% with geo-textile plot, whereas an increase in moisture retention value is observed by 4%–5% only after 1 year, i.e., *kharif* season 1. The harvesting data for the same experiment resulted in higher fodder production but lower grain yield. The lowest values of biological yield were observed in the plot with the highest application of wool, i.e., 1500 GSM. Therefore, it may concluded that the application of geo-textiles increased moisture retention at the initial phase but acted as a barrier for the supply of nutrients to plants from the lower soil to the geo-textile, resulting in a poorer grain yield for the geo-textile applied plots. An experiment to enrich soil fertility by using coarse wool is shown in Fig. 4.5.

Figure 4.5 Geo-textile application of wool for water retention and bio-fertilizer.

Rabi season 1 was the second crop after geo-textile application (after 3 months). Moisture retention in the soil was highest for the W3 (wool 1500 GSM) plot and increased by 10% over the control plot compared with 12% in the first crop. Thus, it appears that aside from being a source of moisture retention, wool start decomposing and may act as a fertilizer. It was found that all plant growth parameters were positive (after 90 days): plant height was observed to be 12% higher and plant fresh and dry weights increased by 22% and 16%, respectively. Plant physiological parameters such as seeds/pods and branches per plant also progressively increased with an increase in wool content with the treatment. The results of harvesting increased fodder production and grain yield with geo-textile application. A maximum increase in fodder and grain yield was observed at 30% and 42%, respectively, for the plot with 1500 GSM of wool.

After 1.25 years and three crops, in *rabi* season 2, decomposition of the wool in the geo-textile applied plots was almost complete. Moisture retention conditions in the control and treatment plots were similar but growth performance was better than in the control plot. This was attributed to wool decomposition, which acted as a bio-fertilizer, although it was observed to be 16%–47% lower than in *rabi* season 1. The results of harvesting were observed to be 8%–11% for the plot with 1000 and 1500 GSM treatment. The fodder had lower value (−1.25%) for treatment at 500 GSM, which shows that the wool decomposition process was completed and almost completed in the other treatment. Overall fodder production was observed to be lower by 5%–26% compared with *rabi* season 1. Thus, the study showed that wool as a geo-textile application initially acts as a moisture reservoir for plant growth, and over the period of decomposition it also works as a natural source of fertilizer for higher crop yields.

4.21 Wool nonwovens

Efforts have been made to incorporate wool fibres into special nonwoven applications. Using nonwoven processes, it is possible to produce low-cost lightweight woollen fabrics with high stretch. Nonwovens have a lot of applications such as home furnishings, military clothing and uses in agriculture and industry. Wool is one of the most widely used animal fibres for the manufacture of nonwovens. The first nonwovens were produced from wool fibres as felt by mechanically interlocking wool fibres under steam, taking advantage of their natural surface scales. The application of nonwoollens depends on various parameters such as the type of material, fibre orientation, packing density and pore size. These key factors determine the various physical and mechanical properties of nonwovens and ultimately their end use.

During the manufacture of nonwovens, staple fibres of similar kinds adhere to each other by mechanical, thermal or chemical methods. The web formation phase of nonwoven manufacturing processes transforms previously prepared or formed fibres, filaments, thermoplastic resins or films into layers of loosely arranged networks, webs, mats or sheets. Multiple fibre composite nonwovens are made of different types of fibres with different characteristics for specific end uses. Some examples are composite nonwovens with both natural and synthetic fibres. Fibres are blended to take advantage of the inherent desirable properties of individual components. Coarse wool can be mixed with cotton to make nonwovens, so that the resulting nonwoven appears to be much softer to handle.

4.22 Applications of wool nonwovens in filtration media

Wool nonwovens have applications in various filter media. Wool has some unique properties for filtration. Wool fibres are naturally crimped and offer bulk and loft. They can easily be electrostatically charged. They have good strength and dimensional stability and are durable. A filter medium is any material that, under the operating conditions of the filter, is permeable to one or more components of a mixture, solution, or suspension, and is impermeable to the remaining components. There are different types of filtration media: solid–liquid, solid–solid, gas–liquid, etc. Wool is used as an absorbent in various filter media; it removes contaminants by absorbing them into the porous structure of the medium, much as a sponge soaks up water from a wet surface. In air filter applications, such as bag house filtration, residential filters, vacuum cleaner bags, automotive filters and industrial respirators, the selection of a filter is depends on the particle size, particle size distribution, and particle type. The advantage of wool fibres the high surface area, which provides high filtration efficiency. It is also has application in the filtration of oils. The filtration efficiency of a fibrous medium can be simulated by single fibre efficiency, based on mechanical mechanisms including inertia impaction, direct interception and Brownian diffusion, as well as electrostatic mechanisms such as coulombic force and image force. Wool nonwovens for filtration and thermal insulation are shown in Fig. 4.6.

4.23 Applications of wool nonwovens in thermal insulation

The thermal properties of nonwoven materials are determined by the type of fibrous materials used and the fibre diameter, as well as the bulkiness and thickness of the mats. When the fabric weight increases, thermal conductivity increases because heat conductivity is transferred through the fibres rather than through the voids. Thus the

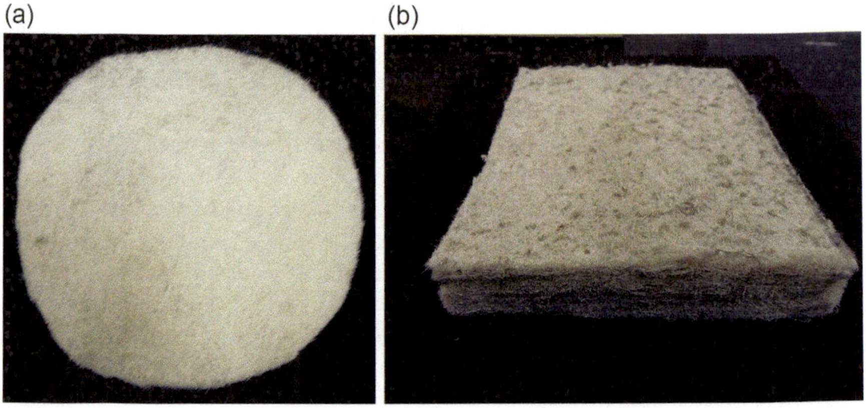

Figure 4.6 (a) Wool nonwoven for filtration; (b) wool felt for thermal insulation.

thermal insulating property of the nonwoven is reduced and the cost increases because more material is used. Thermal blankets produced from wool fibres have excellent insulation and are comfortable. Wool is a suitably stiff and permanently crimped bio-component fibre. The distinct variations in thickness are in the most cases favourable for producing a nonwoven fabric. Wool nonwovens possess high bulkiness and good insulation because of the air trapped between fibres.

4.24 Miscellaneous applications of wool nonwovens

Coarse wool nonwovens can be used to transport fish on ice to pack poultry after coating with superabsorbents to absorb their water and blood. Another possible application for coarse wool nonwovens is floor pads or bath mats. Because wool is hygroscopic and can absorb water, it is well suited for such applications either alone or combined with other natural fibres. Other applications of these coarse wools are in the form of coarse wool felted batts, which may be explored for energy-efficient civil construction, oil spills in the sea, sound absorption and thermal insulation panels for sustainable needs as well as simultaneous wool production. A felt handicraft and its applications are shown in Fig. 4.7.

4.25 Major constraints faced by wool in India

Apparel-grade wool production in India is minimal and meets only 10% of the demand for outer wears such as shawls, etc. To obtain apparel-quality wool, Indian industry mainly depends on imports from major fine wool–producing countries such as Australia, New Zealand, Uruguay and South Africa. The climatic conditions of the Indian subcontinent restrict fine wool production and most sheep breeds produce medium coarse to coarse wool, which goes to the carpet industry. Competition from

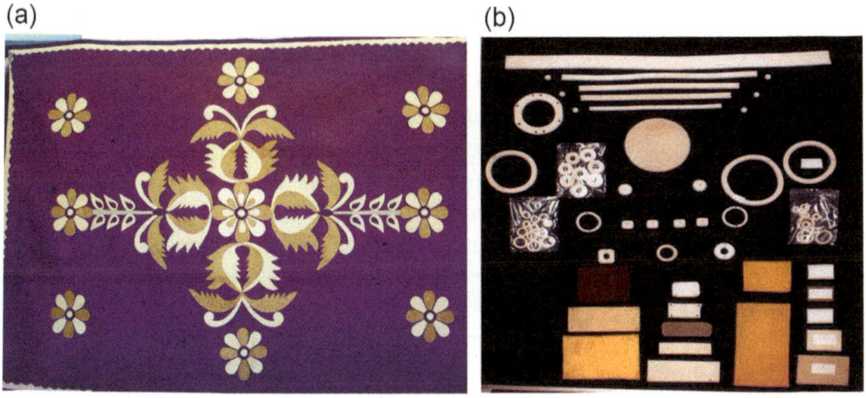

Figure 4.7 (a) Traditional handmade felt; (b) industrial application of nonwoven felt.

the synthetic fibre industry is an important concern. Pure woollens are being replaced by comparatively cheaper wool blends with polyester, acrylic and other synthetic fibres to reduce the cost of the product and customers are attracted by their low price. The coarser quality of wool produced in India is another important issue. Although the domestic availability of wool remains a matter of concern, the quality of the wool also does not meet the needs of the Indian woollen industry. Even for quality carpets two-thirds of quality wool (30–35 μm with 4%–5% medullation) is imported from New Zealand and other medium coarse wool–producing countries, especially to execute their export orders. Another issue faced by farmers is related to the sale of their sheared wool on the local market. The small-herd farmer may harvest only 10–15 kg of wool per season and it is not economical to carry this to the nearest market and sell it expecting a profitable margin. Here comes the role of middleman, who may exploit the pathetic situation of farmers. The middleman collects wool from different farmers and sells it collectively to the industry for a profit. Ultimately sheep farmers do not get the benefit of producing wool. The lack of interest in wool results in the sale of ungraded wool by farmers, although grading before a sale can result in a better price for the wool. Farmers may sell the wool through cooperative societies to industries to obtain appropriate remuneration for the wool. Most farmers still practice age-old hand shearing with scissors, which is a costly affair compared with machine shearing. Hand shearing not only consumes a lot of time and higher shearing costs per animal but also produces inferior-quality wool in terms of short and uneven lengths of fibre and ultimately results in low priced wool. Processing of this coarse wool is mainly done by medium-small or medium-scale industries on indigenously developed, unrecognized machinery with tremendous variability in machine output. Thus the products that result are not competitive enough on the international market. For the economic use of such wool in diverse applications to help the farmers, artisans and small-scale industries to make a profit from their products, a collective effort is needed from the government, scientists, nongovernmental organizations as well as industries to address these issues to realize better prices for coarse wool.

4.26 Conclusion

The low productivity of native sheep is responsible for farmers losing interest in sheep rearing. To meet the heavy demand for carpet wool, it is necessary to increase per-animal productivity. This will be possible only when an integrated approach optimizing nutritional and management inputs with adequate disease control measures are adopted to increase indigenous wool production. Sustainable wool production in terms of both quality and quantity depends mainly on improving farming practices and socioeconomic and ecological considerations. It includes upgrading existing sheep breeds, introducing improver breeds from similar ecological region that can adapt or cross-breeding to evolve new breeds combining the adaptation and hardiness of local breeds and higher productivity of exotic breeds. Adaptation of improver farm management practices and nutritional support are equally important for sustainable growth in global wool production. The development of processes and practices for innovative product manufacturing and improvement in the quality of existing products prepared from of underused coarser wool will increase the demand for such wool and provide better remuneration for sheep farmers.

References

18th Livestock Census, 2007, Department of Animal Husbandry, Dairying & Fisheries, Ministry of Agriculture, India.

Acharya, R.M., 1974. Evaluation of native breeds of sheep for wool and mutton and scope for introduction of exotic inheritance. Indian J. Genet. 34 A, 945 Paper presented at the 2nd General Congress of the Society for the Advancement of Breeding Research in Asia and Oceania.

Acharya, R.M., 1984. In: ICAR in Proceeding of National Seminar on Carpet Wool Production and Carpet Manufacture.

Amble, V.N., Malhotra, J.C., 1968. Statistical studies on Rambouillet x Rampur Bushair cross of sheep at Pipalkoti, U.P. Indian J. Anim. Sci. 38, 10–11.

Amble, V.N., Khandekar, N.C., Garg, J.N., 1967. Statistical studies on breeding data of Deccani and crossbred sheep. Indian J. Vet. Sci. Anim. Husb. 37 (4), 305.

Bhagle, G.D., Whitley, K.J., 1963. Some Rheological properties of Indian carpets wools. Text. Res. J. 33, 161–162.

Bohra, S.D.J., Jain, A., Sharma, S.C., January–March, 1993. Kheri – a New Type of Sheep in Rajasthan (India). Wool and Woollens of India, pp. 23–24.

Burns, R.H., Johnsonand, A., Chen, W.C., 1940. The improvement of Chinese and other carpet wool. J. Text. Inst. 31, T-37.

Chaudhry, A.L., Chaudhary, S.R., 1991. Genetic improvement of small ruminants for fibre, meat and pelt production. In: Golden Jubilee Symposium on Genctic Research and Education: Current Trends and the Next Fifty Years, vol. III, pp. 1455–1466.

Das, M.N., Rajagopalan, V.R., 1956. Studies on the scope of improvement of Bellary sheep. Indian J. Vet. Sci. Anim. Husb. 26, 17–21.

Lemon, H., 1961. Sheep that Produce Carpet Wool–Sheep Farming in Many Lands. Department of Education of the International Wool Secretariat.

Ministry of Textile, Section V, Report, 2012; p. 350.

MSSP Annual Report, 2016. Mega Sheep Seed Project (2015-16). ICAR-CSWRI, Avikanagar.
NWPSI Annual Report, 2016. Network Project on Sheep Improvement (2015-16). ICAR-CSWRI, Avikanagar.
Nagashanker, R., Bohra, S.D.J., 1984, Production of carpet wool with especial reference to India. In: Proceeding of National Seminar on Carpet Wool Production and Carpet Manufacture, pp. 8–18.
Raina, B.L., Desai, R.N., Bhat, P.N., 1973. Studies in crossbred sheep. III. Genetic and phenotypic parameters of body weight and fleece weight in different crosses of Rambouillct x Karnah, Rambouillet x Poonch and Rambouillet x Kashmir Valley. Indian J. Anim. Prod. 4, 99–106.
Shakyawar, D.B., Parthasarathy, S., Swain, N., 2007. Performance of Bharat merino & its cross in Kolar and future prospects. In: Compendium Interactive Meet on Sheep Production for Meat and Wool and its Utilization in the Plains of Southern Region of India 29th May 2007. Organized by Southern Regional Research Centre, CSWRI, Mannavanur, Kodaikanal, TN.
Singh, V.K., Singh, G., 1998. Breeding strategies for improving wool production. In: Paper Presented in National Seminar on "Golden Jubilee Seminar on Sheep, Goat and Rabbit Production and Utilization," from 24–26th April, 1998. Organized by Indian Society for Sheep and Goat Production and Utilization at Jaipur.
Sule, A.D., 1984. Scope of research and development in Indian wools. In: Proceeding of National Seminar on Carpet Wool Production and Carpet Manufacture, pp. 184–194.
Task Force, 1996. Report of the Task Force on Sheep, Goat and Rabbit Production. Ministry of Agriculture, Department of Animal Husbandry & Dairying, Govt. of India.
Terminal Report, 1986. Terminal Report of AICRP on Sheep Breeding, Project Coordinator Sheep Breeding. CSWRI, Avikanagar.
Wool Production and Supply, Trend, Drivers and Prospects, 2015. .
Sherf, B.D. (Ed.), September 1995. World Watch List for Domestic Animal Diversity, second ed. F.A.O, Rome.

Further reading

Acharya, R.M., 1982. Sheep and Goat Breeds of India. F.A.O., Rome, p. 190(FAO Animal Production and Health paper No. 30).
Acharya, R.M., Bhat, P.N., 1984. Livestock and Poultry Genetic Resources in India. IVRI, Piblication.
Annual Report, 2016. Central Sheep & Wool Research Institute, Avikanagar.
Chaudhry, A.L., Arora, A.L., 1989. Research in sheep breeding for fine and carpet wool production in the country. In: National Seminar on 'Genetics Applied to Livestock Production' Held at GAU, Anand from 23–25 October, pp. A7–A18.
Das, S., Mohanty, A., Sahoo, G., 2001. Performance of Malpura sheep and its crosses under hot and humid climatic conditions. Indian J. Anim. Prod. Manage. 17, 20–21.
Report, 1998. 35-Years of Research. CSWRI, Avikanagar, India.
Terminal Report, 2007. Terminal Report (1994-2006) of Network Project on Sheep Improvement, Project Coordinator Sheep Breeding. CSWRI, Avikanagar.

Ecological dyeing of wool with biomordants

5

Shahid-ul-Islam, Faqeer Mohammad
Jamia Millia Islamia (A Central University), New Delhi, India

5.1 Introduction

Over the past few decades, the use of natural products for textile dyeing and finishing has gained importance possibly because of enhanced awareness about contamination associated with synthetic azo dyes (Ghoranneviss et al., 2011; Nasirizadeh et al., 2012; Yusuf et al., 2012). During different textile processing stages including sizing, scouring, bleaching, mercerizing, printing, dyeing and finishing, inefficiencies in dyeing produce wastewater rich in finishing agents, surfactants, inhibitor compounds, active substances, chlorine compounds, salts and dyeing substances which ultimately result in environmental pollution (Farooq et al., 2013; Islam and Mohammad, 2015b; McNeil and McCall, 2011; Tanapongpipat et al., 2008). In this context, natural colourants from plant, animal and mineral sources for textile applications have delivered promising results (Islam et al., 2014a; Islam and Mohammad, 2014; Vankar et al., 2007). Therefore, textile and polymer scientists have intensified efforts in the search for ecological dyes for wool and cotton.

Among different textile fibres, wool has been widely dyed with natural colourants for use in different fields of application (Tsatsaroni et al., 1998; Yusuf et al., 2013; Zhang et al., 2014). To understand the mechanism of wool dyeing effectively, electrostatic forces, hydrogen bonding and hydrophobic and van der Waals interactions must be considered (Fig. 5.1).

Wool dyeing is a multistep process considered to be complex; it requires several stages to turn raw wool into coloured wool for multifunctional uses (Lewis et al., 1992):

- diffusion of dyeing compounds through an aqueous dye bath to a textile substrate
- transfer of dyeing components across the fibre's surface
- diffusion of dyeing materials from the textile's surface to the whole fibre

Wool fibres are currently extensively dyed with synthetic dyes such as azo, anthraquinone, metal complex and indigo. Synthetic dyes offer many advantages such as low cost, global acceptability, a wide range of shades, ease of application and reproducibility (Islam et al., 2013; Yusuf et al., 2015). On the other hand, the production of synthetic dyes is based on nonrenewable petrochemical sources and some of them, particularly azo dyes, produce carcinogenic amines. It is also difficult to remove synthetic colourants once they find their way into water bodies (Yusuf et al., 2015). In the light of these issues, research and development (R&D) is under way worldwide to explore alternative sources of dyes using natural products (Fig. 5.2).

Sustainable Fibres and Textiles. http://dx.doi.org/10.1016/B978-0-08-102041-8.00005-6

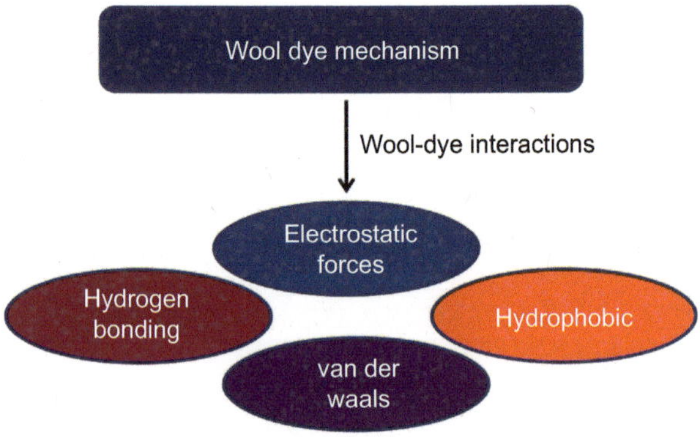

Figure 5.1 Different wool–dye interactions.

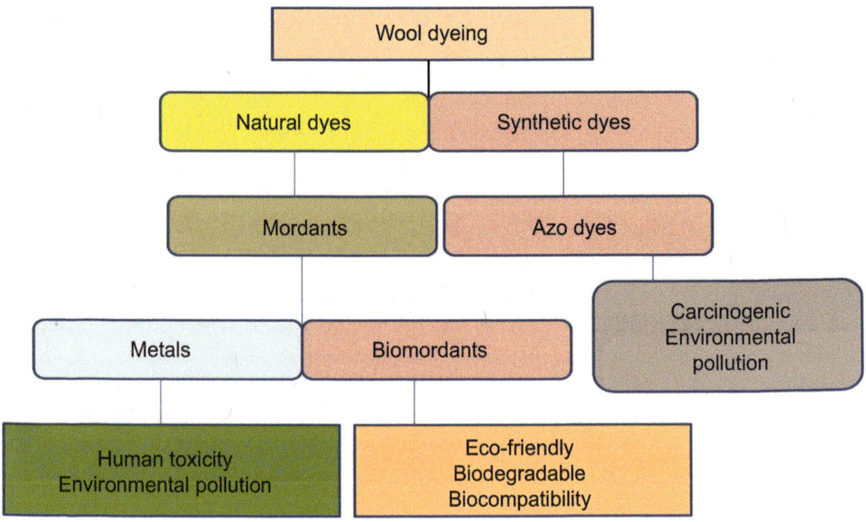

Figure 5.2 Road map of wool dyeing using eco-friendly biomordants.

5.2 Biomordants in wool dyeing

Wool is a suitable substrate for mordanting. Some naturally derived colourants directly dye fibres including wool, cotton and silk, whereas the majority of them require metal salts as anchoring agents for better fixation of dyeing principles with the textile substrate (Shahidi et al., 2014; Vankar and Shanker, 2008; Vankar et al., 2007). These chemicals are known as mordants. Metal salts of Al, Fe and Sn are mainly used in the colouration and functional finishing of textile materials. The schematic complexation of metal mordants with wool and natural dyes is shown in Fig. 5.3.

The use of metallic mordants introduces an additional role in the functional treatment of textile materials (Islam and Mohammad, 2015a; Islam et al., 2013; Rather

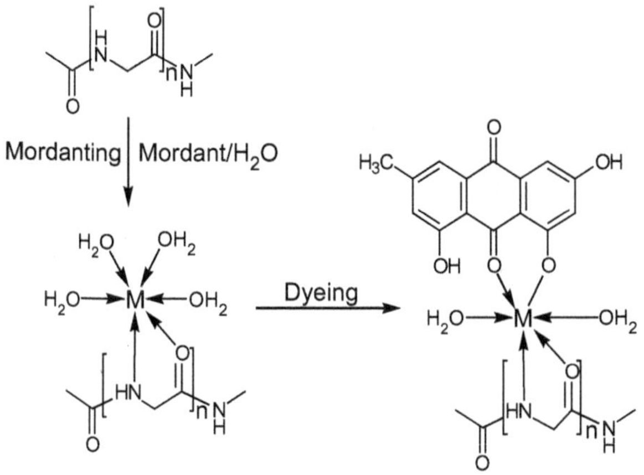

Figure 5.3 Wool–mordant–dye complex formation (Khan et al., 2012).

et al., 2015). It has been observed that mordants may increase dye exhaustion, shade variation, colour fastness and other fibre properties (Samantaa and Agarwal, 2009; Zhou et al., 2015).

Natural dyeing of wool with metal mordants generates effluents rich in toxic metal ions, which has a negative impact on flora and fauna. Therefore the exploitation of natural mordants in place of metallic salts is an active area of research (Erdem İşmal et al., 2014; Rehman et al., 2013; Shahid et al., 2013; Vankar et al., 2008). Some commonly employed biomordants are described next.

5.2.1 Pomegranate

Punica granatum L. is a plant of the Punicaceae family generally cultivated in Spain, Italy, Turkey, Egypt, Iran, California and India (Fischer et al., 2011; Tiwari et al., 2010). Pomegranates are generally rich in tannins, which have applications in fishing nets, ropes, sails and postal bags, and are used as stains and also as mordants (Sinha et al., 2012b; Sivakumar ct al., 2011). The rind of pomegranate is rich in polyphenols such as punicalagin, punicalin, gallagalic acid, gallic acid and ellagic acid (Fig. 5.4) (Madrigal-Carballo et al., 2009; Viuda-Martos et al., 2010). Granatonine present in the alkaloid form *N*-methyl granatonine is the main colouring compound present in pomegranate rind (Erdem İşmal et al., 2014). The phytochemical present in pomegranate has been used to dye wool, silk and cotton (Prabhu and Teli, 2014; Shams-Nateri et al., 2014). The rind extract has been employed in conjunction with many natural dyes as a biomordant to result in faster dyeing. Erdem İşmal et al. (2014) investigated the natural dyeing of wool by valorization of the outer green shell of almond fruit as a novel dye in the presence of pomegranate biomordant and found that pomegranate improved light and wash fastness properties. Rather et al. (2016) studied the dyeing properties of *Adhatoda vasica* on wool pretreated with pomegranate biomordant and found that pomegranate produces a range of beautiful shades with acceptable fastness properties.

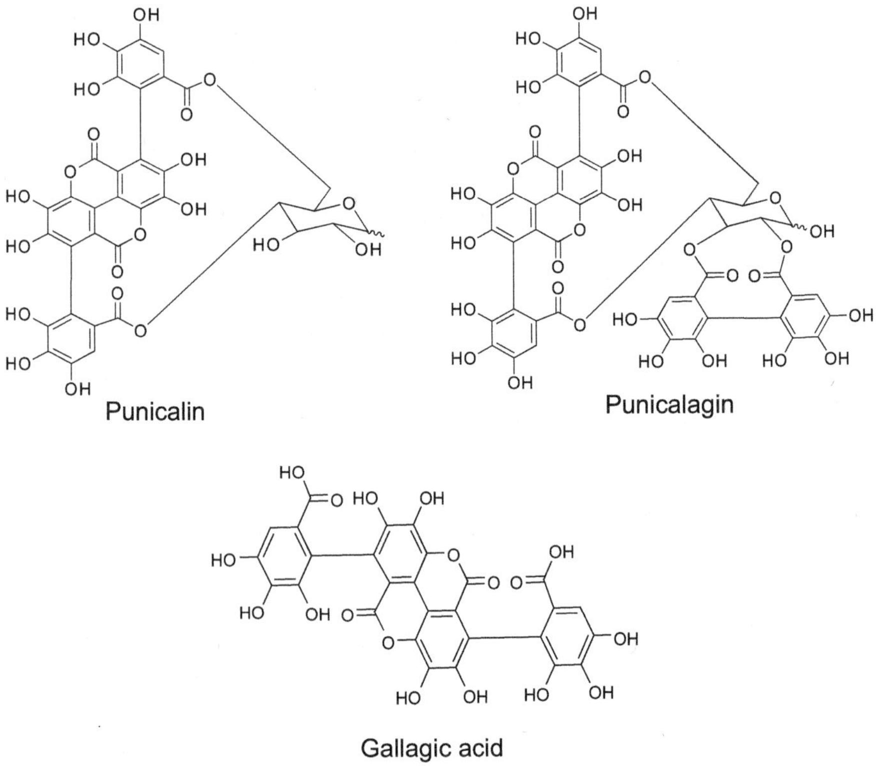

Figure 5.4 Tannins in pomegranate peel extract.

5.2.2 Gallnut

Quercus infectoria belongs to the family Fagaceae. Galls of the *Q. infectoria* tree contain a considerable amount of tannins, mainly gallotannic acid, gallic acid, ellagic acid, starch and glucose (Fig. 5.5) (Lee et al., 2009; Perkin and Everest, 1918; Shahid et al., 2012). The main dyeing principle in gallnut is ellagic acid, which contains auxochrome –OH groups. These hydroxyl groups have affinity with the textile substrates (Lee et al., 2014; Onal et al., 2005). Gallnut has a biomordanting property because of the presence of tannin. Several research studies were undertaken to discover the dyeing and biomordanting properties of gallnut extracts (Hong et al., 2012; Koh and Hong, 2014). Rather et al. (2016) studied the effect of various metal salts (ferrous sulphate, alum and stannous chloride) and natural tannin extracts (gallnut, pomegranate peel and babul bark) as mordants on the colour and fastness properties of wool dyed with natural dye and found that all of these biomordants imparted good colour strength and fastness properties.

5.2.3 Myrobalan

Myrobalan (*Terminalia chebula*) is a large tree grown in many parts of the world, mainly in south India and also in the sub-Himalayan tract of west Bengal and

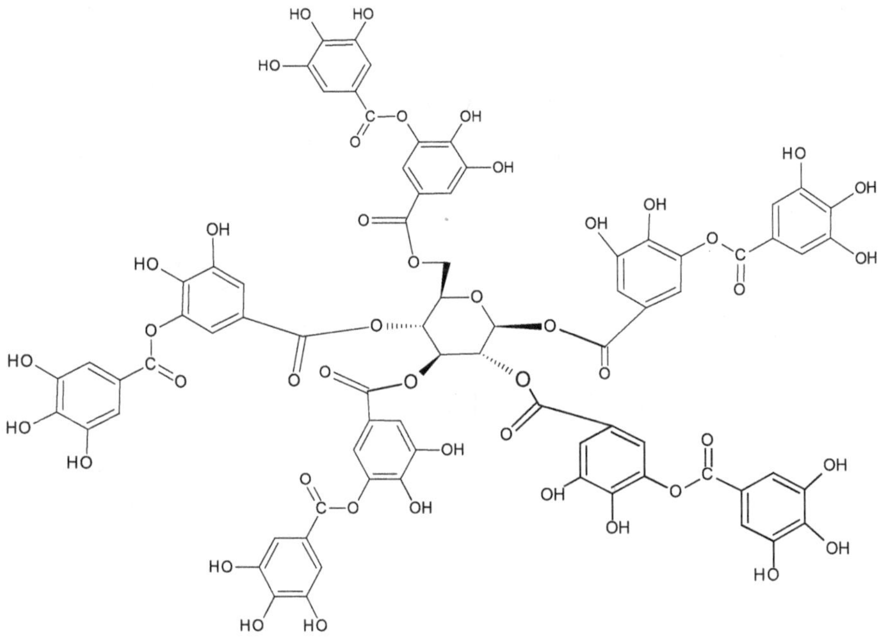

Figure 5.5 Gallotannic acid.

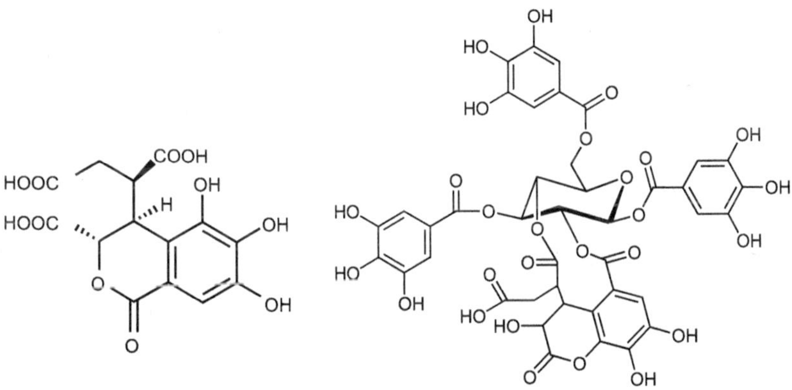

Chebulic acid

Chebulinic acid

Figure 5.6 Tannins in myrobalan.

Assam (Perkin and Everest, 1918). Myrobalan is used in the tanning of leather, as a natural dye for wool and silk, as well as to combat various human infections (Murugajothi and Moses, 2008). Myrobalan is rich in a tannin compounds, mainly chebulagic acid, chebulinic acid, gallic acid and ellagic acid (Fig. 5.6) (Shabbir et al., 2016a). Myrobalan has been used as a natural textile dye mainly because of its tannin compounds (Shabbir et al., 2016b).

5.3 Some natural colourants for wool

5.3.1 Henna

Lawsonia inermis (henna) belongs to the family Lythraceae. It is a shrub-like tree 3–6 m long and is widely cultivated in Oriental countries, India, Pakistan, Egypt, Yemen, Iran and Afghanistan (Gulrajani and Gupta, 1992; Mayer and Cook, 1943). The valuable portion of this plant is its entire leaves, which are small, entire, ovate or lanceolate (Gupta and Gulrajani, 1994). A literature survey on this plant showed the isolation of different classes of biologically active compounds: naphthoquinones, flavonoids, xanthones, triterpenoids, sterols, phenolic glycosides, coumarins and tannins (Semwal et al., 2014). The phytochemicals present in henna extracts are endowed with a rich wealth of biological activities including antibacterial, antifungal, anticancer, anthelmintic, antidiabetic, antioxidant, hepatoprotective, wound healing, antimalarial, anti-inflammatory, hypoglycaemic and antiulcer properties (El-Hag et al., 2007). Chemical structures of some biologically active principles present in henna are shown in Fig. 5.7.

Henna leaves are rich in the colouring compound lawsone (chemically: 2-hydroxy-1,4-naphthoquinone); it is commercially known as Yellow Natural Orange 6 and its Colour Index (C.I.) number 754,800. It has been used since prehistoric times to dye hair, eyebrows and fingernails during religious festivals, marriages etc. (Alam et al., 2007; Ali et al., 2009; Yusuf et al., 2012; 2013; Zarkogianni et al., 2011). Depending on the mordant combination and dyeing conditions, this dye exhibits a wide range of colours on different types of textiles ranging from dark orange-brown to yellowish-brown, greenish-brown and dark brown (Iqbal and Bhatti, 2008; Rehman et al., 2012). Henna dye has imparted beautiful colours on textile substrates previously treated with gamma irradiation (M'Garrech and Ncib, 2009). Henna dye along with chitosan pretreatment has proved markedly effective in imparting antibacterial properties to wool (Dev et al., 2009). Several researchers reported the antimicrobial activity of henna-dyed wool in the presence and absence of metallic salt mordants (Yusuf et al., 2012).

5.3.2 Walnut

Walnut is a popular natural dye-yielding tree 10–15 m height. It belongs to the family Juglandaceae. It is found in temperate regions and is cultivated commercially in Asia, western South America, the United States and central and southern Europe (Tutak and Benli, 2011). Various parts of this plant have been widely used for a myriad of medical conditions (Mayer and Cook, 1943). A variety of compounds representing different classes including polyphenol, steroids, fatty acids and other miscellaneous compounds have been identified from all parts of *Juglans regia*. Phenolic compounds including flavonoids and naphthoquinones are particularly prevalent in juglone extracts (Thakur, 2011). The chemical structures of some active compounds are depicted in Fig. 5.8. The textile colouring property is attributed to the presence of juglone (5-hydroxy-1,4-napthoquinone) (C.I. 75,500) in green walnut shells, bark and leaves (Bechtold and Mussak, 2009; Park et al., 2005). Juglone dye produces a brownish-yellow colour on wool mordanted with salts of aluminium, chromium or iron (Ghaheh et al., 2012; Mirjalili et al., 2011; Sharma

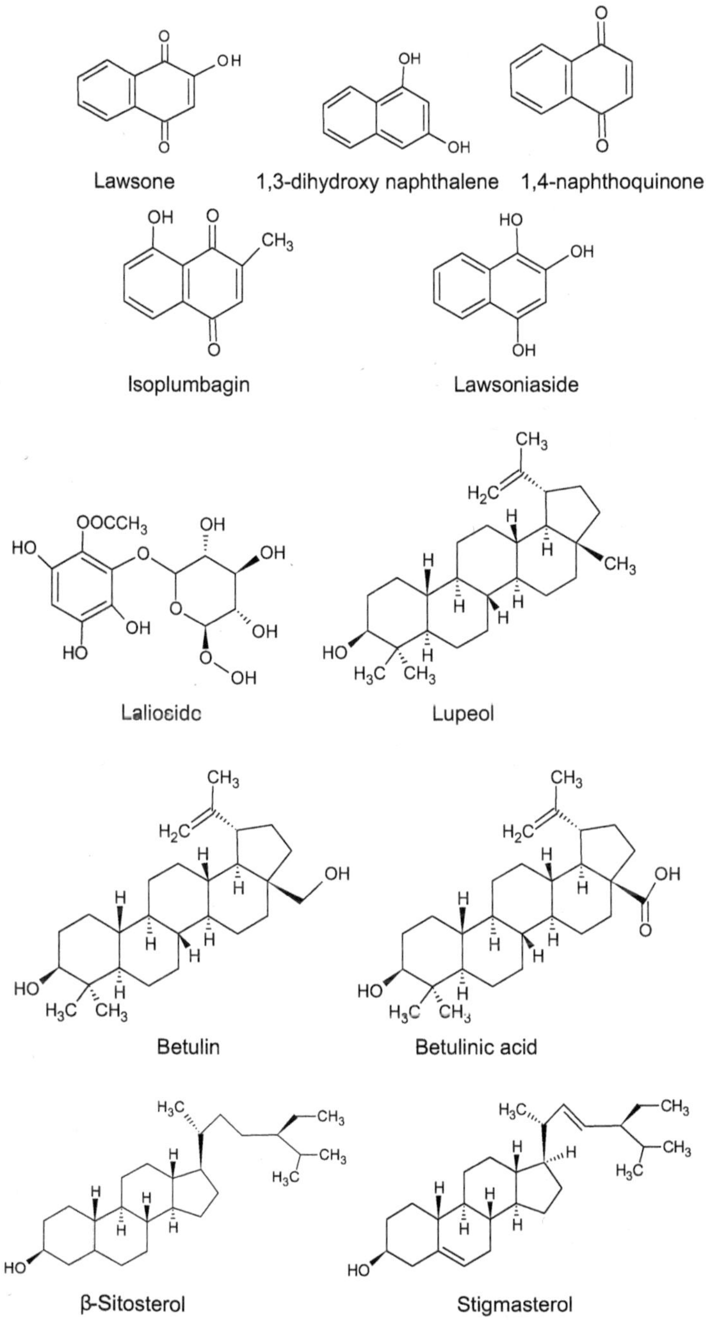

Figure 5.7 Chemical structures of some biologically active compounds present in henna.

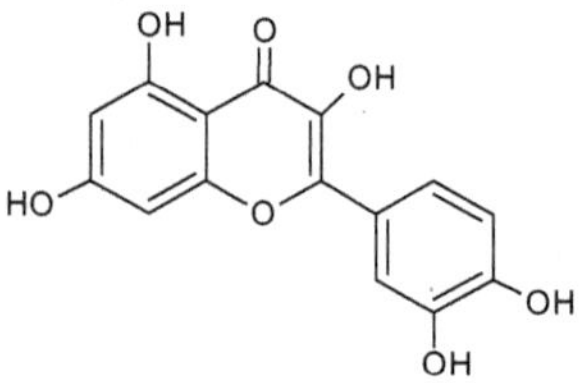

Quercetin

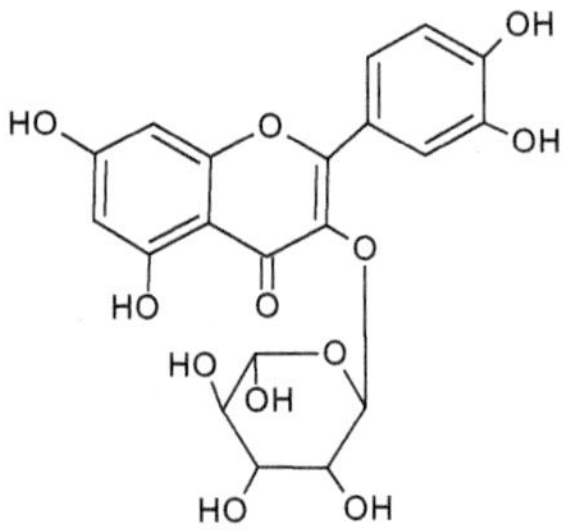

quercetin-3-O-rhamnoside

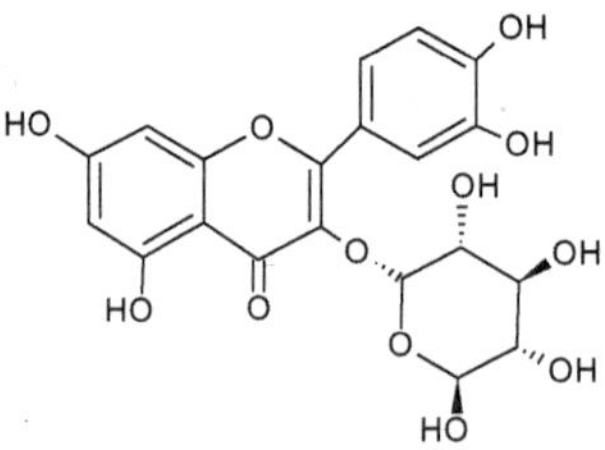

Quercetin-3-O-arabinoside

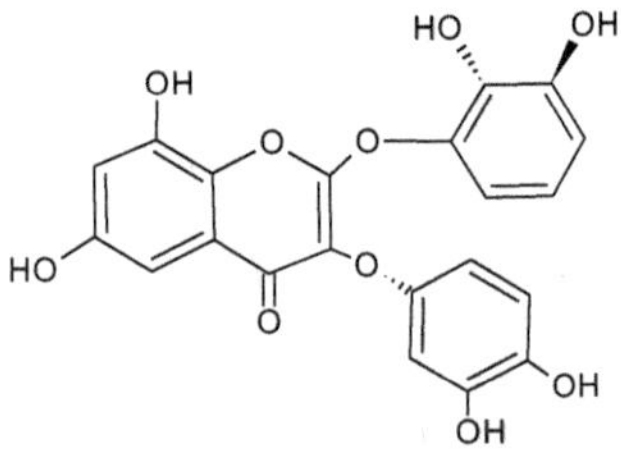

Quercetin-3-O-xyloside

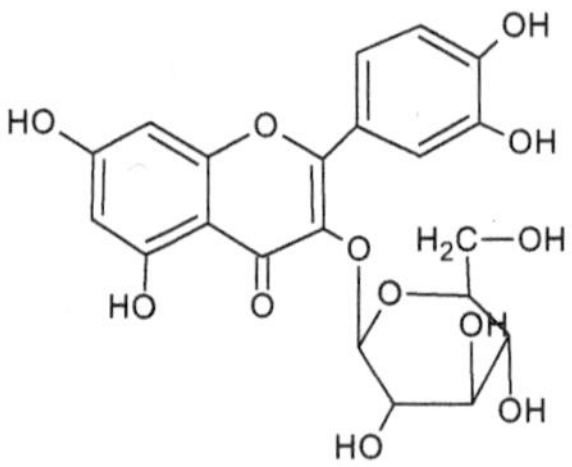

Quercetin-3-O-galactoside

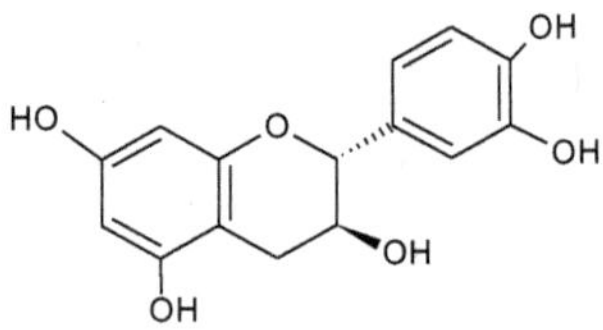

Catechin

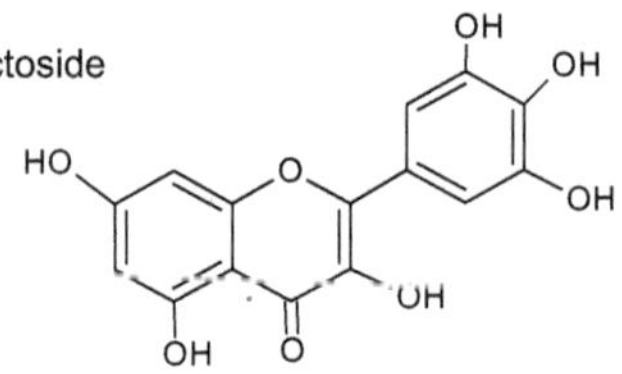

Myricetin

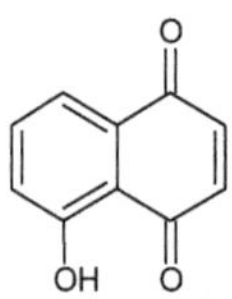

Juglone

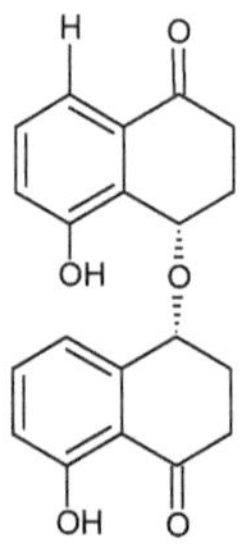

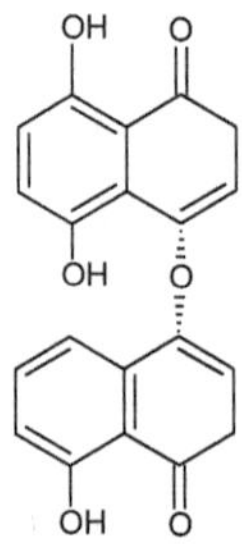

Juglanone A Juglanone B

Figure 5.8 Chemical structures of some compounds present in walnut.

and Grover, 2011). The aqueous extract of walnut dye after application to wool has been found to inhibit the growth of both gram-positive and gram-negative bacteria (Ghaheh et al., 2012; Mirjalili and Karimi, 2013).

5.3.3 Palas

Palas (*Butea monosperma* L.), an attractive medium-sized tree of the family Fabaceae, is native to the tropical and subtropical regions of India (Jharkhand) extending into the northwest Himalayas, Sri Lanka and Malaysia, and is also found in Myanmar (Perkin and Everest, 1918; Yadava and Tiwari, 2007). The leaves are large, pinnate and trifoliate with an 8- to 16-cm petiole and leaflets 10–20 cm long (Patil et al., 2006). The flowers are scentless and 2.5 cm long, with few to many flower clusters or racemes, bright orange-red in colour; the bracteoles and bracts are small and deciduous (Gulrajani and Gupta, 1992; Sinha et al., 2012a). The flowers are attractive for effective pollination by purple sunbirds (*Nectarinia asiatica*). The fruit is a pod 4–5 cm broad; it is pale green, wing-like, and about 15–20 cm long (Bezkorowajnyj and Sequeira, 1998).

The versatile utility of this plant as fuel wood, medicine for tribal and rural people to combat various infections, a source of natural dye for use in textiles, food, in the manufacture of solar cells and as a host species for lac production in India has prompted researchers to deepen investigations into the exploitation of new compounds for the full use of this plant (Sinha et al., 2012a). Investigations have been carried out on the flowers of this plant to isolate and identify its dyeing compounds. The main yellow colouring pigment of the flowers is identified as butein, chemically described as 2′, 3, 4, 4′-tetrahydroxychalcone (Wagner et al., 1986). Other than chalcones, structurally diverse compounds have been isolated and identified from all parts of this plant: flavonoids, terpenoids, imides, steroids and saponins (Fig. 5.9) (Tiwari et al., 2012).

5.3.4 Annatto

Bixa orellana is a medium-sized plant, 3–6 m, commonly found in India and Central and South America. The dye extracted from this plant is used in a wide range of food commodities such as dairy products, flour, confectionery, fish, soft drinks, meat products, snack foods and dry mixes (Rao et al., 2007; Sekar, 2004). A survey of the literature shows that annatto is widely used worldwide to treat a number of health disorders such as constipation, fever, heartburn, asthma, ulcers, diarrhoea, measles, allergy, leprosy, infectious diseases, burns, measles, gonorrhoeae, tumours, and skin problems (Islam et al., 2016). Several investigations have shown the presence of medicinally important phytoconstituents such as carotenoids, polyphenols, terpenoids, terpenes, steroids and tannins in all parts of this plant (Mercadante et al., 1997). These phytochemicals exhibit a wide range of pharmacological activities. The principle colouring compound present in annatto seeds is bixin (*cis*-form) (Sinha et al., 2013). The seeds are also rich in other carotenoid compounds, mainly apocarotenoid bixin, norbixin and other less important compounds such as cryptoxanthin, lutein, zeaxanthin and methylbixin (Galindo-Cuspinera et al., 2002) (Fig. 5.10). More detailed information about its use in textile dyeing was reviewed by Islam et al. (2016).

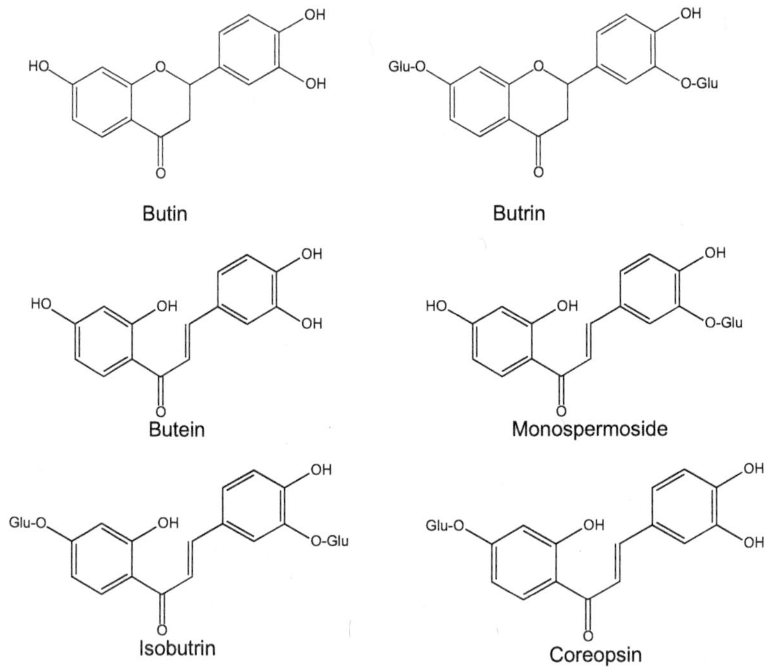

Figure 5.9 Chemical constituents of *Butea monosperma* flowers.

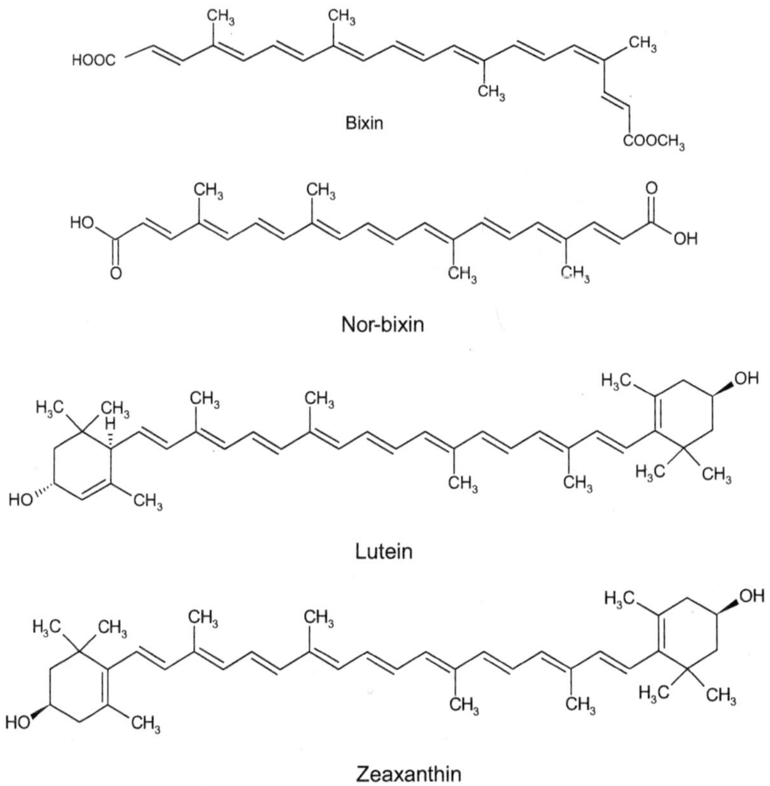

Figure 5.10 Chemical constituents of *Bixa orellana* seeds.

5.3.5 Teak

Tectona grandis, commonly known as teak, is a large deciduous tree belonging to the family Verbenaceae. It is a well-known timber-yielding plant naturally distributed throughout India (Madhya Pradesh, Maharashtra, Tamil Nadu and Kerala), Burma and Thailand (Moya et al., 2014; Nidavani and Mahalakshmi, 2014). It has traditionally been used to treat hyperacidity, swelling, piles, diabetes, asthma, uterine diseases, colic, anaemia and leprosy and has also been used as an expectorant and antiinflammatory (Ramachandran et al., 2011). This plant is mainly used as a source for durable and mite-resistant timber. In folk medicine it is employed to combat various infections, and it is also a promising source for natural dye (Islam et al., 2013). Different classes of natural products such as flavonoids, anthraquinones, steroids and fatty acids have been isolated from all parts of this plant, which have been linked to many pharmacological activities such as antihemolytic anaemic, antibacterial, antioxidant, antiinflammatory, analgesic and free radical scavenging properties (Nidavani and Mahalakshmi, 2014). The leaves are mainly rich in quinonoid compounds (Fig. 5.11) (Prusty et al., 2010). The dye from this plant produces beautiful shades with acceptable wash- and rub-fastness properties on wool (Vankar and Shanker, 2008).

5.4 Current challenges in wool dyeing

The largest use of wool is in apparel; hence colouration is an integral part of any wool-based finished product. As with other fabrics, colouration of wool is a complex physiochemical process involving the use of surfactants, dyes, mordants, salts and other chemical auxiliaries and therefore producing a large quantity of coloured effluent with high concentrations of organic and inorganic chemicals (Shahid et al., 2013). Effluents containing hazardous and toxic compounds released from the wool-dyeing industry are known to be potential carcinogens (Islam et al., 2014a). Considering both the volume and composition of toxic effluents generated from the textile dyeing industries, scientists are constantly searching for alternative ways to reduce their release into wastewaters. Over the years various chemical and physical processes such as ion exchange, membrane filtration, advanced oxidation, biological degradation, photocatalytic degradation, electrocoagulation and adsorption have been applied to remove coloured effluents from textile wastewaters. However, the treatment process of these wastewaters using conventional methods has proved to be markedly ineffective, difficult, and highly expensive (Islam et al., 2014b).

On the other hand, wool dyeing with natural dyes is considered the best route to obtain textiles in an eco-friendly way. In wool dyeing, natural colourants from plants, animals and minerals have been used to develop colour-fast and beautiful shades and as novel agents to produce antibacterial, antifungal, insect-repellant and UV-protective surfaces (Khan et al., 2011; Rather et al., 2016). To reduce the environmental impact of wool colouration and finishing, research in the area of natural dyes has focused on the use of ultrasonics, microwaves, pad dyeing, high-temperature high pressure dyeing, contact dyeing and other optimized methods (Shahid et al., 2013). The main drawback of dyeing wool with plant colourants is the use of metal mordants such

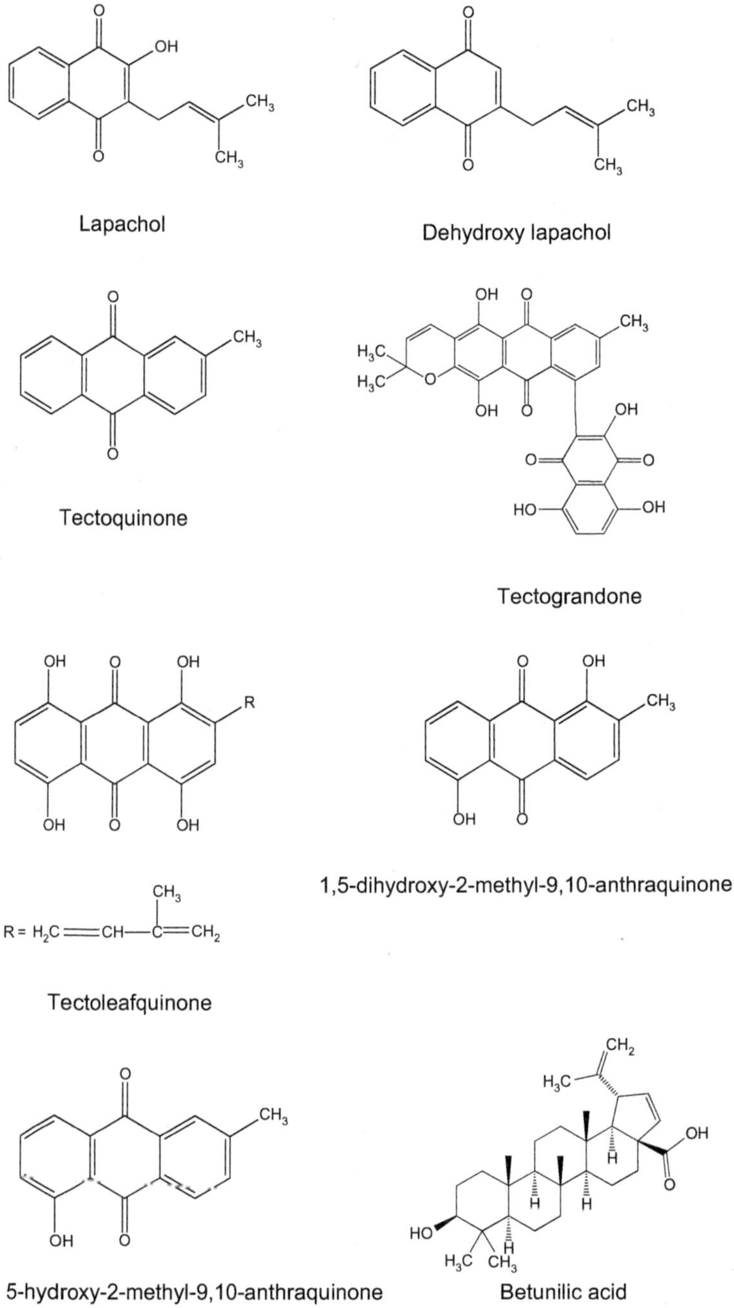

Figure 5.11 Chemical constituents of *Tectona grandis* leaves.

as aluminium potassium sulphate, potassium dichromate, stannous chloride, ferrous sulphate and copper sulphate. These metal salts are used to obtain different shades on wool and to improve the adhesion of natural dyes to wool. As discussed previously in this chapter, mordant salts may have lethal effect on flora and fauna (Islam et al., 2016). In the light of this, biomordants extracted from the plant materials discussed in this chapter offer sustainable anchoring agents instead of metal mordants for the future of the wool dyeing industry. The biomordants selected in this chapter are rich in bioactive compounds which have been shown to have a wide spectrum of biological activities. It is expected that increasing research on biomordants could result in their use in colouration and they may also be developed as potential agents to develop functional textiles with antioxidant, antimicrobial, deodorant and UV protection properties (Rather et al., 2016).

5.5 Conclusion and future aspects

Several efforts over the past few years have been made to exploit various natural dyes and biomordants for wool dyeing and finishing. Because metallic salt mordants used in wool dyeing create environmental pollution, biomordants from natural materials have gained increasing popularity. Use of biomordants and natural dyes for wool colouration has become an important area because of the nontoxic, nonallergic and environmentally friendly natural of plant extracts. In addition, ongoing interest in environmentally friendly colouration of wool will encourage more R & D investigations for the exploitation of biomordants and natural dyes from cheaper sources such as agro-industrial wastes. There is still a long way to go before biomordants and natural dyes can be used on a pilot scale. Future work should focus on the role of biomordants from tannin-bearing plants in the production of antimicrobial, deodorant and insect-repellant textile surfaces. With these additional finishing properties, the ecological dyeing of wool with natural mordants and dyes may introduce many valuable benefits and generate the hope for sustainable textile production in the near future.

References

Alam, M., Rahman, M., Haque, M., 2007. Extraction of henna leaf dye and its dyeing effects on textile fibre. Bangladesh J. Sci. Ind. Res. 42, 217–222.

Ali, S., Hussain, T., Nawaz, R., 2009. Optimization of alkaline extraction of natural dye from Henna leaves and its dyeing on cotton by exhaust method. J. Clean. Prod. 17, 61–66.

Bechtold, T., Mussak, R., 2009. Handbook of Natural Colorants. Wiley, Chichester, UK.

Bezkorowajnyj, P., Sequeira, V., 1998. Improved management of *Butea monosperma* (Lam.) Taub for lac production in India. For. Ecol. Manag. 102, 225–234.

Dev, V.R.G., Venugopal, J., Sudha, S., Deepika, G., Ramakrishna, S., 2009. Dyeing and antimicrobial characteristics of chitosan treated wool fabrics with henna dye. Carbohydr. Polym. 75, 646–650.

El-Hag, A., Al-Jabri, A., Habbal, O., 2007. Antimicrobial properties of *Lawsonia inermis* (henna): a review. Aust. J. Med. Herb. 19, 114.

Erdem İşmal, Ö., Yıldırım, L., Özdoğan, E., 2014. Use of almond shell extracts plus biomordants as effective textile dye. J. Clean. Prod. 70, 61–67.

Farooq, A., Ali, S., Abbas, N., Fatima, G.A., Ashraf, M.A., 2013. Comparative performance evaluation of conventional bleaching and enzymatic bleaching with glucose oxidase on knitted cotton fabric. J. Clean. Prod. 42, 167–171.

Fischer, U.A., Carle, R., Kammerer, D.R., 2011. Identification and quantification of phenolic compounds from pomegranate (*Punica granatum* L.) peel, mesocarp, aril and differently produced juices by HPLC-DAD–ESI/MS(n). Food Chem. 127, 807–821.

Galindo-Cuspinera, V., Lubran, M.B., Rankin, S.A., 2002. Comparison of volatile compounds in water-and oil-soluble annatto (*Bixa orellana* L.) extracts. J. Agric. Food Chem. 50, 2010–2015.

Ghaheh, F.S., Nateri, A.S., Mortazavi, S.M., Abedi, D., Mokhtari, J., 2012. The effect of mordant salts on antibacterial activity of wool fabric dyed with pomegranate and walnut shell extracts. Color. Technol. 128, 473–478.

Ghoranneviss, M., Shahidi, S., Anvari, A., Motaghi, Z., Wiener, J., Slamborova, I., 2011. Influence of plasma sputtering treatment on natural dyeing and antibacterial activity of wool fabrics. Prog. Org. Coat. 70, 388–393.

Gulrajani, M.L., Gupta, D., 1992. Natural Dyes and Their Application to Textiles. Department of Textile Technology, IIT Delhi, New Delhi, India.

Gupta, D.B., Gulrajani, M.L., 1994. Kinetic and thermodynamic studies on 2-hydroxy-1, 4-naphthoquinone (lawsone). J. Soc. Dyers Colour. 110, 112–115.

Hong, K., Bae, J., Jin, S., Yang, J., 2012. Preparation and properties of multi-functionalized cotton fabrics treated by extracts of gromwell and gallnut. Cellulose 19, 507–515.

Iqbal, J., Bhatti, I.A., 2008. Effect of UV radiation on dyeing of cotton fabric with extracts of henna leaves. Indian J. Fibre Text. Res. 33, 157–162.

Islam, S.-U., Mohammad, F., 2015a. High-energy radiation induced sustainable coloration and functional finishing of textile materials. Ind. Eng. Chem. Res. 54 (15).

Islam, S.-U., Mohammad, F., 2015b. Natural colorants in the presence of anchors so-called mordants as promising coloring and antimicrobial agents for textile materials. ACS Sustain. Chem. Eng. 3, 2361–2375.

Islam, S.-U., Shahid, M., Mohammad, F., 2014a. Future prospects of phytosynthesized transition metal nanoparticles as novel functional agents for textiles. In: Syväjärvi, A.T.A. (Ed.), Advanced Materials for Agriculture, Food, and Environmental Safety. John Wiley & Sons, Inc., pp. 265–290.

Islam, S., Rather, L.J., Shahid, M., Khan, M.A., Mohammad, F., 2014b. Study the effect of ammonia post-treatment on color characteristics of annatto-dyed textile substrate using reflectance spectrophotometery. Ind. Crop. Prod. 59, 337–342.

Islam, S.U., Mohammad, F., 2014. Emerging green technologies and environment friendly products for sustainable textiles. In: Muthu, S.S. (Ed.), Roadmap to Sustainable Textiles and Clothing. Springer, Singapore, pp. 63–82.

Islam, S.U., Rather, L.J., Mohammad, F., 2016. Phytochemistry, biological activities and potential of annatto in natural colorant production for industrial applications–a review. J. Adv. Res. 7, 499–514.

Islam, S.U., Shahid, M., Mohammad, F., 2013. Perspectives for natural product based agents derived from industrial plants in textile applications – a review. J. Clean. Prod. 57, 2–18.

Khan, M.I., Ahmad, A., Khan, S.A., Yusuf, M., Shahid, M., Manzoor, N., Mohammad, F., 2011. Assessment of antimicrobial activity of Catechu and its dyed substrate. J. Clean. Prod. 19, 1385–1394.

Khan, S.A., Ahmad, A., Khan, M.I., Yusuf, M., Shahid, M., Manzoor, N., Mohammad, F., 2012. Antimicrobial activity of wool yarn dyed with *Rheum emodi* L. (Indian Rhubarb). Dyes Pigm. 95, 206–214.

Koh, E., Hong, K.H., 2014. Gallnut extract-treated wool and cotton for developing green functional textiles. Dyes Pigm. 103, 222–227.

Lee, Y.-H., Hwang, E.-K., Baek, Y.-M., Kim, H.-D., 2014. Deodorizing function and anti-bacterial activity of fabrics dyed with gallnut (Galla chinensis) extract. Text. Res. J. 85 (10).

Lee, Y.-H., Hwang, E.-K., Kim, H.-D., 2009. Colorimetric assay and antibacterial activity of cotton, silk, and wool fabrics dyed with peony, pomegranate, clove, *Coptis chinenis* and gallnut extracts. Materials 2, 10–21.

Lewis, D.M., Dyers, S.O., Colourists, 1992. Wool Dyeing. Society of Dyers and Colourers.

M'Garrech, S., Ncib, F., 2009. Colorimetric study of effect of gamma-radiation on the color of cotton fabric colored by "henna" dye. Appl. Radiat. Isot. 67, 2003–2006.

Madrigal-Carballo, S., Rodriguez, G., Krueger, C., Dreher, M., Reed, J., 2009. Pomegranate (*Punica granatum*) supplements: authenticity, antioxidant and polyphenol composition. J. Funct. Foods 1, 324–329.

Mayer, F., Cook, A.H., 1943. The Chemistry of Natural Coloring Matters: The Constitutions, Properties, and Biological Relations of the Important Natural Pigments. Reinhold publishing corporation, New York.

McNeil, S.J., McCall, R.A., 2011. Ultrasound for wool dyeing and finishing. Ultrason. Sonochem. 18, 401–406.

Mercadante, A.Z., Steck, A., Pfander, H., 1997. Isolation and structure elucidation of minor carotenoids from annatto (*Bixa orellana* L.) seeds. Phytochemistry 46, 1379–1383.

Mirjalili, M., Karimi, L., 2013. Extraction and characterization of natural dye from green walnut shells and its use in dyeing polyamide: focus on antibacterial properties. J. Chem. 2013, 1.

Mirjalili, M., Nazarpoor, K., Kartmi, L., 2011. Extraction and identification of dye from walnut green husks for silk dyeing. Asian J. Chem. 23, 1055–1059.

Moya, R., Bond, B., Quesada, H., 2014. A review of heartwood properties of *Tectona grandis* trees from fast-growth plantations. Wood Sci. Technol. 48, 411–433.

Murugajothi, K., Moses, J.J., 2008. A study on the characteristic improvement of property of *Terminalia chebula* (myrobalan) on cotton fabric. Oriental J. Chem. 24, 903–910.

Nasirizadeh, N., Dehghanizadeh, H., Yazdanshenas, M.E., Moghadam, M.R., Karimi, A., 2012. Optimization of wool dyeing with rutin as natural dye by central composite design method. Ind. Crops Prod. 40, 361–366.

Nidavani, R.B., Mahalakshmi, A., 2014. Teak (*Tectona grandis* linn.): a renowned timber plant with potential medicinal values. Int. J. Pharm. Pharm. Sci. 6, 48–54.

Onal, A., Sari, A., Soylak, M., 2005. Ellagic acid from gallnut (*Quercus infectoria*): extraction and determination of its dyeing conditions for natural fibres. J. Sci. Ind. Res. 64, 491.

Park, J.H., Gatewood, B.M., Ramaswamy, G.N., 2005. Naturally occurring quinones and flavonoid dyes for wool: insect feeding deterrents. J. Appl. Polym. Sci. 98, 322–328.

Patil, M., Pawar, S., Patil, D., 2006. Ethnobotany of *Butea monosperma* (Lam.) Kuntze in North Maharashtra, India. Nat. Prod. Rad. 5, 323–325.

Perkin, A.G., Everest, A.E., 1918. The Natural Organic Colouring Matters. Longmans, Green and Co, London.

Prabhu, K.H., Teli, M.D., 2014. Eco-dyeing using *Tamarindus indica* L. seed coat tannin as a natural mordant for textiles with antibacterial activity. J. Saudi Chem. Soc. 18, 864–872.

Prusty, A.K., Das, T., Nayak, A., Das, N.B., 2010. Colourimetric analysis and antimicrobial study of natural dyes and dyed silk. J. Clean. Prod. 18, 1750–1756.

Ramachandran, S., Rajasekaran, A., Kumar, K.M., 2011. Antidiabetic, antihyperlipidemic and antioxidant potential of methanol extract of *Tectona grandis* flowers in streptozotocin induced diabetic rats. Asian Pac. J. Trop. Med. 4, 624–631.

Rao, P.P., Satyanarayana, A., Balaswamy, K., Jyothirmayi, T., Nagender, A., Rao, D.G., 2007. Application of annatto dye formulation in bakery and extruded food products. J. Foodserv. 18, 53–58.

Rather, L.J., Shahid, I., Mohammad, F., 2015. Study on the application of Acacia nilotica natural dye to wool using fluorescence and FT-IR spectroscopy. Fibers Polym. 16, 1497–1505.

Rather, L.J., Shahid ul, I., Shabbir, M., Bukhari, M.N., Shahid, M., Khan, M.A., Mohammad, F., 2016. Ecological dyeing of Woolen yarn with Adhatoda vasica natural dye in the presence of biomordants as an alternative copartner to metal mordants. J. Environ. Chem. Eng. 4, 3041–3049.

Rehman, F.-U., Adeel, S., Qaiser, S., Ahmad Bhatti, I., Shahid, M., Zuber, M., 2012. Dyeing behaviour of gamma irradiated cotton fabric using Lawson dye extracted from henna leaves (*Lawsonia inermis*). Radiat. Phys. Chem. 81, 1752–1756.

Rehman, F.-U., Adeel, S., Shahid, M., Bhatti, I.A., Nasir, F., Akhtar, N., Ahmad, Z., 2013. Dyeing of γ-irradiated cotton with natural flavonoid dye extracted from irradiated onion shells (*Allium cepa*) powder. Radiat. Phys. Chem. 92, 71–75.

Samantaa, A.K., Agarwal, P., 2009. Application of natural dyes on textiles. Indian J. Fibre Text. Res. 34, 384–399.

Sekar, N., 2004. Annatto colorants. Colourage 51 (7), 67.

Semwal, R.B., Semwal, D.K., Combrinck, S., Cartwright-Jones, C., Viljoen, A., 2014. *Lawsonia inermis* L. (henna): ethnobotanical, phytochemical and pharmacological aspects. J. Ethnopharmacol. 155, 80–103.

Shabbir, M., Islam, S.U., Bukhari, M.N., Rather, L.J., Khan, M.A., Mohammad, F., 2016a. Application of *Terminalia chebula* natural dye on wool fiber—evaluation of color and fastness properties. Text. Cloth. Sustain. 2, 1.

Shabbir, M., Rather, L.J., Bukhari, M.N., Shahid, M., Khan, M.A., Mohammad, F., 2016b. An eco-friendly dyeing of woolen yarn by *Terminalia chebula* extract with evaluations of kinetic and adsorption characteristics. J. Adv. Res. 7, 473–482.

Shahid, M., Ahmad, A., Yusuf, M., Khan, M.I., Khan, S.A., Manzoor, N., Mohammad, F., 2012. Dyeing, fastness and antimicrobial properties of woolen yarns dyed with gallnut (*Quercus infectoria* Oliv.) extract. Dyes Pigm. 95, 53–61.

Shahid, M., Shahid ul, I., Mohammad, F., 2013. Recent advancements in natural dye applications: a review. J. Clean. Prod. 53, 310–331.

Shahidi, S., Aslan, N., Ghoranneviss, M., Korachi, M., 2014. Effect of thymol on the antibacterial efficiency of plasma-treated cotton fabric. Cellulose 21, 1933–1943.

Shams-Nateri, A., Hajipour, A., Dehnavi, E., Ekrami, E., 2014. Colorimetric study on polyamides dyeing with weld and pomegranate peel natural dyes. Cloth. Text. Res. J. 32, 124–135.

Sharma, A., Grover, E., 2011. Colour fastness of walnut dye on cotton. Indian J. Nat. Prod. Res. 2, 164–169.

Sinha, K., Chowdhury, S., Saha, P.D., Datta, S., 2013. Modeling of microwave-assisted extraction of natural dye from seeds of *Bixa orellana* (Annatto) using response surface methodology (RSM) and artificial neural network (ANN). Ind. Crops Prod. 41, 165–171.

Sinha, K., Saha, P.D., Datta, S., 2012a. Extraction of natural dye from petals of flame of forest (*Butea monosperma*) flower: process optimization using response surface methodology (RSM). Dyes Pigm. 94, 212–216.

Sinha, K., Saha, P.D., Datta, S., 2012b. Response surface optimization and artificial neural network modeling of microwave assisted natural dye extraction from pomegranate rind. Ind. Crops Prod. 37, 408–414.

Sivakumar, V., Vijaeeswarri, J., Anna, J.L., 2011. Effective natural dye extraction from different plant materials using ultrasound. Ind. Crops Prod. 33, 116–122.
Tanapongpipat, A., Khamman, C., Pruksathorm, K., Hunsom, M., 2008. Process modification in the scouring process of textile industry. J. Clean. Prod. 16, 152–158.
Thakur, A., 2011. Juglone: a therapeutic phytochemical from *Juglans regia* L. J. Med. Plants Res. 5, 5324–5330.
Tiwari, H.C., Singh, P., Kumar Mishra, P., Srivastava, P., 2010. Evaluation of various techniques for extraction of natural colorants from pomegranate rind—ultrasound and enzyme assisted extraction. Indian J. Fibre Text. Res. 35, 272.
Tiwari, P., Jain, R., Kumar, K., Mishra, R., Chandy, A., 2012. Antibacterial activity and physicochemical evaluation of roots of *Butea monosperma*. Asian Pac. J. Trop. Biomed. 2, S881–S883.
Tsatsaroni, E., Liakopoulou-Kyriakides, M., Eleftheriadis, I., 1998. Comparative study of dyeing properties of two yellow natural pigments—effect of enzymes and proteins. Dyes Pigm. 37, 307–315.
Tutak, M., Benli, H., 2011. Colour and fastness of fabrics dyed with walnut (*Juglans regia* L.) base natural dyes. Asian J. Chem. 23, 566–568.
Vankar, P.S., Shanker, R., 2008. Ecofriendly ultrasonic natural dyeing of cotton fabric with enzyme pretreatments. Desalination 230, 62–69.
Vankar, P.S., Shanker, R., Mahanta, D., Tiwari, S.C., 2008. Ecofriendly sonicator dyeing of cotton with *Rubia cordifolia* Linn. using biomordant. Dyes Pigm. 76, 207–212.
Vankar, P.S., Shanker, R., Verma, A., 2007. Enzymatic natural dyeing of cotton and silk fabrics without metal mordants. J. Clean. Prod. 15, 1441–1450.
Viuda-Martos, M., Fernández-López, J., Pérez-Álvarez, J.A., 2010. Pomegranate and its many functional components as related to human health: a review. Compr. Rev. Food Sci. Food Saf. 9, 635–654.
Wagner, H., Geyer, B., Fiebig, M., Kiso, Y., Hikino, H., 1986. Isobutrin and butrin, the antihepatotoxic principles of *Butea monosperma* flowers. Planta Med. 2, 77–79.
Yadava, R., Tiwari, L., 2007. New antifungal flavone glycoside from *Butea monosperma* O. Kuntze. J. Enzyme Inhib. Med. Chem. 22, 497–500.
Yusuf, M., Ahmad, A., Shahid, M., Khan, M.I., Khan, S.A., Manzoor, N., Mohammad, F., 2012. Assessment of colorimetric, antibacterial and antifungal properties of woollen yarn dyed with the extract of the leaves of henna (*Lawsonia inermis*). J. Clean. Prod. 27, 42–50.
Yusuf, M., Shahid, M., Khan, M.I., Khan, S.A., Khan, M.A., Mohammad, F., 2015. Dyeing studies with henna and madder: a research on effect of tin (II) chloride mordant. J. Saudi Chem. Soc. 19, 64–72.
Yusuf, M., Shahid, M., Khan, S.A., Khan, M.I., Islam, S.-U., Mohammad, F., Khan, M.A., 2013. Eco-dyeing of wool using aqueous extract of the roots of Indian madder (*Rubia cordifolia*) as natural dye. J. Nat. Fibres 10, 14–28.
Zarkogianni, M., Mikropoulou, E., Varella, E., Tsatsaroni, E., 2011. Colour and fastness of natural dyes: revival of traditional dyeing techniques. Color. Technol. 127, 18–27.
Zhang, B., Wang, L., Luo, L., King, M.W., 2014. Natural dye extracted from Chinese gall – the application of color and antibacterial activity to wool fabric. J. Clean. Prod. 80, 204–210.
Zhou, Y., Zhang, J., Tang, R.-C., Zhang, J., 2015. Simultaneous dyeing and functionalization of silk with three natural yellow dyes. Ind. Crops Prod. 64, 224–232.

Sustainable silk production

6

T. Karthik, R. Rathinamoorthy
PSG College of Technology, Coimbatore, India

6.1 Introduction

Clothing is one of the primary needs of human beings for survival. The textile industry represents the second largest industry after agriculture in India, and hence there are huge demands from consumers with respect to quality, features and green products. However the use of chemicals has increasingly become a subject of environmental concern (Karthik and Gopalakrishnan, 2012a,b,c,d). Therefore, it is essential to develop regulations designed to improve health, safety and the human and natural environments. Living in an environmentally conscious era, consumers have demanding needs in many areas. Because they are aware of environmental issues, nowadays consumers look for eco-friendly products (Karthik and Gopalakrishnan, 2013a,b,c). People have started to search for 'green products' everywhere. Consumer preferences in the purchase of clothing have changed from looking for comfort, aesthetics and style to more sustainable and eco-friendly features (Gardetti and Torres, 2013a,b). As the demand for sustainable products have increased on the consumer side, the textile industries have begun to require eco-friendly fabrics. Several factors, such as the renewability of the raw material, the carbon footprint of resources involved during the production and the product's use, and the use of chemicals to grow and process the materials, are involved in making a product sustainable.

Sustainability can be defined as the property of biological systems to remain diverse and productive indefinitely (Brundtland and Khalid, 1987). Some examples of sustainable biological systems are long-lived forests and healthy wetlands. Sustainability is also known as the endurance of systems and processes. Sustainable development is the organizing principle for sustainability. It includes four domains which are interconnected: ecology, economics, politics and culture (Defra (Department for Environment Food and Rural Affairs), 2008). The study of sustainable development and sustainable science is called sustainability science. Sustainability has other definitions as a socio-ecological process characterized by the pursuit of a common ideal. Sustainability is managed and studied over many levels of space and time and in many contexts of environmental, social and economic organization. If an ecosystem is healthy, it can provide vital goods and good services to every human and other organism (Gopalakrishnan and Karthik, 2012; Fletcher, 2008).

In the first part of this chapter, general aspects related to silk, it types and its conventional way of manufacturing are elaborated upon. Furthermore, sustainability issues related to the production of silk filaments and sericulture sustainability issues such as sustainability in mulberry tree cultivation and the conventional method of silk extraction are briefly discussed. In the second part of the chapter, the sustainable

Sustainable Fibres and Textiles. http://dx.doi.org/10.1016/B978-0-08-102041-8.00006-8

production of silk such as coloured silk, illuminant silk and ahimsa silk are discussed. The sustainable production of materials based on silk materials such as sustainable sensor from silk, nano-imprinting, inkjet printing, chemical-free metallic pattering, sustainable silk material functionalization, optics and photonics applications of silk, and electronics and optoelectronics applications of silk are also discussed in detail. In addition, the aspects of sustainability of spider silk are dealt with briefly.

6.2 Silk

Silk is a vital luxury fibre and raw material for the production of diverse sustainable products. It is the wonder of the haute couture set because of its lavish feel and drape. There are several indigenous ranges of wild silk found in different countries. Among the various silk varieties, the great secrecy and China's dominance over its production and promotion are based on a species of silk known as *Bombyx mori* (Scott, 1993; Rheinberg, 1992). Silk filament is a distinctive and significant category of natural structural proteins which exist in nature. It provides an extensive range of native functions which include protective membranes used to withstand all environmental insults during development (Kaplan, 2014). It has a high-strength web used to capture insects and lifelines to support spiders. Epithelial cells in specialized glands produce silk proteins after biosynthesis, followed by discharge into the lumen of these glands where the proteins are stored before they are spun into filaments.

With remarkable structure and versatility, silk has emerged as a topic of study because of its chemical, physical and biological properties. Silk production starts from polymers produced from amino acid building blocks. Peptide bonds link the 20 different chemical building blocks. To achieve structural function, fibrous protein requires interchain as well as intrachain interactions (Mishra, 2005).

6.2.1 Types of silk

The different kinds of silk are shown in Fig. 6.1. Mulberry silk, called *B. mori*, is very common among all kinds of silk. Mulberry silk itself takes into account about 90% of the global silk supply. *B. mori* silkworms produce this popular kind of silk; they

Figure 6.1 Types of silk: (a) *Bombyx mori*; (b) tussah; (c) muga; (d) eri; (e) mussel; and (f) spider silk.

feed on the mulberry bush and it is abundantly available in countries such as China, Korea and Japan. To maintain its smooth texture special care has to be given, which is a major disadvantage of this kind of silk. In general, silk filament is extracted from the silkworms in an unethical way, by killing them in the cocoon stage to extract it (Mishra, 2005).

The next type is tussah silk, which is produced by tussah silkworms. This kind is special because it has a distinct light golden to dark brown colour. This property results from the consumption of tannin-rich leaves by tussah silkworms. The thick fibre nature of this tussah silk is used for couches, jackets and sweaters.

Muga has a character similar to tussah silk. It is a kind of semidomesticated silkworm, *Antheraea assamensis*. It is from the same family to which tussah silk belongs. Golden brown colour and a glossy texture are the special characters of muga silk, whose origin is in India. Eri and pat silk together with muga are called Assam silk because these silkworms were kept for the exclusive application of royal families in Assam, India, for 600 years. Their production is considered eco-friendly because they do not require delicate care. A disadvantage of their silk is its porosity because of its limitation toward bleach, but that also means that it can be dyed (Scott, 1993; Rheinberg, 1992).

Eri silk is made from *Philosamia rinini*, which is the domesticated silkworm. Its colour is almost as white as the *B. mori* and it also is fine silk. It has a special name, called 'peace silk'. This special name comes from the fact that even though eri silk is spun from the cultivated silk's cocoon, its caterpillars are not killed in the cocoon stage to produce the filaments; they are allowed to become moths to live a full life cycle. Eri silk is not reeled; rather, it is spun. In comparison with the *B. mori*, eri silk is the only completely domesticated silkworm, so it depends on human intervention to be developed. Its durability makes it a great material for clothing and a surprisingly soft furnishing including for couches and curtains. Its only limitation is that it is heavy to wash. It may act as a shelter for microorganisms because it is easy for microorganisms to stick to its thick layers.

Mussel silk is produced by mussels which are found on sea beds; hence it is also called sea silk. It is different from all of the silk mentioned previously because it is not produced by silkworms. It is difficult to source because of pollution and is considered to be the most expensive silk because of its rare production. Byssus is the most common type of mussel (Nguku, 2010).

Spider silk is similar to mussel silk and was long used by ancient cultures. Like silkworms spiders cannot be bred, so the silk cannot easily be produced. Although its production is less and seems difficult, its output is certainly worth the effort. Its special property of durability has significant role in the production of telescopes, wear-resistant clothing and bulletproof vests.

6.2.2 Silk production process

The life cycle of the silk production process is shown in Fig. 6.2.

In a specially prepared paper, silk moths lay eggs. The feeding of mulberry leaves to caterpillars takes place, and then the eggs hatch. The caterpillars are normally heavier by about 10,000 times compared with their weight when they are 35 days old

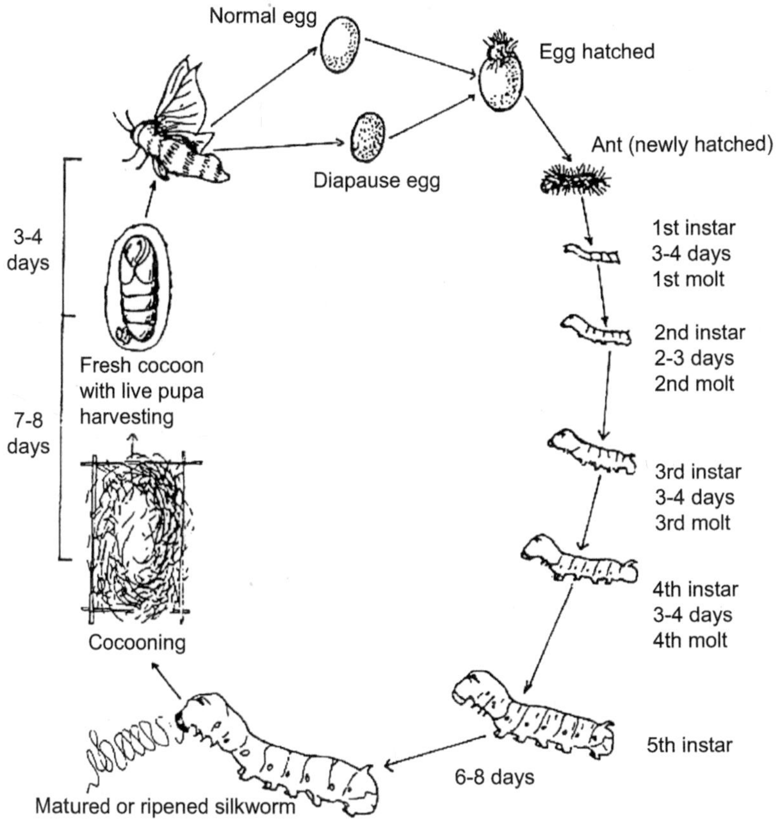

Figure 6.2 Life cycle of the mulberry silkworm.

after hatching and four moultings. The silkworms are sensitive to external conditions such as noise and temperature fluctuations or smells. When the caterpillar is fully grown it is called a pupa. It climbs a stick and starts to spin a cocoon by moving its head in a pattern. The silkworm produces a silk fibroin protein through two salivary glands known as sericteries, which are mixed in the silkworm's mouth with a gummy substance called sericin that is forced out as compound liquid through an opening in the lips (Mishra, 2005; Franck, 2001). As the liquid comes into contact with air, it solidifies and produces a solid continuous filament. The silkworm spins around 1.5 km of filament over 2–3 days and becomes completely enclosed inside a cocoon. When the conversion is finished, the newly formed moth exudes alkali-based fluids which break a hole in the cocoon so that it can come out. Before this can happen, the cocoons are heated to kill them, leaving some amount of transmutation in the moth so that it can breed the next generation of caterpillars.

The cocoons are then soaked in boiling water to mitigate the hardness of the sericin which holding the silk filaments in the shape of a cocoon. To produce a continuous filament the fibres are unwound and three to 10 filaments are spun together for commercial applications to obtain the required strength. The sightless, toothless and flightless moth will mate instantly after coming out of cocoon and lay about 500 eggs during the first 4 or 5 days, and then die (Franck, 2001).

The extraction of cultivated silk as well as the life cycle of the wild silkworms is sustainable in nature itself. Furthermore, when produced by handloom weavers the fabric has almost a zero energy footprint and the framework for sustainable silk is satisfied. A major concern regarding silk production related to sustainability is that the silkworm is killed during production. Domesticated silks are normally produced by boiling the unbroken cocoons and unwinding the silk strand onto reels. For the purpose of breeding, few *B. mori* silk moths are retained during silk production so that more cocoons can be produced.

6.2.2.1 Sustainability issues in silk production

Silk is a renewable material with low environmental impact compared with other fibres. The silkworms feed on mulberry leaves, which do not warrant much application of pesticides or fertilizers to grow. The main sustainability issue in silk production is the killing of the larvae when the cocoon is boiled to extract the filament during the sericulture process; this is criticized heavily by many animal welfare and rights activists. The mulberry silkworm has been completely cultivated and cannot live without humans for their care and feeding. *B. mori* evolved into a blind moth which cannot fly and lives only a few days during which it lays about 500 eggs; afterward it dies within 4 or 5 days. To avoid hurting the insects, Gandhi promoted cotton spinning machines, which do not directly involve hurting any living thing. He also promoted wild silk made from cocoons of wild and semiwild silk moths for those who do not want to wear silk produced by killing silkworms. The organization People for the Ethical Treatment of Animals also campaigned against silk for similar reasons.

Furthermore, processing and transportation of silk lead to energy conservation and sustainability issues. Cleaning of silk necessitates several harsh and intensive treatments and chemicals, which could pollute water. To produce a small quantity of silk requires a large amount of resources. For example, to produce 35 pounds of silk, it takes 1 acre of mulberry trees. In addition, it is a labour-intensive process. Also, mulberry trees have shallow root systems so they are easily blown down. Then it is necessary to cut the tree down and apply an herbicide to kill the roots. The roots of the trees are aggressive and strong; they can cause problems for drainage pipes. The timber is also relatively useless, other to make paper.

6.2.3 Sustainable silk

By nature, the extraction of domesticated silkworms and the life of wild silkworms are sustainable. When produced by weavers on handlooms, silk fabric has a near-zero energy footprint and satisfies most of the guidelines for sustainable silk production. A concern has been that commercial silk kills the silkworm. Conventional silk is made by boiling the intact cocoons and unwinding the silk strand onto reels. Sustainable clothing refers to fabrics derived from eco-friendly resources such as sustainably grown fibre crops or recycled materials. It also refers how these fabrics are used. Silk is the protein fibre spun by silk moth larvae, most commonly *B. mori*, to make the cocoon. For breeding purposes some moths are retained in commercial silk production so more cocoons can be constructed. It is also a biodegradable fibre, which is one of the most important parameters for sustainability. It can go on to produce useful mulch or compost, and hence soil, instead of sticking around for the next 500 years.

6.2.3.1 Organic silk

Organic silk is a more eco-friendly and sustainable way to harvest silk, because there is no need for chemicals or any treatments to produce raw silk. The silkworms are normally permitted to live their full life cycle and die naturally. In this silk production, the silkworms produce cocoons when they are 35 days old and are allowed to follow their natural cycle to morph into the moth stage. After that, they lay eggs and dies naturally about 5 days later. Although it is possible to produce natural coloured silk, the range of colours is limited (Franck, 2001).

6.2.3.2 Peace silk

In the case of conventional silk cultivation, a larger number of silkworms have to be killed to extract the silk filament. A single silk filament about a mile long is extracted from a cocoon, but as the caterpillar emerges from the cocoon, it is broken and ultimately destroys the quality of silk thread. Hence, the cocoons are kept in boiling water to kill the silkworms so that they do not pierce and damage the cocoon when they emerge from it (Cruelty-free silk or peace silk, 2011). In the case of peace silk cultivation, silkworms are not killed in the extraction process, to permit the silkworms to emerge from their cocoons. The silk thread is generally extracted after the silkworm has completed its transformation (metamorphosis) stage and comes out of the cocoon as a moth. The broken cocoons are then collected and the broken sections of silk threads are mended using a spinning process. Every cocoon is inspected to make sure that the moth has emerged before the silk thread is spun.

Because the continuous silk filament secreted by the silkworms is broken as they pierce the cocoon, the length of filament is much smaller less compared with conventional silk. Thus they are degummed to remove the sericin and are spun in the spinning system as with other natural fibres. Because the alkaline solution secreted by the moth to make a hole in the cocoon discolours the silk slightly, it is generally weaker than conventional silk and has a considerably different look and feel compared with *B. mori* silk. Although it is a time-consuming operation to cultivate the silk, the quality of silk is softer and finer compared with conventional silk. Peace silk is more costly because it consumes more time and skill to recover and spin the broken cocoons and fewer people are trained in this type of sericulture (Ahimsa (Peace) Silk; Alicia silverstone).

6.2.3.3 Wild silk

B. mori silk fabrics are generally named wild silks or peace silks because the caterpillars are permitted to live their natural lives in the wild without being sacrificed to make fabric. Other wild silk, tussah or tussur, is also 100% natural; it is organically produced by silkworms that complete their full life cycle. It is also earth-friendly and saves the land from deforestation. The wild silkworms spin a silk that is diverse in colour and texture compared with cultivated *B. mori* (Syiem).

Eri silk is another type of silk cultivated from *P. rinini*. It is a fine silk that is almost as white in colour as the *B. mori* silks. Although it is spun from cultivated silkworms, it is regarded as a peace silk because the caterpillars are not killed in the cocoon and are allowed to emerge as moths and live a full life cycle.

6.2.3.4 *Matka silk*

Rough handloom silks are generally called 'Matka silk' in India. They are produced from thicker yarns spun from damaged cocoon in the weft and organzine in the warp. The threads are obtained from the short ends of silk from mulberry silkworms and hand spun without removing the sericin gum. The silk is strong and lustrous and promotes the use of peace silk and sustainable processes (Syiem; Karthik et al., 2015).

6.2.3.5 *Recycled and vintage silk*

A vintage silk fabric free of stains and rot could be a great investment. When properly handled and cared, silk can last for thousands of years. Vintage silk fabrics provide the perfect platform for recycling. Because silk fabrics rarely degrade, it is the perfect fabric to for reuse and recycling (Karthik et al., 2015).

6.3 Life-cycle assessment of silk production

Sericulture involves the production of cocoons, and then the extraction of the silk filament and its conversion into fabric by the industry. Owing to the social aspects of sericulture and because it is a low-environmental impact process, it contributes to the sustainable development of the nation. Among the various aspects of sustainability, the environmental issue has become vital in the industrial sector. Of the different environmental impact assessment techniques, the life-cycle assessment (LCA) is considered a good tool for assisting in strategic planning and decision making (Barcelos et al., 2013).

An LCA includes the environmental aspects and impacts of all stages involved in the life cycle of a product, from raw materials to the landfill stage; thus it gives a global vision of the product system. An LCA assesses the environmental impacts caused by the particular product through its life cycle, as per the International Organization for Standardization 14040 (2009) standard in four different phases such as the purpose and definition of scope, analysis of inventory, impact assessment and interpretation of data and results (Vollrath et al., 2013). The process flow in the manufacturing of silk fabric is given in Fig. 6.3.

Various factors involved in the production as well as the mass and energy flow in the production of silk are shown in Figs 6.4 and 6.5, respectively. Most energy consumption involving raw silk production is shared between cocoon production (47%) and

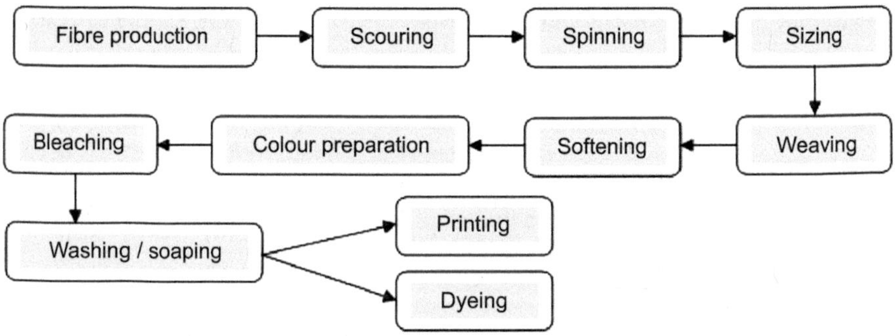

Figure 6.3 Process flow in the manufacturing of silk fabric.

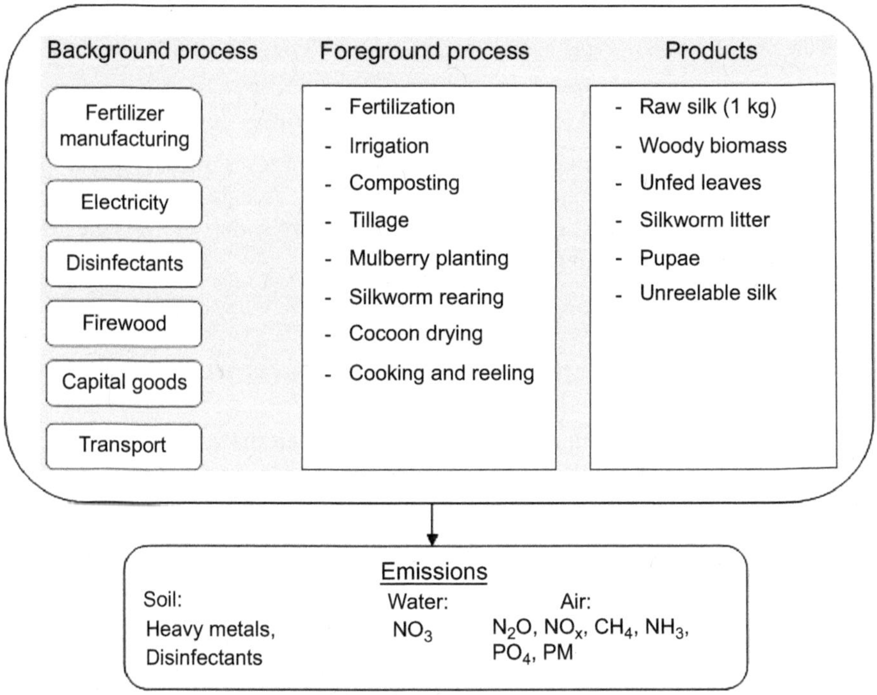

Figure 6.4 System boundaries in life-cycle assessment of silk production. Reproduced with permission from Elsevier.

cocoon cooking (51%). Similar to the other agricultural products, the major energy costs of cocoon production are incurred in the use of fertilizers and irrigation. The quantities of any pesticides used (under 0.04 kg/kg raw silk) make negligible contributions to embodied energy calculations despite their being mainly organophosphates. The energy consumption pattern of irrigation is around 32% of the embodied energy across various irrigation systems.

The average electricity used in irrigation is about 3130 kWh/ha/year. Silk reeling is the most energy-consuming process in the raw silk production process. The temperature

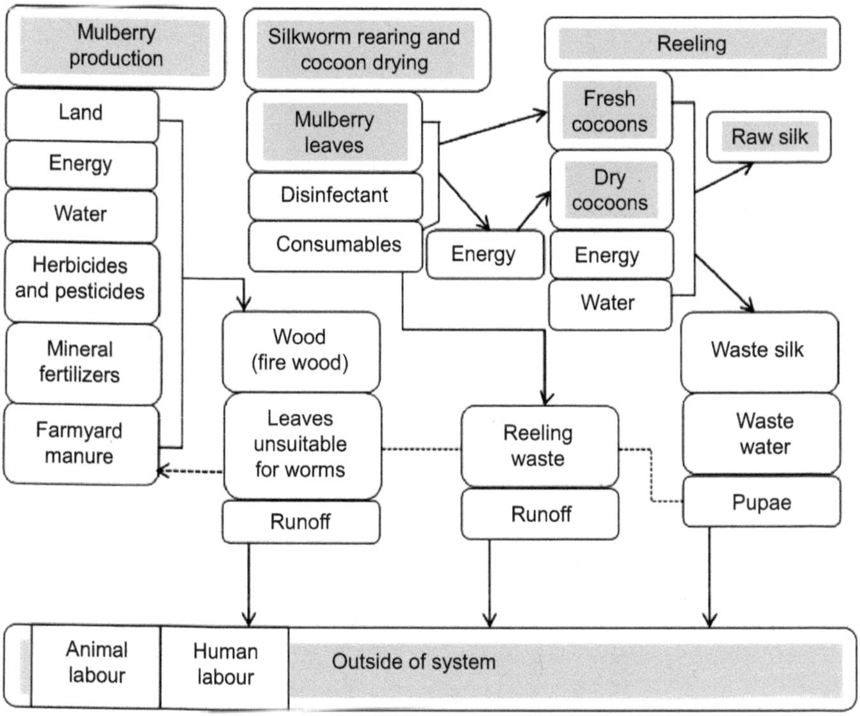

Figure 6.5 Mass and energy flow in the production of raw silk.
Reproduced with permission from Elsevier.

and hardness of water and the pH are important factors in silk reeling. The buoyancy property of cocoons needs more time to submerge in a wire mesh so that it will become soft. The time taken for softening varies around 10 min at temperatures between 60 and 95°C. The degree to which the wood has dried out and its calorific value are important factors which decide the cost of cooking. Two major factors related to the sustainable production of silk are: (1) developing energy efficient technologies and machines for silk production, and (2) optimizing subprocesses such as fertilizer use, irrigation systems and cocoon-rearing processes (Vollrath et al., 2013).

The environmental assessment studies on silk production conducted by Astudillo et al. (2014) are provided in Table 6.1 and Fig. 6.6. The figure shows that most of the environmental impact of sericulture is the result of cocoon production. Embodied solar energy from mulberry leaves and the wood used during the cocoon processing is responsible for renewable cumulative energy demand (CED). On the other hand, electricity used for cocoon drying and irrigation are the main sources of nonrenewable CED. The fertilization process and emissions from composting are major contributors to the global warming potential, and the use of pesticides influences impacts of eco-toxicity.

Fig. 6.6 shows that urban land occupation is insignificant compared with agricultural land occupation. Nearly the entire blue water footprint is a result of the irrigation of mulberry leaves. Main sources of freshwater eutrophication potential are field emissions (50%) and the manufacture of phosphorus fertilizers (18%) (Astudillo et al., 2014).

Table 6.1 Environmental impact of silk compared with other fibres

Category	GWP (kg CO_{eq}/kg)	Renewable CED (MJ/kg)	Nonrenewable CED (MJ/kg)	Eco-toxicity (CTU_e/kg)	ALO (m^2a/kg)	ULO (m^2a/kg)	BWF (m^3/kg)	FE (gP_{eq}/kg)
Farm practices	80.9	1613.6	244.4	1043.1	35.6	1.37	54.0	7.0
Recommended practices	52.5	1350.6	116.7	522.9	19.8	1.13	26.7	4.8
Cotton	3.4	19.7	0.1	71.2	7.8	0.02	7.0	0.8
Nylon 66	8.0	1.3	7×10^{-4}	6×10^{-4}	2×10^{-4}	4×10^{-4}	0.2	0.3
Wool	18.5	81.7	0.1	3.4	53.5	0.36	0.2	0.5

ALO, agricultural land occupation; *BWF*, blue water footprint; *CED*, cumulative energy demand; *FE*, freshwater eutrophication; *GWP*, global warming potential; *ULO*, urban land occupation.
Reproduced with permission from Elsevier.

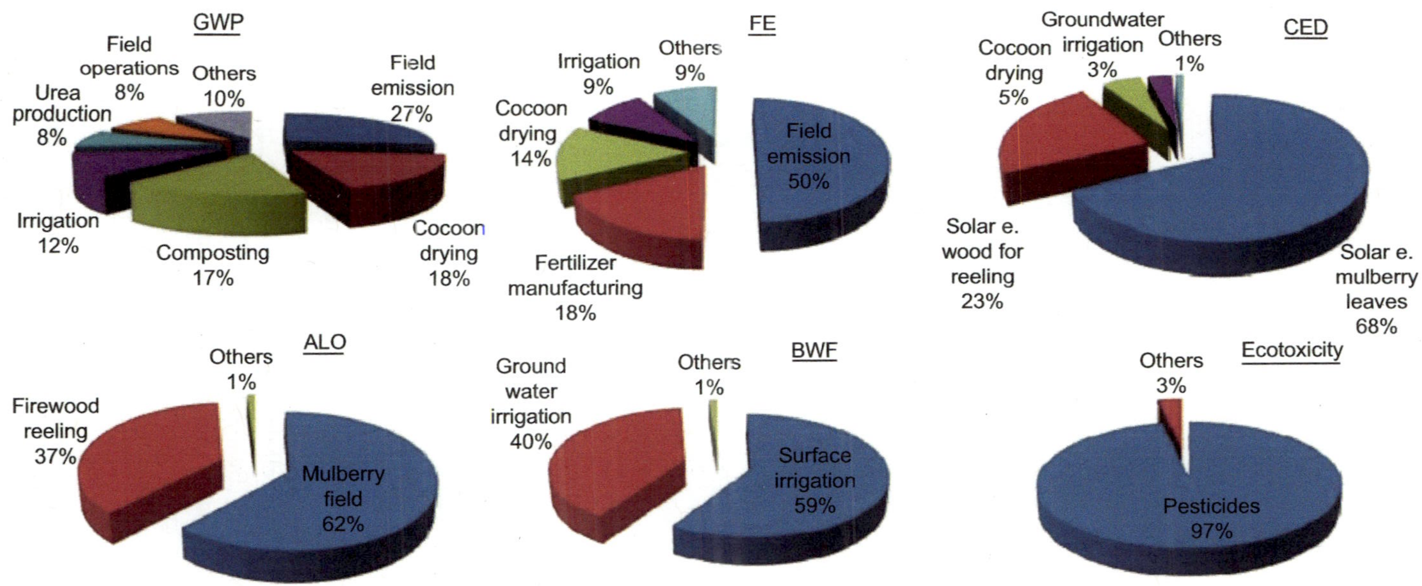

Figure 6.6 Environmental impact of various processes in silk production. *ALO*, agricultural land occupation; *BWF*, blue water footprint; *CED*, cumulative energy demand; *FE*, freshwater eutrophication; *GWP*, global warming potential.
Reproduced with permission from Elsevier.

Animal fibres and animal husbandry normally require higher inputs that plant production and generate a larger amount of co-products. The off-farm processing of silk is complex compared with that for other animal fibres. If these co-products are insufficiently valorized, the result is almost complete attribution of the total impact to reeled silk.

6.4 Sustainable production of mulberry leaves

Organic silk farming promotes the sustainability of mulberry trees, which are the main food source for the silkworms. One acre of mulberry trees could provide 30–35 pounds of raw silk. Mulberry trees are a precious renewable resource for the production of baskets, furniture and even folk remedies (Das et al., 2005).

As a continuous agronomical practice, once they are planted in bimonthly intervals, mulberry trees are maintained for several years. Because of short-term results and for the sake of economy, chemical-based inputs are much preferred. Besides having an adverse effects on mulberry trees' users (silkworms), which is their natural enemy, and complex, beneficial microorganisms, the residues of the chemicals (fertilizers/weedicides/insecticides/fungicides) used in the mulberry garden pose a potential risk of environmental pollution. To avoid the indiscriminate use of chemicals in mulberry gardens, the promotion of organic farming is needed.

6.4.1 *Integrated organic nutrient management practices*

The physical, chemical and biological properties of soil health are improved by organic manure. The water-holding capacity of sandy soils can be improved, while it facilitates aeration and filtration in heavy soils. In alkaline soil the power of the nutrient supply is increased by reducing its pH, making the activities of beneficial microorganisms more fertile besides in addition to adding its own nutrient values (Govindan et al., 2005). To overcome the problem of residues from pesticides, fungicides, herbicides and other heavy metals in the agro-ecosystem, the buffering nature of organic matter is considered advantageous.

6.4.1.1 Farmyard manure

Farmyard manure (FYM) is prepared simply by storing cow and buffalo dung and the droppings of sheep, goats, poultry, etc., along with leftover fodder in pits for a few months for decomposition. Well-decomposed FYM contains approximately 0.5% N, 0.2% P_2O_5, and 0.5% K_2O (Balakrishna et al., 2011). In sericulture, soil test-based FYM application is highly appreciable. The dosage application based on organic carbon in FYM is given in Table 6.2.

6.4.1.2 Compost

Compost is a product of the biodegradation of organic wastes; it is available in different sources such as crop residues, green and dry leaves, weeds, poultry waste and urban waste, and waste from agro-based industries. The principle behind

Table 6.2 **Farmyard manure distribution**

Soil organic carbon (%)	Dosage (MT/ac/year)	Number of splits
<0.35	12	3
0.35–0.65	10	2
0.65–1.00	8	2

composting is to narrow the C:N ratio, aiming for the total destruction of harmful pathogens and weed seeds by high temperature evolved during decomposition and stabilization. Aerobic decomposition will result in end products including carbon dioxide and the release of available plant nutrients and humic substances. Compared with FYM, composting has rich nutrient value (Balakrishna et al., 2011; Anonymous, 2014a).

6.4.1.3 Vermicompost

The bio-oxidation of organic waste which involves the joint action of earthworms and microorganisms is called vermicomposting. In this process, earthworms act as versatile bioreactors converting organic materials into fine granules called vermicast (excreta of earthworms). Vermicompost is rich in plant nutrients, enzymes, antibiotics, plant growth hormones and large beneficial microbial populations which increase the quality and yield of mulberry leaves suitable for the higher productivity of silk (Anonymous, 2014b). Application of 10 MT of vermicompost with a 50% decrease in the application of dose of chemical fertilizers (NPK)/ha/year could produce a mulberry leaf yield equal to the application of 20 MT FYM/ha/year with the recommended dose of chemical fertilizers (Anonymous, 2014c).

6.4.1.4 Poultry manure

Poultry manure is normally used after decomposition and is a better source of organic nutrients. Poultry droppings have phosphorus (2.46%–2.82%), nitrogen (4.55%–5.6%), calcium (4.52%–8.15%) and magnesium (0.52%–0.73%) and appreciable quantities of micronutrients such as Cu, Zn, Fe and Mn. This can be decomposed with suitable organic amendments such as chopped paddy straw or coir pith by adding fungal inoculums of *Pleurotus sajor-caju* (Anonymous, 2014d).

6.4.1.5 Sheep and goat manure

The nutrient content is higher in sheep and goat manure than in farmyard manure and compost. The composition of sheep and goat manure includes 3% N, 2% K_2O and 1% P_2O_5. Application of droppings of sheep and goats is done through two different methods. In the first method, the decomposed droppings of sheep or goats are applied in pits in the field where nutrients in urine reside. In the second method, goats or sheep are kept overnight in the field and urine and faecal matter are incorporated into the soil at a shallow depth using a working harrow or cultivator blade (Anonymous, 2014c).

Figure 6.7 Green manure in mulberry plants with (a) sunn hemp and (b) view of root nodules.

6.4.1.6 Green manure

Green manure can be defined as the practice of incorporating soil with undecomposed green plant tissues to improve its physical structure as well as fertility. It is the best alternative to FYM because the scarcity of FYM and its high cost owing to the decline in livestock farming (Anonymous, 2014c; Chandra, 2005). Green manure crops are grown in interspersed rows of mulberry plants and incorporated into the soil at the preflowering stage because they are grown only for their biomass, which is high in organic nature owing to its ability to fix nitrogen from the air with the help of its root nodule bacteria. Green manure in mulberry plants with sunn hemp is shown in Fig. 6.7.

6.4.1.7 Bio-fertilizers

Bio-fertilizers promote growth by developing and raising the supply or availability of primary nutrients to the plant. Bio-fertilizers are considered as an excellent alternative to chemical fertilizers. Some commonly used bacteria such as *Azotobacter chroococcum* and *Azospirillum* spp. are capable of fixing atmospheric nitrogen. However, *Azospirillum* has an additional role of secreting growth-promoting substance into the soil, inducing disease resistance and drought tolerance in the plants. Similarly, another bacterium, *Bacillus megaterium*, known as phosphorus solubilizing bacteria, is also capable of solubilizing insoluble phosphorus and making it available to the plants (Raja, 2010).

6.4.1.8 Vesicular arbuscular mycorrhizae inoculation for new mulberry plantation

Vesicular arbuscular mycorrhizae (VAM) is a symbiotic fungus also called endomycorrhizae that has a harmonious relationship with mulberry roots (Fig. 6.8). A soil-based inoculum of VAM with a mixed culture of *Glomus mosseae* and *Glomus fasciculatum* is generally used to raise mycorrhizal mulberry saplings in the nursery. Phosphorus can be used at 60 kg/ha/year instead of 120 kg/ha/year after 1 year of practising it. This method ensures a 50% reduction of phosphorus application in the cultivation of mulberry with no loss in the yield or quality of leaves.

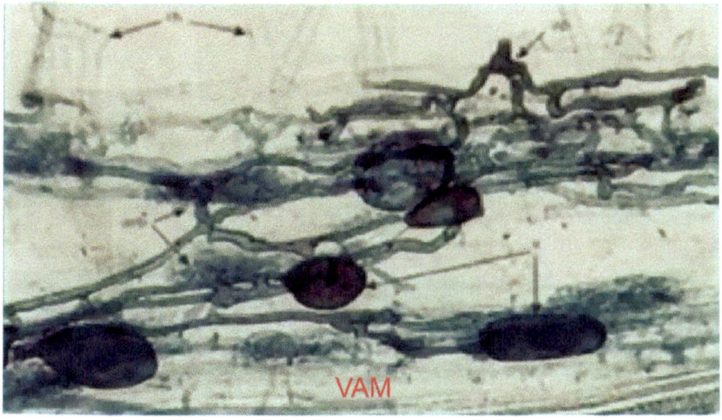

Figure 6.8 Vesicular arbuscular mycorrhizae (VAM) in mulberry roots.

6.4.1.9 Neem oil cake

Inedible oil cakes in general and neem oil cake in particular contain a high amount of plant nutrients and alkaloids which induce immunity against diseases and pests in mulberry in addition to a higher nutrient content than other oil cakes, among different organic sources (Chandra, 2005).

6.4.2 Foliar fertilization

6.4.2.1 Vermin-wash

Vermin-wash is a collection of the secretion of earthworms and excretory products along with micronutrients from soil organic molecules; it is used as a foliar spray. It is composed of plant growth hormones such as auxins and cytokinins together with nitrogen, potash, phosphorus and other micronutrients (Balakrishna et al., 2011). Adult worms of same species measuring 100 g are collected, released into a container with 50 mL lukewarm water (37–40°C) and agitated for 2 min. Then earthworms are taken out and washed in 50 mL water at room temperature for 2 min and released back into tanks. Then 1 L vermin-wash is diluted with 4–5 L water and sprayed as a foliar spray during the late evening hours. A mixture of vermin-wash (1 L) with cow urine (1 L) in 10 L water acts as a bio-pesticide cum liquid manure (Ravikumar et al., 2011).

6.4.2.2 Panchagavya

Panchagavya is an organic product which has a major role in promoting growth and immunity in the plant system. It is prepared using five cow products: cow dung (7 kg), cow urine (10 L), milk (3 L), curd (2 L), ghee (1 kg) and other ingredients, including jiggery (3 kg), well-ripened Ney Poovan banana (1 dozen), tender coconut water (3 L) and plain water (10 L) (Balakrishna et al., 2011).

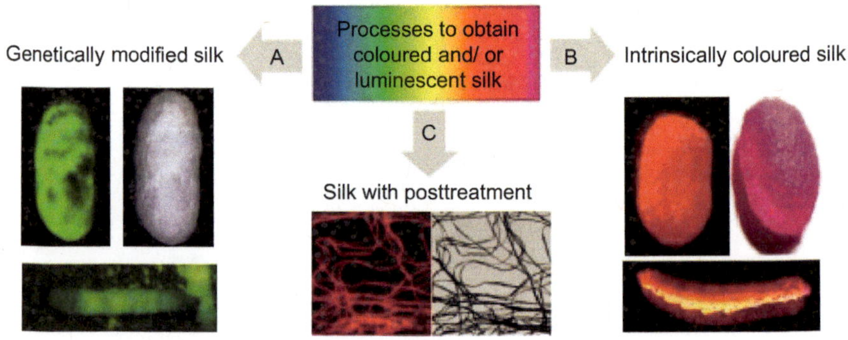

Figure 6.9 Various techniques to produce coloured silk.
Reproduced with permission from Wiley.

6.5 Sustainable dyeing of silk

One of the most polluting industries today is the dyeing of textile fibres. To reduce this pollution there is a need to develop green processes. To apply colour to silk fabric, fibroin is subjected to a dyeing process which involves posttreatments such as washing and drying. This consumes a huge quantity of water, chemicals, dye and electricity (Dandin et al., 2005). There has been research to develop greener methods to manufacture coloured silk. There are three techniques to produce coloured silk, as shown in Fig. 6.9. Besides the natural colour of silk, its luminescence is useful in biomedical applications, where it permits better visualization, monitoring and quantification. Beyond the absorption of dyes to produce colour and luminescence, this concept can be extended to integrate functional molecules such as in drugs, antibiotics, perfumes and nutrients into the silk with therapeutic or nutritional value (Tansil et al., 2011a).

6.5.1 Naturally coloured silk produced by wild silkworms

Silkworms are normally fed with carotenoid-rich mulberry leaves during the larvae stage. In the case of wild silkworms, carotenoid pigments are first absorbed into the epithelia of guts and then transferred to the hemolymph, which are accumulated in the silk gland, resulting in yellow cocoons. These pigments are integrated into the sericin; hence removal of sericin during degumming results in white silk fibroin. Silkworms have been developed to produce white cocoons instead of yellow ones owing to their defect in the transfer of carotenoid (Sangappa et al., 2007). Domesticated silkworms are the largest producers of silk. Colour or luminescence can be added into the resulting white silk for various applications.

6.5.2 Intrinsically coloured and luminescent silk by genetically modified silkworms

Various genetic engineering methods have been explored to optimize the efficient and stable genetic transformation. Numerous studies have been done related to the insertion of fluorescent protein genes into cultivated silkworms to investigate the use of silkworms as bioreactors for the invention of recombinant proteins (Zhao et al., 2008). The fluorescent proteins primarily serve as markers rather than the final end product. Currently, silk and its cocoons with green, orange and red luminescence have been reported.

6.5.3 Intrinsically coloured and luminescent silk by cultivated silkworms using a dye-containing diet

The researchers have developed a green technique to produce colour silk in which the silkworm directly spins a coloured silk cocoon after being fed a diet modified with a dye formulation. The biological incorporation of dyes into silk fibroin is an eco-friendly technique to produce coloured silk because it does not need a further dyeing process later, or the resources (water, energy or additional chemicals) or posttreatments associated with it (Tansil et al., 2011b).

Research toward producing naturally coloured silk started in the early 20th century. A sequence of research demonstrated that orcein in the silkworm's diet was taken up into the silk cocoons whereas other dyes were absorbed by the silkworm guts and migrated into the silkworm's eggs. Several kinds of dyes, such as Berlin blue, gentian violet, safranin, Nile blue, methylene blue, indigo carmine, fuchsin, alizarin and methyl orange, were experimented with and did not appear in silk (Tansil et al., 2012a; Method for Producing Colourful Plane Cocoon and the Produced Colourful Plane Cocoon, 2007; Tolly, 2006; Nisal, 2014).

An emerging technology that can potentially clean up the dyeing of silk fibres widely used for textile application would involve the generation of intrinsically coloured silk cocoons. Naturally coloured silks are produced from silkworms that have natural pigments to produce the colour. Some wild silkworms, such as *Antheraea pernyi*, *Antheraea yamamai* and *Saturniidae*, are known for producing coloured silk cocoons (brown, green and golden yellow). A few varieties of cultivated silkworms (*B. mori*) have also been reported to spin cocoons in a variety of colours such as pink, flesh, salmon, yellow, yellow-green and green. This is accomplished by feeding a modified diet of mulberry leaves composed of a sprayed dye solution for the *B. mori* silkworm (Method for Producing Colourful Plane Cocoon and the Produced Colourful Plane Cocoon, 2007; Tolly, 2006). The production of this method is significantly restricted by the physiochemical properties of the dyes, such as molecular weight, hydrophobic-hydrophilic balance and partition coefficient.

Commercial silk fibre produced by the mulberry silkworm is basically white. Some strains of silkworms can produce pink, yellow, brown, or green silk. Several attempts have been made to develop greener methods to produce coloured silk fabrics. The use of nontoxic natural dyes is included so that the use of hazardous chemicals is minimized. The modified feed is given to silkworms at their fifth instar stage, as shown in Fig. 6.10. The modified feed is composed of mulberry leaves that are dipped into or sprayed with the dye solution, or it can be powdered mulberry leaf feed into which the dye solution is mixed. The large quantities of dye-containing effluent water are eliminated in this process (Nisal, 2014). The dye is taken to the biochemical pathways

Figure 6.10 (a) Fifth instar *Bombyx mori* larvae feeding on mulberry leaves sprayed with direct acid Fast Red dye solution. (b) Silkworm larvae spinning cocoon on a mesh (*chandrike*). (c) Coloured cocoon shells after the larvae have been taken out. (d) Twists of direct acid Fast Red threads after degumming.
Reproduced with permission from ACS Publications.

of the silkworm to produce the coloured cocoon. By controlling the concentration of dye in the feed, the intensity of colour can be controlled.

The special kind of feeding was prepared as a thick green paste prepared by microwaving mulberry powder (100 g, Recorp Inc., Canada) with water (300 mL). Modified feed was prepared by thoroughly mixing an individual dye (50 mg) into normal feed (100 g). Domesticated silkworms (*B. mori*) were cultivated up to the second day of the fifth instar with conventional feed and then were fed the modified feed containing various fluorescent dyes from the third day of the fifth instar until their cocoons were completely spun. These coloured cocoons were then reeled to obtain silk for further processing.

Naturally coloured silk has good lustre and a soft texture similar to conventional silk. The colour of the silkworm is chiefly the result of the flavonoid compounds, but the flavonoids of mulberry leaf should go through several chemical modifications in the silkworm, *B. mori* L. (Lepidoptera: Bombycidae), such as glycosylation.

Tansil et al. (2012b) studied the uptake of dyes such as Rhodamine B into the silk fibroin by feeding of dye-containing diet to silkworms and studied the effect of the molecular properties on the biological inclusion of different molecular materials into fibroin. With this method, the dye is primarily taken up by the fibroin rather than sericin, as in the case of wild silk. They studied various kinds of fluorescent dyes,

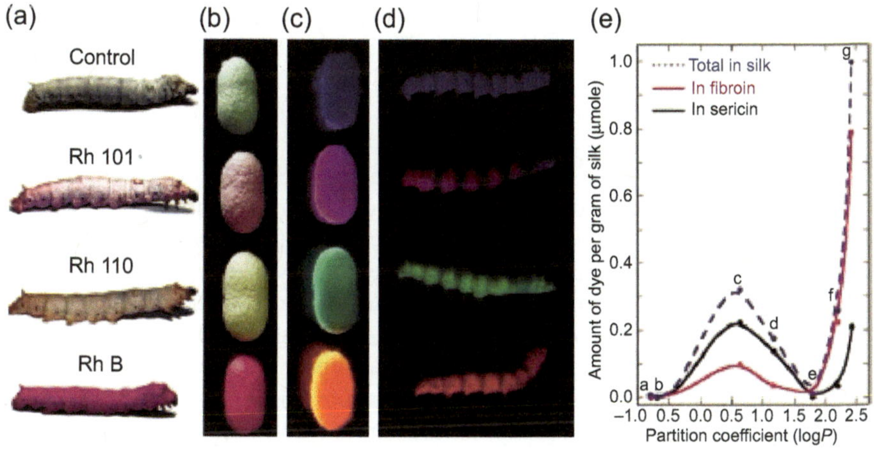

Figure 6.11 Intrinsically coloured and luminescent silks obtained with dye-containing diet. (a) Larvae and cocoons. (b) Colour cocoons viewed under room light. (c and d) Colour cocoons and larvae viewed under UV radiation. (e) Concentration of dye.
Reproduced with permission from Wiley.

such as fluorescein sodium, Rhodamine 116, sulfo Rhodamine 101, Rhodamine 110, Rhodamine 101, acridine orange and Rhodamine B in the silkworm's feed to manufacture intrinsically coloured and luminescent silks (Fig. 6.11).

The study revealed that the uptake of dyes into silk fibroin was determined mainly by their structure-dependent hydrophobicity and self-assembly capability. The feeding method adopted in this technique could be applied to other functional materials to manufacture silk with superior properties (Method for Breeding Color Silkworm as Pet, 2007; Juige et al., 2003; Ma et al., 2016; Zhua and Zhangb, 2014). Dye quantifications in cocoons indicated that lipophilicity is an important parameter governing the association of dye with the two different proteins present in the silk gland. This behaviour is attributed to the fact that sericin is the hydrophilic protein whereas fibroin is more hydrophobic in nature. As the lipophilicity in the dye molecule increases, it preferentially associates with the hydrophobic protein, i.e., fibroin. Dyes with molecular weights more than 400 g/molecule cannot diffuse out of the peritoneal membrane of the alimentary canal. The partition coefficient of the dye also determines the preferential association of dye with either sericin or fibroin protein in the silkworm gland and finally into the cocoon. These insights are extremely important in the development of novel dye molecules that can be used successfully in the search for a green method for producing coloured silk fabrics (Chen et al., 2011).

6.5.4 Extrinsically coloured silk through postprocessing

The conventional method of dyeing using synthetic dyes needs solvents, which could result in microstructural changes in the silk filaments and ultimately affect the tensile strength. Furthermore, dyeing with synthetics dyes and chemicals could lead to environmental pollution. Hence efforts have been made toward developing natural dyes as well as techniques to remove dyes and effluents using advanced methods.

The immobilization of inorganic nanoparticles in silk fibres has a huge potential for practical applications. Silver nanoparticles could be added to silk fibres externally to achieve antibacterial functions. Furthermore, the addition of titanium and silver-coated titanium dioxide nanoparticles exhibits greater anti-UV radiation, antibacterial, and self-cleaning characteristics when coated on silk (Cai et al., 2015).

UV resistance in silk can be incorporated inherently by in vivo uptake of titanium dioxide (TiO_2) nanoparticles by silkworms. Because of the better interactions between TiO_2 and silk fibroin molecules, the nanoparticles can easily be incorporated into the silk gland of silkworms. The modified silks had higher mesophase content, lower crystallinity and higher Herman's orientation functions of the crystalline and mesophase regions than the control group. By adding 1% nano-anatase into the artificial diet, properties such as breaking strength and elongation of the modified silk could be improved up to 548 ± 33 MPa and $16.7 \pm 0.8\%$. UV-resistant properties were obtained by adding 1% TiO_2. This type of insertion of metallic elements could enhance the mechanical properties of certain parts of organisms by incorporating into their inner structure, possibly forming single chelation. It was demonstrated that the toughness of artificial silk spun from regenerated silk fibroin solution was improved by titanium dioxide (TiO_2). Silk fibres could be used as a template to aid in the formation of nanoparticles. The addition of nanoparticles on the surface of silk led to concerns about durability for applications. It is therefore crucial to develop methods that could ensure the much stronger integration of additives within silk fibres before a reliable functional silk can be commercialized (Cai et al., 2015; Chequer et al., 2011).

Transformation silk in other forms has been researched for many years. Research has been reported on the production of regenerated silk fibroin by dissolving it with phosphorus acid, hexafluoroacetone and hexafluoro-2-propanol as well as by self-dialysis. After the dissolution of silk fibroin, additives can be added to the solution to produce functional silk. In addition to proteins, dyes could be added into the silk fibroin solution to prepare dye-doped silk fibroin in the form of nanoparticles or thin films as biosensors.

6.6 Sustainable silk materials

6.6.1 Sustainable material fabrication from silk

Silk fibroin in the form of an aqueous solution can be physically and chemically activated by mixing various organic and/or inorganic dopants into the solution. These doped solutions can be deposited on appropriate materials, and hence the solution can be crystallized through silk fibroin self-assembly upon exposure to air, producing a class of patterned freestanding films (e.g., optical elements) or a biocompatible, robust and bioresorbable substrate for thin-film photonic and electronic devices (Vepari and Kaplan, 2007).

6.6.1.1 Sustainable sensor from silk

The major challenge in the design and development of sustainable methods and techniques concerns the overall aspects of the material's life cycle, such as a greener method of synthesis and fabrication and the use of nontoxic materials; it should degrade to harmless products, as shown in Fig. 6.12. The production of three-dimensional photonic crystals from silk materials provides new ways to design coloured as well as structured biodegradable materials and implantable bio-sensing and targeted therapeutic applications (MacLeod and Rosei, 2013). Photonic crystals produced from cultivated *B. mori* cocoons have the advantage of being produced in abundant quantity compared with production from spider silk fibroin from the dragline of *Nephila clavipes*. The fibroin is extracted from *B. mori* cocoons using sodium carbonate and lithium bromide as reagents.

The extracted silk fibroin aqueous solution is poured over a self-assembled colloidal crystal of closely packed poly(methyl methacrylate) (PMMA) spheres and left to dry so that the PMMA is dissolved, providing an inverse lattice of fibroin. The mixture of inherent structural colour with the biocompatible characteristics of silk fibroin provides silk inverse opals, an excellent contender for application in implantable bio-sensing applications. The biodegradable silk fibroin revealed promising applications in various areas such as for bone growth and environmental sensing, after being functionalized for sensitivity to certain gases (Omenetto and Kaplan, 2008).

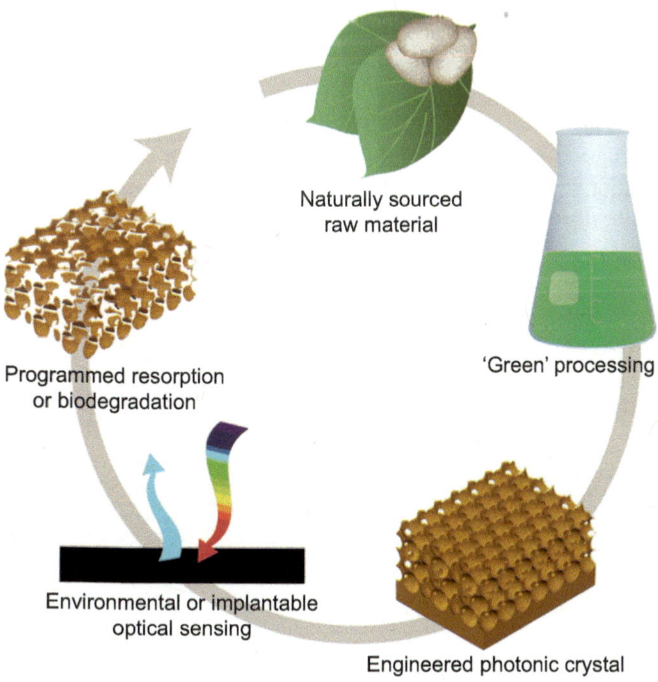

Figure 6.12 Sustainable sensors production from silk fibroin.
Reproduced with permission from Nature Publishing Group.

6.6.1.2 Casting and spinning

An aqueous solution of fibroin could easily be transformed to robust and thermodynamically stable beta sheets of nanometre thickness. Silks are suitable for soft lithography owing to their processing and self-assembly attributes. By casting the silk fibroin in an aqueous form, for any patterned substrates, the topographies can be simulated to characteristic sizes (<10 nm) in the hardened, freestanding silk film (Lawrence et al., 2009). Based on this technique, different kinds of photonic structures could be fabricated, such as diffractive materials, prisms, holograms and multilens arrays.

6.6.1.3 Nano-imprinting

The nano-imprinting of silk films could be done by changing the glass transition temperature (Tg), which is based on its water content and absorbed moisture. Silk films prepared at a humidity of 35% have a Tg of about 100°C, which permits nano-imprinting using a hot embossing process (Tao et al., 2012; Pan et al., 2014; Amsden et al., 2010; Park et al., 2007). The silk films are pressed between 30 and 50 p.s.i. and heated at 100°C on a master with nano-scale patterns for few seconds and then cooled to make nano-imprinted materials, as shown in Fig. 6.13. After imprinting, annealing of silk films could be done by exposing them to methanol, which causes a change in the secondary structure of fibroin from a random coil to a beta-sheet, to make them insoluble in water (Mondia et al., 2010).

6.6.1.4 Inkjet printing

The silk fibroin aqueous solution is used to manufacture optical films can also be used as a starting material for microfluidic printing. Among various possibilities for application, inkjet writing to fabricate silk optical waveguides can be explored (Parker et al., 2009). Silk fibroin inks are treated with polyethylene glycol (15%) for about 14 h

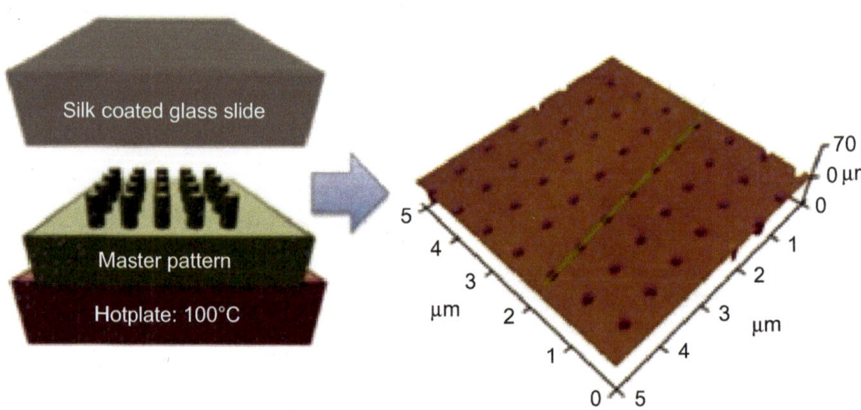

Figure 6.13 Nano-imprinting processes for silk fibroin.
Reproduced with permission from Wiley.

to produce a concentrated silk fibroin ink of (28–30 wt% silk) to attain the viscosity needed for high-quality structures. The printed silk waveguides retained rod-like morphology by crystallization in a methanol-rich coagulation reservoir (Jin et al., 2005).

6.6.1.5 Chemical-free metallic pattering

Using a shadow masking deposition approach, patterning metals on silk has been demonstrated. To deposit metals onto silk films in a dry, chemical-free environment, microfabricated stencils are used. This prevents chemical contamination that might be involved in other photolithography-based metal patterning methods such as lift-off processes and wet etching. By patterning microscale resonators on silk films, the effectiveness of this approach was demonstrated (Tao et al., 2010a). With this method we can gain the advantage of maintaining the integrity and bio-compatibility of the silkworms, which making the resulting materials suitable for incorporation into biological environments.

6.6.2 Sustainable silk material functionalization

To interface ideally with both technological and natural worlds, silk-based materials appear to be suited. The ability to store functional compounds, both organic and inorganic, is considered one of the unique traits of these materials because the inherent block copolymer structure of silk fibroin provides a model to sustain and preserve the biochemical function of biological components such as enzymes and antibodies within the material's structure.

6.6.2.1 Silk as a biologically active substrate

Biological stabilization aids in the complete use of silk material. This functionality of silk films which exist in the bulk material is coupled with the micro- and nano-patterned optical features to function as transducers and monitors or enhancers of the entrained biological functions (Tao et al., 2010b). An example of the transformation of an optical component into an interactive optical component is demonstrated by mixing the blood of a human with silk fibroin solution and then casting or imprinting it into a blood-based diffracted grating structure, as shown in Fig. 6.14.

The haemoglobin in the silk grating remains active and reacting to the oxygenation state of the external atmosphere. This silk–haemoglobin grating was used as the building block of a simple spectrometer using the diffractive characteristics of the doped silk as the transduction mechanism for the oxygenation levels of the entrained haemoglobin.

6.6.2.2 Nanoparticle-doped silk

Silk fibroin films doped with gold nanoparticles could be used as light-activated heating elements which are connected with a thermoelectronic mechanism to power micro-equipment and devices wirelessly (Xu et al., 2007; Perry et al., 2008). Gold-doped silk fibroin films were interfaced to thermally available thermoelectric chips by

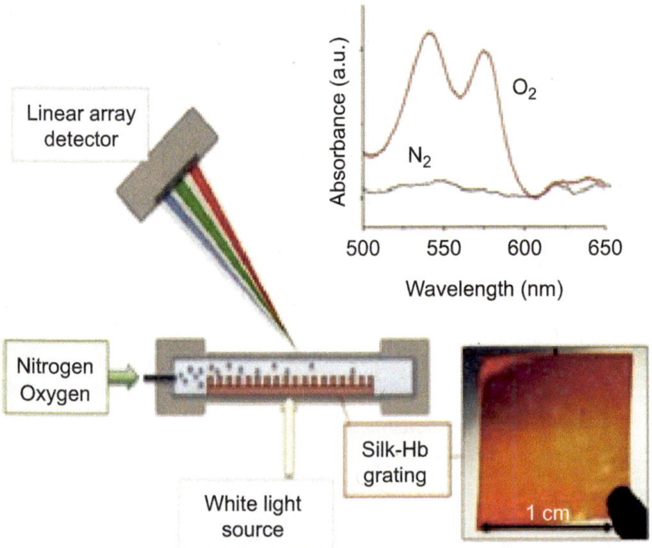

Figure 6.14 Interactive diffracted grating from haemoglobin (Hb)-doped silk grating. Reproduced with permission from Wiley.

casting the solution on the active area of the chip. A continuous wave green laser with an output power up to 450 MW/mm^2 at 532 nm to match the absorption peak of the gold nanoparticles induces an increase in temperature in the silk layer, which in turn is used to generate 20 MW of power (Mills et al., 2006; Boriskina et al., 2010).

6.6.2.3 Fluorescent silk with emission enhancement

Directional and wavelength-specific fluorescent enrichment can be obtained by silk fibroin films doped with fluorophores when they are nano-imprinted with periodic lattices. The environmentally friendly, biodegradable silk fibroin solution allows the incorporation of organic and or inorganic dyes to be used in biomedical applications (Tsioris et al., 2010).

6.6.2.4 Silk microfluidics

Silk can be used as a base substrate for microfluidic devices such as siloxane-based elastomers or glass owing its ability to reform the materials into thick films. Chemical modification is generally carried out to increase the functionality of silk. Chemical bonds available within the fibroin structure are exploited, namely tyrosine residues (>5% of the total amino acids) or acid side chains (1% of the total amino acids). By uniting the utility of facile chemistry to the convenience of a fluid platform that has gained much attention for the convenience of its use in diagnosis, chemically activated microfluidic devices are manufactured (Bettinger et al., 2007; Lu et al., 2010). Hybrid devices which combine azo-benzene functionality within the silk fibroin films and polydimethylsiloxane microfluidics were shown to generate pH-reactive platforms.

6.6.3 Optics and photonics applications of silk

6.6.3.1 Photonic lattices

The success of soft lithography as well as nano-imprinting techniques on silk fibroin films enables more sophisticated photonic components to be produced at the nanoscale. Silk fibroin film-based nano-patterned two-dimensional lattices which were structurally coloured were demonstrated by imprinting periodic lattices of nanoscale holes which display diverse colours as a function of varying lattice spacing under white light illumination (Adato et al., 2009).

6.6.3.2 Metamaterials

Manufacturing of high-quality structures can be explored because of the surface properties of silk, by exploiting microfabrication techniques. Using this technique, metamaterial structures with a larger surface area can be patterned on silk fibroin substrates which could exhibit good resonance response at terahertz frequencies (Wang et al., 2007). Any change in the capacitance or inductance could lead to a change in the resonant response, making metamaterials sensitive to the environment. Thus, these materials could provide an option for incorporation with silk fibroin films for sensing and detection applications by offering a natural as well as biologically active material as the support matrix for split ring resonators.

6.6.3.3 Silk integrated plasmonics

Silk fibroin could be used to make conformal coatings on various kinds of materials with accurate control over the different length scales, ranging from monolayers to the mesoscale. Very thin layers of silk coatings could be achieved dip-coating, spray coating or spin coating the solution (Boriskina et al., 2010; Adato et al., 2009). Creation of precise tunable layers of conformal coating on different surfaces could be used for sensing applications. The capacity of silk fibroin films to stabilize biochemical dopants permits expanded detection schemes exploiting the selectivity and responses of these biological components as active components in these engineered colorimetric chips. Subsequently this approach could push sensitivity limits comparable to surface plasmon sensors within a simpler, cost-effective approach that can cover microfluidics, bioassays and label-free detection of biological methods.

6.6.4 Electronics and optoelectronics applications of silk

6.6.4.1 Silk as an electronic material

The characteristics necessary for photonic applications are mechanical robustness, transparency and surface flatness. Silk fibroin possesses all of these required features, which makes it an ideal product for technological devices such planar microelectronic structures (Wang et al., 2011; Capelli et al., 2011). Silk biomaterials are being explored for the production of various kinds of electronic materials such as low-cost electronic paper devices and organic thin film transistors, and to replace inorganic oxide layers including SiO_2 and PMMA.

6.6.4.2 Mechanically adaptable substrates

The opportunity to tune the mechanical characteristics of silk films by controlling film hydration is an important characteristic of silk films. It is feasible to control the rate of dissolution and phase-transition dynamics by controlling this property by exposing it to either enzymes or water. These kinds of electronics which are capable of being in close contact with the curvilinear surfaces of biological tissues provide significant prospects for monitoring, diagnosing and treating injury or disease (Wang et al., 2005).

6.6.4.3 Bioresorbable electronics

Because of the complex nature of biological responses to many organic and inorganic materials, the task of achieving biocompatibility can be challenging. Electronic devices should be designed from biodegradable materials which can also resorb, such that only small, safe components remain, minimizing adverse biological responses. Silk is considered an attractive biopolymer compared with others such as poly(lactic acid) and collagen, because of its better mechanical properties, the ability to modify the rate of biodegradation, the generation of noninflammatory amino acid degradation products, and the opportunity to set up the materials under ambient conditions to protect sensitive electronic functions (Muller et al., 2011; Kim et al., 2010).

6.7 Sustainable spider silk

Humans have used spider silk as a textile material for a long time before it came into focus as a subject of research. Even during ancient times, natural cobwebs were used to seal bleeding wounds and their threads were used for fishing. Later, spider silks were also used for military purposes, in particular to construct cross-hairs. The range of application of spider silk is the result of its good mechanical properties, smoothness, and biocompatibility compared with other available materials (Gole and Kumar; Karthik and Subramani, 2016; Heim et al., 2009).

Spider silk generally has good mechanical properties compared with cultivated silk, because in the case of cultivated silk the silkworm uses the filament for protection during metamorphosis, whereas spiders use their silk filament to grab the prey. In the 1950s, spider silk, specifically their dragline, was main focus for researchers and scientists owing to its exceptional mechanical properties; it can outperform natural or synthetic fibres. Spider silk is a biological protein fibre in which the hierarchical structure exhibits weak hydrogen bonds at the core, regulating material behaviour on multiple lengths (Karthik and Subramani, 2016; Hisa et al., 2011).

6.7.1 Spider silk production

Silkworms and spiders require less than a second to convert the concentrated and viscous solution of protein into an insoluble solid fibre. Normally silkworms excrete the silk through salivary glands located in the head region, whereas spiders produce silk in the abdominal glands. The main interest of research belongs to spiders which produce silks from orb webs, which are used to grab prey. Female orb-weaving spiders can produce about seven types of silks with different characteristics. Spider silks are characterized by mixture of structures, chemistry and functions, from orb-web structure to adhesives and cocoons (Gole and Kumar; Kang, 2014; Kubik, 2002). The main limitation in the application of spider silk filaments is the difficulty in producing it on a large scale. Efforts to imitate spider silk have found many novel techniques and materials such as genetic engineering, stem cell concepts and recycled silk.

6.7.1.1 Artificial biosynthesis

Direct extraction of spidroins from spiders is not feasible for commercial production. Because the spiders are cannibals and territorial, they normally produce a small quantity of silk, and hence the production of spidroins by biotechnology is considered most efficient and economically viable to supply spidroins. Researchers have invented a procedure and method for the synthetic production of liquid silk spidroin using other organisms. A few of these techniques are reviewed subsequently.

Chimeric silkworms

Generally, silkworms do not reveal cannibalism and territorialism the way spiders do, so they are suitable for mass production. This method of cultivating silk is less expensive to scale up for products, and the hosts are naturally equipped to spin silk fibres; hence less work is needed to process the assembled fibres (Hisa et al., 2011). One basic constraint with this technique is that silkworms will produce endogenous silk fibroin, and hence the output product is a combination of silkworm as well as spider silk. Furthermore, the composite fibres exhibit higher toughness compared with cultivated silkworm fibres, almost close to that of dragline silk fibres. Another downfall is the low level of the incorporation of chimeric spider silk fibres in the overall product.

Transgenic goats

Mammal cells have also been used as a host to produce spider silk monomers. Research in this field was first investigated by Nexia Biotechnogies, which took genes from spider dragline silk, flanked these with regulatory sequences, and then inserted them into mammary gland cells, which are responsible for the production of milk in female goats. As a result, the spider silk proteins were produced only in the milk of lactating goats (Gole and Kumar).

Metabolically engineered *Escherichia coli*

Use of *Escherichia coli* as an expression system to produce spider silk proteins of similar molecular weight and mechanical properties as native spider silk was reported. The vector used was a plasmid, and it contained regulatory sequences to promote the transcription of the gene, a His-tag, to aid in purifying the protein, as well as the spider silk gene sequence. This specific spider silk gene sequence resulted in proteins which were really high in glycine. Thus, to ensure efficient silk protein production, the glycyl-tRNA pool was elevated by inserting a second compatible plasmid into the bacteria, which effectively resulted in overexpression of the glycyl-tRNA (Scott, 2014).

6.7.1.2 Artificial synthesis of spider silk

Injecting the spidroins in appropriate hosts such as bacteria, plants or mammals seems to be attractive from an environmental point of view, because recombinant spidroins correspond to renewable feedstock. On the other hand, based on the host selected, it also has several environmental consequences, such as water and energy consumption, agricultural land or rearing requirements, and the need for physical or chemical treatments of waste and by-products to prevent health hazards (pathogenic, toxic or allergenic contamination).

To facilitate the artificial method of synthesizing spider silk fibres, two major points to be taken into consideration are:

- the synthesis of the feedstock (the unspun silk 'dope' in spiders); and
- mimicry of the spinning apparatus (the funnel, spinning duct, clamp and spigot in spiders).

Synthesis of the spider silk 'dope' can easily be produced by extracting the spider silk gene and using genetically modified organisms to produce the spider silk, such as *E. coli*, caterpillars, silkworms and goats. However, the silk produced via this method is in liquid form owing to the water solubility of the spider silk proteins. The problem is mimicry of the spider's extrusion apparatus to spin a strong fibre. One main characteristic feature of spider silk spidroin is its higher concentration of protein (up to 50% w/v) in the dope with no aggregation, and it can be hastily reassembled into stable fibres when required. The evaluation of self-assembling characteristics and the solubility of recombinant spider proteins is vital to produce commercial spider silk fibres (Hisa et al., 2011; Kubik, 2002).

The main dissimilarity between silkworms and spiders is the way in which the spider spins its silk. It can be found by mimicking the way the spider does this. It requires finding a way to extrude silk monofilaments from a concentrated aqueous solution of silk proteins obtained directly from spiders or silkworms. This is done by a method called biomimetism. Spider silks or proteinaceous filaments that are secreted by specialized abdominal glands of Araneomorphae form one of the major hallmarks of biomimetism. Few spinning techniques have been developed to produce synthetic spider silk, such as microfluidics, electrospinning and the use of a syringe and needle.

Microfluidics is a method currently being developed. The goal is to have a controllable method to test spin very small volumes of unspun fibre, but development costs are expensive. Electrospinning is a production process using an electric field to control the deposition of fibres onto a target material or substrate. A number of proteins such as collagen, silkworm silk, elastin and fibrinogen have been transformed into nanofibres using the electrospinning method successfully. The electrospun protein-based nanofibres can be used in tissue engineering and to develop hemostatic bandages or wound dressings, biological scaffolds for membranes, etc. (Singha et al., 2012).

Spider silk threads produced in vitro are used to 'wet-spin' recombinant spider silk proteins. This is a intricate process which tries to imitate the natural process of spider silk production. The spinning solution is forced through tiny holes in a spinneret (or syringe) to extrude protein fibres. Spinning occurs in an aqueous environment and the resulting fibre is subjected to a subsequent stretching or drawing process. The properties of the resultant fibre depend largely on the spin bath, composition, temperature and concentration of the spinning solution and the force applied during spinning.

Natural fibre formation during the spider's spinning process partly depends on dehydration and protein alignment. 'Wet-spinning' is carried out by forcing a concentrated aqueous spider silk protein through a small needle-like arrangement to imitate the elongational flow and shear forces created by the spider's progressively narrow extrusion duct. The fibre coming from the needle is allowed to enter a dehydrating coagulant bath to reproduce the natural water extraction process. Under a natural spinning environment, spiders connect their dragline to an object with gum from the glands before drawing the silk out by moving away and using their weight to draw the silk. Hence, it is vital to monitor the rate of spinning (Xia et al., 2010). It can also be manufactured by mimicking the biological spinning process using a biomimetic endless fibre spinning machine which operates with conditions as similar as possible to those evolved in nature.

6.7.2 Sustainable aspects of spider silk

Spider silk-inspired materials may help us to make progress toward greater sustainability, in particular to replace oil-based polymers. However, obviously spidroins will not be an adequate substitute in all cases. Despite the finite nature of the resource and their ecological footprint, petroleum-based polymers may be indispensable to manufacture useful devices or materials with specific abilities that cannot be obtained with other compounds. Other abundant proteins may exhibit new or similar characteristics to spidroins, with comparable versatility.

6.7.2.1 Biological properties

Spider silk seems to have superior biocompatibility. Spider silk proteins have already been proven to be appropriate to form scaffolds and promote cell adhesion and development. Templates made of pristine spider silk have allowed the seeding and proliferation of fibroblasts, and the formation of nerve conduits and oriented nucleation of hydroxyl apatite crystals for bone implants. Scaffolds made of several protein constructs appeared to be suitable to support the growth of various cells, including fibroblasts, neural stem cells and chondrocytes (Scott, 2014; Xia et al., 2010). Spider silk-inspired materials show promising potential to replace tissues. The scaffold should exhibit controlled degradation to provide support for cell development and then be replaced by tissues. Resorbability may also be necessary for suture purposes: Surgery filaments must resist over periods of time, so the filaments must have a reasonably slow resorption rate, especially in cases where fatigue is critical, such as for tendon repair. Owing to its biological properties, spider silk is a particularly interesting raw material for various biomedical applications.

6.7.2.2 Potential advantages of spider silk-based materials

Despite various difficulties owing to its mechanical properties and the potential to produce eco-friendly materials, spider silk and its synthetic derivatives represent a promising path for research. Steady developments in the production of regenerated fibres suggest that scientists will succeed in making silk materials with improved properties and a lower ecological footprint. The advantages of spider silk with respect to its properties and aspect of sustainability are given in Fig. 6.15.

Benefits and shortcomings may depend on the chosen application and requirements. Many of these specifications will be determined by the hierarchical organization of the material, which in turn will be dictated by the protein sequence and the process by which it will be structured (Xia et al., 2010). Thus, the design of a protein intended to prepare a specific biomaterial should simultaneously incorporate various and perhaps antagonistic constraints such as appropriate mechanical, optical, biocompatible and biodegradable properties, while taking into account the ecological footprint of its

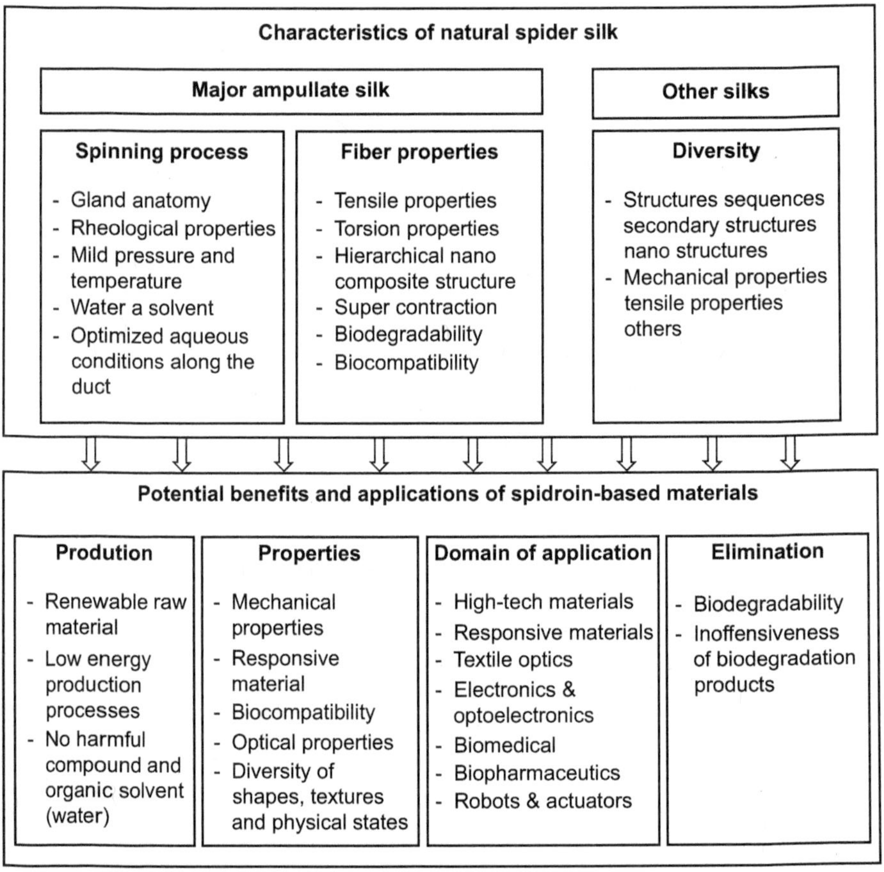

Figure 6.15 Benefits and applications of spider silk for spidroin-based materials.

life cycle. These requirements should be considered from the beginning of the design process of the material.

From a sustainable point of view, apart from having the required properties, silk-based products will have to minimize their environmental impact after a consideration of their complete life cycle, such as product durability, the production of recombinant spidroins, processing, transportation, use, maintenance and recycling, degradation into harmless compounds or disposal at the end of the lifespan. This is required predominantly for the production of recombinant spidroin using biotechnology, because it has several environmental concerns regarding water and energy consumption based on the host.

6.8 Silk in sustainable luxury fashion

Silk has always been connected with luxury and wealth. Silk makes individuals feel important, and when people wear a genuine silk material, their confidence soars. Silk is generally used in a range of fashion-related pieces such as skirts, dresses, scarves, pyjamas, blouses and lingerie.

Samant has adapted grainy, tactile peace silk to high-end couture, producing stylish *sherwanis*, boucle-like jackets and ethereal dresses. The waste incurred during harvesting and cocoon processing offered an opportunity for Samant to develop a new variety of silk with a much coarser weave yet a feel like wool, and it is used in men's jackets. Waste generated during the fabric cutting process becomes the textured layers of appliqué and decoration (Cocooned in luxury: Samant Chauhan's wild silk couture, 2016). In addition to being organic, an 'Ahimsa silk' produced from cocoons without killing the larvae is mostly used to make shawls and stoles. Syiem's initial challenge in the production of fashionable silk products is convincing weavers to supply the fabric in new dimensions. Because the fabric is dyed with vegetable dyes, the brand's colour palette is limited to shades of white and natural earthy tones, which could be a second challenge for the fashion industry. To complement the handlooms, machine-made embellishments or fasteners could be avoided, with a reliance only on drawstrings, knots and buttons made from natural materials.

If the principal concern from the consumer point of view is organic and sustainable silk, consideration can be given to products made from waste silk, muga or eri silks that could be dyed with natural or low-impact, fibre-reactive dyes. On the other hand, if the consumer's main concern is about the ethics of raising silk, consideration can be given to the production of wild silk, spun silk or eri silk, which do not destroy the silkworm to harvest the silk cocoons.

6.9 Conclusion and outlook

Sustainability can be achieved starting from cultivation to the end products. The development of sustainable, biodegradable materials that can be used in a technological manufacturing context could revolutionize the semiconductor and optoelectronics industry. The global implications for the conservation of energy and materials and the sustainable manufacture of optoelectronic devices for high-tech, medicinal and environmental applications are truly paradigm changing. Among the promising cadre of biopolymers, the particular attributes of silk as a sustainable, biocompatible material can provide an important step toward replacing existing photonic and electronic devices with environmentally benign, water-based systems that match the high performance of contemporary devices, while eliminating toxic by-products and energy-intensive fabrication methods. These approaches could be developed into an efficient and low-environmental impact process covering the cycle from material fabrication to disposal, at which point natural processes such as enzymatic degradation break the materials down into their natural building blocks.

References

Adato, R., Yanik, A.A., Amsden, J.J., Kaplan, D.L., Omenetto, F.G., Hong, M.K., Erramilli, S., Altug, H., 2009. Ultra-sensitive vibrational spectroscopy of protein monolayers with plasmonicnanoantenna arrays. Proc. Natl. Acad. Sci. U.S.A. 106, 19227.

Ahimsa (Peace) Silk. http://www.wormspit.com/peacesilk.htm.

Alicia Silverstone. Peace Silk: What Is It? http://thekindlife.com/blog/2012/09/alicia-silverstone-what-is-peace-silk.

Amsden, J.J., Domachuk, P., Gopinath, A., White, R.D., Dal Negro, L., Kaplan, D.L., Omenetto, F.G., 2010. Rapid nanoimprinting of silk fibroin films for biophotonic applications. Adv. Mater. 22, 1746.

Anonymous, 2014a. http://www.agritech.tnau.ac.in/org_farm/orgfarm_composting.html.

Anonymous, 2014b. Preparation of Composting Utilizing Sericulture Waste. http://agritech.tnau.ac.in/sericulture/seri_waste%20product.html.

Anonymous, 2014c. www.sugarcane.res.in/index.php/faqs/crop-production/322?showall=1.

Anonymous, 2014d. http://www.agritech.tnau.ac.in/org_farm/orgfarm_poultry.html.

Astudillo, M.F., Thalwitz, G., Vollrath, F., 2014. Life cycle assessment of Indian silk. J. Clean. Prod. 81, 158–167.

Balakrishna, R., Sakthivel, N., Samuthiravelu, P., Kirsur, M.V., Ravikumar, J., 2011. Comparative efficacy and economics of black polythene mulching, thermal and manual weeding in mulberry. In: Golden Jubilee National Conference on "Sericulture Innovations: Before and Beyond". Central Sericultural Research & Training Institute, Mysore; India.

Barcelos, S.M.B.D., Luz, L.M., Vasques, R.S., Piekarski, C.M., Francisco, A.C., 2013. Introductory background for life cycle assessment (LCA) of pure silk fabric. Indep. J. Manag. Prod. 4 (1).

Bettinger, C.J., Cyr, K.M., Matsumoto, A., Langer, R., Borenstein, J.T., Kaplan, D.L., 2007. Silk fibroin microfluidic devices. Adv. Mater. 19, 2847.

Boriskina, S.V., Lee, S.Y.K., Amsden, J.J., Omenetto, F.G., Dal Negro, L., 2010. Formation of colorimetric fingerprints on nano-patterned deterministic a periodic surfaces. Opt. Express 18, 14568.

Brundtland, G.H., Khalid, M., 1987. Our Common Future Report of the World Commission on Environment and Development. Oxford University Press, UK.

Cai, L., Shao, H., Hu, X., Zhang, Y., 2015. Reinforced and ultraviolet resistant silks from silkworms fed with titanium dioxide nanoparticles. ACS Sustain. Chem. Eng. 3 (10), 2551–2557.

Capelli, R., Amsden, J.J., Generali, G., Toffanin, S., Benfenati, V., Muccini, M., Kaplan, D.L., Omenetto, F.G., Zamboni, R., 2011. Integration of silk protein in organic and light-emitting transistors. Org. Electron 12, 1146.

Chandra, K., 2005. Organic Manures. Regional Centre of Organic Farming, Hebbal, Banglaore, India, p. 24.

Chen, X., Yang Liu, X., Han, M.-Y., 2011. Intrinsically colored and luminescent silk. Adv. Mater. XX, 1–4.

Chequer, F.M.D., Dorta, D.J., de Oliveira, D.P., 2011. Azo dyes and their metabolites: does the discharge of the azo dye into water bodies represent human and ecological risks? In: Hauser, P.J. (Ed.), Advances in Treating Textile Effluent. InTech Publisher, p. 2.

Cocooned in luxury: Samant Chauhan's wild silk couture. https://fashionethnographer/wordpress.com/2009/11/30/cocooned-in-luxury-samant-chauhan%E2%80%99s-wild-silk-couture.

Cruelty-free silk or peace silk, 2011. http://www.ecofashionworld.com/Glossary/Cruelty-Free-Silk-or-Peace-Silk.html.

Dandin, S., Kumar, N., KanikaTrivedi, S.S., 2005. Coloured silk: a preliminary report. Indian Silk 23–24.

Das, P.K., Bhogesah, K., Katiyar, R.S., Rajanna, L., Dassappa, Dandin, S.B., 2005. Sources of organic inputs for sustainable mulberry production. In: Govindan, R., Naika, R., Sannappa, B., Chandrappa, D. (Eds.), Progress of Research in Organic Sericulture and Seri-byproduct Utilization. Seri Scientific Publishers, pp. 11–17.

Defra (Department for Environment, Food and Rural Affairs), 2008. Sustainable Clothing Roadmap Briefing Note. December 2007 (updated March 2008) http://www.defra.gov.uk/environment/business/products/roadmaps/clothing/documents/clothing-briefing-Dec07.pdf.

Fletcher, K., 2008. Sustainable Fashion and Textiles. Earthscan, London.

Franck, R.R., 2001. Silk, Mohair, Cashmere and Other Luxury Fibres. Woodhead Publishing, UK.

Gardetti, M.A., Torres, A.L. (Eds.), 2013a. Sustainability in Fashion and Textiles: Values, Design. Production and Consumption, Greenleaf Publishing, UK.

Gardetti, M.A., Torres, A.L., 2013b. Entrepreneurship, innovation and luxury: the Any Savoirs Des Peuple case. J. Corp. Citizensh. 52, 55–75.

Gole, R.S., Kumar, P. Spider Silk: Investigation of Spinning Process, Web Material and Its Properties, Biological Sciences and Bioengineering, IIT Kanpur, India.

Gopalakrishnan, D., Karthik, T., 2012. Eco-friendly fibres – textile contribution to minimize global warming. Text Rev. 7 (9), 13–15.

Govindan, R., Naika, R., Sannappa, B., Chandrappa, D., 2005. Progress of Research in Organic Sericulture and Seri-byproduct Utilization. Seri Scientific Publishers.

Heim, M., Keerl, D., Schiebel, T., 2009. Spider silk from soluble protein to extraordinary fiber. Angew. Chem. Int. Ed. 48, 3584–3596.

Hisa, Y., Gnesa, E., Jeffrey, F., Tang, S., Vierra, C., 2011. Spider silk composites and applications. In: Cuppoletti, J. (Ed.), Metal, Ceramic and Polymeric Composites for Various Uses. InTech, Rijeka.

Jin, H.J., Park, J., Karageorgiou, V., Kim, U.J., Valluzzi, R., Kaplan, D.L., 2005. Water-insoluble silk films with reduced β-sheet content. Adv. Funct. Mater. 15, 1241–1247.

Juige, F., Wenrui, T., Tao, F., 2003. Technical Procedure for Producing Color Cocoon Silk of Silkworms by Use of Feeding Dye Coloring Matter. Chinese Patent 1430904A.

Kang, S., 2014. Biomimetics: Engineering Spider Silk. University of Southern California.

Kaplan, D.L., 2014. Kirk-othmer Encyclopedia of Chemical Technology. Wiley, New York.

Karthik, T., Gopalakrishnan, D., 2012a. Eco-friendly fibres of the future. Asian Text J. 21, 67–71.

Karthik, T., Gopalakrishnan, D., 2012b. Impact of textiles in environmental issues and environmental legislation, part – I. Text Rev. 7 (3), 14–21.
Karthik, T., Gopalakrishnan, D., 2012c. Impact of textiles in environmental issues and environmental legislation, part – II. Text Rev. 7 (4), 16–21.
Karthik, T., Gopalakrishnan, D., 2012d. Impact of textiles on environmental issues, part – I. Asian Dyer 9, 52–58.
Karthik, T., Gopalakrishnan, D., 2013a. Impact of textiles on environmental issues, part – II. Asian Dyer 9, 45–51.
Karthik, T., Gopalakrishnan, D., 2013b. Impact of textiles on environmental issues, part – III. Asian Dyer 9, 40–43.
Karthik, T., Gopalakrishnan, D., 2013c. Impact of textiles on environmental issues, part – IV. Asian Dyer 9, 43–46.
Karthik, T., Subramani, K., 2016. Sustainable bio-polymer fibers – production, properties and applications. In: Muthu, S.S., Gardetti, M.A. (Eds.), Sustainable Fibres for Fashion Industry, vol. I. Springer, New York, pp. 109–140.
Karthik, T., Rathinamoorthy, R., Ganesan, P., 2015. Sustainable luxury natural fibers—production, properties, and prospects. In: Gardetti, M.A., Muthu, S.S. (Eds.), Handbook of Sustainable Luxury Textiles and Fashion, vol. I. Springer, Newyork, pp. 59–100.
Kim, D.H., Viventi, J., Amsden, J.J., Xiao, J., Vigeland, L., Kim, Y.S., Blanco, J.A., Panilaitis, B., Frechette, E.S., Contreras, D., Kaplan, D.L., Omenetto, F.G., Huang, Y., Hwang, K.C., Zakin, M.R., Litt, B., Rogers, J.A., 2010. Dissolvable films of silk fibroin for ultrathin conformal bio-integrated electronics. Nat. Mater. 9, 511.
Kubik, S., 2002. High-performance fibers from spider silk. Angew. Chem. Int. 41 (15), 2721–2723.
Lawrence, B.D., Marchant, J.K., Pindrus, M.A., Omenetto, F.G., Kaplan, D.L., 2009. Silk film biomaterials for cornea tissue engineering. Biomaterials 30, 1299.
Lu, Q., Wang, X.Q., Hu, X., Cebe, P., Omenetto, F.G., Kaplan, D.L., 2010. Stabilization and release of enzymes from silk films. Macromol. Biosci. 10, 359.
Ma, M., Hussain, M., Dong, S., Zhou, W., 2016. Characterization of the pigment in naturally yellow-colored domestic silk. Dyes Pigm. 124, 6–11.
MacLeod, J., Rosei, F., 2013. Sustainable sensor from silk. Nat. Mater. 12, 98–100.
Method for Breeding Color Silkworm as Pet. Chinese Patent CN1319441C, 2007.
Method for Producing Colourful Plane Cocoon and the Produced Colourful Plane Cocoon. Chinese Patent CN101081017B, 2007.
Mills, C.A., Navarro, M., Engel, E., Martinez, E., Ginebra, M.P., Planell, J., Errachid, A., Samitier, J., 2006. Transparent micro- and nanopatterned poly(lactic acid) for biomedical applications. J. Biomed. Mater. Res. 76, 781.
Mishra, S.P., 2005. A Text Book of Fibre Science and Technology. New Age International Publishers, New Delhi, India.
Mondia, J.P., Amsden, J.J., Lin, D., Negro, L.D., Kaplan, D.L., Omenetto, F.G., 2010. Rapid nanoimprinting of doped silk films for enhanced fluorescent emission. Adv. Mater. 22, 4596.
Muller, C., Hamedi, M., Karlsson, R., Jansson, R., Marcilla, R., Hedhammar, M., Inganas, O., 2011. Woven electrochemical transistors on silk fibers. Adv. Mater. 23, 898.
Nguku, E., 2010. *Bombyx mori* Silk Fibre and Fabric Properties. VDM Verlag, Kenya.
Nisal, A., 2014. Uptake of azo dyes into silk glands for production of colored silk cocoons using a green feeding approach. ACS Sustain. Chem. Eng. 2, 312–317.
Omenetto, F.G., Kaplan, D.L., 2008. A new route for silk. Nat. Photonics 2, 641.
Pan, H., Zhang, Y.P., Shao, H.L., Hu, X.C., Li, X.H., Tian, F., Wang, J., 2014. Nanoconfined crystallites toughen artificial silk. J. Mater. Chem. B 2, 1408–1414.

Park, I., Cheng, J., Pisano, A.P., Lee, E.S., Jeong, J.H., 2007. Low temperature, low pressure nanoimprinting of chitosan as a biomaterial for bio-nano technology applications. Appl. Phys. Lett. 90, 093902.

Parker, S.T., Domachuk, P., Amsden, J., Bressner, J., Lewis, J.A., Kaplan, D.L., Omenetto, F.G., 2009. Biocompatible silk printed optical waveguides. Adv. Mater. 21, 2411.

Perry, H., Gopinath, A., Kaplan, D.L., Dal Negro, L., Omenetto, F.G., 2008. Nano-and micro-patterning of optically transparent, mechanically robust, biocompatible silk fibroin films. Adv. Mater. 20, 3070.

Raja, S.A., 2010. Management of powdery mildew, *Phyllactiniacorylea* (Pers.) Karst of mulberry (*Morus* sp.) using chosen biocontrol agents. J. Biopestic. 3 (2), 483–486.

Ravikumar, J., Ambiga, M., Radhakrishnan, S., Balakrishna, R., Sakthivel, N., Anbazhagan, R., 2011. Bio-efficacy of some botanicals against spiraling white fly, *Aleurodicusdispersus* (Russell) on mulberry. In: Golden Jubilee National Conference on "Sericulture Innovations:Before and Beyond". Central Sericultural Research & Training Institute, Mysore, India, p. 23.

Rheinberg, L., 1992. The Romance of Silk. Textile Institute, Manchester.

Sangappa, S., Dandin, S.B., Trivedy, K., Kumar, S.N., 2007. Coloured cocoons to coloured silk. Indian Silk 22–24.

Scott, P., 1993. The Book of Silk. Thames & Hudson, London.

Scott, A., 2014. Spider silk poised for commercial entry. Chem. Eng. News 92 (9), 24–27.

Singha, K., Maity, S., Singha, M., 2012. Spinning and applications of spider silk. Front. Sci. 2 (5), 92–100.

Syiem, D. Silk root. http://www.livemint.com/Leisure/V9is7nTLM0Qgb3JmVCd7oK/Silkroot.html?utm_source=ref_article.

Tansil, N.C., et al., 2011a. Intrinsically colored and luminescent silk. Adv. Mater. 23, 1463–1466.

Tansil, N.C., et al., 2011b. The use of molecular fluorescent markers to monitor absorption and distribution of xenobiotics in a silkworm model. Biomaterials 32, 9576–9583.

Tansil, N.C., Koh, L.D., Han, M.Y., 2012a. Functional silk: colored and luminescent. Adv. Mater. 24, 1388–1397.

Tansil, N.C., Han, M.Y., Liu, X.Y., Soh, R., Li, Y., 2012b. Intrinsically Colored, Luminescent Silk Fibroin and a Method of Producing the Same. US20120039813A1.

Tao, H., Amsden, J.J., Strikwerda, A.C., Fan, K.B., Kaplan, D.L., Zhang, X., Averitt, R.D., Omenetto, F.G., 2010a. Metamaterial silk composites at terahertz frequencies. Adv. Mater. 22, 3527.

Tao, H., Siebert, S.M., Brenckle, M.A., Averitt, R.D., Cronin-Golomb, C., Kaplan, D.L., Omenetto, F.G., 2010b. Gold nanoparticle-doped biocompatible silk films as a path to implantable thermo-electrically wireless powering device. Appl. Phys. Lett. 97–102.

Tao, H., Kaplan, D.L., Omenetto, F.G., 2012. Silk materials – a road to sustainable high technology. Adv. Mater. 24 (21), 2824–2837.

Tolly, A., 2006. Method for Producing Colorful Silk Using Silkworm. Chinese Patent CN1545874A.

Tsioris, K., Tilburey, G.E., Murphy, A.R., Domachuk, R., Kaplan, D.L., Omenetto, A.G., 2010. Functionalized-silk-based active optofluidic devices. Adv. Funct. Mater. 20 (7), 1083.

Vepari, C., Kaplan, D.L., 2007. Silk as a biomaterial. Prog. Polym. Sci. 32, 991–1007.

Vollrath, F., Carter, R., Rajesh, G.K., Thalwitz, G., Astudillo, M.F., 2013. Life cycle analysis of cumulative energy demand on sericulture in Karnataka, India. In: BISERICA 2013-6th BACSA International Conference: Building Value Chains in Sericulture, pp. 1–12.

Wang, X.Y., Kim, H.J., Xu, P., Matsumoto, A., Kaplan, D.L., 2005. Biomaterial coatings by stepwise deposition of silk fibroin. Langmuir 21, 11335.

Wang, X.Y., Hu, X., Daley, A., Rabotyagova, O., Cebe, P., Kaplan, D.L., 2007. Nanolayer biomaterial coatings of silk fibroin for controlled release. J. Control Release 121, 190.
Wang, C.H., Hsieh, C.Y., Hwang, J.C., 2011. Flexible organic thin-film transistors with silk fibroin as the gate dielectric. Adv. Mater. 23, 1630.
Xia, X.X., Qian, Z.G., Ki, C.S., Park, Y.H., Kaplan, D.L., Lee, S.Y., 2010. Native-sized recombinant spider silk protein produced in metabolically engineered *Escherichia Coli* results in strong fiber. PNAS 107 (32), 14059–14063.
Xu, T., Zhang, N., Nichols, H.L., Shi, D.L., Wen, X.J., 2007. Modification of nanostructured materials for biomedical applications. Mater. Sci. Eng. C Biomim. Supramol. Syst. 27, 579.
Zhao, Y.P., et al., 2008. SSR based linkage and mapping analysis of C, a yellow cocoon gene in the silkworm, *Bombyx mori*. Insect Sci. 15, 399–404.
Zhua, L., Zhangb, Y.Q., 2014. Identification and analysis of the pigment composition and sources in the colored cocoon of the silkworm, *Bombyx mori*, by HPLC-DAD. J. Insect Sci. 14, 31.

Further reading

Domachuk, P., Perry, H., Amsden, J.J., Kaplan, D.L., Omenetto, F.G., 2009. Bioactive 'self-sensing' optical systems. Appl. Phys. Lett. 95, 253702.

Sustainable production of regenerated cellulosic fibres

7

Mohd Shabbir, Faqeer Mohammad
Jamia Millia Islamia (A Central University), New Delhi, India

7.1 Introduction

At the start of 21st century, as technology quickly accelerated, researchers in every field began to have concerns about sustainability. Development that meets the needs of the present without compromising the ability of future generations is defined as sustainable development required in this era. Sustainability is composed of social, economical and environmental factors; these factors have their own subsets, including social, economical and environmental sustainable developments. To benefit both the present and future, these three factors must be taken into consideration so that their cumulative effect can make development sustainable (Lehtonen, 2004).

The textile industry heavily influences the economy, society and environment of the region (Islam et al., 2017). Primary base materials for the textile industry are fibres, which are of various kinds according to their origin and chemical composition (Fig. 7.1). Fibres of natural origin, such as wool, silk and cotton, have been used from the beginnings of clothing to cover the human body. Wool and silk are composed of amino acid units as monomers in long polymeric chains, whereas cotton is made of cellulose, i.e., glucose units as monomers. With the passage of time, synthetic materials evolved in the textile industry and synthetic fibres such as nylon, Dacron and polyester were developed to meet the needs of modern society (Cook, 1984). As the world population and global demands for multifunctional textile materials increasing, there arose a need for alternative textile materials with characteristic features that were cost-effective and used abundant raw materials. Cellulose, which was discovered in 1839, is one of the most abundant materials in the world; it is used in the textile and paper industries. Cotton, a natural cellulose fibre, inspired modern research to use abundant cellulose resources from plants for textile fibre manufacturing. Natural cellulose fibres such as cotton and regenerated cellulose fibres (solvent spun from cellulose pulp) are widely used in clothing textiles (Stanković et al., 2008). One of the natural fibres, silk, which is obtained from silkworms, has exclusive properties that attract people toward it, in addition to a high cost. A tremendous amount of research was proposed and designed to obtain cost-effective silk with its patent properties. However, in the 19th century a fibre material similar to silk, with all of the properties of silk, was discovered and was obtained from cellulose material. Since then, it was an alternate to natural silk with low production costs. In 1855, a Swiss chemist, Georges Audemars, discovered the manufacturing process for cellulose nitrate, and later in 1884, Count Hilaire de Chardonnet made the first man-made fibres from

Sustainable Fibres and Textiles. http://dx.doi.org/10.1016/B978-0-08-102041-8.00007-X

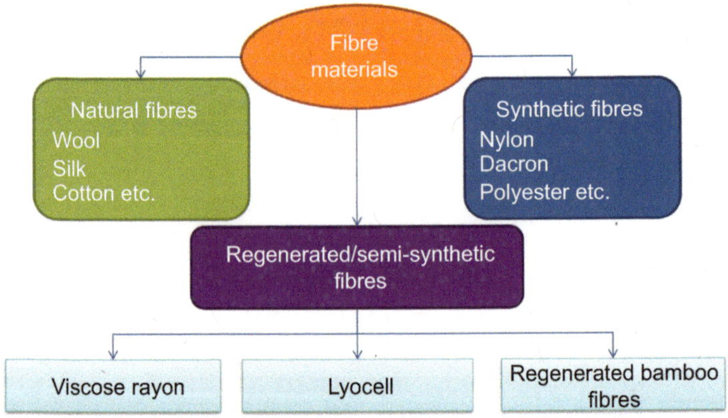

Figure 7.1 Fibre materials classification array.

nitrocellulose. The last century also evidenced a tremendous increase in research and developments in regenerated cellulosic fibre manufacturing to compete with textile fibres of natural and synthetic origin (Haule et al., 2016a).

Research then evolved the manufacturing methods and use of raw materials to create a variety of regenerated cellulose fibres with specific characteristics. On the basis of the variation in properties, rayon developed into many types of fibres such as viscose rayon, high wet modulus (HWM) rayon and Lyocell (Santamala et al., 2016). Bamboo was used as a raw material for cellulose, and fibres manufactured from this evolved as regenerated bamboo fibres with inherent antimicrobial and UV protection properties.

The purpose of this chapter is to review the sustainable production of regenerated cellulosic fibres manufactured with cost-effective and environmentally friendly processes. Sustainability issues with fibre manufacturing, the effluents generated and the recovery of the chemicals used are discussed, as well as the comparative structure and applications of these fibres. Future perspectives and the related scope of development in manufacturing these cellulosic fibres are also discussed for an awareness of the best use of these fibres.

7.2 Key sustainability issues in regenerated fibre production

As discussed earlier regarding sustainable development and its important pillars of economy, society and the environment (Fig. 7.2), there is a need to fulfil each aspect. Sustainable production demands cost-effective and eco-friendly raw materials and processes. Cotton is a natural fibre composed of cellulose polymer obtained from cotton crops grown in fields of black soil. As far as the formation of cellulose fibre is concerned, cotton crops may not be a sustainable option because there is a shortage of land for the production of cotton crops, according to the demands of the times.

Figure 7.2 Sustainable development pillars.

Other than cotton, cellulosic fibres can be manufactured from cellulose sources such as various softwood and hardwood species. These cellulose-producing species are grown all over the world and include *Pinus silvestris*, *Larix sibirica*, *Picea abies*, *Picea glauca*, *Abies balsamea*, *Pinus banksiana*, *Pinus taeda*, *Populus tremula*, oak, maple and beech species (Lonnberg, 2001).

Environmental concerns with regard to textile manufacturing wastewater and waste are basic and the most important concerns to be considered to obtain functional and sustainable textile materials. Toxic chemicals used in the textile industry such as some synthetic dyes may be replaced by products such as natural dyes extracted from plants and animals which are considered biodegradable and eco-friendly (Shabbir et al., 2016; Yusuf et al., 2016). A lot of research has focused on environmental safety and wastewater treatments, which may be a major task to be handled, and a lot of money can be involved in this. Various physical, chemical and biological treatments of wastewater released from the rayon industry were attempted and developed to make the environment safe (Coia-Ahlman and Groff, 1990). Rayon attained a unique position as a textile fibre because it can be tailored to fit numerous end-use applications. Rayon (a regenerated cellulosic fibre) offers superior performance alone and in blends with other fibres such as polyester, wool and cotton. Considering the availability of raw materials and the demands of the modern growing population, viscose rayon will be one of the most continuously versatile engineered fibres in the foreseeable future. To achieve sustainability in viscose production primarily with respect to the economy depends on the reduction of raw material consumption, such as lower carbon disulphide, and lower caustic and higher cellulose in viscose, so these factors have been principal objectives for research. High-performance rayon production using low-cost pulp generally requires more costly steps in processing, so use of cost effective pulp is a failure to sustainability and not to be considered a hot topic of research in rayon production (Dyer and Daul, 1981).

Effluent generation from regenerated cellulose fibre manufacturing plants causes sustainability issues regarding environmental sustainability primarily from the three pillars of sustainable development.

Textile industries that manufacture rayon yarn from wood pulp or cotton linters generate waste sludge containing zinc. The rayon manufacturing process requires wood pulp treatment with caustic soda to produce alkali cellulose, which is converted into cellulose xanthate with carbon disulphide. Xanthate is subsequently dissolved in dilute caustic soda to produce a solution known as viscose, which is then extruded under pressure into fine jets through numerous spinnerets into a spinning bath containing dilute sulphuric acid (8%–10% m/v), zinc sulphate (0.5%–1.0% m/v) and other chemicals essential to the spinning process. The alkaline and acidic wash solutions are generated during the manufacture of rayon, and are mixed and neutralized with lime or alkali to precipitate zinc hydroxide as waste sludge (Jha et al., 2004). Zinc in the wastewater of rayon industry effluent is a major threat to the water ecosystem; its recovery should be discussed. Some measures or developments attempted to eliminate zinc in the past, parallel to rayon manufacturing developments (De Jong, 1976; Jha et al., 2007; Ghosh et al., 2011).

Side reactions of various processes of regenerated cellulose fibre manufacturing are other sustainability issues. The Lyocell process involves dissolving cellulose material into amine oxide solvents, which can lead to various side reactions producing effluents in the air and environment. The gradual decomposition of solvents and the degradation of cellulose material 'pulp' during the process make manufacturing economically ineffective, producing chromophores and other side products that adversely affect the quality of fibres and the environment. Control of these issues related to side reactions can be optimized through technical measures, material management and the use of some additives (Rosenau et al., 2001). Thus effluents generated from manufacturing processes, raw material waste via degradation and side reactions, the establishment of wastewater treatment plants, and the cost of research and development are primary issues regarding the sustainable production of these fibres.

7.3 Production of regenerated fibres

With the world's population explosion, there is an increased demand for textile materials, because clothing is one of the basic needs for humans. Thus, apart from natural fibres (wool, silk and cotton), with limited resources, and because synthetic fibres (nylon, Dacron and acrylic fibres) harm the environment during their processing, regenerated cellulose fibres from cellulose (Fig. 7.3) have arisen as an abundant resource to meet the needs of the modern era. Based on the need for fibres, regenerated cellulosic fibres are manufactured by various methods to achieve the desired properties, and on this basis they are categorized into various classes. Modern manufacturing methods and their evolution are devoted to minimizing problems with sustainability in all aspects. A vast amount of research has been done to achieve the maximum output of regenerated fibres without disturbing the environment and the ecosystem, while optimizing the process (Kim et al., 2006; Bernt et al., 2016; Zhu et al., 2016; Wang et al., 2016).

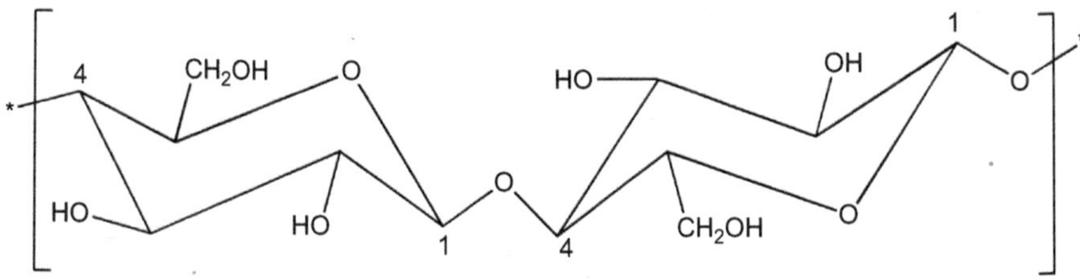

Figure 7.3 Glucose units 1–4 linkage in cellulose structure.

7.4 Closed-loop manufacturing of high wet modulus rayon and Lyocell

The production of regenerated cellulosic fibres involves steps from the selection of cellulose resources up to the recovery of side reactants and effluents in water. A number of patents were registered in the late 1990s regarding production methods and ways to optimize the production of regenerated cellulosic fibres (Nicolas and Paul, 1955; Kimura et al., 1970; Kubota et al., 1971; Siclari and Rossi, 1976; Bockno, 1980). The textile industry, which requires a higher sustainable production of these multifunctional and versatile textile fibres, needs to invent eco-friendly and human-compatible ways to produce these fibres abundantly and sustainably. Closed-loop manufacturing to reduce costs and improve the quality and accuracy of the manufactured products is the aim of current textile industry research. For this production, a systematic approach needs to be followed with stepwise observation and feedback. Steps such as proper planning, the use of resources, recovery of residues and their remanufacturing, estimation of production of target value and measurement of residual and product value are the basics of closed-loop manufacturing. This type of manufacturing can make products cost-effective and eco-friendly, and conserve energy; the proper use of time and personnel are other main benefits. One of such knowledge-based study of rayon productivity management was carried out by Arunkumar and Janakiram in 1991.

Rayon is the first regenerated cellulosic fibres among man-made cellulose fibres, developed and evaluated in the last century. Developments in manufacturing methods and the selection of raw materials and chemicals revealed various kinds of cellulosic fibres with their characteristic features. The early developed viscose rayon fibres were similar to silk and were named artificial silk owing to their resemblance. Later, in the further development of their manufacture, cotton-like fibres developed and emerged as a cost-effective alternative to natural fibres. Various methods were created to produce fibres such as viscose and cuprammonium, among which the viscose method (Fig. 7.4) is most common (Woodings, 2003; Chen and Burns, 2006).

The viscose method involves first producing alkali cellulose by steeping cellulose pulp sheets in warm caustic soda and then pressing them to remove the extra solution. Then the treated cellulose is broken up in a shredder to form powdery crumbs. The crumbs are aged in the reaction of caustic soda with cellulose to form alkali cellulose. Aged crumbs of alkali cellulose are mixed with carbon disulphide

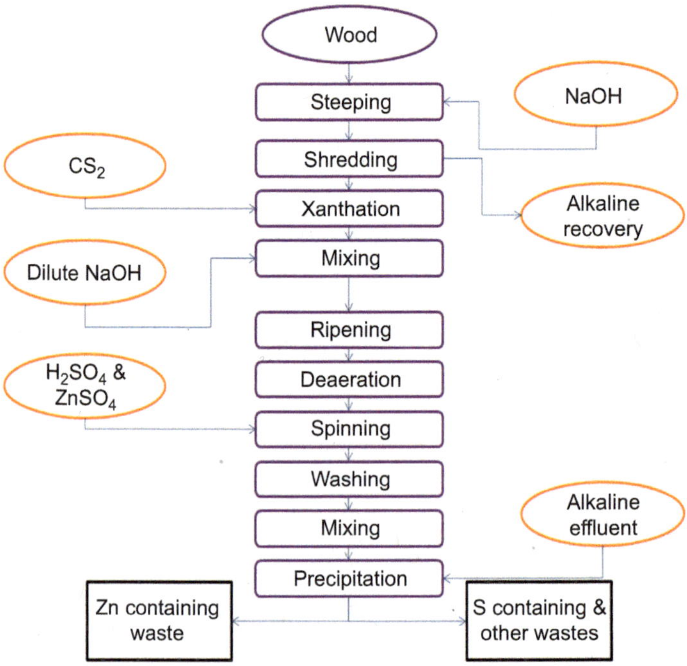

Figure 7.4 Viscose rayon manufacturing steps (Jha et al., 2004).

in a revolving drum and the white crumbs gradually turn yellow and then orange as sodium cellulose xanthate is formed (Fig. 7.5). The batch is tipped into a dilute solution of caustic soda, forming a viscous orange-brown solution. Sodium cellulose xanthate and sodium hydroxide are in a loose association and the lustre of the rayon is controlled at this stage.

The viscose solution (sodium cellulose xanthate) is allowed to stand and ripen for up to about 24 h at a controlled temperature, during which it is filtered repeatedly. The viscosity of the solution initially falls; later the viscosity of the solution begins to rise again as cellulose is regenerated by the breakdown of some of the sodium cellulose xanthate upon further standing. Ripening is allowed to continue until the solution reaches a state suitable for spinning, because ripening for too long duration can lead to cellulose deposition.

Then deaeration is carried out to remove air bubbles or other gases which can interfere with the smooth flow of the solution during spinning. Ripened viscose solution is passed through a final filtering stage, and then is forced through the spinneret. After it emerges from the spinneret holes, the jet of viscose enters a coagulating bath (mixture of H_2SO_4, Na_2SO_4 and $ZnSO_4$). Sodium cellulose xanthate is converted back into cellulose in the coagulating bath (Jacobs and Renn, 1953; Wilkes, 2001; Woodings, 2003).

The cross-section of rayon fibre resembles a distorted circle with a serrated contour and the fibre surface is smooth but striated longitudinally, as shown in Fig. 7.6. The lustre of viscose rayon fibre can be bright, semidull, or dull (Chen, 2015). Manufactured

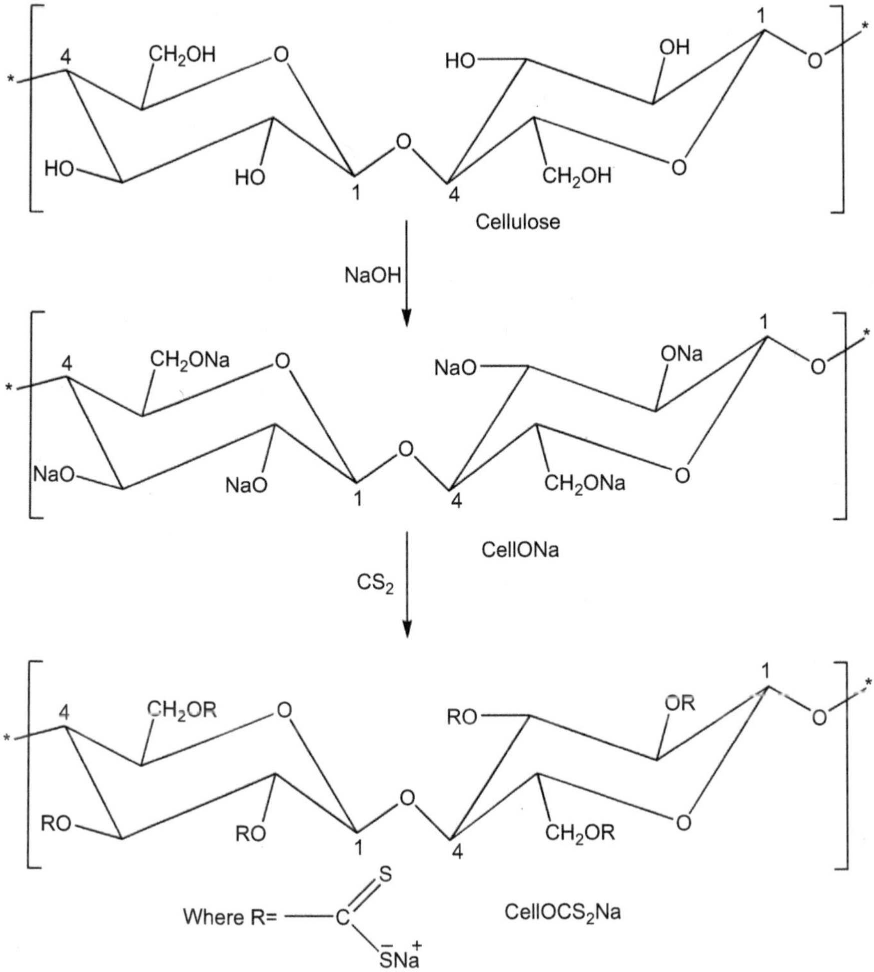

Figure 7.5 Chemical reaction steps of viscose rayon formation.

viscose rayon has some advantages and disadvantages over natural fibres such as cotton. The tenacity of viscose rayon decreases under wet conditions as a result of the absorption of water and the swelling of fibre. The modulus of rayon falls and extensibility increases under wet conditions, and the recovering ability of rayon is less. Thus during washing, the stability of fibres decreases, and this behaviour is reflected in the fabrics. This drawback of viscose rayon has motivated a lot of investigations to improve the quality of manufactured cellulose fibres and HWM rayon evolved to overcome these drawbacks and achieve higher wet stability of the fibres and fabrics. Some processing steps of rayon manufacturing were investigated and modified accordingly to obtain HWM rayon (Tachikawa, 1956).

Figure 7.6 Viscose rayon fibre cross-section and longitudinal view.
Adapted from Chen, J., 2015. Synthetic textile fibres: regenerated cellulose fibres. In: Sinclair, R. (Ed.), Textiles and Fashion. Woodhead Publishing in Association with the Textile Institute, copyright 2015, with permission from Elsevier.

7.5 High wet modulus rayon

The production of HWM rayon with varying fibre properties needs some changes to the viscose reaction conditions and control parameters. These modified viscose rayon fibres (HWM viscose and high-tenacity viscose) are most important for a wide range of apparel and other applications. The production of HWM viscose fibre no longer requires aging in cellulose alkalization and ripening in xanthation, and chemical concentrations are reduced, so that the speed of cellulose coagulation is reduced, allowing more time for the fibre to stretch. For the production of high-tenacity viscose fibre, the speed of cellulose regeneration needs to be reduced by increasing the concentration of zinc sulphate in the spin bath. 'Modal' and 'Polynosic' are two commonly used brand names for HWM viscose fibre (Gray, 1967).

Tensile strength and breaking elongation are two of the most important mechanical properties for textile fibre materials. Fibre tensile strength is generally expressed by tenacity with a unit of force per denier or tex. HWM rayon fibre has a higher tensile strength under wet conditions such as during washing than under dry conditions. However, elongation is higher for viscose rayon up to 25% (dry) and 30% (wet). Thus viscose rayon or rayon-blended fabrics tend to have good stretch ability. HWM viscose fibre has a higher tensile strength under both dry and wet conditions but a lower elongation at break (Chen, 2015). Average comparative characteristics of cotton, rayon, HWM rayon, Lyocell and regenerated bamboo fibres are displayed in Table 7.1 (Woodings, 2003; Chavan and Patra, 2004).

Use of these chemicals in processing steps such as cellulose alkalization, xanthation and spinning may cause environmental and sustainability issues in the production of the fibres. This can be overcome with optimized conditions of processing with respect to problematic chemicals and through the proper planning of resources or chemicals and the steps involved in the process of manufacturing. Proper recycling

Table 7.1 Characteristic differentiations of various regenerated and natural cellulose fibres

Fibres	Dry tenacity (cN/tex)	Wet tenacity (cN/tex)	% Elongation at break	% Moisture absorption
Cotton	20–43	27–56	6–10	7–8
Rayon	18–26	9–15	15–25	10–16
High wet modulus rayon	24–36	12–24	13–25	12–13
Lyocell	37–45	30–39	12–16	11–12
Regenerated bamboo fibre	22–25	13–17	14–24	13

techniques and the evolution of eco-friendly or recyclable chemicals should be the focus in future research on sustainable production. Cellulose resources (wood pulp) are primary materials supplied for rayon production, which may be selected with the consideration of sustainability parameters. Trees for wood pulp, such as beech and eucalyptus, are harvested quickly and without artificial irrigation or the need to use pesticides. Industrial use of these plants without disturbing the environment or agricultural systems can be an important sustainable step for HWM rayon production.

7.6 Lyocell manufacturing

Lyocell manufacturing requires fewer steps compared with viscose rayon fibre manufacturing. Steps such as the derivatization, alkalization and xanthation of rayon are omitted in the Lyocell process. Few chemicals are used, among which *N*-methyl morpholine-*N*-oxide (NMMO) and water are almost recycled, which makes the process economically favourable (Rosenau et al., 2001; Dawson, 2012). The starting material for Lyocell manufacturing, wood pulp, is directly dissolved using an organic cyclic polar solvent. NMMO is considered nontoxic and is easily regenerated. NMMO has higher cellulose dissolving capacity than the other organic polar solvent, such as dimethyl sulphoxide or *N,N*-dimethylformamide. The molecular level of dissolution of cellulose in NMMO is unknown. It is thought that the swelling and dissolution of cellulose disrupt the hydrogen bonding networks in cellulose by forming new H-bonds between the solvents and cellulose. However, these processes are extremely complicated and no individual step has been identified so far. A better understanding of the interactions between cellulose and solvent molecules would be highly important to improving the current Lyocell process (Zhao et al., 2007). NMMO, which is commonly called amine oxide, is used in a 50:50 (w/w) mixture with water. The melting point of the monohydrate solvent is about 76°C. NMMO can be produced from *N*-methyl morpholine and hydrogen peroxide according to the reaction in Fig. 7.7 (Fink et al., 2001).

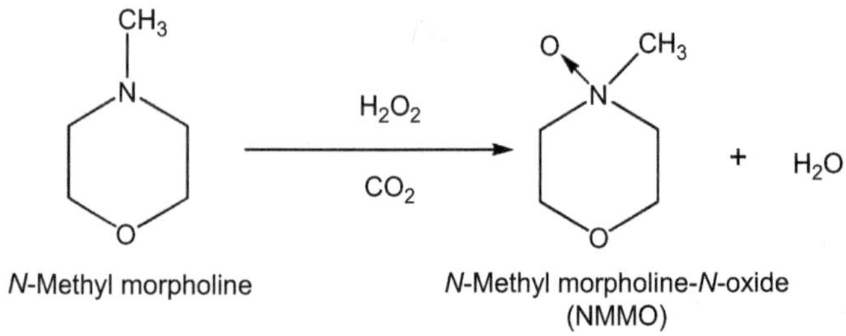

Figure 7.7 *N*-Methyl morpholine-*N*-oxide production chemical reactions.

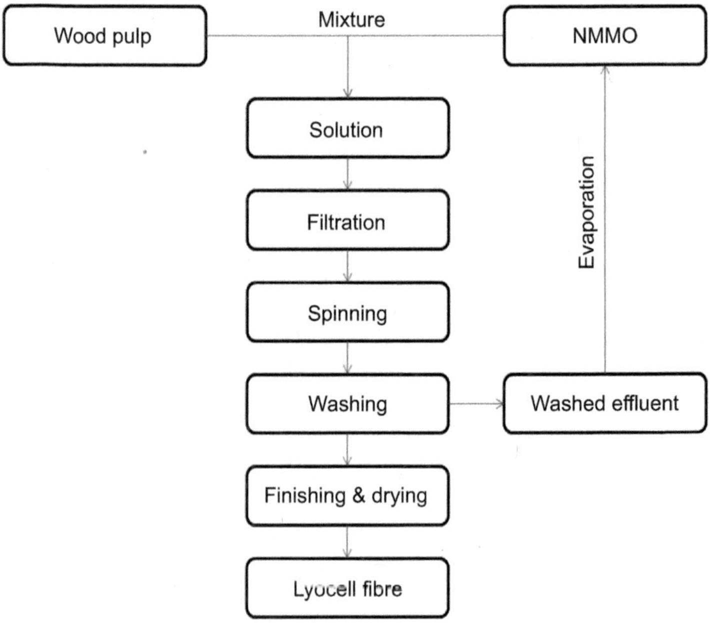

Figure 7.8 Closed loop manufacturing flow chart of Lyocell fibre. *NMMO*, *N*-methyl morpholine-*N*-oxide.

Industrial dissolving pulp with 96% cellulose is dispersed in concentrated aqueous NMMO and dissolved under the exertion of intensive shear forces and the simultaneous evaporation of water. The starting point of the process is a suspension of approximately 13% cellulose, 20% water and 67% NMMO, and dissolution of cellulose in NMMO is done at 120°C (a safe temperature for NMMO), resulting in a highly viscous solution. The solution is filtered and then extruded into a water bath through fine jets. As the solvent is washed out, the fibres formed into fine filaments are collected as tow, from which the staple fibre is produced. Water is evaporated off and the remaining concentrated NMMO is recycled into the process (Fig. 7.8).

Recycling of more than 99% of the solvent within the process makes the fibre production eco-friendly. Very small quantities of NMMO emitted via wastewater are readily degraded in biological wastewater treatment plants. Therefore, nontoxicity of

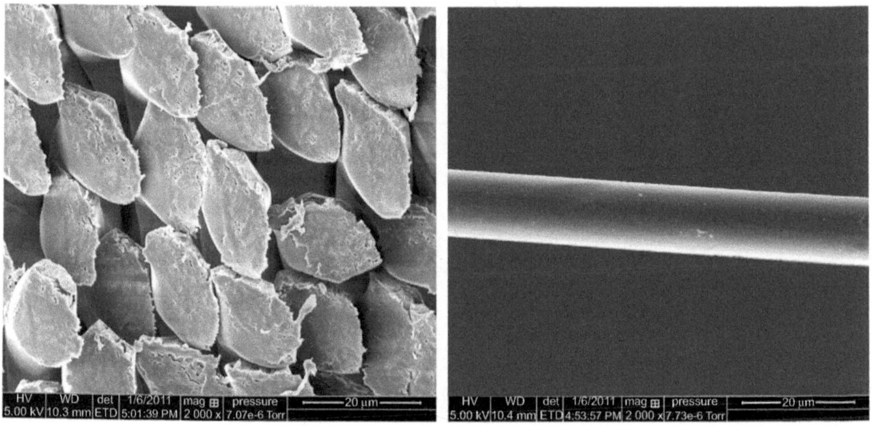

Figure 7.9 Lyocell rayon fibre cross-section and longitudinal view.
Adapted from Chen, J., 2015. Synthetic textile fibres: regenerated cellulose fibres. In: Sinclair, R. (Ed.), Textiles and Fashion. Woodhead Publishing in Association with the Textile Institute, copyright 2015, with permission from Elsevier.

the solvent can be assumed for all practical purposes. According to the various scientific tests, it is also not mutagenic. The quantity of chemicals used in the process of manufacturing Lyocell does not pose an environmental concern. When purified by a biological wastewater treatment plant, wastewater contains only small amounts of organic chemicals and salt, mainly sodium sulphate (Chavan and Patra, 2004). The flexibility of the cellulose resources in the Lyocell process is another advantage, because paper-grade pulp, cotton, rayon and waste paper can be employed as raw materials (Rosenau et al., 2001; Haule et al., 2016b).

Lyocell fibre has close to circular cross-section and its longitudinal surface is smooth and cylindrical with no striation (Fig. 7.9). Lyocell rayon fabrics exhibit better fabric feel and drape compared with viscose rayon owing to the fibre shape and appearance of Lyocell. The regenerated cellulose fibre Lyocell is known for a higher dry and wet tensile strength than others because of the properties of the regenerated cellulose of Lyocell, such as high crystallinity, long crystallites, the high degree of orientation and well-oriented amorphous regions (Chanzy et al., 1980; Woodings, 2000). The dry tenacity value of Lyocell fibre is larger than that of viscose and HWM rayon fibres. Lyocell fibre is capable of keeping almost 85% of its dry tenacity under wet conditions. Compared with viscose rayon fibre, Lyocell fibre has a significantly reduced elongation, slightly more than that of HWM rayon fibre (Chen, 2015).

Up to the point of manufacturing, Lyocell fibres are considered to be an eco-friendly and sustainable closed-loop process with most of the solvent recycled, but the transformation of the fibres into fabrics requires some toxic or hazardous chemicals which can cause environmental disturbance to some extent. Lyocell fibres also have less surface energy, because of which there is less substantivity of these fibres toward dyes, requiring the use of chemicals for finishing and a loss of sustainability. The inherent property of fibrillation (longitudinal splitting of a single fibre into microfibres) in Lyocell fibres also affects the dyeing behavior of these fibres, which then require chemical pretreatment for finishing. Alkali treatment of Lyocell fibres before dyeing is carried out to enhance dyeing abilities successfully (Goswami et al., 2009). Use of eco-friendly or less toxic pretreatment chemicals to solve these issues and their recovery are steps toward sustainable production. Flexibility of the Lyocell process with respect to cellulose resources is an important factor for environment and sustainability. Use of the resources of cellulose such as from the paper and timber industry, which otherwise would be discarded and lead to environmental issues, would be sustainable steps for production. Proper optimization and planning in the production steps are further important factors for sustainability in production.

7.7 Production of regenerated bamboo fibres

With the evolution in textile fibre manufacturing, the latest development is the use of bamboo fibre in various textile products generally used for construction materials, decorating items, furniture, etc. An extremely soft feel, antimicrobial properties, moisture-wicking properties, an antistatic nature, moisture absorption and breaking tenacity are common advantages of these fibres which are picking up speed in research in the bamboo textile industry. The bamboo plant, which is an abundant and cheap natural resource, is used to obtain regenerated bamboo fibre. Bamboo grows in tropical climates widespread throughout Asian countries and is harvested after 3–4 years. *Phyllostachys edulis* is the most common species of bamboo used in the bamboo rayon textile industry (Organic Blanket Info, 2016). Regenerated cellulosic bamboo fibre was first manufactured in 2002 by Hebei Jigao Chemical Fibre Co., Ltd. in China (Majumdar and Arora). Bamboo fibre is obtained from bamboo pulp, which is extracted from the bamboo stem and leaves by wet spinning, including a process of hydrolysis-alkalization and multiphase bleaching that is similar to that of viscose rayon fibre (Oda, 2006; Erdumlu and Ozipek, 2008).

Manufacturing of regenerated bamboo fibres can be done by mechanical as well as chemical processing. In mechanical processing, initially cut and crushed wood of the bamboo is treated with natural enzymes that break the bamboo into a soft and mushy substance. Subsequently, the natural fibres can be combed out mechanically to obtain individual fibres, followed by spinning yarn from them. The fabric manufactured via this process is often termed bamboo linen and this process is considered to be eco-friendly because no harmful chemicals are used (Haule et al., 2016a; Kaur et al., 2013a,b).

Chemical processing (Fig. 7.10) requires the use of chemicals such as carbon disulphide and sodium hydroxide, which can introduce pollutants into the environment via side reactions in processing. These are not considered environmentally friendly, but some methods such as Lyocell processing use solvents (NMMO) which can be recycled and further used in the process. Also, some pollution-creating chemicals are used, such as zinc, which makes the process harmful to the environment. Thus research is ongoing to find greener and more sustainable methods to obtain better regenerated bamboo fibres. The cross-sectional shape and longitudinal surface of bamboo viscose

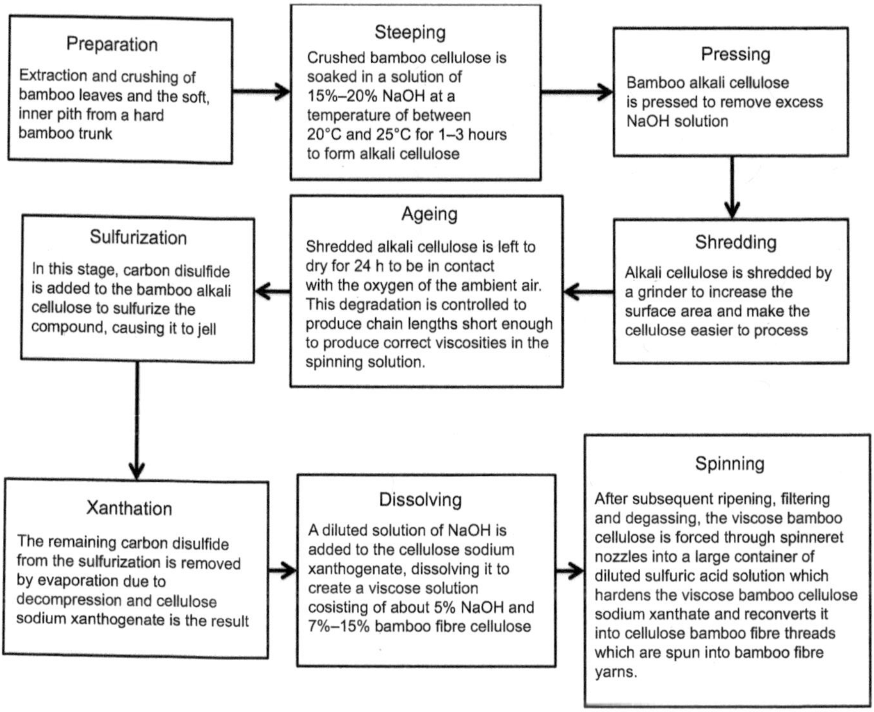

Figure 7.10 Stepwise manufacturing process of regenerated bamboo fibres (Oda, 2006; Haule et al., 2016a; Erdumlu and Ozipek, 2008).

fibre (Fig. 7.11) are more or less similar to those of conventional viscose fibres with a cross-sectional shape that irregular and toothed (Xu et al., 2007).

Bamboo grass, a cellulose resource for regenerated bamboo fibre, is considered easy to grow and organic farming takes the place of nonorganic farming of crops such as cotton. The development of bamboo harvesting is an evolutionary subject for future research to make regenerated bamboo fibre manufacturing more sustainable and environmentally compatible. The Lyocell process for manufacturing, rather than the rayon process, could be more sustainable and eco-friendly. The use of chemicals in processing, such as in the Lyocell process, are factors which can lead to side reactions and the introduction of hazardous chemicals into the environment. Mechanical processing using enzymes, which are eco-friendly in nature, would be an alternative to the chemical processing of regenerated bamboo fibre production.

Distinctive characteristics of regenerated bamboo fibres, such as natural antibacterial and biodegradable properties, a high moisture absorption capacity, softness, brightness and UV protection, have found use in the textile market. Regenerated bamboo fibre is used in underwear, sports clothes, T-shirts and socks. It is preferred for summer clothing because of its UV protection abilities, especially in garments for pregnant women and children, to protect them from UV radiation. These fibres are also considered suitable for hygienic products and sanitary

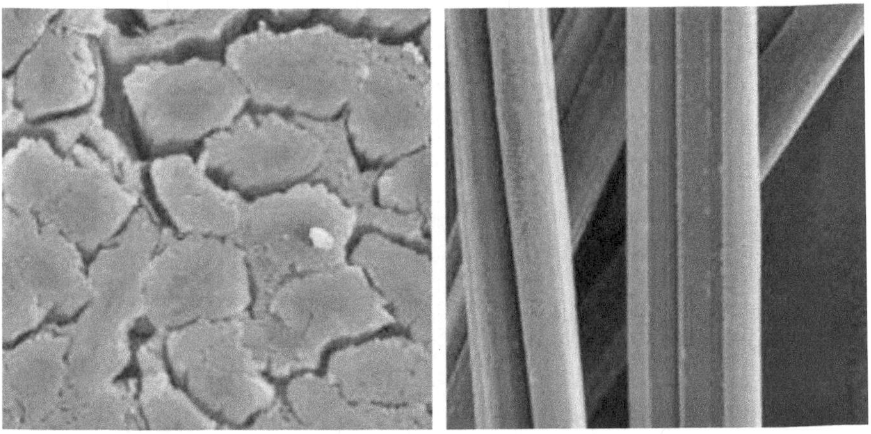

Figure 7.11 Regenerated bamboo fibre cross-section and longitudinal views.
Adapted from Xu, Y., Lu, Z., Tang, R., 2007. Structure and thermal properties of bamboo viscose, tencel and conventional viscose fibre. J. Therm. Anal. Calorim. 89 (1), 197–201, copyright 2007, with permission from Springer.

materials such as sanitary napkins, absorbing pads, masks, mattresses, bandages, surgical clothes, food packaging, etc. owing to their natural antimicrobial properties with no involvement of chemical additives that cause allergies. Other than these applications, regenerated bamboo fibres are useful for home furnishings such as towels, bathrobes, bedding sets, blankets and curtains (Nayak and Mishra, 2016; Haule et al., 2016a).

7.8 Future trends

As far as John Dyer and George C. Daul reviewed the recent developments and future prospects regarding viscose rayon production without losing sustainability in all aspects whether it is economy, environment or society; a special attention like this is highly needed in this sector (Dyer and Daul, 1981). The study focused on the developments in the manufacture of regenerated cellulose fibres with a special emphasis on the optimization of properties and resources in the manufacturing processes. Sustainable issues regarding economy and environment were discussed with the evolution of cost-effective and low-effluent generation methods. A lot of research has been done to investigate various methods for regenerating cellulose fibre on an industrial as well as laboratory scale with characteristic evaluations. In this era of overpopulation and overpollution there is a great need for sustainable production manufacturing methods, optimization of the use of resources and minimization of the spillage of effluents into water bodies. The US Environmental Protection Agency issued a proposed rule to control hazardous air pollutants (hydrogen sulphide and carbon disulphide) during cellulose manufacturing processes. An activated carbon-based process was developed to remove both of these hazardous compounds with greater efficiency. An impregnated activated carbon is used to remove hydrogen sulphide by converting it to sulphur on the interior surface of the activated carbon. The resulting minimal waste can be landfilled at a relatively low cost. The use of a unique type of activated carbon can remove carbon disulphide from the air stream and recover carbon disulphide for reuse (Enneking, 2002). A comparative analysis of these various kinds of regenerated fibres with their characteristic features can be used for a large range of applications by studying their blending abilities with other natural or synthetic fibres (Majumdar et al., 2010). Majumdar et al. (2010) studied the properties of ring-spun yarns made from cotton and regenerated bamboo cellulosic fibres and their blends. Three blend proportions (100% cotton, 50:50 cotton-bamboo and 100% bamboo) were used to produce yarns of three different counts (20, 25 and 30s); these yarns were tested for their diameter, tensile, evenness and hairiness properties. Blending of different fibres is a common practice in the spinning industry. Generally blending is carried out to enhance the properties of the resultant fibre mix and optimize the cost of the raw material. The properties of blended yarns primarily depend on the properties of the constituent fibres and their compatibility. Moreover, the proportion of fibres in the blend has a significant role (Majumdar et al., 2011). The thermal comfort properties of single jersey knitted fabric structures made from cotton, regenerated bamboo and cotton-bamboo blended yarns were also studied by Chidambaram et al. (2012). Cotton, bamboo fibre and blends of the two fibres (100% cotton, 100% bamboo, 50:50 cotton–bamboo, 67:33 cotton–bamboo and 33:67 cotton–bamboo) were spun into yarns of identical linear density (20 tex). The thermal conductivity of the fabrics was generally found to decrease with an increase in the proportion of bamboo fibre. The water vapor permeability and air permeability of the fabrics were observed to increase with an increase in the bamboo fibre content. Statistical analysis also indicated that the results were significant for air permeability, thermal resistance, thermal conductivity and water vapor permeability of the fabrics (Chidambaram et al., 2012). Some modifications in the fibre properties were carried out through a synthesis of graft copolymers of rayon with acrylonitrile by a chemical method using ceric ammonium nitrate as a redox initiator and a gamma radiation mutual method (Kaur et al., 2013a,b). Different raw materials and solvents in regenerated cellulose fibre manufacturing also need to be studied to evaluate various properties of fibres or knitted fabrics. Chen et al. (2015) investigated the combined effects of materials and solvents on the preparation and structural and mechanical properties of regenerated cellulose fibres; four cellulosic materials (microcrystalline cellulose, cotton linter pulp, bamboo pulp and bleached softwood sulphite dissolving pulp) and six nonderivative solvents (NaOH/urea aqueous solution, *N,N*-dimethylacetamide/lithium chloride, NMMO, 1-butyl-3-methylimidazolium chloride, 1-allyl-3-methylimidazolium chloride and 1-ethyl-3-methylimidazolium acetate) were used to prepare fibres with the wet spinning method. Results showed that the dissolvability of the solvent was the determining factor in cellulose dissolution, and the dissolving time was influenced by the raw materials' properties, such as the molecular weight, exposed area and hemicellulose content. The crystallinity and elongation at break of the fibres were almost fixed and not affected by the materials and solvents.

However, the tensile strength of the fibres was directly proportional to the molecular weight of the raw materials, and varied with the type of solvents through cellulose degradation (Chen et al., 2015). Process optimization, chemical recovery and waste treatment, new raw material investigation and evaluation of characteristics with blending of various natural and synthetic fibres may be important factors for future developments in regenerated cellulosic fibre production.

7.9 Conclusion

Regenerated cellulosic fibres are finding applications everywhere in the textile industry, alone or blended with other natural or synthetic fibres. From sportswear to health care textiles, these are used abundantly owing to their characteristic properties such as tensile strength, smoothness and antimicrobial qualities. Suits, jackets and blazers with high-quality appearance and moisture permeability are also manufactured from these fibres blended with wool, polyester or acrylic fibres. To be able to fulfil such a vast array of applications, there needs to be the large-scale production of regenerated cellulosic fibres with green technologies and sustainability. Evaluation of the production techniques and procedures has been ongoing since the development of the first such man-made fibres.

Sources of further information and advice

Woodings, C., 2001. Regenerated Cellulose Fibres, vol. 18. Woodhead Publishing.

Woodings, C., 2003. Fibers, Regenerated Cellulose. Kirk-Othmer Encyclopedia of Chemical Technology.

Wallenberger, F.T., Weston, N. (Eds.), 2003. Natural Fibers, Plastics and Composites. Springer Science & Business Media.

Kalia, S., Kaith, B.S., Kaur, I. (Eds.), 2011. Cellulose Fibers: Bio-and Nano-polymer Composites: Green Chemistry and Technology. Springer Science & Business Media.

Needles, H.L., 1986. Textile Fibers, Dyes, Finishes, and Processes. Noyes Publications, Park Ridge, New Jersey.

Graenacher, C., Sallmann, R., Soc of Chemical Ind, 1939. Cellulose Solutions and Process of Making Same. U.S. Patent 2,179,181.

Schrell, A., Meier, S., Hoechst Aktiengesellschaft, 1998. UV-active Regenerated Cellulose Fiber. U.S. Patent 5,770,110.

Acknowledgements

The author Mohd Shabbir is highly grateful to the University Grants Commission, Government of India, for financial support provided through a BSR Research Fellowship in Science for Meritorious Students.

References

Arunkumar, S., Janakiram, N., 1991. Knowledge based approach to productivity management in rayon industry. Decis. Support Syst. 7 (3), 199–219.

Bernt, I., Roggenstein, W., Woodings, C., Harms, H., Kelheim Fibres Gmbh, 2016. Regenerated Cellulose Fiber. U.S. Patent 9,279,196.

Bockno, G.C., Avtex Fibers Inc, 1980. Viscose Rayon and Method of Making Same. U.S. Patent 4,242,405.

Chanzy, H., Peguy, A., Chaunis, S., Monzie, P., 1980. Oriented cellulose films and fibers from a mesophase system. J. Polym. Sci. 18 (5), 1137–1144.

Chavan, R.B., Patra, A.K., 2004. Development and processing of lyocell. Indian J. Fibre Text. Res. 29 (4), 483.

Chen, H.L., Burns, L.D., 2006. Environmental analysis of textile products. Clothing Text. Res. J. 24 (3), 248–261.

Chen, J., Guan, Y., Wang, K., Zhang, X., Xu, F., Sun, R., 2015. Combined effects of raw materials and solvent systems on the preparation and properties of regenerated cellulose fibers. Carbohydr. Polym. 128, 147–153.

Chen, J., 2015. Synthetic textile fibers: regenerated cellulose fibers. In: Sinclair, R. (Ed.), Textiles and Fashion. Woodhead Publishing in Association with the Textile Institute.

Chidambaram, P., Govindan, R., Venkatraman, K.C., 2012. Study of thermal comfort properties of cotton/regenerated bamboo knitted fabrics. Afr. J. Basic Appl. Sci. 4 (2), 60–66.

Coia-Ahlman, S., Groff, K.A., 1990. Textile wastes. Res. J. Water Pollut. Control Fed. 62 (4), 473–478.

Cook, J.G., 1984. Handbook of Textile Fibres: Man-made Fibres. Elsevier.

Dawson, T., 2012. Progress towards a greener textile industry. Color. Technol. 128 (1), 1–8.

De Jong, H.C., 1976. Elimination of Zn from wastewater of rayon plants by precipitation. Resour. Recovery Conserv. 1 (4), 369–379.

Dyer, J., Daul, G.C., 1981. Viscose rayon: recent developments and future prospects. Ind. Eng. Chem. Prod. Res. Dev. 20 (2), 222–230.

Enneking, J.C., 2002. Control of carbon disulfide emissions from viscose processes. Environ. Prog. 21 (3), 169–174.

Erdumlu, N., Ozipek, B., 2008. Investigation of regenerated bamboo fibre and yarn characteristics. Fibres Text. East. Eur. 16 (4), 69.

Fink, H.P., Weigel, P., Purz, H.J., Ganster, J., 2001. Structure formation of regenerated cellulose materials from NMMO-solutions. Prog. Polym. Sci. 26 (9), 1473–1524.

Ghosh, P., Samanta, A.N., Ray, S., 2011. Reduction of COD and removal of Zn^{2+} from rayon industry wastewater by combined electro-Fenton treatment and chemical precipitation. Desalination 266 (1), 213–217.

Goswami, P., Blackburn, R.S., Taylor, J., White, P., 2009. Dyeing behaviour of lyocell fabric: effect of NaOH pre-treatment. Cellulose 16 (3), 481–489.

Gray, K.L., 1967. HWM-fibres: their structure, properties and fabric performance. Pure Appl. Chem. 14 (3–4), 337–352.

Haule, L.V., Carr, C.M., Rigout, M., 2016a. Investigation into the supramolecular properties of fibres regenerated from cotton based waste garments. Carbohydr. Polym. 144, 131–139.

Haule, L.V., Carr, C.M., Rigout, M., 2016b. Preparation and physical properties of regenerated cellulose fibres from cotton waste garments. J. Cleaner Prod. 112, 4445–4451.

Islam, S., Shabbir, M., Mohammad, F., 2017. Insights into the functional finishing of textile materials using nanotechnology. In: Textiles and Clothing Sustainability. Springer, Singapore, pp. 97–115.

Jacobs, H.L., Renn, C.E., 1953. Rayon waste recovery and treatment [with discussion]. Sewage Ind. Wastes 296–304.

Jha, M.K., Kumar, V., Maharaj, L., Singh, R.J., 2004. Studies on leaching and recycling of zinc from rayon waste sludge. Ind. Eng. Chem. Res. 43 (5), 1284–1295.

Jha, M.K., Kumar, V., Bagchi, D., Singh, R.J., Lee, J.C., 2007. Processing of rayon waste effluent for the recovery of zinc and separation of calcium using thiophosphinic extractant. J. Hazard. Mater. 145 (1), 221–226.

Kaur, I., Sharma, N., Kumari, V., 2013a. Modification of fiber properties through grafting of acrylonitrile to rayon by chemical and radiation methods. J. Adv. Res. 4 (6), 547–557.

Kaur, V., Chattopadhyay, D.P., Kaur, S., 2013b. Study on extraction of bamboo fibres from raw bamboo fibres bundles using different Retting techniques. Text. Light Ind. Sci. Technol. 174–179.

Kim, I.S., Kim, J.P., Kwak, S.Y., Ko, Y.S., Kwon, Y.K., 2006. Novel regenerated cellulosic material prepared by an environmentally-friendly process. Polymer 47 (4), 1333–1339.

Kimura, Y., Yamamura, T., Kawai, A., Ikeda, M., Katsuyama, T., Oda, T., Kondo, S., Mitsubishi Rayon Co, 1970. Process for Producing Polynosic Fibers. U.S. Patent 3,539,679.

Kubota, M., Yamamura, T., Kawai, A., Katsuyama, T., Ikeda, M., Omoto, S., Mitsubishi Rayon Co, 1971. Process for Producing Polynosic Fibers. U.S. Patent 3,574,812.

Lehtonen, M., 2004. The environmental–social interface of sustainable development: capabilities, social capital, institutions. Ecol. Econ. 49 (2), 199–214.

Lonnberg, B., 2001. Industrial Cellulose. Woodhead Publishing Ltd., Cambridge, pp. 22–36.

Majumdar, A., Arora, S., Bamboo Fibres in Textile Applications. 285–304. http://www.frienvis.nic.in/WriteReadData/UserFiles/file/Content-Page/Books/Bamboo/Bamboo-Fibres-in-Textile.pdf.

Majumdar, A., Mukhopadhyay, S., Yadav, R., 2010. Thermal properties of knitted fabrics made from cotton and regenerated bamboo cellulosic fibres. Int. J. Therm. Sci. 49, 2042–2048.

Majumdar, A., Mukhopadhyay, S., Yadav, R., Mondal, A.K., 2011. Properties of ring-spun yarns made from cotton and regenerated bamboo fibres. Indian J. Fibre Text. Res. 36, 18–23.

Nayak, L., Mishra, S.P., 2016. Prospect of bamboo as a renewable textile fiber, historical overview, labeling, controversies and regulation. Fash. Text. 3 (1), 1–23.

Nicolas, D., Paul, H., Textile & Chemical Res Company, 1955. Process for the Production of Rayon Products. U.S. Patent 2,705,184.

Oda, T., Katsuyama Technos Ltd, 2006. Yam and Cloths Made Mainly from Bamboo Sheaths and Methods for Manufacturing the Same. U.S. Patent 7,060,211.

Organic Blanket Info.2016 http://organicblanket.info/2013/02/17/analysis-of-bamboo-textile-properties/.

Rosenau, T., Potthast, A., Sixta, H., Kosma, P., 2001. The chemistry of side reactions and byproduct formation in the system NMMO/cellulose (Lyocell process). Prog. Polym. Sci. 26 (9), 1763–1837.

Santamala, H., Livingston, R., Sixta, H., Hummel, M., Skrifvars, M., Saarela, O., 2016. Advantages of regenerated cellulose fibres as compared to flax fibres in the processability and mechanical performance of thermoset composites. Compos. A Appl. Sci. Manuf. 84, 377–385.

Shabbir, M., Islam, S.U., Bukhari, M.N., Rather, L.J., Khan, M.A., Mohammad, F., 2016. Application of *Terminalia chebula* natural dye on wool fiber—evaluation of color and fastness properties. Text. Clothing Sustainability 2 (1), 1–8.

Siclari, F., Rossi, P.P., 1976. Method for the Production of Cellulose-based Fibres and Polynosic Fibres Having a High Resistance to Combustion, and Fibres and Textile Articles Obtained Thereby. U.S. Patent 3,947,276. Snia Viscosa Societa'Nazionale Industria Applicazioni Viscosa SpA.

Stanković, S.B., Popović, D., Poparić, G.B., 2008. Thermal properties of textile fabrics made of natural and regenerated cellulose fibers. Polym. Test. 27 (1), 41–48.

Tachikawa, S., 1956. Process for Manufacturing of Artificial Fiber by Viscose Method. U.S. Patent 2,732,279.

Wang, S., Lu, A., Zhang, L., 2016. Recent advances in regenerated cellulose materials. Prog. Polym. Sci. 53, 169–206.

Wilkes, A.G., 2001. The viscose process. Regener. Cellul. Fibers 37–61.

Woodings, C., 2000. Lyocell staple fibre in industrial applications (part 2) Manufacturers of lyocell fibres remain optimistic about their future. Tech. Text. Int. 9 (9), 15–18.

Woodings, C., 2003. Cellulose fibers, regenerated. In: Encyclopedia of Polymer Science and Technology.

Xu, Y., Lu, Z., Tang, R., 2007. Structure and thermal properties of bamboo viscose, tencel and conventional viscose fiber. J. Therm. Anal. Calorim. 89 (1), 197–201.

Yusuf, M., Mohammad, F., Shabbir, M., 2016. Eco-friendly and effective dyeing of wool with anthraquinone colorants extracted from *Rubia cordifolia* roots: optimization, colorimetric and fastness assay. J. King Saud University-Science. http://dx.doi.org/10.1016/j.jksus.2016.06.005.

Zhao, H., Kwak, J.H., Wang, Y., Franz, J.A., White, J.M., Holladay, J.E., 2007. Interactions between cellulose and N-methylmorpholine-N-oxide. Carbohydr. Polym. 67 (1), 97–103.

Zhu, C., Richardson, R.M., Potter, K.D., Koutsomitopoulou, A.F., van Duijneveldt, J.S., Vincent, S.R., Wanasekara, N.D., Eichhorn, S.J., Rahatekar, S.S., 2016. High modulus regenerated cellulose fibers spun from a low molecular weight microcrystalline cellulose solution. ACS Sustainable Chem. Eng. 4 (9), 4545–4553.

Sustainable synthetic fibre production

8

T. Karthik, R. Rathinamoorthy
PSG College of Technology, Coimbatore, India

8.1 Introduction

The major concern in today's world is about the environment, i.e., global warming, climate change, sustainability. Renewability is one of the aspects of sustainable development. Sustainability is defined by the United Nations World Commission as 'the continual development which satisfies the needs of the present without compromising the capacity of future generations to meet their own needs'. Sustainable development implies the minimization of adverse impacts on the quality of air, water and other natural elements so as to sustain the ecosystem's overall integrity (Brundtland and Khalid, 1987; Joy et al., 2012; Gardetti and Torres, 2013). In general, the product or process should be made by keeping in mind the entire life cycle (cradle to grave) of the product. The use of renewable biomass has to be understood to reduce the carbon footprint of textile materials. Burning of fossil fuels and deforestation resulted in an imbalance in the carbon cycle, leading to the release of a massive quantity of CO_2 into the atmosphere which cannot be completely compensated for by photosynthesis or dissolution into the oceans (Defra (Department for Environment, Food and Rural Affairs) Sustainable clothing roadmap briefing note, 2008; Prakash et al., 2009). The consequence of this activity is a massive buildup of CO_2 in the atmosphere, contributing to global warming. Consumers are now more aware about greenhouse gas (GHG) emissions and the environment. The use of renewable resources should be considered for sustainable development.

Sustainability is an important concern in the textile manufacturing sector. The textile industry, one of the major contributors to environmental pollution, is working relentlessly toward producing environmentally friendly and sustainable products for its customers as well as to preserving the environment for the future (Gardetti and Torres, 2013a,b). The challenge for textile manufacturers is in the areas of fibre, yarn and fabric production, textile chemical processing, garment making-up, transportation of goods, product use, recycling of textiles and disposal of clothing. Deliberations about global warming have a huge impact on big players in the global market to reshape their policies. Textile companies have understood the competitive disadvantage if they not willing to focus on producing sustainable products; hence they are continuously striving to produce sustainable textile products. This critical situation has led them to search for more renewable raw materials such as cellulosic fibres, in which wood from trees is a raw material which has a very low impact on the environment. Synthetic fibres are dominant over natural

Sustainable Fibres and Textiles. http://dx.doi.org/10.1016/B978-0-08-102041-8.00008-1

ones in textile clothing and other similar applications owing to their superior properties such as tenacity, abrasion resistance, and resistance to chemicals and mildew, as well as durability and cost. By considering these factors, the low cost of synthetic fibres cannot be assured in the future owing to the increasing costs of raw materials, energy and transportation (Gopalakrishnan and Karthik, 2012). Unfortunately, most synthetic polymers are produced from petrochemicals raw materials which are nonbiodegradable. Hence, the major challenge in the synthetic fibre manufacturing sector is environmental degradation and the rising demand for eco-friendly fibres. Synthetic polymers are responsible for environmental pollution because of the emission of CO_2 during their product life cycle, the depletion of fossil resources, and their harm to aquatic and wild-life ecosystems (Karthik & Gopalakrishnan, 2012a,b,c, 2013). For instance, aquatic life will be affected as a result of the disposal of normal nonbiodegradable plastic materials, and the burning of plastic waste leads to emissions of toxic substances and gases. Furthermore, recycling of plastics creates a negative eco-balance owing to the need to wash plastics during the recycling phase as well higher energy consumption during the recycling process (Karthik and Gopalakrishnan, 2013b,c). In the case of the synthetic fibre manufacturing industry, sustainability aspects primarily focus on lessening GHG emissions during production and increasing the lifetime of synthetic textile products by means of recycling and reuse. According to this aspect, the prospective of biopolymers has been documented for a long time because they provide a solution to conquer the drawbacks of fossil oil resources in the manufacturing of synthetic polymers (Karthik and Gopalakrishnan, 2012d).

8.2 Environmental impact of textiles

Fig. 8.1 outlines the social, environmental and economic impacts at each stage in the life of an item of clothing. Generally, the textile sector uses huge quantities of water and fuel and different classes of chemicals in a long process chain which produces a substantial quantity of waste (Karthik and Gopalakrishnan, 2014). The low-efficiency textile processes lead to the generation of a significant amount of waste as well as severe environmental problems. The major environmental concern related to the textile sector is characteristically related to water pollution caused by the discharge of untreated textile effluents. Other environmental issues of equal importance are air emissions, notably volatile organic compounds (VOCs) and excessive noise or odours as well as workspace safety (Vijayaraaghavan and Karthik, 2004; Fletcher, 2008; Climate Change and the Textile Industry, 2014).

A life-cycle assessment (LCA) is the generally adopted approach by industries for assessing the environmental impact of products and processes. Four indicators have been devised under this approach to assess the environmental impact:

- eco-costs
- CO equivalent
- cumulative energy demand (CED)
- RIVM and Radboud University, CML and PRé Consultants (ReCiPe) method

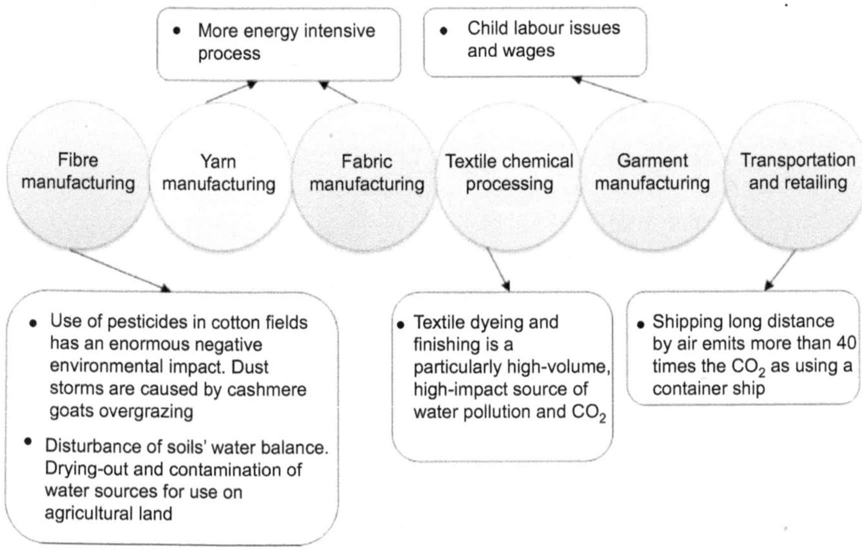

Figure 8.1 Ecological and social impacts of textile processes.

Eco-costs are commonly an indicator of the volume of environmental burden of a product on the basis of prevention of that burden. Otherwise, eco-costs are the costs incurred to minimize environmental pollution and materials depletion to a level in line with the carrying capacity of the earth. The other two single indicators commonly used in an analysis of LCA are CO equivalent and CED, because they are easy to understand and interpret, but the level of toxicity and material depletion is not considered in this. The eco-cost and ReCiPe indicators include human toxicity, eco-toxicity, materials depletion and land use.

Van der Velden et al. (2014) studied the environmental impact of various synthetic fibres and compared them with cotton fibre; the results are given subsequently. The influence of various processes involved in synthetic fibre fabric manufacturing on eco-costs, CO equivalent, CED and ReCiPe indicators are given in Fig. 8.2. The figure shows that nylon fibre fabric has a higher negative impact on the environment, followed by elastane, acrylic and polyester. An LCA assessment of various synthetic fibres along with cotton, using a cradle-to-gate approach from raw materials to consumer use and end of life in terms of eco-cost indicators, is shown in Fig. 8.3. Fig. 8.3 demonstrates that polyester and acrylic have a minimal negative impact on the environment, followed by elastane and nylon, and cotton fibres have the highest negative environmental burden.

8.3 Biodegradable polymers

Biodegradable polymers are a growing field in many industrial sectors. Biopolymers are polymers derived from living organisms or plastic materials produced from renewable resources, polymers which are fragmented by means of microorganisms or polymers which support sustainable environment during their life cycle. They have a

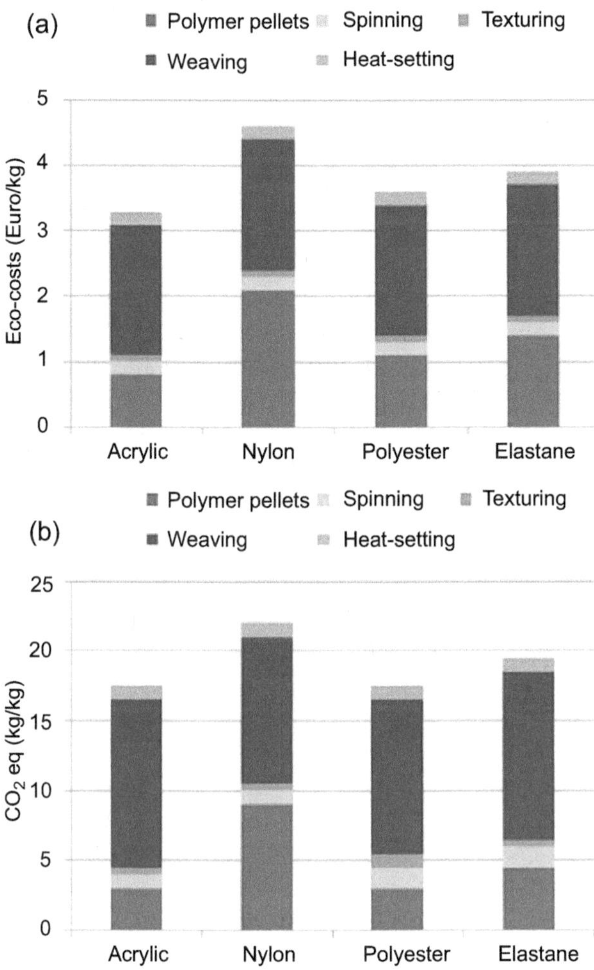

Figure 8.2 Environmental impact of various synthetic fibres during their production process (a) eco-costs (b) carbon footprint (Van der Velden et al., 2014).

significant positive impact on the environment because they diminish the use of fossil fuels, resulting in a lower carbon footprint (Karthik and Subramani, 2016; Steinbuchel, 2003; Avérous and Pollet, 2012; Gandini, 2011; Van de Velde and Kiekens, 2002; Babu et al., 2013). A huge number of biodegradable polymers such as cellulose, starch, polylactide, chitin and collagen have been synthesized or formed in the natural environment during the growth cycles of organisms. The first generations of biopolymers are polymers derived from agricultural feedstocks such as corn, potatoes and other carbohydrate sources (Mohanty et al., 2002; Rydz et al., 2015; Averous and Pollet, 2012; Vroman and Tighzert, 2009). However, focus has shifted from food-based resources owing to significant breakthroughs in biotechnology. Biopolymers can be produced by bacterial fermentation processes by synthesizing the building blocks from renewable resources including lignocellulosic biomass, fatty acids and organic waste.

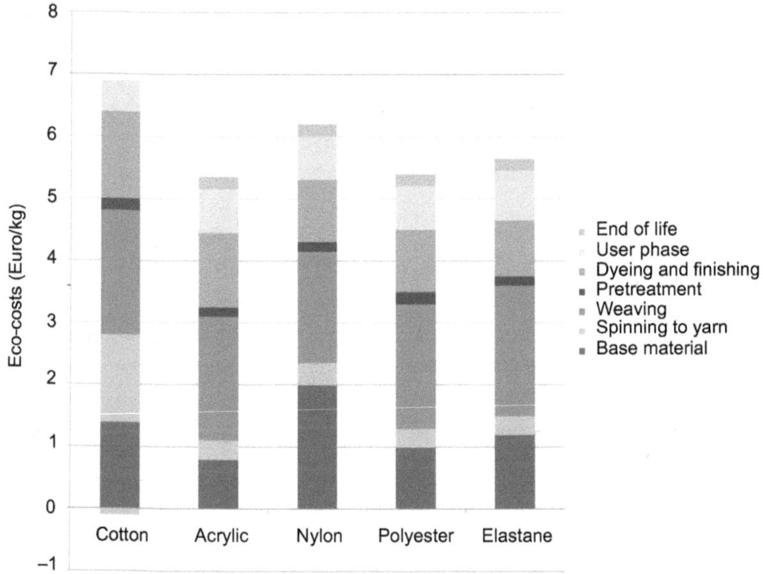

Figure 8.3 Eco-costs over the life cycle of various synthetic fibres compared with cotton for a woven textile product, 70 dtex (Van der Velden et al., 2014).

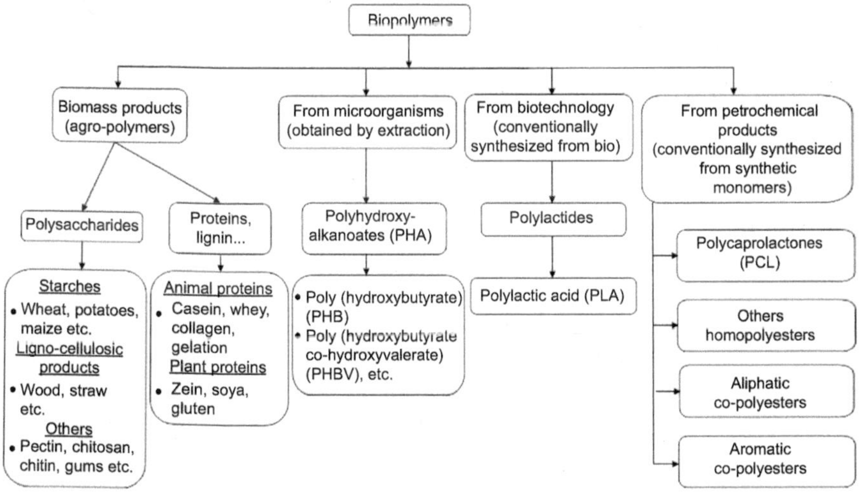

Figure 8.4 Classification of biopolymers (Avérous and Pollet, 2012).

The classification of biopolymers based on the process of synthesis is shown in Fig. 8.4:

- polymers from biomass: for example, agro-polymers derived from agro-resources such as cellulose or starch
- polymers obtained from microorganisms such as polyhydroxyalkanoates (PHAs)
- polymers obtained by a conventional synthesis process using monomers derived from agro-resources, e.g., polylactic acid (PLA)
- polymers obtained from fossil resources

In a daily life, polymers produced from fossil resources such as polyethylene, polypropylene, polyester and polyvinylchloride are used in large quantities, and other polymers in smaller quantities (Shen et al., 2009; Karlsson and Albertsson, 1998; Kaplan et al., 1993; Aeschelmann and Carus, 2015). In this chapter, the sustainable production of main synthetic fibres such as polyester, nylon and acrylic by way of biopolymers is discussed in detail.

8.4 Sustainable production of polyester fibres

Polyesters are one of the most promising families of polymers owing to their renewability, range of good properties, potential biodegradability and easy accessibility of bio-based monomers for polyesters. Polyesters consist of several materials with different monomer structures and characteristics, and have a huge range of end-use applications such as fibres, bottles, recording tape and photographic films; thus they represent a highly versatile and important class of materials. Several biomass-based chemicals such as sugars and their derivatives, vegetable oils, organic acids, suberin, cutin and glycerol can be used as building blocks for the production of polyester (Sousa et al., 2014; Pellis et al., 2016; Steinbuchel and Doi, 2002; Jiang and Loos, 2016).

Because polyester is a synthetic fibre, it is generally produced from petroleum resources, which are a carbon-intensive nonrenewable resource. Petroleum products are used as feedstock (raw material is used to make the fibre) and also to generate the energy needed for manufacture. Over 70 billion barrels of oil are used to make polyester each year. It is not biodegradable and will persist in the ecosystem even as it eventually breaks apart. Although it is less energy intensive compared with the production of nylon, it still requires more than double the energy of conventional cotton to produce. The production of polyester uses harmful chemicals including carcinogens and if emitted untreated into the water and air, it can cause significant environmental damage. The water intensity of production is much lower than for natural fibres. However, polyester cannot be dyed using low-impact and natural dyes. This means that the detrimental impact on water supplies is potentially far greater. The environmental impacts of polyester are significant. Cheap polyester is probably the biggest offender when it comes to the high environmental cost of fashion. Because of the impact of microfibres on oceans, I would avoid polyester garments as much as possible.

Polyesters are generally polymers in which the monomers are linked to one another by means of ester groups. Some naturally occurring polyesters are shellac, cutin and poly(hydroxybutyrate); synthetic polyesters are poly(ethylene terephthalate) (PET), poly(butylene succinate) (PBS) and PLA. Based on the chemical composition of the main chain, polyesters are classified as aliphatic, semiaromatic and aromatic. Aliphatic polyesters are generally sustainable because they are mostly produced from biomass feedstocks (Amass et al., 1998; Sousa et al., 2015). Furthermore, they are biodegradable materials which can be recycled, disposed of or incinerated with minimum negative impact on the environment. PLA is a well-known example of an aliphatic polyester which can be used as fibres, food packaging materials and durable goods. Semiaromatic polyesters have better thermal and mechanical properties than aliphatic polyesters and are generally used as a commodity as well as a thermal engineering plastic. Poly(trimethylene terephthalate) (PTT), PET and poly(butylene terephthalate) (PBT) are a few examples of semiaromatic polyesters, of which PET is the most commonly used fibre. Aromatic polyesters generally possess excellent mechanical, thermal and chemical properties compared with aliphatic and semiaromatic polyesters. Hence they are used in high-end application industries such as chemical, electronic, aviation and automobile (Okada, 2002).

In general, polyester fibres can be manufactured by two methods:

- step-growth polycondensation of diols and diacid/diesters, or hydroxyacids/hydroxyesters
- ring-opening polymerization of cyclic monomers and cyclic oligomers

Both of these methods have merits and demerits. Generally, the building blocks for step-growth polycondensation can easily be found at a comparatively low cost. However, higher reaction temperatures, high vacuum conditions, longer reaction times, heavy metal catalysts and accurate stoichiometric balance between monomers are needed for polycondensation. Conversely, removal of by-products is not required in ring-opening polymerization, and hence under mild conditions very high-molecular weight products can be obtained. Nevertheless additional synthesis steps and heavy metal catalysts are frequently necessary to prepare cyclic monomers and oligomers. Few bio-based polyesters are available on the commercial market including fully bio-based PLA, PHA, and poly(ethylene furanoate) (PEF); partially bio-based PBS, PET, PTT and poly(butylenesadipate-co-terephthalate) (PBAT), etc. Yet, polyesters and polyamides are mainly obtained from petroleum oils. The next section gives a brief summary of bio-monomers used to manufacture bio-based polyesters (Steinbuchel and Doi, 2002; Rydz et al., 2015).

8.4.1 Renewable bio-monomers for polyester manufacturing

Biomass-derived monomers are generally categorized based on their natural molecular biomass origins. They are classified into four main types:

1. oxygen-rich monomers such as carboxylic acids, polyols, furans and di-anhydro-alditols
2. hydrocarbon-rich monomers such as fatty acids, vegetable oils, terpenes, terpenoids and resin acids
3. hydrocarbon monomers such as bio-propene, bio-ethene, bio-butene and bio-isoprene
4. nonhydrocarbon monomers such as carbon monoxide and carbon dioxide

Among the numerous sources of building blocks for the synthesis of polyester fibres, lignin, carbohydrates and vegetable oils and, to a lesser extent, suberin, are the most interesting.

8.4.1.1 Bio-based aliphatic diacids

Polysaccharides, in particular cellulose, which is the most abundantly available biomass from the plant, are considered to be the most vital raw materials for the synthesis of sustainable chemicals, to prepare a variety of polymers such as polyesters, polyamides, polyethers, polyurethanes and polycarbonates owing to their biodegradability and biocompatibility. Alditols, aldonic and furans, which are derived from monosaccharides through chemical or biological routes, are needed for the synthesis of aliphatic or aromatic monomers for polyester.

Itaconic acid

Itaconic acid (IA), an unsaturated dicarboxylic acid, was first recognized during the thermal decomposition of citric acid in 1837. It is normally produced by fermentation using fungi, which is more economical compared with thermal decomposition or chemical methods of production (Lockwood and Nelson, 1946; Willke and Vorlop, 2001; Mondala, 2015). In 1932, Kinoshita isolated IA from the cultivation media of the osmophile eukaryotic *Aspergillus itaconicus*. Later, various *Aspergillus terreus* strains were found to be more suitable for the biosynthesis of IA using fermentation (Gyamerah, 1995). To achieve a higher yield rate, glucose or sucrose as substrates can be used, but by considering the economic sustainability of the process, complex carbon sources such as starch, molasses and hydrolysates of wood or corn syrup were also experimented with and found to be suitable for biosynthesis (Klement and Büchs, 2013; Devin et al., 2010).

Adipic acid

Higher-grade and high-purity adipic acid (AA) is normally used in the production of nylon 66, and low-purity AA is used to produce polyurethanes. In most synthetic fibre manufacturing industries, AA is normally produced by the oxidation of cyclohexane or other petroleum-based feedstock (Polen et al., 2013). AA is normally found in nature in the juice of sugar and red beets, and the synthesis of AA from renewable carbon sources such as glucose derived from starch or cellulose is a likely alternative to the synthesis of AA (Zorn and Czermak, 2014). On the other hand, biosynthesis of AA can also be accomplished through the α-amino adipate pathway or starting from long-chain carbon substrates (Draths and Frost, 1994; Niu et al., 2002).

Succinic acid

Succinic acid (SA) was first obtained from amber distillation in 1546 by Georgius Agricola. During the process of fermentation, cornstarch, cane molasses, whey, wood hydrolysates and glycerol are generally converted into SA by the action of particular strains of bacteria. *Actinobacillus succinogenes*, *Mannheimia succiniciproducens*, *Anaerobiospirillum succiniciproducens* and recombinant *Escherichia coli* are the bacterial strains used to prepare SA with higher yields and lower amounts of by-products (Cheng et al., 2012; Song and Lee, 2006; Kurzrock and Weuster-Botz). The synthesis of SA through fermentation is sustainable because it uses CO to convert the substrates into the product (Beauprez et al., 2010). The production of SA through fermentation is shown in Fig. 8.5. The high attention to SA is because this dicarboxylic acid is a main component in the production of many industrial adhesives, solvents, printing inks, coating resins, magnetic tapes, plasticizers, deicing compounds, emulsifiers and chemical and pharmaceutical intermediates (Kang et al., 2010; McKinlay et al., 2007). Furthermore, SA can be hydrogenated to obtain 1,4-butanediol (which in turn can be carbonylated to obtain AA) (Fig. 8.5).

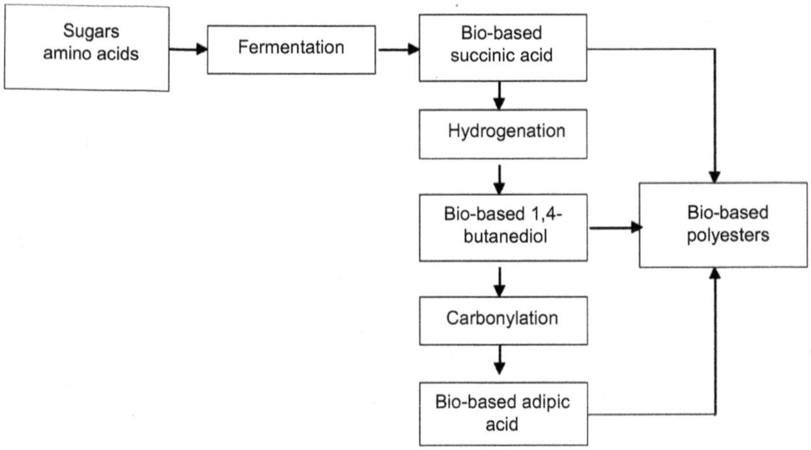

Figure 8.5 Production of bio-based succinic acid and its derivatives 1,4-butanediol and adipic acid (Kurzrock and Weuster-Botz, 2010).

Fumaric acid

Fumaric acid is a naturally occurring dicarboxylic acid and was first isolated from *Fumaria officinalis*. Fumaric acid is produced from the isomerization of maleic anhydride; interest has shifted toward a sustainable process through fermentation (Xu et al., 2010). Fungi from *Cunninghamella, Rhizopus, Mucor* and *Circinella genera* are capable of producing fumaric acid during the fermentation of renewable raw materials at a reasonable yield (Xu et al., 2012; Goldberg et al., 2006). *Rhizopus* species have been demonstrated to be the most favourable, owing to their capability of producing fumaric acid under aerobic as well as anaerobic conditions (Engel et al., 2008). Although fungi are the widely used microorganisms in the production of fumaric acid, a few bacterial strains such as *Lactobacillus* ssp. and metabolically engineered yeasts such as *Saccharomyces cerevisiae* have also been used to produce fumaric acid (Xu et al., 2012).

Terephthalic acid

Terephthalic acid is the key monomer for the synthesis of polyester along with ethylene glycol (EG) for the production of PET. Currently, terephthalic acid is produced by the catalytic oxidation of petrochemical p-xylene; however, three methods are available for the synthesis of bio-based p-xylene: (1) pyrolysis of biomass; (2) yeast fermentation of sugars into isobutanol; and (3) chemical conversion of carbohydrates (Collias et al., 2014). One method to produce bio-based terephthalic acid is the cyclo-addition of isoprene and acrylic acid. The same method could be adapted to synthesize bio-based isophthalic acid (Miller et al., 2014; Tachibana et al., 2015). Fig. 8.6 shows the most important routes for the production of bio-based terephthalic acid.

Furandicarboxylic acid

The dehydration of sugars in bio-refineries could result in a set of products such as furans, dehydro sugars and levulinic acid. Furandicarboxylic acid (FDCA) is a

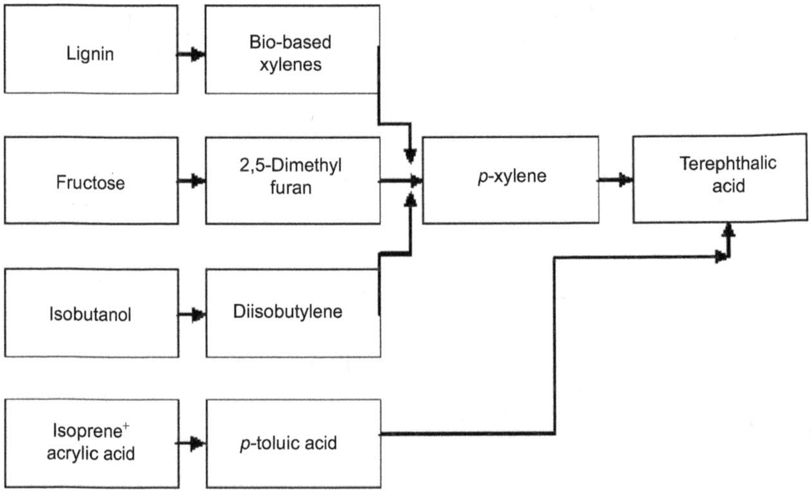

Figure 8.6 Important routes for the production of bio-based terephthalic acid.

member of the furan family and is normally derived by oxidative dehydration of glucose using oxygen or by oxidation of 5-hydroxymethylfurfural (Moreau et al., 2004; Werpy and Petersen, 2004). Owing to the comparable properties of PET produced from FDCA, it can substitute terephthalic acid in the production of various bio-based polyesters. The major limitation in commercial production of FDCA from renewable materials is the uncontrolled process of sugar dehydration (Sousa et al., 2013). Furthermore, an effective esterification reaction for the synthesis of polyester cannot be obtained because of the control of FDCA reactivity. Avantium (Geleen, The Netherlands) publicized a technology that involves a highly reactive catalyst along with an efficient separation technology for the commercially viable production of FDCA (Gubbels et al., 2013).

8.4.1.2 Bio-based aliphatic diols and polyols

1,3-Propanediol

The wide range of applications of propanediol (PDO), such as in the synthesis of polymers, solvents, lubricants and cosmetics, has attracted huge attention to this monomer. It is also used in the production of a polymer with excellent stretch recovery known as PET (Biebl et al., 1999; Zeng and Biebl, 2002).In a nonsustainable way it can be synthesized using acrolein or ethylene oxide, but it requires harsher conditions and nonrenewable resources, and mainly releases toxic by-products. Hence, biological routes have been attempted to overcome these limitations using several kinds of bacteria such as *Klebsiella pneumoniae*, *Lactobacillus brevis* and *Enterobacter agglomerans* to produce 1,3-PDO during anaerobic growth on glycerol (van Gelder et al., 2012). An aerobic process has also been attempted for bioconversion of glycerol into PDO because it is more economical than an anaerobic process (Lee et al., 2004).

Butanediol

Fermentation to synthesize 3-butanediol (2,3-BDO) is an old method. The sugar or carbon substrates are the main carbon source for the fermentation process of 2,3-BDO and more costly (Ji et al., 2011). Hence, biomass-derived sugars such as lignocellulosic or noncellulosic, which are cheaper, have been attempted to produce it. Another monomer in this category is 1,4-BDO and its production is mostly based on fossil carbon resources (Anvari et al., 2009). The fermentation process has been successfully attempted to produce bio-based 1,4-BDO by *S. cerevisiae* under anaerobic conditions. Bacterial production is known for its slow fermentation rate and relatively low yields and therefore has received limited attention (Celinska and Grajek, 2009).

Glycerol

Owing to the increase in the cost of propylene and its scarcity in developing countries, glycerol has become an attractive substrate for the production of several chemicals. The production of this chemical by *S. cerevisiae* under anaerobic conditions in which sugars are converted into ethanol in a redox-neutral process leads to glycerol as a by-product (Wang et al., 2001). It can also be produced by bacteria and algae, which is known for its slow fermentation rate and low yields.

Sorbitol and isosorbide

Although sorbitol can be derived by the chemical reduction of glucose, numerous bio-based methods have been explored for its synthesis. A bio-based fermentation with sucrose or glucose–fructose mixtures using *Zymomonas mobilis* could produce sorbitol with comparative yields (van Haveren et al., 2008; Silveira and Jonas, 2002). On the other hand, enzymatic hydrolysis using cereal-extracted starch and mannans followed by the production of D-sorbitol and D-mannitol through hydrogenation have been reported.

Ethylene glycol

The huge increase in biodiesel production over the years with glycerol as a by-product has led to a drastic reduction in cost, making this monomer a striking raw material for the synthesis of EG and 1,2-propylene glycol. Ethylene glycol is synthesized by the oxidation of ethylene followed by the addition of water. It is also possible to produce bio-based EG, but the yields of oxygen and carbon are lower (Ladero et al., 2007).

8.4.1.3 Bio-based aliphatic diamines

1,4-Butanediamine (1,4-BDA) can be synthesized naturally by the decomposition of amino acids in dead or living organisms. Bio-based 1,4-BDA can be synthesized through the chemical conversion of biomass-derived SA or by means of fermentation of sugars using *E. coli*. 1,5-Pentanediamine (1,5-PDA), a natural compound, is synthesized by the hydrolysis of protein during the tissue putrefaction of animals (Kang et al., 2010). The bio-based synthesis of 1,5-PDA is carried out by the decarboxylation of lysine using numerous microorganisms or the fermentation of sugars using metabolic engineering. 1,6-hexanediamine (1,6-HDA) is a raw material for the synthesis of nylon 6,6, nylon 6,10 and PA6,T, which is currently produced industrially from petrol-based butadiene or propylene (Steiger et al., 2013). Developments show that bio-based 1,6-HDA can be produced by chemical-catalytic conversion of AA or 1,6-hexanediol derived from carbohydrates, or by fermentation. 1,8-Octanediamine can potentially be derived from biomass. It can be produced by amination of suberic acid, which can be converted from plant oils. 1,10-Decanediamine can be chemically converted from sebacic acid derived from castor oils. They are interesting bio-based monomers for the synthesis of fully bio-based nylon 10,10, which is commercially available on the market (Xie and West, 2009).

8.4.1.4 Bio-based aromatic monomers

Regarding the aromatic monomers, lignin and carbohydrates are considered to be the major sources, with FDCA and vanillic acid significant examples to be emphasized. Lignin is one of the main noncarbohydrate components in a lignocellulosic biomass which is composed by oxidative coupling of three phenylpropane compounds such as *p*-coumaryl alcohol, sinapyl alcohol and coniferyl alcohol (Hatakeyama and Hatakeyama, 2010). Owing to its distinctive chemical structures and properties, it offers a wide range of possibilities for the production of several bio-based aromatic materials which are not possible through chemical or biological modifications of other biomass feedstocks, as shown in Scheme 8.1. Yet, it is a big task to develop an efficient method for the recovery of aromatic chemicals (Pandey and Kim, 2011).

Vanillin is an aromatic aldehyde extracted from the genus *Vanilla planifolia*. Only vanillic acid is synthesized by oxidizing lignosulphonates. It is obtained by air oxidizing lignin in the presence of metal catalysts. It is also possible to produce vanillic acid by a bio-based process using fungi, bacteria or genetically modified microorganisms. Vanillic acid, a kind of hydroxyl acid, also can be synthesized by means of bioconversion of ferulic acid (Pearl, 1942).

Terephthalic acid, which is currently synthesized by the oxidation of *p*-xylene, is used as a main precursor for the production of aromatic polyesters and polyamides. Bio-based methods to produce terephthalic acid and its precursors are shown in Scheme 8.2.

FDCA is considered to be the most capable monomer for the synthesis of bio-based polyesters compared with chemical-based monomers. FDCA is synthesized from biomass feedstocks and can be synthesized through a bio-catalytic approach starting from hydroxymethylfuran (HMF), as shown in Scheme 8.3. Other interesting bio-based furan monomers for polyester or polyamide synthesis include 2,5-*bis*(hydroxymethyl) furan, 2,5-*bis*(hydroxymethyl)-tetrahydrofuran, 2,5-*bis*(aminomethyl)furan and 2,5-*bis*(aminomethyl)tetrahydrofuran (Scheme 8.3). The conversion process of HMF into 2,5-furandicarboxylic acid consists of two main steps: (1) conversion of HMF to a dialdehyde derivative, and (2) oxidation of the two aldehyde groups to carboxylic acid groups (Scheme 8.4). Earlier, oxidation of HMF to FDCA normally was done using nitric acid, silver oxide or potassium permanganate in the presence of organic solvents such as dimethyl sulphoxide (Becker et al., 2015). More eco-friendly solvents and oxidants are now used in the oxidation reactions using heterogeneous metal catalysts

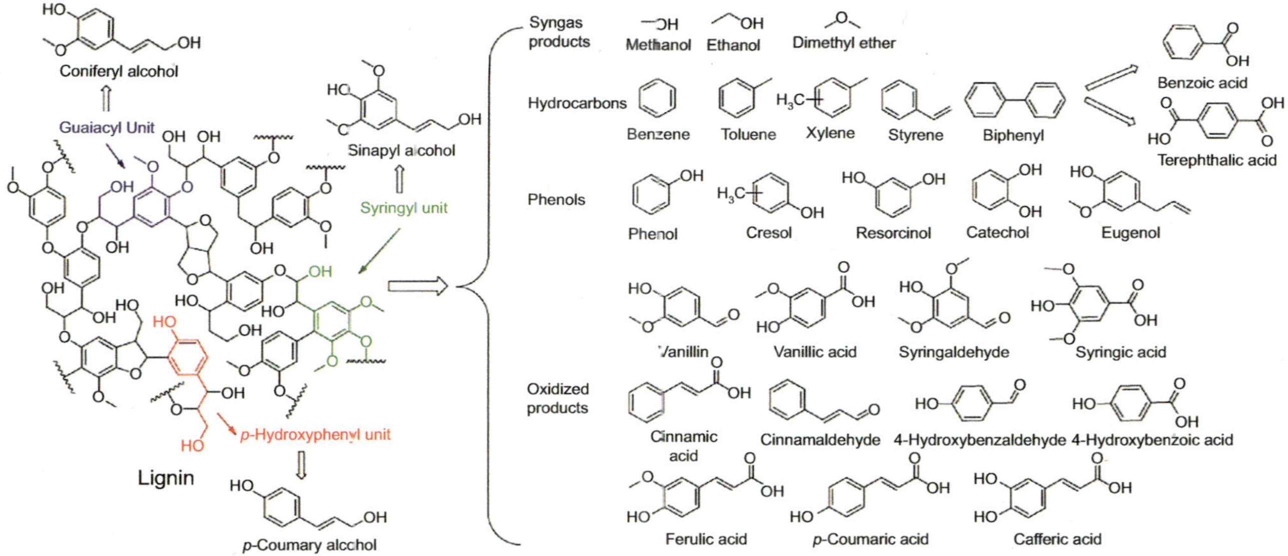

Scheme 8.1 Favorable bio-based chemicals derived from lignin (Hatakeyama and Hatakeyama, 2010).

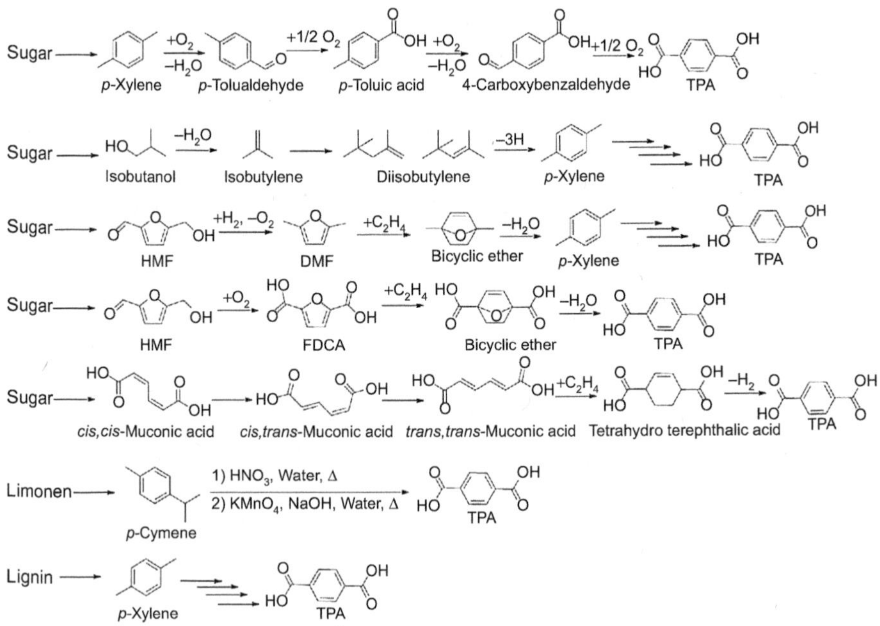

Scheme 8.2 Bio-based routes for the production of terephthalic acid and its precursors (Becker et al., 2015).

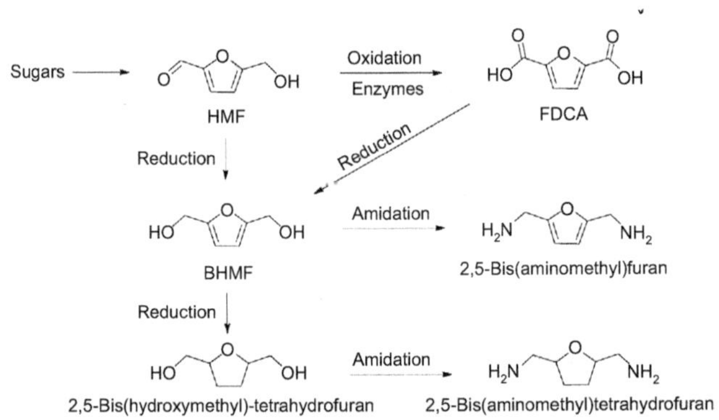

Scheme 8.3 Bio-based furan monomers for the synthesis of polyester or polyamide (Becker et al., 2015).

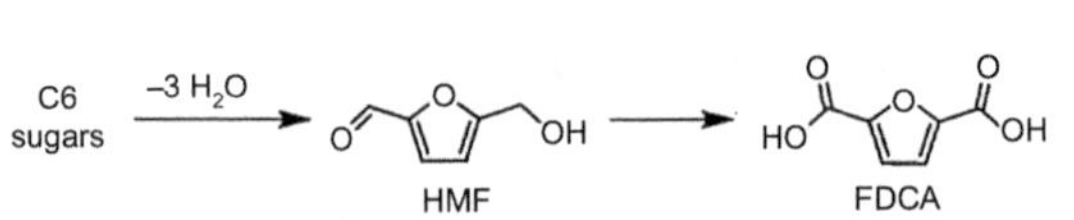

Scheme 8.4 Conversion of C6-sugars into 2,5-furandicarboxylic acid (Becker et al., 2015).

such as platinum supported on alumina (Pt/Al) or carbon (Pt/C) in water under an atmospheric oxygen pressure. Alternatively, enzymes could also be used for the bioconversion of HMF into FDCA (Dijkman et al., 2014; Jong et al., 2012).

8.4.1.5 Other bio-based monomers

Lactic acid, a kind of organic acid found in several carbohydrates, is identified as a main metabolic intermediate. It is currently produced in a chemical synthesis method through the hydrolysis of lacto-nitrile using concentrated hydrochloric acid. Although this method looks simple, it leads to a racemic mixture of L- and D-lactic acid (Datta et al., 1995). It can also be produced by the bacterial fermentation of sugars, which necessitates the use of biomass feedstocks and requires lower energy consumption and low temperature. Although it can be synthesized from bio-based glycerol, because of its high cost it is not considered a feasible method (Onda, 2014).

3-Hydroxypropionic acid, a bio-based platform building block, can be synthesized through a chemical method using 1,3-PDO, 3-hydroxypropionaldehyde or acrylic acid, but is costlier. Bio-based methods employing the fermentation of sugars using genetically modified microorganisms have been initiated (John et al., 2007). Several long chain fatty acids such as oleic acid, erucic acid, ricinoleic acid and vernolic acid can be synthesized from renewable resources such as plant oils and fats, which are considered good building blocks for the production of polyesters and polyamides. Furthermore, there are several other bio-based building blocks for the synthesis of polyesters or polyamides, such as EG, 11-amino-undecanoic acid and polycarboxylic acids (citric acid and tartaric acid) (Okano et al., 2010).

8.4.2 Sustainable aliphatic polyesters

Aliphatic polyesters are favourably obtained through polycondensation reactions of alkanediols with α,ω-alkanedioic acids (or) are derived from hydroxyl-alkanoic acids which are readily available from inexpensive renewable resources.

8.4.2.1 Poly(lactic acid)

Among the family of aliphatic polyesters, PLA is one of the most promising industrial products for plastics with the prospective to substitute for polymers such as PET, polystyrene and polycarbonate. The high-molecular weight PLA can be produced by ring opening polymerization (ROP) of the corresponding lactide, as shown in Scheme 8.5. The process is based on removing water under milder conditions, without a solvent to produce a cyclic intermediate dimer referred to as a lactide. Production creates three different stereoisomeric forms: L-lactide, D-lactide and meso-lactide, and the final properties of PLA produced depend on the ratio of the forms. The production of PLA through the lactide route allows the possibility of modification for superior control of the polymer's properties by controlling the optical sequence of the polymer backbone (Avinc and Khoddami, 2009; Gruber and

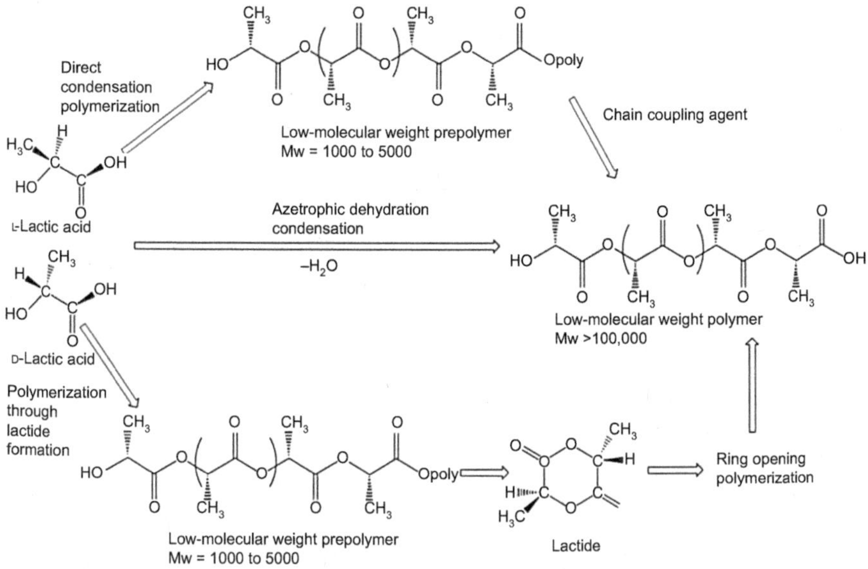

Scheme 8.5 Ring-opening method of production of polylactic acid (Averous, 2008).

O'Brien, 2005). The ratio of D and L isomers and their distribution along the polymer backbone influence the molecular weight, crystallinity and melting point of the end product.

Cargill Dow LLC developed a patented, low continuous process for the production of lactic acid-based polymers. The process starts with a continuous condensation reaction of lactic acid to produce low-molecular weight PLA prepolymer. Prepolymer is converted into a mixture of lactide stereoisomers using tin catalysts to enhance the rate of intramolecular cyclization reaction (Averous, 2008). Finally, PLA polymer is produced using a ring-opening lactide polymerization, eliminating the use of solvents. After the polymerization is complete, any remaining monomer is removed under vacuum and is recycled for use as a raw material (Gupta et al., 2007).

PLA's main applications are in short-term packaging, owing its biodegradability, and in biomedical applications such as implants, sutures and drug encapsulations owing to its biocompatibility in contact with living tissue (Farrington et al., 2005; Hongu et al., 2005). Currently, Nature Works LLC (Blair, Nebraska, United States) is the leading supplier of PLA biopolymer under the brand name Ingeo, with a production capacity of 140,000 metric tons per year.

8.4.2.2 Polyhydroxyalkanoates

PHA, an important class of polyesters, is a polymer produced naturally by means of several kinds of microorganism; it is considered a potential substitute for conventional hydrocarbon-based polymers. These kinds of biopolymers belong to the family of intracellular biopolymers synthesized by fermentation using bacteria (Fig. 8.7). They can also be produced using plants and transgenic microorganisms. The

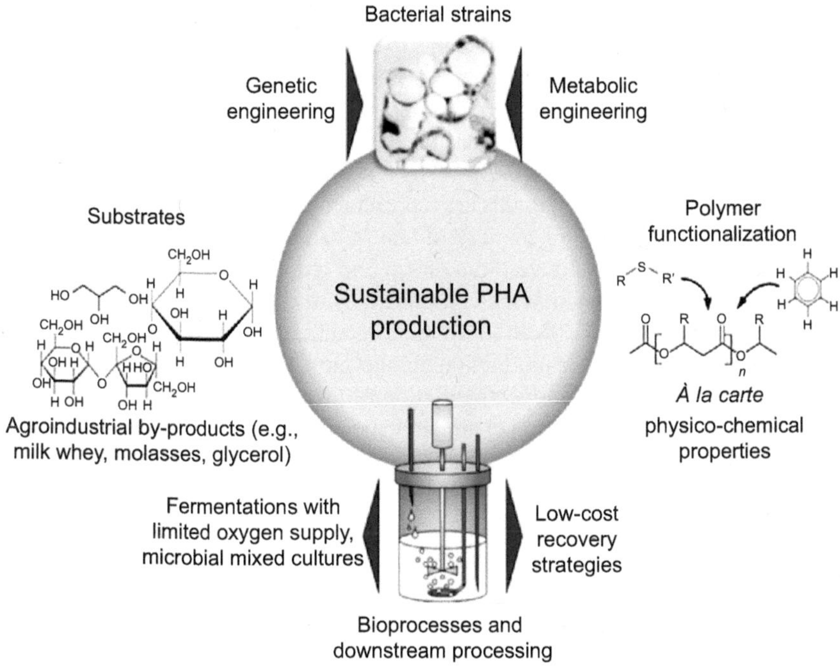

Figure 8.7 Bio-based production of polyhydroxyalkanoates (PHA) using fermentation (Madison and Huisman, 1999).

important kind of biopolymer in this family is the polyhydroxybutyrate homopolymer, but different poly(hydroxybutyrate-cohydroxyalkanoates)co-polyesters exist such as poly(hydroxybutyrate-co-hydroxyvalerate), poly(hydroxybutyrate-co-hydroxyoctanoate), poly(hydroxybutyrate-co-hydroxyhexanoate) and poly(hydroxyl butyrate co-hydroxy octadecanoate) (Vink et al., 2004; de Koning, 1993). They are suitable for various applications such as short-term packaging, are biocompatible in contact with living tissue and can be used for biomedical applications. Worldwide, more than 24 companies are engaged in their production and applications. However, compared with PLA, their production is lower.

PHA polymers are synthesized in the bodies of bacteria fed with glucose in a fermentation plant. In the late 1980s, ICI Zeneca commercialized PHAs under the trade name Biopol. However their high cost, the small difference between their melting and thermal degradation temperatures and their low-impact resistance prevented their large commercial application. Nevertheless many companies are currently producing PHA because their development looks promising (Zinn et al., 2001; Kotnis et al., 1995; Ebata et al., 2007).

8.4.2.3 Vegetable oil-based polyesters

Amongst several vegetable oil-based monomers, ricinoleic acid (12-hydroxy-9-cis-octadecenoic acid) extracted from castor oil has an important role owing to its bifunctionality. The use of ricinoleic acid to produce homopolyesters is rare. Indeed, there is only a reference to poly(ricinoleic acid) prepared by lipase-catalysed condensation of methyl ricinoleate, which was subsequently cross-linked to produce thermosetting elastomers. This work has the merit of having used both a renewable monomer and an eco-friendly synthesis approach, i.e., bulk enzymatic polytransesterification using several lipases. Furthermore, it was also demonstrated that poly(ricinoleic acid) can easily be recycled by enzymatic hydrolysis followed by repolymerization.

Other developments related to this are represented by the preparation of long chain aliphatic polyesters, which are semicrystalline in nature, using fatty acids, in which they are converted into a saturated α,ω-diacid and reduced to its diol counterpart to obtain the monomer from which the consequent synthesis of polyester is carried out by esterification or transesterification. In another kind of study, erucic acid was used to synthesize monomer for the production of renewable long chain aliphatic polyesters using olefin self-metathesis. Polycondensation of the resultant 1,26-hexacosanedioic acid with α,ω-diols produced long chain aliphatic polyesters with higher crystallinity and melting and degradation properties (Kotnis et al., 1995; Ebata et al., 2007). In another context, alternative methods such as the use of thiolene and metathesis polymerization for the synthesis of aliphatic polyesters were reported by several researchers. Glycerol is abundantly available because it is produced from vegetable oils. Hence, it is used as a better intermediate for the synthesis of several industrial chemicals. In addition, glycerol-based polyesters and their synthesis route and properties have been reported (Shogren, 1995; Petrovicn et al., 2010; Slivniak and Domb, 2005).

8.4.2.4 Suberin and cutin-based polyesters

Another method for producing bio-based aliphatic polyesters is to use suberin and cutin monomers. Little research has been done on the use of long chain hydroxyl-alkanoic acids isolated from suberin or cutin or, instead, a mixture of monomers to synthesize polyesters (Sousa et al., 2012; Sousa et al., 2008; Heredia-Guerrerom et al., 2009). Depolymerized suberin extracted from cork, which consists mainly of long chain α,ω-dicarboxylic acids and ω-hydroxy acids, was used to synthesize bio-based polyesters by emulsion polymerization or using *Candida antarctica* lipase B (CALB) (Pintom et al., 2009). In other research, depolymerized suberin extracted from cork as well birch outer bark along with some functional groups such as diols or diacids to balance the OH/COOH stoichiometry for the production of several interesting bio-based polyesters was reported. The resulting linear or semicross-linked polyesters had a reasonable yield and showed properties similar to those of conventional petroleum-based aliphatic polyesters (Sousa et al., 2008; Kricheldorf et al., 2003).

8.4.2.5 Sugar diol-based polyesters

Owing to advantages such as rigidity, nontoxicity and sustainability of the resultant polyesters using the 1,4:3,6-dianhydrohexitols (DAHs) monomers such as isosorbide, isomannide and isoidide, these monomers have gained importance for the synthesis of polyesters. The main limitation of using this group of monomers for the synthesis of polyesters is that the ensuing polyester has low molecular weight, owing to the low reactivity of hindered hydroxyl groups (Garaleh et al., 2010). Research in the production of bio-based aliphatic polyesters from DAHs is based on melt polymerization using a renewable source of isohexide, in which the isoiodide dicarboxylic acid, diacid, and the isoidide dimethanol, diol, were prepared by a multistep process from isomannide (Wu et al., 2013). On the other hand, several interesting research works were carried out on the production of bio-based polyesters from diols derived from sugars (Alla et al., 2011). For example, aliphatic polyesters possessing high wettability and solubility were produced from dimethyl 2,3:4,5-di-O-methylene-galactarate.

8.4.2.6 Other aliphatic polyesters

Among the wide range of monomers available from renewable resources, carboxylic acids such as SA and IA have received much attention and produced interesting materials. Other carboxylic acids that are worth mentioning but are less common include tartaric acid and citric acid (Albertsson and Varma, 2002; Cameron and Shaver, 2011).

8.4.3 Renewable aromatic polyesters

The aliphatic polyesters that were discussed earlier are the most promising biopolymers but they do not possess the important properties needed to substitute completely for conventional plastics. On the contrary, aromatic polyesters have good properties such as high molecular weight and higher melting and glass transition temperatures compared with aliphatic polyesters. High-performance thermoplastic PET and PBT are the main commercial examples of aromatic polyesters. However, the precursors used to produce these fibres are fossil-based rather than renewable feedstock, and hence they are not degradable under normal environmental conditions. Thus, considerable research has been done in aromatic polyester fibre production to imitate their excellent properties using renewable monomers and to make them biodegradable (Xie and West, 2009; Houck et al., 2001).

Many big players in the polymer market have commercialized biodegradable or partially renewable-based aromatic polyesters. BASF has commercialized PBAT with the brand name Ecoflex and have carried out biodegradation studies on PBAT using several degradation enzymes. DuPont pioneered a way to produce PDO from corn sugar using a patented fermentation process; the result was a PDO termed Bio-PDO (Kurian, 2005; Shen et al., 2009). They introduced a PTT, also known as 3 GT, under the trade name Sorona. It is a corn-based fibre made from 1,3-PDO and a petrochemical-based monomer. The fermentation process consists of two steps: (1) yeasts ferment glucose to glycerol, and (2) microbes ferment glycerol to PDO. DuPont developed a patented process in collaboration with Genencor consisting of the metabolization of glucose derived from wet-milled corn by the genetically engineered microorganism *E. coli* (Fig. 8.8). PTT can be produced either by transesterification of dimethyl terephthalate with PDO or by an esterification route starting with purified terephthalic acid and PDO. After the organism ingests glucose, it produces PDO, which is separated by filtration and is concentrated by evaporation, followed by purification by distillation. Several kinds of aromatic monomers have been recognized for the bio-based production of aromatic polyesters, such as polysaccharide-based substrates including lignin-based compounds and furan derivatives (Wolf et al., 2005).

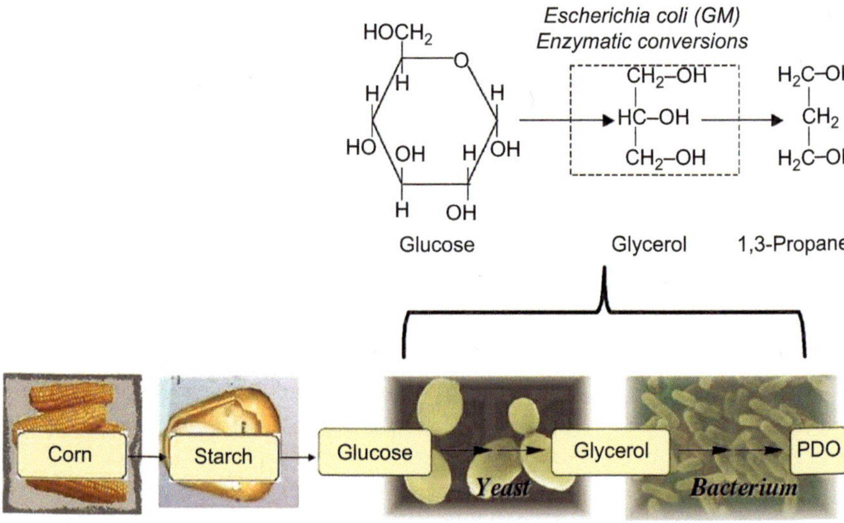

Figure 8.8 Production of Bio-PDO by the fermentation route (Kurian, 2005). *PDO*, propanediol.

8.4.3.1 Furan-based polyesters

Furan polymers are not new. Poly(hexamethylenefuranoate), a polyester based on furan monomer was produced by Moore and Kelly in 1978b. Subsequently, several forms of furan polyesters were successfully synthesized by many researchers (Moore and Kelly, 1978a; Gandini et al., 2009a,b; Ma et al., 2012a,b). Gandini et al. (2009a,b) confirmed that FDCA can be used in polycondensation reactions to produce PEF and other FDCA-based (co)polyesters. In the case of PEF, the similarity with PET in terms of thermal properties was unambiguously demonstrated. The realm of FDCA can be extended to a more vast number of polyesters besides PEF, such as the ones prepared by using FDCA and three isomers of 1,4:3,6-dianhydrohexitol, PDO or hydroquinone. Avantium has demonstrated YXY for the production of fully bio-based PEF through an innovative chemical, catalytic process that allows for the economic production of furanics based on a wide range of carbohydrates (Ma et al., 2012a). The YXY process converts biomass from various sources into the building blocks of alkoxy-methyfurfural (RMF) using a dehydration process in the presence of alcohols (R-OH) and catalysts. The process then transforms the RMF to FDCA via an oxidation process (Ma et al., 2012b). FDCA is polymerized in the presence of EG forming PEF, yielding a high-performance engineering thermoplastic capable of multiple applications, as shown in Fig. 8.9.

Research in the field of furan-based polyesters includes Zhuon polyester [poly(butylene 2,5-furandicarboxylate)] based on FDCA and BDO; it possess a crystal structure, mechanical and thermal properties similar to those of PBT. A new way to produce an FDCA-based polyester derived from furfural and with total carbon use has been proposed. It consists of four main steps: (1) oxidation of furfural

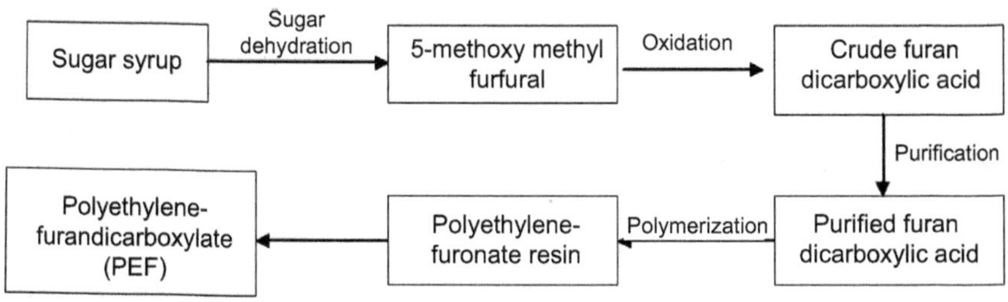

Figure 8.9 Bio-based production of poly(ethylene furanoate) using furan monomer (Ma et al., 2012a).

into furoate; (2) disproportionation of furoate to furan and 2,5-furandicarboxylate; (3) hydrogenation and hydrolysis of furan to 1,4-butanediol; and (4) polymerization of 1,4-butanediol with FDCA to poly(butylenes 2,5-furandicarboxylate).

8.4.3.2 Lignin-based polyesters

In one research project, acetic anhydride and vanillin extracted from lignin was used to synthesis 4-acetyldihydroferulic acid. This was employed to produce bio-based polyester, poly(dihydro ferulic acid), with mechanical and thermal properties comparable to conventional PET, providing a new route of lignin-derived monomers for the preparation of aromatic polymers (Mialon et al., 2010; Firdaus and Meier, 2013). In other research work, aromatic hydroxyl acids extracted from the lignin were used for the production of biorenewable aromatic polyesters. In this method, poly(alkylene4-hydroxybenzoate), poly(alkylene syringate) and poly(alkylene vanillate) were prepared by means of bulk polycondensation of the resultant alkylene hydroxyl benzoates (Mialon et al., 2010; Lovell et al., 2013). In a kind of research, thermoplastic aromatic polyesters with better thermal stability were produced with fatty acid derivates along with vanillin through three different polymerization reactions such as added thiolene, acyclic diene metathesis and polycondensation (Dong et al., 2011).

8.4.4 Sustainable routes to polyesters

This section of the chapter deals with research towards the sustainable synthesis of polyester by means of gentle polymerization conditions, green catalytic systems and/or atom-efficient approaches. One drastic change in the sustainable production of polyester is the replacement of highly efficient (poly)transesterification antimony-based catalysts by less toxic catalysts based on titanium. For example, in the synthesis of furanic-based polyesters, transesterification is followed by polytransesterification using titanium(IV) tetraisopropoxide, tin(II) 2-ethyl hexanoate and titanium(IV) but also oxide or zirconium(IV) butoxide. The main challenge faced with this technique is the need for a high temperature and low pressure system to achieve a high molecular weight and higher yield (Mathers and Meier, 2011). These limitations are partially nullified by the ring opening polymerization method, which does not involve the vacuum removal of by-products; the reaction occurs at atmospheric pressure.

In other work, the production of aliphatic polyester by polycondensation is carried out in water in the presence of a Bronsted acid surfactant catalyst, in which polycondensation is achieved at the interface of the emulsion (Loos, 2010). The advantages of this route are it requires a low reaction temperature, avoids undesirable side reactions, uses cheaper solvents and is environmentally friendly. Metal trifluoro-methane-sulphonates known as Lewis acid catalysts have gained much attention as polycondensation catalysts for polyester production. Thus, bismuth(III) triflate-mediated polycondensation reactions of aliphatic monomers have attracted some interest throughout scientific communities owing to their low toxicity, ease of handling, low cost, and stability (Takasu et al., 2006).

Enzymatic polymerization is a promising technique for the production of sustainable polyesters because the enzymes have a highly selective reaction and require mild and neutral conditions for reaction at low temperatures. Much research work has been carried out on the use of lipase enzymes for mild polycondensation or polytransesterification. Lipases can be extracted from a range of sources including porcine pancreas and many microorganisms (Manabe et al., 2002). The enzymes are generally used as a biocatalyst for the synthesis of various kinds of polymers. Esterases, in particular lipases, catalyse the hydrolysis of fatty acid esters in aqueous environments. Most hydrolases of these kinds were found to be stable in organic solvents and are capable of catalysing reverse reactions such as esterification and transesterification. Thus, these kinds of enzymes were explored for the synthesis of aliphatic polyesters to a larger extent, and aromatic polyesters to a lesser extent.

Hydrolases for the production of polyester are stable, easily extractable and commercially available. Among the various kinds of lipases, the most commonly used biocatalyst for the synthesis of polyester is lipase B from CALB, an immobilized catalyst. It is synthesized by means of the genetically modified *Aspergillus oryzae* microorganism and is immobilized on poly(methylmethacrylate-*co*-butyl methacrylate) resin. This enzyme is active for polycondensation as well for ROP in bulk or organic media, and results in a reasonably higher yield of polyesters (Manabe et al., 2001). The most important merits of CALB-catalysed reactions is their selectivity towards primary hydroxyl groups, which permits the one-pot synthesis of polyesters of 'diols' with extra secondary OH groups without the need for tedious protection/deprotection steps.

Several varieties of lipase enzymes were explored by various researchers to produce polyesters. From various research works, it was observed that another promising biocatalyst for the synthesis of bio-based polyester is lipase from *Mucormiehei.* Using this enzyme, polymers of AA and 1,4-BDO were obtained with higher molecular weight. The biocatalyst generally exhibits different ranges of activity and selectivity based on the reaction medium, and hence supercritical carbon dioxide was also explored as a reaction solvent to replace organic solvent. The urge to widen the substrate specificity for the synthesis of polyesters resulted in research towards various hydrolases, particularly the cutinase family (Yashiro et al., 2009). The structure of CALB and cutinase from *Humicola insolens* was used for the biosynthesis of polyesters. Although these kinds of enzyme-based bio-catalytic production of polyester are sustainable, they have not yet been commercialized owing to the long reaction time and higher cost involved. Hence, even in such a promising strategy for polyester, industry and scientists have to find a balance to discover cheap and viable processes to enable the use of enzymes on a large scale in polyester synthesis.

8.5 Sustainable nylon

Nylon is a synthetic like polyester derived from petrochemicals, It is versatile and used for various kinds of applications such as stockings and parachutes, carpets, packaging and even car parts. Around 70 billion barrels of oil are consumed every year to produce polyester. Although more polyester than nylon is produced each year, nylon requires more energy (almost three times that of cotton) to manufacture. Nylon production results in the release of GHGs such as nitrous oxide, which contribute significantly to global warming. Dyeing of nylon is not environment friendly because it is not suited to natural or low-impact chemical dyes. Further, nylon is not biodegradable and will persist in the environment indefinitely. The two main sources of micro-plastic pollution in the aquatic zone are nylon fishing nets and synthetic textile fibres that wear off during washing. This means that the impact of nylon on the aquatic environment is high.

Like polyester, nylon is a manufactured fibre that became accessible to the overall population around World War II. Nylon had a few vital parts in wartime. First, because of its quality and strength, nylon was widely used for military items, including parachutes, tents, rope and tires. Furthermore, nylon supplanted everything that was once produced using silk, e.g., silk stockings, because silk imports from Asia experienced critical deficiencies and vacillations in value. Fundamentally, nylon is a kind of plastic derived from unrefined petroleum. This plastic is then put through a concentrated substance procedure, bringing about the solid, stretchy filaments that make it as valuable as a fabric (Anon, 2016a).

Nylon is a manufactured man-made fibre from petrochemicals, a thermoplastic widely used throughout the design business. It is flexible to a great degree and is employed for everything from stockings and parachutes, floor coverings, bundling and even auto parts. The sheer broadness of nylon's uses implies that it is hard to reveal what the general commitment of nylon in design is to environmental change and natural impacts. Nevertheless, we can consider the effect of polyester to get a general feeling for this. About 70 billion barrels of oil are used each year to create polyester. Although more polyester than nylon is delivered every year, nylon requires more vitality to produce. Nylon also takes three times more concentrated energy than cotton to deliver. Considering this, we can begin to get a feeling for the size of the effect.

In particular, nylons are a group of materials called polyamides. They are produced using responding carbon-based chemicals found in coal and petroleum in a high-pressure, warmed environment. This compound response, known as buildup polymerization, shapes a substantial polymer as a sheet of nylon. To make nylon fabric for clothing, this nylon sheet is then broken into chips, liquefied and drawn through a mechanical spinneret to deliver singular filaments that are woven into fabric. Various types of nylon have distinctive properties; however what is consistent ideas about each is their quality, sturdiness and capacity to be formed into shape. On the other hand, no type of nylon is biodegradable; so once you no longer need your torn leggings or old toothbrush, it sits in a landfill for no less than 30 years.

Nylon was introduced to fashion in the 1930s as a dependable contrasting option to silk stockings. Currently low-quality nylon is the standard, and numerous nylon items do not last as well as they ought to. High-quality nylon is hard-wearing and keeps going admirably. The creation of nylon results in the release of nitrous oxide, a GHG that makes a huge contribution to global warming. Discharges from a single United Kingdom nylon plant in the 1990s were thought to have had a worldwide effect on altering temperature comparable to more than 3% of the United Kingdom's whole carbon dioxide outflow (Wen et al., 2012).

Nylon is also not suited to natural dyes and lowest-impact chemical dyes, which implies that the way to colour fibre also creates a huge amount of water contamination. With more nylon being produced in nations with weaker environmental protections set up, this makes nylon a huge contributor of water contamination and mutagenicity. However, nylon requires less water to deliver than regular filaments, so a portion of the strands' effect on water is relieved by this. Nylon is not biodegradable and will remain in nature indefinitely. Two of the biggest sources of micro-plastic contamination in the sea are nylon fishing nets and manufactured material filaments that wear off with washing. This implies that the effect of nylon on the aquatic environment is huge. On the positive side, nylon is recyclable polymer, and hence we could stop the virgin nylon production drastically and still meet our craving for nylon. For this to happen we would need to expand the availability of nylon reusing plans to the normal customer. Some programmes for reusing nylon work, but these are a long way from being implemented everywhere. In addition to supporting some of the world's dirtiest industries, the manufacture of nylon has several other direct environmental impacts (Anon, 2016a):

- GHGs: Producing nylon creates nitrous oxide, a GHG that is 310 times more potent than carbon dioxide.
- Water: Manufacturing nylon is a thirsty process; large amounts of water are used to cooling the fibres, which can be a source of environmental contamination and pollution.
- Energy: Manufacturing nylon is an energy-hungry process which contributes to environmental degradation and global warming.

8.5.1 Sustainable production of nylon

8.5.1.1 Sustainable adipic acid production

AA is one of the most important chemicals in the manufacturing industry. This chemical serves as a major precursor for the nylon 6,6 synthesis process. In 2013 alone, nearly 3.5 million metric tons of AA was produced with an average annual growth rate of 5% per year (Wen et al., 2012; Anon, 2014b).

Kuo Chu Hwang reported an N_2O-free process for AA synthesis as a sustainable step for nylon manufacturing. In their study they mentioned different steps:

1. First, there is air oxidation of cyclohexane under high temperatures (125–165°C) and high pressure (8–15 atm) to produce KA oil (i.e., a mixture of cyclohexanone and cyclohexanol) (Hwang, 2014; Castellan et al., 1991; Chauvel and Lefebvre, 1989).
2. Second, nitric acid is applied as an oxidant and the conversion is approximately 100%. Selectivity for AA is 93%–95% with some other short-chain acids as side products.

In the existing industrial production process, to produce 1 kg AA, approximately 0.3 kg N_2O gas is formed. Other than this, existing AA production emits NO, which is an ozone-depleting GHG. It consumes more energy, the process makes the reaction vessels corrosive and above all, the manufacturing process yields only a low amount of AA (Schneider et al., 2010; Cavani and Teles, 2009). In the current industrial process, after the application of all possible recovery, recycling and energy conservation efforts are made, the industry still polluting the environment with approximately 5%–8% of anthropogenic N_2O emissions worldwide annually (Reimer et al., 1994). AA can also be produced via the multistep enzymatic conversion of D-glucose via an *o*-hydroxy phenol intermediate. Despite the high overall yield of 97%, this process requires a large quantity of enzymes and is too costly for commercial production (Draths and Frost, 1994).

In a simple experiment with no metal catalysts or solvent, only simple ozone gas was bubbled through neat cyclohexane with concurrent UV irradiation at room temperature. This process yielded a solid precipitation at the bottom of the reaction vessel after 2–8 h. A portion of the liquid cyclohexane evaporated as a result of ozone gas bubbling. The AA was obtained by researchers as a yield after the treatment of neat cyclohexane, cyclohexanol or cyclohexanone with ozone at room temperature and 1 atm pressure. Addition of acidic water, exposure to UV light irradiation or a combination of both dramatically enhanced the oxidative conversion of cyclohexane to AA (Hwang, 2014).

8.5.1.2 From microemulsions to adipic acids

Microemulsions are dispersions of oil and water which are transparent and thermodynamically stable. Microemulsions are can be of nonaqueous structured polar solvents (formamide, glycerol and their mixture with water). Microemulsions are distinguished from other emulsions by characteristics such as higher thermodynamic stability and their nanostructure. Direct microemulsions are usually expressed as oil-in-water microemulsions in which nanoscopic oil droplets are dispersed in water with the help of a surfactant and a cosurfactant. These microemulsions are suitable as reaction media for sustainable development because of their numerous properties (Blach et al., 2010); they allow organic synthesis in water, the specific localization of the reactants, interfacial orientation of the molecules and an increase in local concentrations and the reaction rate (Ravishankara et al., 2009).

Philippe Blach et al. proposed a recyclable and environmentally friendly process for the oxidation of cyclohexene by hydrogen peroxide in microemulsions (Lapisardi et al., 2005). They developed nano-structured media based on the molecular-economy principle. In the current process used in the manufacturing industries, the oxidation of a cyclohexanol/cyclohexanone mixture or cyclohexane with nitric acid is used to produce AA. This process is considered as harmful owing to the higher amount of toxic gas emissions into the atmosphere. These gases contribute greatly to ozone depletion and global warming. Hence, sustainable developments in the process are vital for the nylon manufacturing industries (Bolm et al., 1999; Che et al., 2006; Deng et al., 1999; Karimi et al., 2005). Current research by Philippe Blach et al. mentioned a sustainable way to develop AA by oxidizing cyclohexene by hydrogen peroxide in the presence of a catalyst. The reaction scheme is shown in Scheme 8.6.

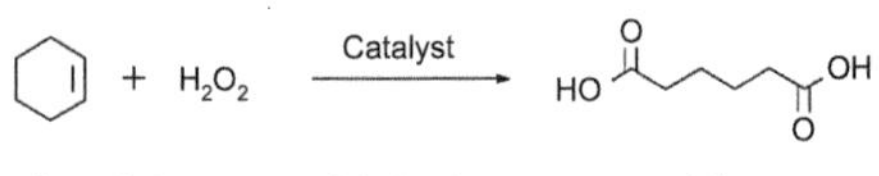

Scheme 8.6 Oxidation of cyclohexene with hydrogen peroxide.

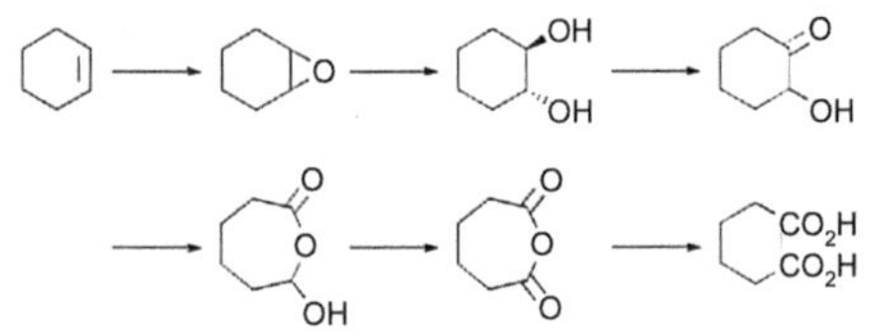

Scheme 8.7 Pathway for the formation of adipic acid (Blach et al., 2010).

In this process, hydrogen peroxide (H_2O) is used owing to its wide acceptance as a green oxidant, because it is relatively nontoxic and its decomposition leads to benign by-products such as water and oxygen. Hydrogen peroxide is an important chemical for production because of its low cost and easy handling ability and storage; it also possesses a high amount of oxygen (Jiang et al., 2002; Phillips and Romão, 1999; Saladino et al., 2009; Nardello-Rataj et al., 2008; Sato et al., 1998). The researchers also detailed that for this process to achieve conversion, hydrogen peroxide alone is insufficient: it is essential to use catalysts. Oxidation of cyclohexene by H_2O_2 typically involves epoxidation of the double bond, its opening and transformation to a diol and then a Baeyer-Villiger oxidation and multiple hydrolysis steps resulting in AA (Scheme 8.7).

A microemulsion was developed with a high concentration of cyclohexene with a relatively low concentration of surfactant (to limit the viscosity of the microemulsion) in respect to the stoichiometry of the reaction. The composition of this point was the following: cyclohexene 8 wt% (0.975 zmol L1); BenzCl 15 wt%; water 77 wt%. Oxidation of cyclohexene was performed in the selected microemulsion and surfactant, hydrogen peroxide (30 wt%), catalyst, cyclohexene and sulphuric acid were mixed (homogeneous solution) at 70°C for 18 h. After reaction and cooling at room temperature, AA precipitated and was recovered by filtration. Pure AA (analysed by gas chromatography and nuclear magnetic resonance) was obtained after washing the crude solid with cold water (Chauvel and Lefebvre, 1989).

Recycling ability of the microemulsion system

The study conducted by Philippe Blach et al. confirmed that it is possible to obtain a good yield in the production of pure AA without using extreme conditions of stirring, high temperatures or fastidious purification steps. The researchers also mentioned the recyclability of a newly developed system to showcase the sustainability of the process. In the synthesis process, after precipitation and recovery of the AA (formed during the first cycle), the team analysed the presence of peroxide in the solution and found it to be absent. Then they evaporated the remaining water under low pressure until precipitation (solubilized AA and surfactant) occurred. After this process, as in the first cycle, the same amounts of hydrogen peroxide and cyclohexene were reintroduced and another reaction was started under the same conditions as in the first cycle. After this second cycle, AA was again recovered by filtration after cooling to room temperature. The yield rapidly reached extremely interesting values, up to 90% after five successive cycles.

The AA was also synthesized by other researchers in water emulsion by hydrogen peroxide oxidation of cyclohexene without surfactant and under strong stirring. In this system, the addition of an organic acid as co-catalyst and under severe reaction conditions was required to obtain a good yield of pure AA (Vijayaraaghavan and Karthik, 2004). The AA was also produced with the use of a phase transfer agent (methyl-tri-octylammonium-hydrogen-*o*-sulphate) in water with a good yield, but reuse of the system required an additional amount of phase transfer agent to be effective (Timofeeva et al., 2008). More recently, another method for producing AA was discussed by Timofeeva et al. The potential of transition metal-mesoporous materials as heterogeneous systems have been described in water or organic solvent, but the long reaction time and/or low AA production yield are not advantageous for an industrial process (Draths and Frost, 1994).

8.5.1.3 Microbe-based conversion of D-glucose into adipic acid

Draths and Frost detailed the biocatalytic microbe-based conversion of D-glucose into *cis,cis*-muconic acid and its subsequent hydrogenation into AA (Van de Velde and Kiekens, 2002). This process is an environmentally friendly process because of its renewable raw material for synthesis by means of an intrinsically safe process.

To synthesis the raw material, *cis,cis*-muconic acid, *E. coli* [WN1/pWN2.248] was developed and cultured in glucose for 48 h of culturing under fed-batch fermentor conditions. This process yields 36.8 g L of *cis,cis*-muconic acid in a 22% (mol/mol). The researchers also performed optimization of the carbon flow directed into *cis,-cis*-muconic acid biosynthesis and manipulated the enzyme to avoid the accumulation of biosynthetic intermediates. The derived muconic acid could then be hydrogenated to AA using supported Rh or Pt catalyst (Niu et al., 2002). The reaction mechanism of the process is shown in Scheme 8.8.

8.5.1.4 Bio-adipic acid

cis,cis-Muconic acid is a polyunsaturated dicarboxylic acid which can be developed from the biological conversion of sugars and lignin-derived aromatic compounds. Afterwards, the developed muconic acid can be catalytically converted into AA, the precursor for nylon production. This is because of the specific oxygen-rich characteristics of muconic acid. Thus it can be produced from both the carbohydrate and lignin fractions of lingo-cellulosic biomass. Catechol is one of the ortho isomer of the three

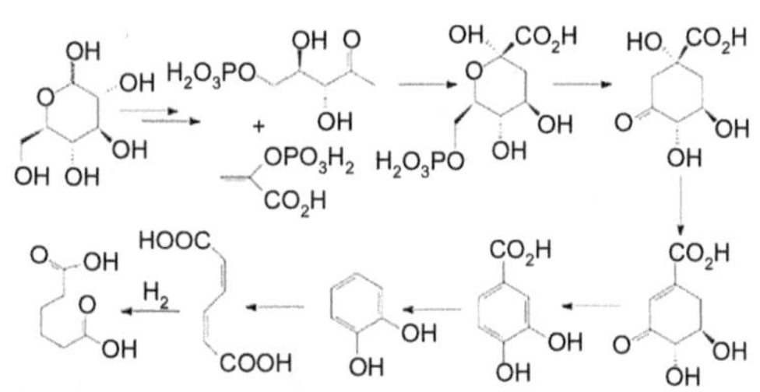

Scheme 8.8 Enzymatic transformation of D-glucose to *cis,cis*-muconic acid and subsequent hydrogenation to adipic acid (Niu et al., 2002).

isomeric benzenediols and is synthesized by rerouting of aromatic amino acids. After this process, the catechol is oxidatively ring opened to produce *cis,cis*-muconate as the terminal metabolite (Draths and Frost, 1994; Niu et al., 2002; Joo et al., 2017; van Duuren et al., 2012).

Alternatively, other researchers mentioned that certain microorganisms can be used for catabolization of lignin-derived aromatic compounds using upper pathway reactions. This process will convert the aromatic compounds into central intermediates such as catechol, which is the direct intermediate before oxidative ring opening for muconate (Fuchs et al., 2011). Lignin has some unique properties such as its aromatic polymer structure and relative abundance in biomass. Hence, the valorization of lignin to fuels and chemicals has significant economic and environmental potential. With the help of a chemo-catalysis, muconic acid can be converted into myriad high-value petrochemical substitutes such as AA immediately after biological production from sugar- or lignin-derived monomers (Vardon et al., 2015).

In their research, Vardon et al. generated muconic acid for downstream processing. With the help of the engineered strain *Pseudomonas putida* (KT2440-CJ102), the benzoate was biologically converted. After this process, the muconic acid was recovered from cell-free culture media, as shown in Fig. 8.10.

- The first treatment step for the process, as mentioned by the researchers, is microfiltration, in which the proteins were removed.
- In the second step, purified muconic acid was recovered from the broth by pH/temperature shift crystallization at pH 2 and 5°C.

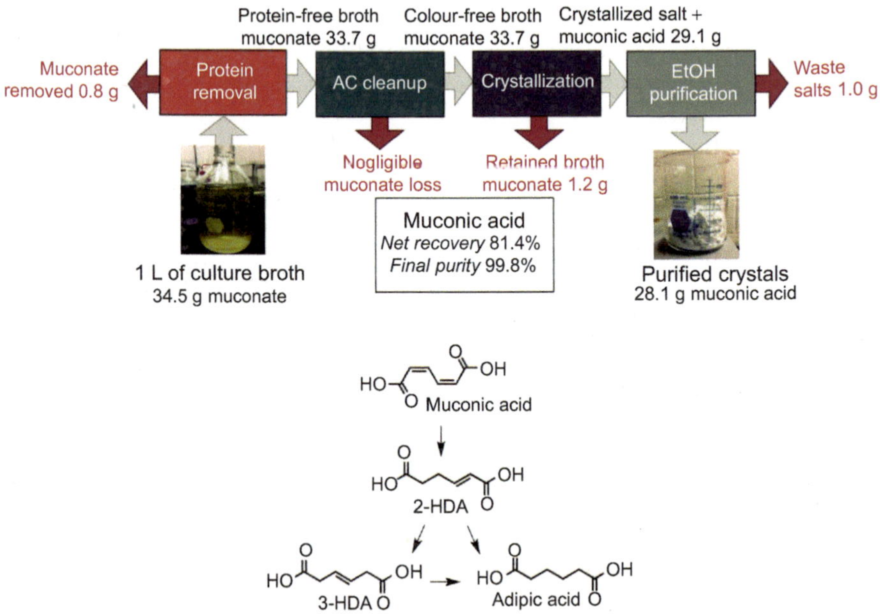

Figure 8.10 Reaction network for the hydrogenation of muconic acid via the intermediate hexenedioic acid to produce adipic acid (Vardon et al., 2016). *AC*, acid crystal; *EtOH*, ethyl alcohol.

- In the third step, the precipitated muconic acid crystals were vacuum filtered [0.2 μm polyethersulphone (PES)] and dried in a vacuum oven for 48 h.
- As a final treatment step, muconic acid crystals were dissolved in ethanol and filtered through a 0.2-μm PES membrane.

The acid obtained after all of these purification steps is 99.8% pure. The acid is then converted into bio-adipic acid in the trickle bed reactor operating with a higher catalyst loading at an elevated bed temperature and reduced liquid flow rate to facilitate complete conversion (Vardon et al., 2016).

8.5.1.5 Metabolic engineering

In industrial practice, nylon 6 is produced by ring opening of caprolactam, which is a product from the benzene value chain. This process is considered to be an environmentally polluting operation because of its higher amount of GHG emissions. Therefore, research has focused on alternative resources for nylon production. It was found that a significant reduction in these emissions can be achieved by the production of the bulk intermediates from biorenewables (Herman et al., 2007).

One way to produce nylon 6,6 is through cadaverine, which is similar to hexamethylenediamine, one of the building block polymers of nylon 6,6. A metabolically engineered *Corynebacterium* strain is used for the fermentative production of nylon building blocks. In this process, a *Corynebacterium* strain expresses the L-lysine decarboxylase gene (cadA) from *E. coli* and is polymerized with SA or castor oil to produce two novel 100% bio-nylons (Mimitsuka et al., 2007; Kind et al., 2014). Another way to produce nylon is through putrescine, also known as 1,4-diaminobutane. Metabolically engineered *E. coli* strains were used to produce putrescine. After fermentation, putrescine used chemical modifications have been proposed and developed for several nylon building blocks (Beardslee and Picataggio, 2012; Li et al., 2013). For AA, a chemo-enzymatic production process was developed by Frost et al. using a metabolically engineered *E. coli* strain for the fermentative production of *cis,cis*-muconic acid from glucose followed by a chemical conversion to AA (Frost, 2005).

Frost et al. also developed a chemo-enzymatic route for the production of caprolactam. In that process, the researchers converted the lysine, derived from fermentation, into an ε-caprolactam route. They had also discussed another route for the production of caprolactam, in which lysine is chemically converted into 6-aminohex-2-enoic acid.

Turk et al. mentioned the production of 6-aminocaproic acid (6-ACA) by two different ways. One possible way is the adipyl-coenzyme A (CoA) route, in which the condensation of succinyl-CoA and acetyl-CoA produces 3-oxoadipyl-CoA. In the subsequent process, it is converted into adipyl-CoA. With the adipyl-CoA-dependent aldehyde dehydrogenase, adipate semialdehyde is formed, which can subsequently be converted to 6-ACA by transamination (Turk et al., 2016). The second way is referred as the α-ketopimelate (AKP) route. In this method, intermediate AKPs are used for decarboxylation of their corresponding α-ketoacid. The fundamental principle used for the process is branched-chain amino acid degradation. Here, aldehydes such as adipate semialdehyde produced by decarboxylation could potentially allow for the

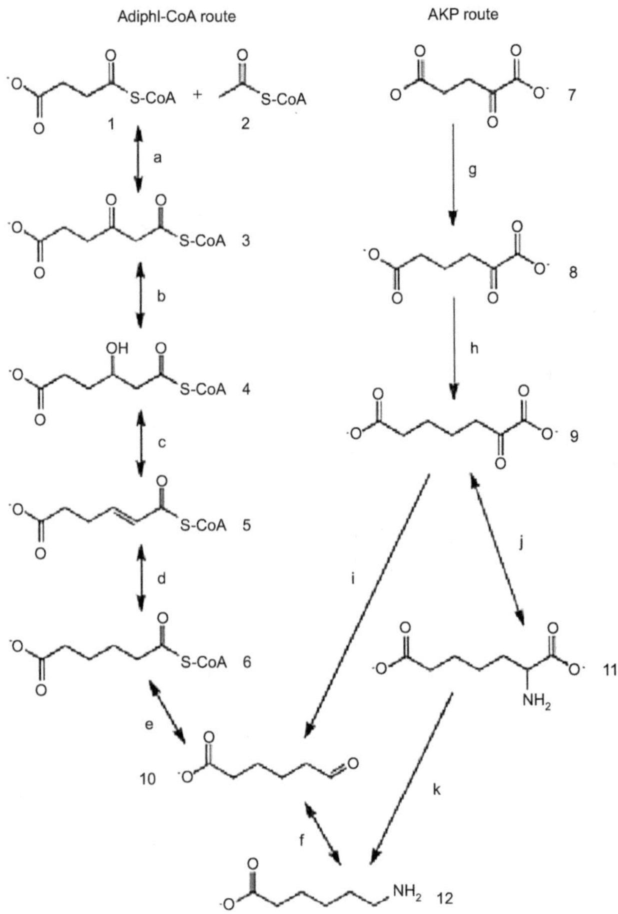

Scheme 8.9 α-Ketopimelate and adipyl-coenzyme A method of 6-aminocaproic acid production (Turk et al., 2016).

formation of 6-ACA (Anon, 2012). The schematic representation of both methods is provided in Scheme 8.9.

8.5.1.6 *Scrap wood*

As a major breakthrough in sustainability in nylon production, researchers found that nylon can be made from scrap wood without creating waste. Bouwman and Raoufmoghaddam developed a new path to synthesize caprolactam out of scrap wood without producing harmful by-products. The researchers mentioned that cellulose from waste wood can be converted into levuline acid, which is similar to the caprolactam molecule. They mentioned that this acid can be derived from materials such as waste products from maize (Anon, 2013a).

8.5.2 Sustainable nylon on the market

8.5.2.1 Econyl: sustainable nylon

Econyl is a certified recycled nylon textile manufacturer. It produces recycled nylon textile products made from used fishing lines and other postconsumer waste collected from the oceans. Aquafil, a leading recycled producer and marketer of recycled nylon fibres, entered into a partnership with Outerknown, a new menswear apparel brand (Sousa et al., 2015). Outerknown's first collection launched on July 15, 2015 and features the Evolution Series jackets and boardshorts made with Econyl brand regenerated nylon yarn. The company claims that, Econyl offers quality and performance indistinguishable from traditionally produced nylon yet with staggering ecological advantages. The fibre is produced from recovered waste and can be reused an uncountable number of times with no change in quality, according to Aquafil. Econyl occupies worldwide waste streams from landfills and seas through the recuperation of abandoned fishing nets and other cast-off nylon waste materials. Econyl has developed an eco-friendly nylon made from recycled plastics in a closed-loop system, radically lessening waste and discharges. Their technology has been embraced by brands such as Levi's and Rapanui (Anon, 2013b).

8.5.2.2 Rennlon

Rennlon, another commercial polymer, is made using Rennovia's bio-based Rennlon AA and hexamethylenediamine (HMD) monomers that are produced via the company's proprietary chemical-catalytic process technology using widely available renewable glucose feedstocks (Okada, 2002). The manufacturers mentioned that, based on the pilot run production, they refined its glucaric acid-based AA process. Rennovia is confident that its bio-based AA will be highly cost-competitive against not only petroleum-based AA (adipic acid) but also other bio-based AA that are currently being developed, as shown in Scheme 8.10.

The manufacturing company mentioned that there is plenty of glucose from which to source their feedstock and that there are opportunities for corn wet mill capacity to shift from high-fructose corn syrup production to renewable chemical production without threatening the food supply. However, the company also mentioned that it is planning to diversify its feedstock source from existing carbohydrate supply chains into non-food based raw materials once after it is stabilized and mature enough in the production process. This company claims that the production costs for its monomers will be 20%–25% lower than costs for petroleum-based AA and HMD production on the market. The company also anticipates Rennlon AA and HMD production will generate 85%–50% less GHG emissions, respectively, compared with petroleum-based AA and HMD production (Anon, 2013c).

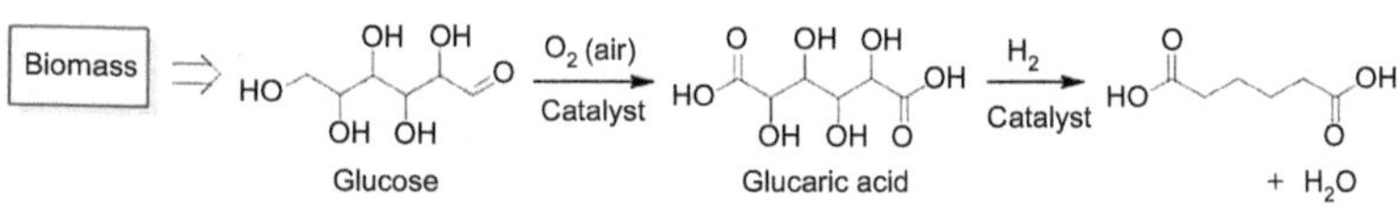

Scheme 8.10 Renewable adipic acid production.

8.6 Sustainable carbon fibre production

Carbon fibre is one of the most important technical fibres with unique chemical, electrical and mechanical properties (Barrero et al., 2010). The need for carbon fibre is expected to increase many fold in the forthcoming decade because numerous application areas have been found in consumer end products (Alway-Cooper et al., 2013). In industrial practice, carbon fibres are produced using high-purity polyacrylonitrile precursor fibre. Existing carbon fibre production is mostly influenced by two factors: (1) the cost of the process, and (2) low-quality polyacrylonitrile fibre. Other than this, the manufacturing process is considered highly energy-intensive because the operation requires a high temperature. This factor constrains the scope of feasible applications for carbon fibre (Endoc et al., 2002).

However, the production process of carbon fibre is highly regulated by the environmental community. Manufacturers are also constantly seeking for a low-cost carbon fibre manufacturing process that reduces the energy demands for this heat-intensive process. In the carbon fibre manufacturing process, the furnaces used, such as oxidation and carbonization furnaces, have the potential to emit hydrogen cyanide, ammonia and VOCs. These pollutants are highly toxic to the environment and human health, even in very small quantities. Other than these pollutants, GHGs such as carbon monoxide and nitrogen oxide not only contribute to environmental and global warming but have a direct correlation to manufacturers' consumption of energy. (Anon, 2013d)

8.6.1 Carbon fibre from wool fibre

In previous studies, mostly high-purity polyacrylonitrile (PAN), pitch, cellulose, regenerated cellulose and lignin were investigated as a carbon fibre precursor. The carbon content of PAN, lignin, wool and cotton is 67.91%, 63.4%, 50% and 44.4%, respectively. The wool fibres are a source of sulphur; hence, the rich sulphur content helps the wool generate higher yield of carbon fibre. This can be understood from the carbon content; even though wool has a lower carbon content than lignin and PAN, pyrolysis provides a higher carbon fibre yield than cotton. Wool fibres are renewable, have good mechanical properties and are abundantly available at cheap prices (Hassan and McLaughlin, 2013).

Carbonization of untreated and cross-linked wool fibre can be used to produce carbon fibres. Hassan et al. found that the carbon yield of the resulting fibres was a function of the type of cross-linking agents applied to the wool fibre. They also monitored the strong relationship between the type of cross-linking treatment used and the surface topology of the resulting carbon fibre. The results also revealed that the cross-linked wool fibres were able to produce a strong carbon fibre with higher tensile strength than untreated wool fibre. The researchers noticed that carbon fibres derived from the wool were strongly hydrophilic; this appeared to result from the increase in hydrophilic groups introduced to the surface of carbon fibre (Hassan et al., 2015). Wool fibre could be a sustainable and renewable carbon fibre precursor, and this high-value application of waste wool will also mitigate current disposal problems of wool waste, thereby improving our environment.

8.6.2 Carbon fibre from lignin

The most important alternative precursors for carbon fibre production with the highest potential for the automotive industry are polyethylene and lignin. The production of carbon fibre through a lignin precursor is comparatively 50% cheaper than PAN as a precursor (Baker, 2010). Lignin, which is second to cellulose in natural abundance, is readily available and relatively inexpensive. Lignin is a high-molecular weight aromatic biopolymer with a reported total worldwide production of approximately 50 million tons per year (Gellerstedt et al., 2010)

Lignin is a by-product of 'wood-free' papermaking. Lignin is commercially used as a fuel in the pulping chemical recovery process. Commercial lignins such as those isolated from alkaline, acidic or organic solvent-based processes are highly degraded because of extensive fragmentation during processing. Because of this fermentation process, the lignins obtained vary in their structure, chemical properties, molecular weight, functional group and degree of condensation. Moreover, the type of intermolecular linkages and type of monomer differ among processes based on the type of wood and the type and length of chemical process. Main advantages to lignin as a precursor for carbon fibre production are that it:

- is a sustainable, renewable resource material
- is the third most abundant organic substance (polymer) on earth after cellulose and chitin
- accounts for 70% of CO_2 sequestered by plants
- is readily available as a co-product of pulping processes; about 200 million metric tons annually pass through pulp and paper mills worldwide (2005 data)
- has future availability from biorefineries as a by-product of cellulosic ethanol production

Hardwood lignin is mainly used for carbon fibre production process as a precursor. This is because it is as a waste product of paper mills and is cheap and available in huge amounts. Hardwood lignin is also a sustainable, renewable resource. The use of hardwood lignin offers significant cost savings potential in the production of carbon fibre and makes it possible to produce a carbon fibre based on renewable resources (Mainka et al., 2015).

In the manufacture of carbon fibre from hardwood, two steps are added to prepare the fibre spinning process, compared with the conventional PAN-based carbon fibre production process: the washing of lignin powder and drying as the first step, and pelletizing of the lignin powder as the second step. These result in five main steps which illustrate the use of hardwood lignin as a precursor for carbon fibre production:

- washing of lignin powder and drying
- pelletizing of the lignin powder
- melt spinning of the lignin fibre
- oxidation of the fibre
- oarbonization of the fibre

The sequence of the process of fibre manufacturing is:

1. **The lignin powder**: Nearly 128 million tons of lignin wastes is produced by the pulping industry alone every year around the world. At the pulp mill, lignin is dissolved from cellulose into black liquor. Then the lignin is precipitated out of the black liquor. After the washing and drying process, hardwood lignin powder is available (Eberle et al., 2013).

2. **Compounding and pelletizing**: The pelletizing of the hardwood lignin powder is done mainly for three reasons: (1) to create better flow ability; (2) to reduce moisture in the wood; (3) to remove other volatile compounds for the melt spinning process.
3. **Precursor fibre production**: In carbon fibre manufacturing, the most important aspect is quality. The quality of the carbon fibre mainly depends on the quality of the precursor fibre.
4. **Fibre stabilization**: This is an oxidation process to make the fibre stable. Fibre stabilization is performed as in the conventional carbon fibre production process; it is the most critical step during conversion. The oxidation process must be handled carefully to prevent damage to the fibre.
5. **Fibre carbonization**: The stabilized material is carbonized in a furnace under nitrogen atmosphere. It typically operates at 500–1500ºC for 5–10 min. Approximately 65% of the material is vaporized during carbonization with gases exhausted through an incineration system.

8.6.3 Carbon microtubes from chicken feathers

Chicken feathers were used as a source for the production of carbon microtubes by Melissa E.N. Miller and Richard P. Wool of the University of Delaware, Newark. Chicken feathers are a major waste in the poultry industry. Approximately 5 billion pounds of poultry feathers are generated annually in the United States alone. However, not all of the waste is disposed of safely into the environment, which creates landfill disposal and potential health problems. The research conducted by Miller and Wool used this waste as a valuable feedstock for high-performance, bio-based materials. Because they contain 47.83% carbon, chicken feathers are hollow and strong in nature owing to their 91% keratin content. In this research, the chicken feathers were carbonized to create strong, inexpensive, hollow fibres. Current research shows that the carbonization method of heating chicken feathers at 225°C for 26 h, followed by 2 h at 450°C in a nitrogen environment results in carbonized fibres which maintain their hollow structure. These fibres were added to a soybean oil-based resin to make biocomposites. Their results were promising and resulted in a 236% increase in the composite storage modulus at 35°C, from 0.639 GPa with 0% fibre to 2.145 GPa with 3.45 wt% fibre. The extracted fibre modulus is on the order of 142 GPa, comparable to low/medium modulus carbon fibre. The authors listed potential applications for their fibre, ranging from use in the aerospace industry to airbags, catalysts, ligament repairs, batteries, hydrogen storage and fuel cells (Anon, 2007).

8.6.4 High-performance sustainable carbon fibres from Australian spinifex grass

Darren Martin et al. came up with a precursor for carbon fibre production. Spinifex grass is a type of grass which covers approximately 30% of the Australian continent including the driest regions. Using this grass, the researchers found that it was uniquely easy to break down into ultralong, thin cellulose nanofibrils. By the application of novel catalysts and advanced processing techniques, the cost-effective production of high-strength, sustainable carbon fibres from nanocellulose could be achieved (Martin et al., 2015).

8.6.5 Carbon fibre from microwave-assisted plasma processing

A system to produce fully carbonized or graphitized carbon fibres continuously using microwave-assisted plasma processing was developed by White et al. This process employs an elongated chamber in which microwave plasma is excited in a selected gas atmosphere. In this chamber, the fibre is allowed to pass through gradient microwave power, in which the higher power is at the fibre's exit region and lower power is at the fibre's entry region. The feedstock for carbon fibre production may be of any conventional precursors such as PAN, pitch or any other suitable organic/polymeric fibres. The oxidized polymeric fibre precursors are allowed to run continuously through a microwave-assisted plasma reactor in an inert, nonoxidizing atmosphere to heat the fibres, drive off unwanted elements such as oxygen, nitrogen and hydrogen, and produce carbon or graphite fibres faster than conventionally produced carbon fibres (Mainka et al., 2015; Eberle et al., 2013; Anon, 2007).

Each precursor used in carbon fibre manufacturing brings specific characteristics to the carbon fibres.

8.6.6 Carbon fibre recycling: a step towards sustainability

In general, carbon fibre manufacturing is an energy-intensive process, but after their life cycle most products go to the landfill. Recycling of carbon fibres has been carried out successfully at the Nottingham Innovative Manufacturing Research Centre, United Kingdom, in the form of fluffy individual filaments which are eco-friendly and cost-effective. The researchers also developed nonwoven mats with the help of recycled carbon fibre, which can be used as heating elements for garments as well as insulated food bags. In fact, recycling of carbon fibres has opened up the potential for the sustainable manufacturing of lightweight high-strength materials for aerospace applications and more. Several studies reported different methods to improve the overall properties of reclaimed fibres and improve the processing of these materials (US Patent 7,824,495 B1, 2010; Carbon Fibre Technology Facility). Following are the some common methods used in the carbon fibre recycling process:

- **Pyrolysis**: One of the most used techniques for carbon fibre recycling. The pyrolysis process is generally carried out with or without the aid of a catalyst. In this process the composites or carbon fibre materials are placed in an oven in the absence of air and at a temperature of 400–500°C. This process cleans the carbon fibres; they maintain 90%–95% of their original properties.
- **Fluidized bed**: In this process, at a higher temperature, approximately at 550°C, a stream of fluid (for example, a gaseous airstream) is forced through the material. The crushed composite scrap material or the material with carbon content which needs to be recycled is kept on a bed of sand. Organic material, including the composite's resin matrix and any contaminants present in the scrap material, is oxidized. The process is designed in such a way that organic compounds present in the composite scraps will be burned up and the lightweight fibre will be forced upwards while heavier matter, such as metallic material, remains in the reactor 'bed'. Because the process of the fluidized bed is performed at an elevated temperature of 550°C, the carbon fibres obtained by the process have a degradation in strength of 20%; however, research identified that the original stiffness of the fibre was retained after

recycling. In addition, an analysis of the surface after recycling showed only a small reduction in oxygen content, indicating good bondability to a polymeric matrix if it is reused in a composite application (Anon, 2014a).

- **Microwave**: In this process, the microwave energy is passed onto a material which needs to be recycled and the energy is absorbed by the carbon fibres, which subsequently heat the matrix resin internally rather than externally. It can result in a more rapid resin decomposition and recovery of fibres without the formation of char, and a shorter overall processing time. It requires smaller-scale equipment than other pyrolysis methods (Wise and Girard, 2016).
- **Supercritical fluids**: For this method, supercritical fluids such as *n*-propanol are used for the carbon fibre recycling process. Here, supercritical fluids are used to recycle composites to enable the collection of pure PAN-based carbon fibres at the end of the process. The reclaimed fibres reportedly retain up to 97% of tensile strength of the virgin material, with no change in modulus (Pickering, 2006).
- **Combination of methods**: Several methods are combined to recycle the carbon fibre. The first method was developed by the Massachusetts Institute of Technology: this technique combines both dry and wet processes. Here, low temperature was used to remove the bulk matrix and some contaminants, followed by thermal posttreatment through vacuum pyrolysis to eliminate any remaining resin. The research team reported that this method was able to produce 99% pure carbon fibre. Another innovative method was developed by Adherent Technologies Inc., ATI (Albuquerque, New Mexico); it combined three methods: vacuum pyrolysis, a dry process operated at around 500°C and the recovery of resins as marketable liquids (Job, 2010; McConnell, 2010).

8.6.7 Recycled carbon fibre products on the market

To date, the most successful use of recycled carbon fibres in a commercial product is reclaimed carbon fibre material in BMWs i3 and 7-Series model cars (Munich, Germany). The company used the reclaimed carbon fibre from manufacturing waste, in which the dry fibre processes were converted into new product forms and used in closed-loop recycling solutions. This was achieved because of the joint venture between the BMW and SGL Groups, resulting in an integrated supply chain system. The main advantage of recycled carbon fibre is that it can reduce the cost of lightweighting, used either on its own or in conjunction with virgin carbon fibres (Pickering et al., 2009; CompositesWorld, 2010). Practically speaking, making lightweight products at an affordable cost is a primary goal if manufacturing companies are looking for the increased use of carbon fibre in high-volume applications. This recycling process helps industry in this aspect by reducing the cost of carbon fibre manufacturing. From an environmental standpoint, and also from a social responsibility point of view, reducing waste and reusing materials that have high embedded energy, and which can be recycled in an energy efficient manner, is the right thing to do. It also is a legislative requirement in some countries (Barnes, 2016).

8.6.7.1 RECAFIL

The recycling of carbon fibre material is of increasing importance for producing fibre composite components suitable for mass production. The Joint Venture of RECAFIL with SGL Automotive Carbon Fibres developed an innovative recycling process which makes it possible to recirculate valuable carbon fibres into the production process and thus closes the material cycle. The development of RECAFIL recycled carbon fibre material simultaneously establishes a new material class. The company sells the product on the market as carbon fibre roving which is based on a PAN precursor and is supplied as single tows off spools. The company also sells carbon fibre flocks with about 92% carbon fibre content and an average fibre length of 230 μm. They mentioned that these materials can be used as an additive to produce antistatic or electrically conductive material (Anon, 2016b).

8.6.8 Carbon fibre remanufacturing

Carbon Fibre Remanufacturing (CFR) develops new products for industrial, commercial and retail product applications using reclaimed carbon fibre. The company developed an alternative to landfilling of excess postindustrial carbon fibre materials. It was estimated that 30% of carbon fibre from the composites industry becomes excess or part of the waste stream in the manufacturing process. The company developed carbon fibre fire-retardant fabric, insulation, friction/brakes, filtration, thermal barriers and laminate compounds using reclaimed carbon fibre waste from composite industries. Remanufactured carbon fibres retain 100% of their virgin properties. They can be cut to specified lengths; be incorporated into nonwoven rolled cloth, veil, mat/felt product; and compounded to specifications for bulk and sheet-moulded product applications (Van de Velde and Kiekens, 2002). CFR is in various stages of moving excess postindustrial carbon fibres into various carbon fibre material forms as well as working with an increasing number of fabric and composite firms in the development of final whole goods (Anon, 2015). Reclaimed carbon fibres are optimized when produced in various broad goods, including:

- veil/cloth for fire-retardant benefits for pultruded products
- fabric-in blend with other fibres, producing fire-retardant and self-extinguishing results
- felts/mats to preform for friction/brakes and rigid insulation product for high-temperature industrial furnaces
- carbonized: advancing broad goods to designed carbon content levels optimizes their use in bulk and sheet-moulded composite laminates

8.7 Recycling of synthetic fibres

Recycling of textiles is the technique of reusing or reprocessing used clothing or materials from the manufacturing process. Textiles in municipal solid waste are found mainly in discarded clothing; other sources include furniture, carpets, tires, footwear and nondurable goods such as sheets and towels. Major players in textile retailing such as H&M and C&A have taken the initiative towards energy efficiency in their stores and increased recycling of clothing hangers. One major drawback in recycling textile materials is that they contain fibre blends that are different in nature, such as cotton–polyester blended clothing. Some materials such as cotton and linen can be composted, but petroleum-based fibres such as polyester have little chance for reuse.

There are basically two kinds of textile recycling: fibre and polymer recycling. In fibre recycling the garment is shredded into the fibre stage and then blended with new fibres to produce a yarn and further garments or carpets. Polymer recycling involves polyesters that are shredded and ultimately melted down and turned into plastic pellets that are then are respun into fibres for new polyester applications. There are two types of polymer recycling: mechanical and chemical. In mechanical recycling, plastics are melted and reextruded to produce yarns. This can be carried out only a few times before the molecular structure breaks down and makes the yarn suitable only for the landfill. Chemical recycling involves breaking the polymer into its molecular parts and reforming the molecule into a yarn. Depolymerization and then repolymerization is expensive and almost nonexistent.

The textile manufacturing process is generally chemically intense, and residues from the processes might stay in the textile. Physical aging causes an increase in the ordering of noncrystalline polymer chains, which could decrease the molecular mobility of the polymer. Chemical aging cause changes in molecular bonds including thermal degradation, hydrolysis and various oxidation reactions. Chemical aging generally causes chain scissoring of the polymer chain, which causes a rapid decrease in the degree of polymerization. The net decrease in molecular weight causes a decrease in tensile strength and elasticity of the fibre, and it may also affect its chemical reactivity by introducing more sites of reaction. To produce new high-value textiles from textile waste there is a need to develop an automatic technology for sorting of textile wastes and separating fibre blends into pure fibre fractions. Based on some estimates, 65% of fabrics manufactured worldwide are synthetic and mostly made of polyester (70%). Synthetic fibres such as polyester, acrylic, nylon and elastane are all made from organic compounds derived from fossil oil. Specific monomers are reacted to form a polymer with the desired properties.

8.7.1 Recycled polyester

Polyester is produced from the same petrochemicals as gasoline. It is an energy-intensive process and relies on finite, nonrenewable natural resources. However, while taking into consideration the entire life cycle of polyester from the raw material stage to the use phase and the end of the life cycle, it is not as environmentally damaging as other synthetic fibres such as nylon. It has lower energy impacts during the washing and cleaning phase and is also completely recyclable at the end of its life. About 65%–70% of global polyester production is used for textiles and the remaining 25%–30% is used to manufacture PET bottles. Of the 30% of PET which is used to make bottles, only a small percentage is recycled into fibres. From a survey on the use of petrochemical oils for polyester production, it was observed that around 104 million barrels of oil are used to produce PET, out of which 70 million barrels of oils are used to produce polyester fabrics. Hence, most petrochemical fossil oils are used to produce textiles and not bottles. The recycled polyester (rPET) could be considered a sustainable option for to two reasons: (1) the energy required to produce rPET is less than for virgin polyester, and (2) bottles and other plastics are kept out of landfills owing to recycling.

The most common form of textiles produced from recycled polyester is fleece, a knitted pile fabric often used by outdoor clothing companies to make jackets. Patagonia is the most well-known promoter of polyester recycling. It partnered with Teijin, a Japanese company that developed its own closed-loop polyester recycling system. In general, recycling of polyester uses 33%–53% less energy compared with the production of virgin polyester (125 MJ/kg). Furthermore, it emits a low 54% carbon footprint (5.19 kg CO_2) compared with virgin polyester (952 kg CO_2).

The major limiting factor in producing recycled PET is that it is not produced under a closed-loop system; i.e., when the plastic is heated it degenerates, so subsequent iterations of the polymer are degraded and cannot meet quality standards for the production of bottles such as soda and milk bottles and are used to making low-quality products. This kind of second-generation plastic can be recycled again for further low-quality applications such as park benches, speed bumps, carpet or other products with very low value. The cycle is completed when the plastic is no longer stable enough to be used for any product, so it is sent to the landfill where it is incinerated. Another limitation in recycling polyester is that antimony, which is present in 80%–85% of all virgin PET, is converted into antimony trioxide at high temperatures, which leads to the release of this carcinogenic matter from the polymer. Practical problems in recycling PET fibres are related to its dyeing. The base colour of recycled polyester chips range from white to creamy yellow, which makes it difficult to achieve consistency in shades. Once the recycled fibres are woven into fabrics, most fabrics are rendered nonrecyclable because the fabrics have lamination or other kinds of finishers that are blends of different synthetics. Either of these renders the fabric unsuitable for the mechanical method of recycling, which cannot separate the various chemicals to produce recycled yarn.

8.7.2 Recycled nylon

Like polyester, nylon fibre is made from petroleum resources and is not sustainable. Nylon does not break down easily and is more difficult to recycle than polyester. Hence it accounts for about 10% of waste in the ocean in the form of fishing nets. According to the records of the World Society for the Protection of Animals, more than 600,000 tons of nylon fishing gear, including nets, is dumped into oceans every year.

The recycling of nylon is not economically effective, and polymers are cheap to buy new. This is why most of industries have chosen PET for the manufacture of soda and water bottles. Another reason could be contamination. Because nylon is melted at a lower temperature, some kinds of microbes, bacteria or nonrecyclable contaminants can survive, which necessitates a thorough cleaning of nylon before the recycling process.

Hyosung, a Korean-based company, developed a method for recycling discarded nylon items into a textile-grade fibre called Regen. Because textile-grade yarn is the finest achievable form of the fibre, the recycled material is not at all inferior compared with virgin nylon, and it can be used for a number of applications. Although the recycling method is not completely environmentally sustainable, recycling nylon keeps a large percentage of petrochemical waste from going into the landfill or being incinerated and releasing toxic emissions into the atmosphere. It also uses 27% fewer natural resources than the production of virgin nylon, reduces GHG emissions by 28% and can be processed over and over again. Aquafil's a product Econyl is a recycled nylon produced from fishing nets.

8.8 Conclusions and outlook

Several sectors including agriculture, textiles, automotives, medicine and packaging need eco-friendly polymers. Environmental responsibilities by consumers as well as industries are constantly increasing. In particular, industries that produce biodegradable material are at an advantage because they emit lower CO_2 emission during manufacturing and further processes, and degrade it into organic matter after disposal. Although biodegradable materials are not economically competent compared with synthetic ones, the availability of biodegradable materials will allow many consumers to choose them on the basis of their environmentally responsible disposal.

Processes which use renewable feedstocks are most promising in the future for the development of biopolymer products. Biodegradable plastics composed of a starch or cellulose fibres seem to be the most likely to have potential growth in use. Time is of the essence for the development of biodegradable polymer because society's current views on environmental responsibility make this an ideal for the further growth of biopolymers. Challenges that should be take into consideration with reference to the production of bio-based polyesters and bio-plastics in the future are: (1) the sustainable way to manage raw material resources, (2) the sustainable production of chemicals as well as polymeric materials from bio-based materials, (3) the performance of bio-based materials in the long run and (4) economy of production and commercialization.

Biodegradable plastics manufactured from microorganisms are future technology, but their economical feasibility has to be taken into account. Apart from an economical point of view, the yield of the fermentation process is very low according to a biotechnological perspective. Hence, enhancement of the fermentation process is a major challenge for enlarging the range of monomers available in industrial production, such as PLA. The commercialization of bio-based polymers is still in the early stages and expansion of bio-based polymers is also hindered mainly by economical aspects, such as that currently available bio-based chemicals and polymers are not economical compared with petroleum products. Hence, two methods seem to be logical in the short to medium term to encourage the growth of bio-based products: (1) to produce bio-versions of existing fossil-based polymers, such as with Sorana, a PTT from DuPont, which is produced from renewable 1,3-propanediol and petrochemical terephthalic acid; and (2) to replace petroleum-based homologues gradually with their renewable-based homologues: e.g., Avantium, which produces FDCA, announced the development of the first PEF bottle.

In the near future, polymer chemistry could introduce bio-based monomers within existing productive processes and create promising building blocks for the production of monomers. In that respect, the possibility of fine-tuning polymer functional properties and biodegradability by means of optimized biocatalysts will guide the development of new effective polymers. Significant issues that are not yet resolved and which require further research are the development of an agreed-upon methodology for the evaluation of emissions from direct and indirect land use change and the quantification of the impacts of biomass production on regional biodiversity. Such assessments should be complemented by further approaches such as environmental risk assessment, or certification approaches for good and sustainable agricultural practices including eco-management and audit schemes. Currently, no single approach gives a complete and balanced picture.

References

Aeschelmann, F., Carus, M., 2015. Bio-Based Building Blocks and Polymers in the World-Capacities, Production and Applications: Status Quo and Trends Towards 2020. nova-Institut GmbH, Hürth, Germany, pp. 1–500.

Albertsson, A.C., Varma, I.K., 2002. Aliphatic polyesters: synthesis, properties and applications. Adv. Polym. Sci. 157, 1–40.

Alla, L.A., de Ilarduya, A.M., Benito, E., Garcia-Martin, M.G., Galbis, J.A., Munoz-Guerra, S., 2011. Carbohydrate-based polyesters made from bicyclic acetalized galactaric acid. Biomacromolecules 12, 2642.

Alway-Cooper, R.M., Anderson, D.P., Ogale, A.A., 2013. Carbon black modification of mesophase pitch-based carbon fibers. Carbon 59, 40–48.

Amass, W., Amass, A., Tighe, B., 1998. A review of biodegradable polymers: uses, current developments in the synthesis and characterization of biodegradable polyesters, blends of biodegradable polymers and recent advances in biodegradation studies. Polym. Int. 47 (2), 89–144.

Anon., 2014a http://www.jeccomposites.com/news/composites-news/researchers-university-northtexas-have-created-new-carbon-fiber-plants.

Anon, 2012. Producing Environmentally Friendly Nylon from Scrap Wood. http://news.leiden.edu/news-2014/producing-environmentally-friendly-nylon.html.

Anon, 2013a. Outerknown Uses Econyl Recycled Nylon Fibres in New Line. http://www.fibre2fashion.com/news/textile-news/newsdetails.aspx?news_id=173638.

Anon, 2013b. Rennovia Develops Rennlon™ Biobased Nylon 6,6 Polymer. http://www.textileworld.com/textile-world/fiber-world/2013/12/rennovia-develops-rennlon-biobased-nylon-66-polymer/.

Anon, 2013c. The Polymer Is Made Using Rennovia's Biobased Rennlon AA and HMD Monomers. http://resources.made-in-china.com/article/industry-view/jEtnhvKHHQiN/Rennovia-Develops-Rennlon-Biobased-Nylon-6-6-Polymer/.

Anon, 2013d. Addressing the Environmental Impact of Carbon Fiber. http://www.sailingscuttlebutt.com/2013/10/17/addressing-environmental-impact-carbon-fiber/.

Anon, 2014b. Adipic Acid (ADPA): World Market Outlook and Forecast up to 2018. Merchant Research and Consulting, Birmingham, UK.

Anon, 2016a. How Sustainable Nylon Is? http://goodonyou.org.au/material-guide-nylon/.

Anon., 2007 Carbon microtubes from chicken feathers. http://acs.confex.com/acs/green07/techprogram/P41563.HTM.

Anon., 2015 CFR – Global Leader in Carbon Fiber Remanufacturing. http://carbonfiberremanufacturing.com/pages/product/landing.php.

Anon., 2016b RECAFIL® Recycled Carbon Fibres. http://www.sglgroup.com/cms/international/products/product-groups/cf/recafil/index.html?__locale=en.

Anvari, M., Pahlavanzadeh, H., Vasheghani-Farahani, E., Khayati, G., 2009. In situ recovery of 2,3-butanediol from fermentation by liquid-liquid extraction. J. Ind. Microbiol. Biotechnol. 6, 313–317.

Averous, L., 2008. Polylactic acid: synthesis, properties and applications. In: Belgacem, N., Gandini, A. (Eds.), Monomers, Oligomers, Polymers and Composites from Renewable Resources. Elsevier, Amsterdam, pp. 433–450.

Avérous, L., Pollet, E., 2012. Biodegradable polymers. In: Avérou, L., Pollet, E. (Eds.), Environmental Silicate Nano-biocomposites, Green Energy and Technology. Springer-Verlag, London (Chapter 2).

Averous, L., Pollet, E., 2012. Environmental Silicate Nano-biocomposites, Green Energy and Technoloy. Springer-Verlag, London.

Avinc, O., Khoddami, A., 2009. Overview of poly(lactic acid) fibers. Fiber Chem. 41 (6), 391–401.

Babu, R.P., O'Connor, K., Sreeram, R., 2013. Current progress on bio-based polymers and their future trends. Prog. Biomater. 2 (8), 1–16.

Baker, F., 2010. Low Cost Carbon Fiber from Renewable Resources. Oral Presentation – U.S. Department of Energy, Washington.

Barnes, F., 2016. Recycled Carbon Fiber: Its Time Has Come. http://www.compositesworld.com/columns/recycled-carbon-fiber-its-time-has-come.

Barrero, A., Bedia, J., Cordero, T., Lallave, M., Loscertales, I.G., Rodriguez-Mirasol, J., Ruiz-Rosas, R., 2010. The production of submicron diameter carbon fibers by the electro-spinning of lignin. Carbon 48, 696–705.

Beardslee, T., Picataggio, S., 2012. Bio-based adipic acid from renewable oils. Lipid Technol. 24 (10), 223–225.

Beauprez, J.J., de Mey, M., Soetaert, W.K., 2010. Microbial succinic acid production: natural versus metabolic engineered producers. Process Biochem. 45 (7), 1103–1114.

Becker, J., Lange, A., Fabarius, J., Wittmann, C., 2015. Top value platform chemicals: bio-based production of organic acids. Curr. Opin. Biotechnol. 36, 168–175.

Biebl, H., Menzel, K., Zeng, A.P., Deckwer, W.D., 1999. Microbial production of 1,3-propanediol. Appl. Microbiol. Biotechnol. 52, 289–297.

Blach, P., Bostrom, Z., Franceschi-Messant, S., Lattes, A., Perez, E., Rico-Lattes, I., 2010. Recyclable process for sustainable adipic acid production in microemulsions. Tetrahedron 66, 7124–7128.

Bolm, C., Beckmann, O., Dabard, O.A.G., 1999. The search for new environmentally friendly chemical processes. Angew. Chem. Int. Ed. 38, 907–909.

Brundtland, G.H., Khalid, M., 1987. Our Common Future, Report of the World Commission on Environment and Development. Oxford University Press, UK.

Cameron, D.J.A., Shaver, M.P., 2011. Aliphatic polyester polymer stars: synthesis, properties and applications in biomedicine and nanotechnology. Chem. Soc. Rev. 40, 1761–1776.

Carbon Fibre Recycling: Ongoing Research. , March 1, 2010. www.compositesworld.com.

Carbon Fibre Technology Facility, Oak Ridge National Laboratory, Tennessee. www.ornl.gov/science-discoveryclean-energy/facilitie-and-centres/carbon-fibre-technology-facility.

Castellan, A., Bart, J.C.J., Cavallaro, S., 1991. Industrial production and use of adipic acid. Catal. Today 9, 237–254.

Cavani, F., Teles, J.H., 2009. Sustainability in catalytic oxidation: an alternative approach or a structural evolution? ChemSusChem. 2, 508–534.

Celinska, E., Grajek, W., 2009. Biotechnological production of 2,3-butanediol–current state and prospects. Biotechnol. Adv. 27, 715–725.

Chauvel, A., Lefebvre, G., 1989. Petrochemical Processes: Major Oxygenated, Chlorinated and Nitrated. Editions Technip, Paris.

Che, C.M., Yip, W.P., Yu, W.Y., 2006. Ruthenium-catalyzed oxidation of alkenes, alkynes, and alcohols to organic acids with aqueous hydrogen peroxide. Chem. Asian J. 1, 453–458.

Cheng, K.K., Zhao, X.B., Zeng, J., Zhang, J.A., 2012. Biotechnological production of succínico acid: current state and perspectives. Biofuel Bioprod. Bior. 6, 302–318.

Climate Change and the Textile Industry, 2014. https://oecotextiles.wordpress.com/2014/10/15/climate-change-and-the-textile-industry/.

Collias, D.I., Harris, A.M., Nagpal, V., Cottrell, I.V., Schultheis, M.W., 2014. Biobased terephthalic acid technologies: a literature review. Ind. Biotechnol. 10, 91–105.

Datta, R., Tsai, S.P., Bonsignore, P., Moon, S.H., Frank, J.R., 1995. Technological and economic potential of poly(lactic acid) and lactic acid derivatives. FEMS Microbiol. Rev. 16, 221–231.

Defra (Department for Environment, Food and Rural Affairs) Sustainable Clothing Roadmap Briefing Note, 2008. http://www.defra.gov.uk/environment/business/products/roadmaps/clothing/documents/clothing-briefing-Dec07.pdf 2007.

Deng, Y., Wang, Z.M., Chen, J., 1999. Clean synthesis of adipic acid by direct oxidation of cyclohexene with H_2O_2 over peroxytungstate–organic complex catalysts. Green Chem. 1, 275–276.

Devin, G.B., Timothy, J.M., Christopher Luft, J., Yousaf, M.N., 2010. One-step syntheses of photocurable polyesters based on a renewable resource. Macromolecules 43 (23), 9660–9667.

Dijkman, W.P., Groothuis, D.E., Fraaije, M.W., 2014. Enzyme-catalyzed oxidation of 5-hydroxymethylfurfural to furan-2,5-dicarboxylic acid. Angew. Chem. Int. Ed. 53, 6515–6518.

Dong, W., Li, H., Chen, M., Ni, Z., Zhao, J., Yang, H., Gijsman, P., 2011. Biodegradable bio-based polyesters with controllable photo-crosslinkability, thermal and hydrolytic stability. J. Polym. Res. 18, 1239–1247.

Draths, K.M., Frost, J.W., 1994. Environmentally compatible synthesis of adipic acid from D-Glucose. J. Am. Chem. Soc. 116, 399–400.

van Duuren, J.B., Wijte, D., Karge, B., dos Santos, V.A., Yang, Y., Mars, A.E., Eggink, G., 2012. pH-stat fed-batch process to enhance the production of cis, cis-muconate from benzoate by *Pseudomonas putida* KT2440-JD1. Biotechnol. Prog. 28 (1), 85–92.

Ebata, H., Toshima, K., Matsumura, S., 2007. Lipase-catalyzed synthesis and curing of high-molecular-weight polyricinoleate. Macromol. Biosci. 7 (6), 798–803.

Eberle, C., Albers, T., Chen, C.D., 2013. Webb Commercialization of Newcarbon Fiber Materials Based on Sustainable Resources for Energy Applications. (ORNL/TM-2013/54).

Endoc, H., Fana, T., Fujinob, T., Hirosea, T., Okabe, T., Yoshimura, M., 2002. Effect of carbonization temperature on the structural changes of wood-ceramics impregnated with liquefied wood. Carbon 40, 761–765.

Engel, C., Straathof, A., Zijlmans, T., van Gulik, W., van der Wielen, L., 2008. Fumaric acid production by fermentation. Appl. Microbiol. Biotechnol. 78, 379.

Farrington, D.W., Davies, J.L., Blackburn, R.S., 2005. Poly(lactic acid) fibers. In: Blackburn, R.S. (Ed.), Biodegradable and Sustainable Fibers. Woodhead Publishing Series, pp. 191–219.

Firdaus, M., Meier, M.A.R., 2013. Renewable co-polymers derived from vanillin and fatty acid derivatives. Eur. Polym. J. 49, 156–166.

Fletcher, K., 2008. Sustainable Fashion and Textiles. Earthscan, London.

Frost, J.W., 2005. Synthesis of Caprolactam from Lysine. Patent application. WO2005123669.

Fuchs, G., Boll, M., Heider, J., 2011. Microbial degradation of aromatic compounds—from one strategy to four. Nat. Rev. Microbiol. 9, 803–816.

Gandini, A., 2011. The irruption of polymers from renewable resources on the scene of macromolecular science and technology. Green Chem. 13, 1061–1083.

Gandini, A., Silvestre, A.J.D., Neto, C.P., Sousa, A.F., 2009a. The furan counterpart of poly (ethylene terephthalate): an alternative material based on renewable resources. J. Polym. Sci. Part A Polym. Chem. 47, 295–298.

Gandini, A., Coelho, D., Gomes, M., Reis, B., Silvestre, A., 2009b. Materials from renewable resources based on furan monomers and furan chemistry: work in progress. J. Mater. Chem. 19, 8656–8664.

Garaleh, M., Yashiro, T., Kricheldorf, H.R., Simon, P., Chatti, S., 2010. Macromol. (Co-) Polyesters derived from isosorbide and 1,4-cyclohexane dicarboxylic acid and succinic acid. Chem. Phys. 211, 1206–1214.

Gardetti, M.A., Torres, A.L., 2013a. Sustainability in Fashion and Textiles: Values, Design, Production and Consumption. Greenleaf Publishing, UK.

Gardetti, M.A., Torres, A.L., 2013b. Entrepreneurship, innovation and luxury: the Any Savoirs Des Peuple case. J. Corp. Citizensh. 52, 55–75.

van Gelder, A.H., Aydin, R., Alves, M.M., Stams, A.J.M., 2012. 1,3-Propanediol production from glycerol by a newly isolated Trichococcus strain. Microb. Biotechnol. 5 (4), 573–578.

Gellerstedt, G., Sjöholm, E., Brodin, I., 2010. The wood-based biorefinery: a source of carbon fiber? Open Agric. J. 3, 119–124.

Goldberg, I., Rokem, J.S., Pines, O., 2006. Organic acids: old metabolites, new themes. J. Chem. Technol. Biotechnol. 81, 1601–1611.

Gopalakrishnan, D., Karthik, T., 2012. Eco-friendly fibres – textile contribution to minimize global warming. Text Rev. 7 (9), 13–15.

Gruber, P., O'Brien, M., 2005. Polylactides "Natureworks® PLA". Biopolymers 6 (8), 235–239.

Gubbels, E., Jasinska-Walc, L., Koning, C.E., 2013. Synthesis and characterization of novel renewable polyesters based on 2,5-furandicarboxylic acid and 2,3-butanediol. J. Polym. Sci. Part A Polym. Chem. 51, 890.

Gupta, B., Revagade, N., Hilbron, J., 2007. Poly(lactic acid) fiber: an overview. Prog. Polym. Sci. 32 (4), 455–482.

Gyamerah, M.H., 1995. Oxygen requirement and energy relations of itaconic acid fermentation by *Aspergillus terreus* NRRL 1960. Appl. Microbiol. Biotechnol. 44, 20–26.

Hassan, M.M., McLaughlin, J.R., 2013. Formation of poly(methyl methacrylate) thin films onto wool fiber surfaces by vapor deposition polymerization. ACS Appl. Mater. Interf. 5, 1548–1555.

Hassan, M.M., Schiermeister, L, Staiger, M.P. Sustainable production of carbon fiber: effect of crosslinking in wool fiber on carbon yields and morphologies of derived carbon fiber. ACS Sustain. Chem. Eng. **3** (11), 2015, 2660–2668.

Hatakeyama, H., Hatakeyama, T., 2010. Lignin structure, properties, and applications. In: Abe, A., Dusek, K., Kobayashi, S. (Eds.). Abe, A., Dusek, K., Kobayashi, S. (Eds.), Biopolymers, vol. 232. Springer-Verlag, Berlin/Heidelberg, Germany, pp. 1–63.

van Haveren, J., Scott, E.L., Sanders, J., 2008. Bulk chemicals from biomass. Biofuel Bioprod. Bior. 2, 41–57.

Heredia-Guerrerom, J.A., San-Miguel, M.A., Sansom, M.S.P., Heredia, A., Benitez, J.J., 2009. Chemical reactions in 2D: self-assembly and self-esterification of 9(10), 16-dihydroxypalmitic acid on mica surface. Langmuir 25, 6869–6874.

Herman, B.G., Blok, K., Patel, M.K., 2007. Producing bio-based bulk chemicals using industrial biotechnology saves energy and combats climate change. Environ Sci. Technol. 41, 7915–7921.

Hongu, T., Philips, G.O., Takigami, M., 2005. New Millennium Fibers, the Textile Institute. Woodhead publishing, Cambridge, England.

Houck, M.M., Menold II, R.E., Huff, R.A., 2001. Poly(trimethylene terephthalate): a new type of polyester fiber. Probl. Forensic Sci. XLVI, 217–221.

Hwang, K.C., Sagadevan, A., 2014. One-pot room-temperature conversion of cyclohexane to adipic acid by ozone and UV light. Science 346 (6216), 1495–1498.

Ji, X.J., Huang, H., Ouyang, P.K., 2011. Microbial 2,3-butanediol production: a state-of-the-art review. Biotechnol. Adv. 29, 351–364.

Jiang, Y., Loos, K., 2016. Enzymatic synthesis of biobased polyesters and polyamides. Polymers 8 (7), 243.

Jiang, H., Gong, H., Yang, Z., Zhang, X., Sun, Z., 2002. Clean synthesis of adipic acid by direct oxidation of cyclohexene in the absence of phase transfer agents. React. Kinet. Catal. Lett. 75, 315–321.

Job, S., September 2010. Composite Recycling, Materials KTN Report.

John, R., Nampoothiri, K.M., Pandey, A., 2007. Fermentative production of lactic acid from biomass: an overview on process developments and future perspectives. Appl. Microbiol. Biotechnol. 74, 524–534.

Jong, E.D., Dam, M.A., Sipos, L., Gruter, G.J.M., 2012. Furandicarboxylic acid (FDCA), a versatile building block for a very interesting class of polyesters. In: Biobased Monomers, Polymers, and Materials, vol. 1105. American Chemical Society, Washington, DC, USA, pp. 1–13.

Joo, J.C., Khusnutdinova, A.N., flick, R., Kim, T., Bornscheuer, U.T., Yakunin, A.F., Mahadevan, R., 2017. Alkene hydrogenation activity of enoate reductases for an environmentally benign biosynthesis of adipic acid. Chem. Sci. http://dx.doi.org/10.1039/C6SC02842J.

Joy, A., Sherry, J.F., Venkatesh, A., Wang, J., Chan, R., 2012. Fast fashion, sustainability, and the ethical appeal and luxury brands. Fashion Theory 16 (3), 273–296.

Kang, Z., Gao, C.J., Wang, Q., Liu, H.M., Qi, Q.S., 2010. A novel strategy for succinate and polyhydroxybutyrate co-production in *Escherichia coli*. Bioresour. Technol. 101 (19), 7675–7678.

Kaplan, D.L., Mayer, J.M., Ball, D., McCassie, J., Allen, A.L., Stenhouse, P., 1993. Fundamentals of biodegradable polymers. In: Ching, C., Kaplan, D.L., Thomas, E.L. (Eds.), Biodegradable Polymers and Packaging. Technomic Pub Co, Lancaster, pp. 1–42.

Karimi, B., Ghoreishi-Nezhad, M., Clark, J.H., 2005. Selective oxidation of sulfides to sulfoxides using 30% hydrogen peroxide catalyzed with a recoverable silica-based tungstate interphase catalyst. Org. Lett. 7, 625–628.

Karlsson, S., Albertsson, A., 1998. Biodegradable polymers and environmental interaction. Polym. Eng. Sci. 38 (8), 1251–1253.

Karthik, T., Gopalakrishnan, D., 2012a. Impact of textiles in environmental issues and environmental legislation, part – I. Text Rev. 7 (3), 14–21.

Karthik, T., Gopalakrishnan, D., 2012b. Impact of textiles in environmental issues and environmental legislation, part – II. Text Rev. 7 (4), 16–21.

Karthik, T., Gopalakrishnan, D., 2012c. Impact of textiles on environmental issues, part – I. Asian Dyer 9, 52–58.

Karthik, T., Gopalakrishnan, D., 2012d. Eco-friendly fibres of the future. Asian Text. J. 21, 67–71.

Karthik, T., Gopalakrishnan, D., 2013a. Impact of textiles on environmental issues, part – II. Asian Dyer 9, 45–51.

Karthik, T., Gopalakrishnan, D., 2013b. Impact of textiles on environmental issues, part – III. Asian Dyer 9, 40–43.

Karthik, T., Gopalakrishnan, D., 2013c. Impact of textiles on environmental issues, part – IV. Asian Dyer 9, 43–46.

Karthik, T., Gopalakrishnan, D., 2014. Environmental analysis of textile value chain: an overview. In: Muthu, S.S. (Ed.), Roadmap to Sustainable Textiles and Clothing. Springer, New York, pp. 153–188.

Karthik, T., Subramani, K.B., 2016. Sustainable biopolymer fibres – production, properties and applications. In: Muthu, S.S., Gardetti, M.A. (Eds.), Sustainbale Fibres for Fashion Industry, vol. 1. Springer, New York, pp. 109–140.

Kind, S., Neubauer, S., Becker, J., Yamamoton, M., Volkert, M., von Abendroth, G., Zelder, O., Wittmann, C., 2014. From zero to hero-production of bio-based nylon from renewable resources using engineered *Corynebacterium glutamicum*. Metab. Eng. 25, 113–123.

Klement, T., Büchs, J., 2013. Itaconic acid – a biotechnological process in change. Bioresour. Technol. 135, 422–431.

de Koning, G.J.M., 1993. Prospects of Bacterial Poly[(R)-3-hydroxyalkanoates]. Eindhoven University of Technology, Eindhoven.

Kotnis, M.A., O'Brien, G.S., Willett, J.L., 1995. Processing and mechanical properties of biodegradable poly(hydroxybutyrate-co-valerate)-starch compositions. J. Environ. Polym. Degrad. 3 (2), 97–105.

Kricheldorf, H.R., Chatti, S., Schwarz, G., Kruger, R.P., 2003. Macrocycles 27: cyclic aliphatic polyesters of iso-sorbide. J. Polym. Sci. Part A Polym. Chem. 41, 3414–3424.

Kurian, J.V.A., 2005. New polymer platform for the Future-Sorona ® from corn derived 1,3 propanediol. J. Polym. Environ. 13 (2), 159–165.

Kurzrock, T., Weuster-Botz, D., 2010. Recovery of succinic acid from fermentation broth. Biotechnol. Lett. 32, 331–339.

Ladero, V., Ramos, A., Wiersma, A., Goffin, P., Schanck, A., Kleerebezem, M., Hugenholtz, J., Smid, E.J., Hols, P., 2007. High-level production of the low-calorie sugar sorbitol by *Lactobacillus plantarum* through metabolic engineering. Appl. Environ. Microbiol. 73, 1864–1872.

Lapisardi, G., Chiker, F., Launay, F., Nogier, J.P., Bonardet, J.L., 2005. Preparation, characterisation and catalytic activity of new bifunctional Ti–AlSBA15 materials. Application to a "one-pot" green synthesis of adipic acid from cyclohexene and organic hydroperoxides. Microporous Mesoporous Mater. 78, 289–295.

Lee, S.Y., Hong, S.H., Lee, S., Park, S.J., 2004. Fermentative production of chemicals that can be used for polymer synthesis. Macromol. Biosci. 4, 157–164.

Li, X., Wu, D., Lu, T., Yi, G., Su, H., Zhang, Y., 2013. Highly efficient chemical process to convert mucic acid into adipic acid and DFT studies of the mechanism of the rhenium-catalyzed deoxyhydration. Angew. Commun. 53, 4200–4204.

Lockwood, L.B., Nelson, G.E., 1946. Some factors affecting the production of itaconic acid by *Aspergillus terreus* in agitated cultures. Arch. Biochem. 10, 365–374.

Loos, K., 2010. Biocatalysis in Polymer Chemistry. Wiley-VCH Verlag GmbH & Co. KGaA, Weinheim, Germany.

Lovell, C.S., Ries, M.E., Ward, I.M., Montes de Oca, H., Farrar, D., 2013. Morphology, phase behavior, and relaxation processes in a series of aromatic-aliphatic thermotropic polymer biomaterials. Macromolecules 46, 1201–1211.

Ma, J., Pang, Y., Wang, M., Xu, J., Ma, H., Nie, X., 2012a. The copolymerization reactivity of diols with 2, 5-furandicarboxylic acid for furan-based copolyester materials. J. Mater. Chem. 22, 3457.

Ma, J., Yu, X., Xu, J., Pang, Y., 2012b. Synthesis and crystallinity of poly(butylene 2,5-furandicarboxylate). Polymer 53, 4145–4151.

Madison, L.L., Huisman, G.W., 1999. Metabolic engineering of poly(3-hydroxyalkanoates): from DNA to plastic. Microbiol. Mol. Biol. Rev. 63 (1), 21–53.
Mainka, H., et al., 2015. Lignin – an alternative precursor for sustainable and cost-effective automotive carbon fiber. J. Mater. Res. Technol. 4 (3), 283–296.
Manabe, K., Sun, X.M., Kobayashi, S., 2001. Dehydration reactions in water. Surfactant-type brønsted acid-catalyzed direct esterification of carboxylic acids with alcohols in an emulsion system. J. Am. Chem. Soc. 123, 10101–10102.
Manabe, K., Iimura, S., Sun, X.M., Kobayashi, S., 2002. Dehydration reactions in water. brønsted acid-surfactant-combined catalyst for ester, ether, thioether, and dithioacetal formation in water. J. Am. Chem. Soc. 124, 11971–11978.
Martin, D., Annamalai, P.K., Laycock, B., 2015 High Performance Sustainable Carbon Fibres from Australian Spinifex Grass. http://researchers.uq.edu.au/research-project/23303.
Mathers, R.T., Meier, M.A.R., 2011. Green Polymerization Methods: Renewable Starting Materials, Catalysis and Waste Reduction. Wiley-VCH Verlag GmbH & Co. KGaA, Weinheim, Germany.
McConnell, V., March 29, 2010. Reinforced Plastics Magazine. www.reinforcedplastics.com/view/8116/launching-the-carbon-fibre-recyclinindustry.
McKinlay, J.B., Vieille, C., Zeikus, J.G., 2007. Prospects for a bio-based succinate industry. Appl. Microbiol. Biotechnol. 76 (4), 727–740.
Mialon, L., Pemba, A.G., Miller, S.A., 2010. Biorenewable polyethylene terephthalate mimics derived from lignin and acetic acid. Green Chem. 12, 1704–1706.
Miller, K.K., Zhang, P., Nishizawa-Brennen, Y., Frost, J.W., 2014. Synthesis of biobased terephthalic acid from cycloaddition of isoprene with acrylic acid. ACS Sustain. Chem. Eng. 2, 2053–2056.
Mimitsuka, T., Sawai, H., Hatsu, M., Yamada, K., 2007. Metabolic engineering of *Corynebacterium glutamicum* for cadaverine. Fermentation. Biosci. Biotechnol. Biochem. 71, 2130–2135.
Mohanty, A.K., Misra, M., Drzal, L.T., 2002. Sustainable bio-composites from renewable resources: opportunities and challenges in the green materials world. J. Polym. Environ. 10 (1/2), 19–26.
Mondala, A.H., 2015. Direct fungal fermentation of lignocellulosic biomass into itaconic, fumaric, and malic acids: current and future prospects. J. Ind. Microbiol. Biotechnol. 42, 487–506.
Moore, J.A., Kelly, J.E., 1978a. Thermally initiated crosslinking of an unsaturated heterocyclic polyester. J. Macromol. 11 (3), 568–573.
Moore, J.A., Kelly, J.E., 1978b. Polyesters derived from furan and tetrahydrofuran nuclei. Macromolecules 11, 568–573.
Moreau, C., Belgacem, M.N., Gandini, A., 2004. Recent catalytic advances in the chemistry of substituted furans from carbohydrates and in the ensuing polymers. Top Catal. 27, 11–30.
Nardello-Rataj, V., Caron, L., Borde, C., Aubry, J.M., 2008. Oxidation in three-liquid-phase microemulsion systems using "balanced catalytic surfactants". J. Am. Chem. Soc. 130, 14914–14915.
Niu, W., Draths, K.M., Frost, J.W., 2002. Benzene-free synthesis of adipic acid. Biotechnol. Prog. 18, 201–211.
Okada, M., 2002. Chemical syntheses of biodegradable polymers. Prog. Polym. Sci. 27, 87–133.
Okano, K., Tanaka, T., Ogino, C., Fukuda, H., Kondo, A., 2010. Biotechnological production of enantiomeric pure lactic acid from renewable resources: recent achievements, perspectives, and limits. Appl. Microbiol. Biotechnol. 85, 413–423.
Onda, A., 2014. Production of lactic acid from sugars by homogeneous and heterogeneous catalysts. In: Jin, F. (Ed.), Application of Hydrothermal Reactions to Biomass Conversion. Springer-Verlag, Berlin/Heidelberg, Germany, pp. 83–107.

Pandey, M.P., Kim, C.S., 2011. Lignin depolymerization and conversion: a review of thermochemical methods. Chem. Eng. Technol. 34, 29–41.

Pearl, I.A., 1942. Vanillin from lignin materials. J. Am. Chem. Soc. 64, 1429–1431.

Pellis, A., Acero, E.H., Gardossi, L., Ferrario, V., MGuebitz, G., 2016. Renewable building blocks for sustainable polyesters: new biotechnological routes for greener plastics. Polym. Int.

Petrovicn, Z.S., Milic, J., Xu, Y., Cvetkovic, I., 2010. A chemical route to high molecular weight oil based polyhydroxyalkanoate. Macromolecules 43, 4120–4125.

Phillips, A.M.F., Romão, C., 1999. Synthesis of γ-butyrolactones by a Baeyer–Villiger oxidation with hydrogen peroxide, catalysed by methyltrioxorhenium. Eur. J. Org. Chem. 8, 1767–1770.

Pickering, S.J., 2006. Recycling technologies for thermoset composite materials – current status. Composites Part A 37, 1206–1215.

Pickering, S.J., et al., 2009. Characterisation of carbon fibres recycled from carbon fibre/epoxy resin composites using supercritical n-propanol. Composites Sci. Technol. 69 (2), 192–198.

Pintom, P.C.R.O., Sousa, A.F., Silvestre, A.J.D., Neto, C.P., Gandini, A., Eckerman, C., Holmbom, B., 2009. *Quercus suber* and *Betula pendula* outer barks as renewable sources of oleochemicals: a comparative study. Ind. Crops Prod. 29, 126–132.

Polen, T., Spelberg, M., Bott, M., 2013. Toward biotechnological production of adipic acid and precursors from biorenewables. J. Biotechnol. 167, 75–84.

Prakash, C., Maruthavanan, T., Parvathi, C., 2009. Environmental impacts of textile industries. Indian Text. J. 23, 22–26.

Ravishankara, A.R., Daniel, J.S., Portmann, R.W., 2009. Nitrous oxide (N_2O): the dominant ozone-depleting substance emitted in the 21st century. Science 326, 123–125.

Reimer, R.A., Slaten, C.S., Seapan, M., Lower, M.W., Tomlinson, P.E., 1994. Abatement of N_2O emissions produced in the adipic acid industry. Environ. Prog. 3, 134–137.

Rydz, J., Zawidlak-Węgrzyńska, B., Christova, D., 2015. Degradable polymers. In: Mishra, M.K. (Ed.), Encyclopedia of Biomedical Polymers and Polymeric Biomaterials. CRC Press, Boca Ratón, FL, USA.

Rydz, J., Sikorska, W., Kyulavska, M., Christova, D., 2015. Polyester-based (Bio)degradable polymers as environmentally friendly materials for sustainable development. Int. J. Mol. Sci. 16, 564–596.

Saladino, R., Bernini, R., Neri, V., Crestini, C., 2009. A novel and efficient catalytic epoxidation of monoterpenes by homogeneous and heterogeneous methyltrioxorhenium in ionic liquids. Appl. Catal. A 360, 171–176.

Sato, K., Aoki, M., Noyori, R., 1998. A "Green" route to adipic acid: direct oxidation of cyclohexenes with 30 percent hydrogen peroxide. Science 281, 1646–1647.

Schneider, L., Lazarus, M., Kollmuss, A., 2010. Industrial N_2O Projects Under the CDM: Adipic Acid – A Case of Carbon Leakage? Working Paper, WP-US-1006. Stockholm Environment Institute, Somerville, MA.

Shen, L., Haufe, J., Patel, M., 2009. Pro-BIP Product Overview and Market Projection of Emerging Bio-based Plastics, Report Commissioned by European Polysaccharide Network and Excellence (EPNOE) at European Bioplastics. .

Shogren, R.L., 1995. Poly(ethylene oxide)-coated granular starch-poly(hydroxybutyrate-cohydroxyvalerate) composite materials. J. Environ. Polym. Degrad. 3 (2), 75–80.

Silveira, M.M., Jonas, R., 2002. The biotechnological production of sorbitol. Appl. Microbiol. Biotechnol. 59, 400–408.

Slivniak, R., Domb, A.J., 2005. Lactic acid and ricinoleic acid based copolyesters. Macromolecules 38, 5545–5553.

Song, H., Lee, S.Y., 2006. Production of succinic acid by bacterial fermentation. Enzym. Microb. Technol. 39 (3), 352–361.

Sousa, A.F., Gandini, A., Silvestre, A.J.D., Neto, C.P., 2008. Synthesis and characterization of novel biopolyesters from suberin and model comonomers. ChemSusChem 1, 1020–1025.

Sousa, A.F., Silvestre, A.J.D., Gandini, A., Neto, C.P., 2012. Synthesis of aliphatic suberin-like polyesters by ecofriendly catalytic systems. High Perform. Polym. 24, 4–8.

Sousa, A.F., Matos, M., Freire, C.S.R., Silvestre, A.J.D., Coelho, J.F.J., 2013. New copolymers derived from terephthalic and 2,5-furandicarboxylic acids: a step forward in the development of biobased polyesters. Polymer 54, 513.

Sousa, F., Vilela, C., Fonseca, A.C., Matos, M., Freire, C.S.R., Gruter, C.J.M., Coelho, J.F.J., Silvestre, A.J.D., 2014. The quest for sustainable polyesters – insights into the future. Polym. Chem. 5, 3119–3141.

Sousa, F., Vilela, C., Fonseca, A.C., Matos, M., Freire, C.S.R., Gruter, C.J.M., Coelho, J.F.J., Silvestre, A.J.D., 2015. Bio-based polyesters and other polymers from 2,5furandicarboxylic acid: a tribute to furan excellency. Polym. Chem. 6, 5961–5983.

Steiger, M.G., Blumhoff, M.L., Mattanovich, D., Sauer, M., 2013. Biochemistry of microbial itaconic acid production. Front. Microbiol. 4, 23.

Steinbuchel, A., 2003. Biopolymers, General Aspects and Special Applications. Wiley-VCH, Weinheim, p. 10.

Steinbuchel, A., Doi, Y., 2002. Biopolymers: Polyesters III—Applications and Commercial Products, vol. 4. Wiley-VCH, Weinheim.

Tachibana, Y., Kimura, S., Kasuya, K.I., 2015. Synthesis and verification of biobased terephthalic acid from furfural. Sci. Rep. 5, 8249–8253.

Takasu, A., Takemoto, A., Hirabayashi, T., 2006. Polycondensation of dicarboxylic acids and diols in water catalyzed by surfactant-combined catalysts and successive chain extension. Biomacromolecules 7, 6–9.

Timofeeva, M.N., Kholdeeva, O.A., Jhung, S.H., Chang, J.S., 2008. Titanium and cerium-containing mesoporous silicate materials as catalysts for oxidative cleavage of cyclohexene with H_2O_2: a comparative study of catalytic activity and stability. Appl. Catal. A 345, 195–200.

Turk, S.C.H.J., Kloosterman, W.P., Ninaber, D.K., Kolen, K.P.A.M., Knutova, J., Suir, E., Schürmann, M., Raemakers-Franken, P.C., Muller, M., De Wildeman, S.M.A., Raamsdonk, L.M., van der Pol, R., Wu, L., Temudo, M.F., van der Hoeven, R., Akeroyd, M., van der Stoel, R.E., Noorman, H.J., bovenberg, R.A.L., Trefzer, A.C., 2016. Metabolic engineering towards sustainable production of nylon-6. ACS Synth. Biol. 5 (1), 65–73.

US Patent 7,824,495 B1. System to Continuously Produce Carbon Fibre via Microwave Assisted Plasma Processing. 2010.

Vardon, D.R., et al., 2015. Adipic acid production from lignin. Energy Environ. Sci. 8, 617–628.

Vardon, D.R., Rorrer, N.A., Johnson, C.W., Ciesielski, P.N., Menart, M.J., Steirer, X., Salvachúa, D., Settle, A.E., Cleveland, N.S., Dorganand, J.R., Beckham, G.T., 2016. *cis,cis*-Muconic acid: separation and catalysis to bio-adipic acid for nylon-6,6 polymerization. Green Chem. 18, 3397–3413.

Van de Velde, K., Kiekens, P., 2002. Biopolymers: overview of several properties and consequences on their applications. Polym. Test 21 (4), 433–442.

Van der Velden, N.M., Patel, M.K., Vogtländer, J.G., 2014. LCA benchmarking study on textiles made of cotton, polyester, nylon, acryl, or elastane. Int. J. Life Cycle Assess. 19, 331–356.

Vijayaraaghavan, N.N., Karthik, T., 2004. Multi-component fiber technology for medical and other filtration applications. Synth. Fibres 33 (1), 5–8.

Vink, E.T.H., Rabago, K.R., Glassner, D.A., Springs, B., O'Connor, R.P., Kolstad, J., Gruber, P.R., 2004. The sustainability of NatureWorks™ polylactide polymers and Ingeo™ polylactide fibers: an update of the future. Macromol. Biosci. 4, 551–564.

Vroman, I., Tighzert, L., 2009. Biodegradable polymers. Materials 2 (2), 307–344.
Wang, Z.X., Zhuge, J., Fang, H., Prior, B.A., 2001. Glycerol production by microbial fermentation: a review. Biotechnol. Adv. 19, 201–223.
Wen, Y., Wang, X., Wei, H., Li, B., Jin, P., Li, L., 2012. A large-scale continuous-flow process for the production of adipic acid via catalytic oxidation of cyclohexene with H_2O_2. Green Chem. 14, 2868–2875.
Werpy, T., Petersen, G. (Eds.), 2004. Top Value Added Chemicals from Biomass. US Department of Energy.
Willke, T., Vorlop, K.D., 2001. Biotechnological production of itaconic acid. Appl. Microbiol. Biotechnol. 56, 289–295.
Wise, R., Girard, P., 2016. Carbon Fibres: A Review of Technology and Current Market Trends. The Knowledge Transfer Nework, London. https://connect.innovateuk.org/documents/2998699/3676121/Carbon_Fibre_Report_2016.pdf/.
Wolf, O., Patel, M., Weidemann, F.M., Schleich, J., Husing, B., Angerer, G., 2005. Techno-economic feasibility of large scale production of bio-based polymers in Europe. Eur. Comm.
Wu, J., Eduard, P., Jasinska-Walc, J., Rozanski, A., Noordover, B.J.A., Van Es, D.S., Koning, C.E., 2013. Fully isohexide-based polyesters: synthesis, characterization, and structure-properties relations. Macromolecules 46, 384–394.
Xie, G., West, T.P., 2009. Citric acid production by *Aspergillus niger* ATCC 9142 from a treated ethanol fermentation co-product using solid-state fermentation. Lett. Appl. Microbiol. 48, 639.
Xu, Q., Li, S., Fu, Y., Tai, C., Huang, H., 2010. Two-stage utilization of corn straw by *Rhizopus oryzae* for fumaric acid production. Bioresour. Technol. 101, 6262–6264.
Xu, Q., Li, S., Huang, H., Wen, J., 2012. Key technologies for the industrial production of fumaric acid by fermentation. Biotechnol. Adv. 30, 1685–1696.
Yashiro, T., Kricheldorf, H.R., Huijser, S., 2009. Synthesis of polyesters from succinic anhydride and various diols catalyzed by metal triflates. Macromol. Chem. Phys. 210, 1607–1616.
Zeng, A.P., Biebl, H., 2002. Bulk chemicals from biotechnology: the case of 1,3-propanediol production and the new trends. Adv. Biochem. Eng. Biotechnol. 74, 239–259.
Zinn, M., Witholt, B., Egli, T., 2001. Occurrence, synthesis and medical application of bacterial polyhydroxyalkanoate. Adv. Drug Deliv. Rev. 53 (1), 5–21.
Zorn, H., Czermak, P., 2014. Biotechnology of food and feed additives. Springer 143, 301.

Further reading

Doi, Y., Steinbuchel, A., 2002. Biopolymers, Applications and Commercial Products – Polyesters III. Wiley-VCH, Weiheim.

Part Two

Sustainable textile and apparel production

Low water-consumption technologies for textile production

9

Shanmugasundaram O. Lakshmanan[1], *Guruprasad Raghavendran*[2]
[1]K.S. Rangasamy College of Technology, Tiruchengode, India; [2]ICAR-Central Institute for Research on Cotton Technology, Mumbai, India

9.1 Introduction

Our planet Earth, with its diverse and abundant life forms, including over seven billion humans, is facing a serious water crisis. The number of people on our planet is growing quickly and our water use is growing even faster. Water is the most widely occurring substance on earth; however, only 2.53% is fresh water whereas the remainder is salt water. Some two-thirds of this fresh water is locked up in glaciers and permanent snow cover. Hence critical challenges lie ahead in coping with progressive water shortages and water pollution. By the middle of this century, at worst 7 billion people in 60 countries will be water-scarce, at best 2 billion people in 48 countries. Freshwater resources are further reduced by pollution caused by disposal of industrial wastes and chemicals, human waste and agricultural wastes. The global overview of water availability versus the population shows the continental disparities, and in particular the pressure put on the Asian continent, which supports more than half the world's population with only 36% of the world's water resources (Water for People, 2003).

The textile industry is one of the largest users of water for its production system. The industry uses vast amounts of water throughout all processing operations. Almost all dyes, specialty chemicals and finishing chemicals are applied to textiles in water baths. Most fabric preparation steps, including desizing, scouring, bleaching and mercerizing, use water. And each one of these steps is followed by thorough washing of the fabric to remove all chemicals used in that step before moving on to the next step. A cotton T-shirt requires approximately 2650 L of water to grow, produce and transport. A substantial proportion of this water use (almost 20% and more) is used in the dyeing process alone. Up to 100 L of fresh water and a very high amount of energy are required to dye just 1 kg of cotton fabric (Hulme, 2015).

The source of water for industries is from surface sources such as rivers and shallow wells, deep wells or municipal or public water systems. Many mills have high water costs, especially when the water is purchased from a municipal system. The quality of textiles produced by any manufacturing operation which employs wet processes, such as preparation, dyeing and/or finishing, is profoundly affected by water quality. The quality of water to be used for efficient processing is 300 parts per million. Many mills have softening plants to soften the hard water for use, which adds to the cost.

Adding to the problem of water scarcity is the problem of pollution when the water used for processing is discharged without treatment. The waste water that flows in the

Sustainable Fibres and Textiles. http://dx.doi.org/10.1016/B978-0-08-102041-8.00009-3

drains corrodes and clogs the sewerage pipes. If allowed to flow in drains and rivers, it affects the quality of drinking water and making it unfit for human consumption. Such polluted water can be a breeding ground for bacteria and viruses.

The interest of industries towards eco-friendly wet-processing textile techniques has increased because of increased awareness about environmental issues. The move towards more sustainable water use is becoming a priority for organizations across all sectors of life. With the global demand for water increasing every day, it is necessary to look into the sustainable aspects of textile chemical processing. There is a need to adopt measures that will reduce the consumption of water in the chemical processing of textiles. New production methods that use no water or a lesser quantity of water have to be developed and adopted on a large scale. By developing suitable production methods and adopting appropriate measures, the dependence of processing industries on water can be reduced. With the improvement in water efficiency, a range of benefits can be gained by industries, including an improved financial position, reduced carbon footprint and effective environmental leadership. This chapter covers various aspects of water consumption in the processing of textiles, strategies to control waste water and methods to reduce the water footprint in textiles industries, such as plasma technology, ultrasound-assisted dyeing and supercritical carbon dioxide dyeing.

9.1.1 Water consumption in textile processing

Woven and knitted fabrics cannot be processed into finished goods without several water-intensive finishing operations to enhance the appearance, durability and serviceability of fabrics. Woven fabric finishing includes preparation of the cloth (desizing, scouring, bleaching and mercerizing), dyeing, printing and finishing (resin treatment, waterproofing or flame-proofing and special finishes). The knit fabric finishing sector produces knit fabric goods and hosiery outerwear and inner wear. The processes for knitted fabrics include bleaching, dyeing, printing and special treatments. Desizing and mercerizing operations are not required for knitted materials. Important areas of water use in processing are given in Table 9.1 (Water and Chemical Use, 1997).

Table 9.1 Important areas of water use

Process water use	Ancillary water use
Scouring processes	Evaporation: usually small unless cooling load is large
Bleaching processes	Domestic/toilets
Dyeing process	Boilers: can be excessive if condensate is not returned
Finishing processes	Factory washing: may be high if hose pipes are used
Process cooling	Air conditioning
Process rinsing/washing	Utility cooling: can easily be recycled

9.1.2 Specific water consumption

Specific water consumption is calculated using the following formula:

$$\text{Specific water consumption} = \text{Amount of water consumed}/\text{Amount of product produced}$$

For example, the specific water consumption of woven fabric is 113 L/kg, and for knitted fabric is 83 L/kg. These values have to be considered as only indicative and not standard values. The amount of water required in textile processing varies widely even between similar types of wet processing at different sites. The information given in Table 9.2 indicates values for liquor ratios for various types of dyeing machinery. In general, heating of dye baths constitutes the major portion of energy consumed in dyeing; therefore, using low-liquor ratio dyeing equipment can result in major energy savings. However, the washing efficiency of many types of low-liquor ratio dyeing machines, such as jigs, is poor. Washing efficiency depends on mechanical factors such as bath and fabric turnover rate or 'contacts', liquor ratio, turbulence and other mechanical considerations and physical flow characteristics. Thus, lower liquor ratio and reduced total water use do not always correlate closely (Smith and Rucker, 1987).

9.1.3 Reasons for more consumption of water

The excess use of water occurs as a result of the following factors:

- overuse of water in the winches
- as a result of high temperature, flash steam escapes from the dye liquor
- leakage of water in pipelines and equipment
- dead storage in the sump and feed water tank, and also leakage in these systems
- excessive use of water for cleaning, gardening and labour
- improper process sequencing
- reprocessing operations

Normally excessive water use, considering waste alone in most of the units, works out to be 10%–20% on average (Balachandran and Rudramoorthy, 2008).

Table 9.2 Liquor ratios of dyeing machines

Dyeing machine	Material to liquor ratio
Continuous	1:1
Winch	1:15
Jet	1:4–1:10
Jigger	1:3–1:6
Beam	1:8–1:10
Package	1:6–1:8
Paddle	1:40
Stock	1:4–1:12
Hank	1:12–1:25

9.1.4 Minimizing water consumption in textile processing

9.1.4.1 Reduction in number of process steps

With continual improvements in chemical performance, processes should be reviewed regularly to ensure that every stage remains necessary. Many firms have dramatically reduced rinse water by reducing the number of process steps involved. Strategies involve combining desizing, scouring and bleaching; dyeing two fibres in one bath for blends whenever possible; and combining the scouring and dyeing of cotton or synthetics when strict shade requirements do not need to be met. These are widely practiced and vary greatly with end-use requirements, shade, fastness, specific blend and equipment.

9.1.4.2 Reduction in process water use

Washing and rinsing are both important for reducing impurity levels in the fabric to predetermined levels. Because water and effluent disposal costs have been low, there has been a tendency to overuse water. Because the cost to use water is now increasing, the optimization of water use could pay dividends. One possible option is to reduce the use of rinse water for lighter shades. Examples of water reduction in batch and continuous operations are:

Winch dyeing
By dropping the dye batch and avoiding overflow rinsing, water consumption can be reduced by 25%.
High and low
By replacing the overflow with pressure-jet dyeing and batch-wise rinsing, water consumption may be cut by approximately 50%.
Beam dyeing
Preventing overflow during soaking and rinsing can reduce water consumption by about 60%. Automatic controls proved to be beneficial.
Jig dyeing
Reductions in water consumption ranging from 15% to 79% were possible by switching from overflow to step-wise rinsing.
Cheese dyeing
A reduction in water consumption of around 70% proved possible with intermittent rinsing.
Continuous dyeing
A 20%–30% savings was realized by introducing automatic water stops. An effective method of washing is to use a countercurrent system. Horizontal washing equipment delivered double the performance of vertical washing machines using the same amount of water.

9.1.4.3 Recycle cooling water

Many cooling water systems are operated on a once-through basis. The resulting hot water is generally uncontaminated and can be reused in the process as makeup or rinse water.

9.1.4.4 Reuse process water

It is sometimes possible to reuse certain waste streams, e.g., dilute wash water, in other parts of the processes:

- as process water in other textile operations, with or without the addition of chemicals

- as rinse water for another process in which low-grade rinse water is acceptable
- as rinse water for direct use in a continuous countercurrent washing system in which dilute rinses are reused in successively dirtier washing bowls

Other options for process water reuse are to:

- use scouring rinse water for desizing or machine cleaning
- use mercerizing water to prepare baths for scouring, bleaching and wetting fabric
- recover the condensate from the indirect use of steam and use it as process water

As water and effluent costs continue to rise, new technologies for treating and recycling water for process use have to be explored.

9.1.4.5 Countercurrent washing/rinsing

Washing/rinsing is the most common operation in textile processing. Many processes involve washing and rinsing stages, and therefore optimization of these processes can conserve significant amounts of water. Washing and rinsing require about 70%–75% more water than the other stages such as bleaching and dyeing. Countercurrent washing/rinsing is an established technique common on continuous ranges. This system of operation can significantly reduce water use.

9.1.5 General measures to reduce water consumption

- Monitoring of water consumption in the various processes
- Turning off running taps and hoses when machines are not in operation
- Leak control: audits should be carefully conducted for broken and leaking pipes, drums, pumps and valves
- Use of automatic controllers for control of fill volume and liquor temperature in batch machines
- Optimization of scheduling in production (e.g., dyeing dark shades after dyeing light shades will reduce water and chemical consumption for machine cleaning)
- Adjustment of pretreatment processes to suit downstream processing (bleaching not necessary for dark shades)
- Education/training of employees about the importance of water conservation

Technologies to reduce water footprint of textile products

9.2 Plasma technology

9.2.1 Introduction

Plasma is the fourth state of matter, invented by Langmuir in 1926. The three other states are solid, liquid and gas. Plasma is a cloud of protons, neutrons and electrons in which all of the electrons have come loose from their respective molecules and atoms, giving the plasma the ability to act as a whole rather than as a bunch of atoms.

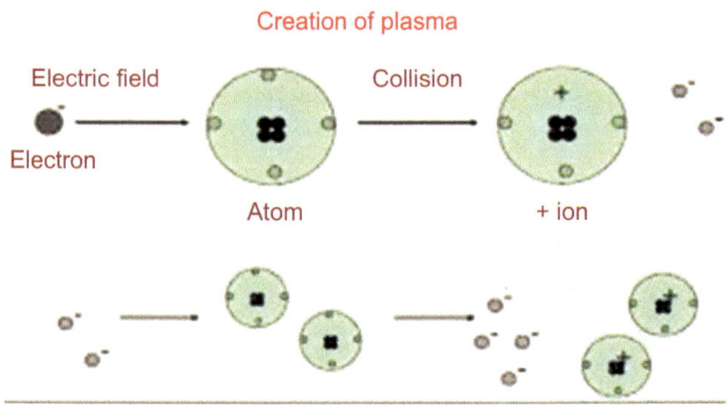

Figure 9.1 Creation of plasma.

Plasma is more like a gas than any of the other states of matter because the atoms are not in constant contact with each other, but it behaves differently from a gas. Plasma gas is a mixture of electrons, ionized atoms and molecules, photons and neutral species. It can be generated by applying current across a dielectric gas, as shown in Fig. 9.1. Conventional wet treatments used in textile processing for fibre surface changes have many limitations. These treatments mainly have the problems of higher energy consumption and environmental issues.

The use of plasma technology at low temperature in textile processing can prove to be the best alternative for these problems. In plasma application, the plasma does not penetrate deeply into fibre but reacts only with the fabric surface. Hence the internal structure remains unaffected (Shah and Shah, 2013). Plasma technologies can modify the surface properties of textile materials by depositing chemicals by plasma polymerization to improve functionality or remove substances by (plasma etching) from the textile materials for better use.

The functional properties of the fabric can be changed by nanoscale etching of surface by plasma gas particles. In textile processing, this technology can be examined in various areas such as pretreatment, dyeing and finishing through different discharge methods including glow, corona and dielectric barrier to improve the functionality and change of surface properties of textile materials (Poll and Schladitz, 2001; Wang et al., 2008).

Plasma technology is used for surface treatment and is beneficial over the conventional process, because it does not alter the inherent properties of the textile materials. It is carried out at the dry stage, and hence no cost is involved in treating effluent. Different methods used to establish the ionization of gases are: (1) glow discharge, (2) corona discharge and (3) dielectric barrier discharge. Plasma technology reduces the processing cost involved in preparatory processes by limiting the use of chemicals and energy and reducing the pollution load.

A schematic diagram of an atmospheric pressure plasma unit is shown in Fig. 9.2. Plasma treatment has been used to establish both surface and bulk property enhancements of textile materials, resulting in improvements in textile products from conventional fabrics to advanced composites. It can improve the functionality of textile

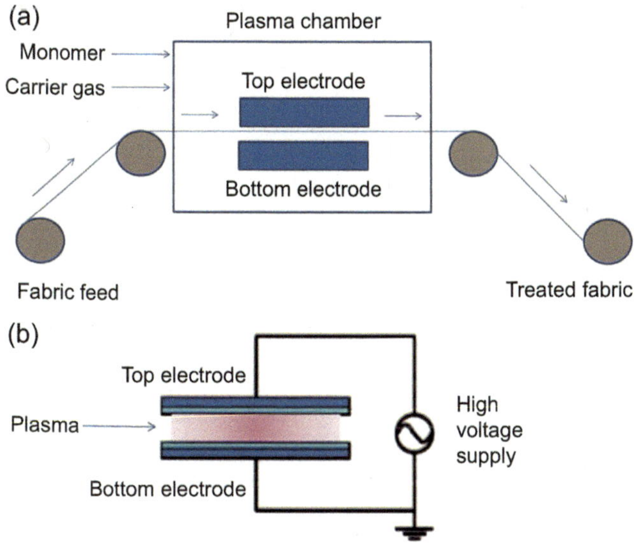

Figure 9.2 (a, b) Schematic diagram of atmospheric pressure plasma treatment.

materials such as the wettability, hydrophobic nature, adhesion and product quality (Shah and Shah, 2013).

Plasma treatment can enhance the ability of polyester, wool and polypropylene fibres to retain moisture on their surface. Cotton fabric treated with hexamethyldisiloxane gas gave a smooth surface and increased the contact angle of water. It can improve the adhesion property of chemical coating and increase the dye affinity of textile materials. Oxygen plasma can be used for antifelting finishing of wool fabric. Plasma gas may be used to give UV protection, an antibacterial quality, waterproofing, a flame resistance finish, etc.

In the desizing process, a mixture of oxygen and helium gas can effectively remove poly(vinyl alcohol) (PVA) sizing material from cotton fabric. Plasma treatment makes PVA sizing agent more soluble by breaking down the chains of PVA. X-ray photoelectron microscopy results show that plasma adds oxygen and nitrogen groups to the surface of PVA, which provides greater polarity, thus increasing the solubility of PVA in water.

9.2.2 Treatment of wool material by plasma technology

Kan et al. (2001) achieved an good antifelting effect on wool fabric by plasma treatment. Plasma treatment reduces the curling effect, enhancing wettability and introducing surface polar groups by etching off the exocuticle and hydrophobic epicuticle. From the atomic force microscopy study, it was found that the surface area of fibre increased from $0.1\,m^2/g$ to $0.35\,m^2/g$. Millard and Klausen achieved an increase in hydrophilic groups and fibre–fibre friction, a decrease in directional frictional effect, a change from cystine to cysteic acid and an improvement in dyeing and shrink-proofing effect on wool fabric by plasma treatment.

9.2.3 Dyeing of natural and synthetic fibres

Fabric properties, such as wettability, capillary action, enhancement of the surface area, creation of reaction sites on fibres, reduction in crystallinity and many more, can be improved by plasma treatment. The following changes take place in cotton fibre when treated with either argon or air plasma: (1) an increase in fabric surface area per unit volume; (2) a removal of surface fibre impurities (3) introduction of carbonyl and carboxyl groups in the fibre surface; and (4) the formation of free radicals in chains of cellulose. The actions of plasma gas change the surface properties, which leads to an increase in the dye uptake of cotton fabric (Desai, 2008). Plasma-pretreated cotton materials required a smaller amount of heat energy and shorter duration during dyeing processes. The dye uptake of plasma-treated wool fabric increased to 90% in 30 min compared with 60 min in untreated wool fabric (Kan et al., 2010). The cationic dye can be used to dye plasma-treated polyester fabric, which leads to low energy consumption, environmentally friendly results and low temperature for dyeing by introducing polar groups to the polyester fibre surface (Shah and Shah, 2013).

Use of plasma is eco-friendly for the processing of textile materials. It reduces the processing duration and heat energy in dyeing processes by increasing the hydrophilicity of certain materials such as hemp, wool and polypropylene (Radetic et al., 2007). Inkjet printing is a method of printing textile materials which can print materials possessing a hydrophilic property. Polypropylene materials are hard to dye and print because of their hydrophobic nature. Plasma-pretreated polypropylene material possesses good colour fastness and results in better fixation by inkjet printing.

9.3 Ultrasonic-assisted dyeing technology

9.3.1 Introduction

Sound waves with frequencies above the audible range (20 kHz) are ultrasonic waves, or ultrasound or ultrasonics. These waves are high-frequency sound waves and highly energetic, and are not capable of being heard by humans. The wavelengths of ultrasound waves (Fig. 9.3) are small and found in many applications such as for cutting and matching hard materials, ultrasonic soldering and welding, ultrasonic flaw detection, measurement of flow devices, thermal effects, communications, dyeing and finishing of textile materials and medicines (Bruno Pollet, 2012).

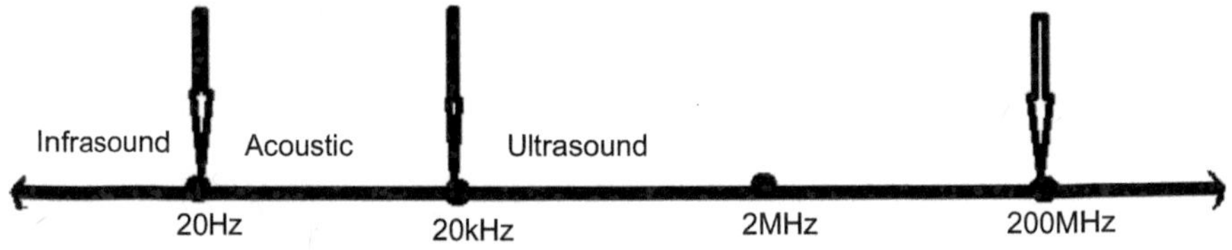

Figure 9.3 Wavelengths of different sounds.
Reproduced from https://en.wikipedia.org/wiki/Ultrasound.

9.3.2 Production of ultrasonic sound waves

Ultrasonic waves are have frequencies above 20,000 Hz and are produced by mechanical generators, magnetostriction generators or oscillators and piezo-electric oscillators.

9.3.3 Applications in dyeing and finishing of textile materials

Dyeing cotton fabric with reactive, direct and vat dyes requires a huge quantity of sodium chloride as an exhausting agent to increase dye uptake and causes serious environmental problems. As a result, a large volume of effluent containing a significant quantity of electrolyte, dyes and chemicals are discharged from the cotton dyeing industry (Hauser and Tabba, 2001). Lewis and Lei (1989) suggested a method to avoid this problem, i.e., to cationize the material by treatment with cationic agents and, as a result of this process, increase the colour yield and improve the wash-fastness properties of the dyed material.

Ultrasound is used to dye textile materials with anionic dyes such as reactive and direct dye, which increases the colour strength of the dyeing process.

Kamel et al. (2009) studied ultrasonic dyeing of cationized cotton fabric using cochineal dye. The dye is extracted from cochineal insects using conventional and ultrasound techniques. They stated that various factors such as particle size, temperature, ultrasound power and duration affect the quantity of dye extraction from cochineal powder and achieve maximum yield by ultrasound at 300 W, at low temperature and short duration. They also investigated the effect of dyeing conditions such as the percentage of salt, pH, ultrasound power, dyeing temperature and duration on colour strength and fastness properties of the dyed fabric.

Montazer et al. (2007), Kamel et al. (2007, 2009) and Nazaria et al. (2011) reported a salt-free dyeing technique of cationized cotton fabric with direct and reactive dyes; they obtained a high exhaustion rate of dye on cotton material. Hashem et al. (2003) achieved high reaction efficiency and maximum dye exhaustion in cationized cotton fabric by 3-chloro-2-hydroxypropyltrimethyl ammonium chloride.

Much research has contributed to study the use of ultrasonic waves in various stages of textile wet processing including desizing, scouring, bleaching, dyeing and finishing of cellulosic materials. Thakore et al. (1990), Moholkar et al. (2004), Vajnhandl and Le Marechal (2005) and Sun et al. (2010) achieved high dye uptake efficiency in ultrasound-assisted dyeing processes owing to cavitation phenomena and mechanical effects such as diffusion, agitation and dispersion.

The use of ultrasound in textile wet processing enhances the wash-fastness properties, improves dye uptake, and reduces energy, processing duration and water consumption (Carr et al., 1997; Vouters et al., 2004; Ma et al., 2008).

Ilker Mistik and Yükseloğlu (2005), Karaboğa et al. (2007) and Basto et al. (2007a,b) reported a savings of energy (20%) and water consumption while using ultrasonic waves as a mechanical catalyst in desizing, scouring and bleaching of cotton materials. Perincek et al. (2009) achieved high desizing efficiency and good water absorbency of cotton fabric when ultrasonic waves were applied in desizing with amylase enzyme and scouring with pectinase enzyme. Application of ultrasound agitation in chlorination treatment reduces felting and shrinkage during the laundering of wool fabrics (Khajavi and Azari, 2007; Hurren et al., 2008). An increased chemical reaction between the dye and fibre and the rate of diffusion of the dye inside the fibre was observed in the dyeing of polyester material (Lee et al., 2003), cotton fabric (Kamel et al., 2003, 2007), nylon fabric (El-Shishtawy et al., 2003; Merdan et al., 2004) and wool material (Kamel et al., 2005). A combination of ultrasonic waves and electro-oxidation was used to decolourize textile effluents (Vajnhandl and Le Marechal, 2005). Parvinzadeh et al. (2010) obtained better softener finishing efficiency, 10%–20% increased adsorption and fixation of softener on cotton fabric in an ultrasonic-assisted finishing process.

Vankar et al. (2007) observed enhanced dye uptake, high colour yield and good fastness properties of cotton fabric dyed with *Eclipta* natural dye in an ultrasonicator dyeing method. Oner et al. (1995) and Khatri et al. (2011) achieved good colour strength, increased dye uptake, good fastness property, less water and chemical consumption, shorter duration and less energy consumed for heating and agitation and reduced effluent load using ultrasonic-assisted dyeing of cellulosic material with reactive dyes. The outcomes of research done by Thakore (2011) were improved colour strength, short dyeing time and increased wash-fastness of cotton fabric dyed with reactive dyes in an ultrasonic-assisted dyeing method. Lei et al. (2013) reported good colour yield and increased dye uptake and fixation of dye on cotton fabric dyed with reactive dye using a microwave heating technique.

Wang et al. (2010) reported that ultrasound energy increases the colour strength value of poly(trimethylene terephthalate) fabric dyed with disperse dyes and reduces the particle size of dye in solution and slightly decreases the crystallinity of fibre. Moreover, adsorption isotherm well fitted with the Langmuir model and improve the colour strength, reducing the dyeing duration as well as saving heat energy. Ultrasound with a frequency range of 20–100 kHz commonly used to accelerate chemical reactions and enhance physical processes, such as cleaning, emulsification and degassing extraction, improved results over other techniques under less optimal conditions such as lower temperatures and chemical concentrations (Sivakumar and Rao, 2003); in addition, mechanical effects are produced, including dispersion, degassing and diffusion, inside the fibre (Vajnhandl and Le Marechal, 2005). Better results observed in ultrasound-assisted dyeing processes are generally attributed to the cavitation phenomenon. Many researchers have investigated the dyeing behaviour of poly(trimethylene terephthalate) fabric with the use of ultrasound energy (Luo et al., 2002; Wang et al., 2004; Zhu and Cui, 2008).

Much research has been carried out to study ultrasonic effects on dye uptake on natural and synthetic fibres such as cotton (Padma et al., 2007), silk (Shukla and Mathur, 1995) and poly(ethylene terephthalate) fibre (Lee et al., 2003). Ferrero and Periolatto (2012) observed a positive effect of sonication on the dyeing of wool fabric with acid levelling dye, such as mass transfer, uniform dye dispersion, dye diffusion inside the fibre, with no damage to the fibre's structure or surface. Shukla and Mathur (1995) studied the effect of ultrasound energy on the dyeing characteristics of silk fabric with cationic, acid and metal–complex dyes at low temperatures. They observed an increase in dye uptake in all classes of dye at low temperatures, with less dyeing time compared with conventional dyeing methods and no fibre damage caused by cavitation phenomena.

9.4 Supercritical carbon dioxide technology

9.4.1 Introduction

The textile industry is one of the major industries in the world; it consumes a huge volume of fresh water to colour textile materials. Water is used as a medium (solvent) to dissolve dyestuff, chemicals and other auxiliaries in various processes of the textile industry, such as desizing, scouring, bleaching, dyeing, printing and finishing. After a large volume of fresh water is used in many pretreatment and finishing processes, it is discharged as waste which contains unreacted and unfixed dyestuff and other chemicals. Moreover, the effluent discharged from the textile wet processing industry causes pollution to the environment (Lablu Miah et al., 2013). A technology which can colour textile materials without using water is highly desirable. One such technology is 'supercritical carbon dioxide dyeing' of textile materials.

Supercritical carbon dioxide is used as a dyeing medium to dye synthetic fibres and fabrics. Various research activities have been carried out extensively because of environmental issues and the rise in the cost of water in normal dyeing processes (Bach et al., 1998a; Bach et al., 1998b; Beltrame et al., 1998a,b; De Giorgi Cadoni et al., 2000; Gebert et al., 1994; Guzel and Akgerman, 1999; Ozcan et al., 1998a,b; Poulakis et al., 1991; Saus et al., 1993a,b; West et al., 1998; Youssef, 2000).

9.4.2 Properties and applications of supercritical carbon dioxide

The dissolving power of supercritical fluids has been found in many applications such as dissolving organic molecules (Brogle, 1985) and to extract natural products (Hirata and Ishikawa, 1987; Paulaitis et al., 1983; Schneider et al., 1980; Squires and Paulaitis, 1987), extract caffeine from coffee beans (Zosel and Ger, 1970) and tobacco (Philip morris, 1971) and remove lanolin from raw wool (Zhuse et al., 1958) and for chromatographic applications (Cowper and DeRose, 1988). Many scientists have used supercritical fluids as transport media for water-insoluble dyes to dye synthetic and natural fibres (Poulakis et al., 1997; Saus et al., 1993a,b; Schollmeyer et al., 1985).

A supercritical fluid can be obtained by increasing the temperature and vapour pressure of a liquid and pressure in a closed system. A supercritical state is a state at which the fluid reaches critical value where no boundary between the liquid and gaseous state can be notable. Furthermore, an increase in the pressure of a system will increase the dielectric constant, as a result imparting dissolving power to a system under normal conditions of temperature and pressure (Saus et al., 1993a,b). The critical data for some substances are listed in Table 9.3.

Carbon dioxide is the best choice for producing supercritical fluids because it is nontoxic and nonflammable, it is used in the food and beverage industry and it is supplied in large amounts from combustion processes. Carbon dioxide can be recycled in a closed system (Beutler et al., 1988). The essential properties of supercritical fluids are low viscosity, high dissolving power and high diffusion property even in small pores without the need for vigorous treatment.

The dyeing procedure involved in supercritical fluids is explained by Saus et al. (1993a,b). The sample is wrapped around a perforated stainless-steel tube and mounted

Table 9.3 **Critical value of some compounds (Saus et al., 1993a,b)**

Substances	Boiling temperature (°C)	Critical value Temperature (°C)	Pressure (atm.)	Density (g/cm³)
Ethane	−88	32.4	48.3	0.203
Chlorotrifluoromethane	−81.4	28.8	39.0	0.58
Carbon dioxide	−78.5	31.3	72.9	0.448
n-Propane	−44.5	96.8	42	0.220
Ammonia	−33.4	132.3	111.3	0.24
Trichlorofluoromethane	23.7	196.6	41.7	0.554
Diethylether	34.6	193.6	36.3	0.267
Water	100	374.4	226.8	0.334

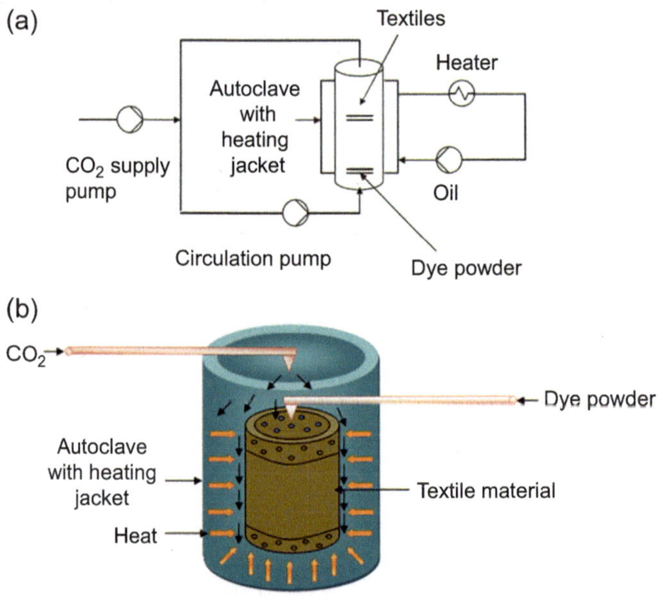

Figure 9.4 (a, b) Schematic diagram of supercritical carbon dioxide dyeing technique. Reproduced from van der Kraan, M., Fernandez Cid, M.V., Woerlee, G.F., Veugelers, W.J.T., Witcamp, G.J., 2007. Dyeing of natural and synthetic textiles in supercritical carbon dioxide with disperse reactive dyes. J. Supercrit. Fluids 40 (3), 470–476.

inside the autoclave, as shown in Fig. 9.4. Disperse dye in a powder form is placed in the bottom of the vessel and the autoclave is sealed, then purged with gaseous CO_2 and preheated. Critical temperature and pressure are maintained during the dyeing period of 60 min. Finally the sample is rinsed using acetone to remove residual dye and dye uptake is determined spectroscopically. The authors reported the advantages of using supercritical fluids (carbon dioxide) as a dyeing medium for synthetic fibres, including avoiding the use of water during dyeing, eliminating water pollution, saving heat energy in drying textiles, no need for dispersing agents and surfactants, a high degree of exhaustion and fixation, recovery of residual dye, little duration for dyeing, no need for after-treatment (reduction clearing) and savings of a huge volume of water.

According to the report by Dyecoo, on average, 28 billion kilograms of various textile materials are dyed annually with an approximate use of 140–150 L water per kilogram of textile material. Hence, eliminating the use of water and other chemicals would be a real breakthrough for the textile processing industry. The Dutch company and Dyecoo textile systems launched the world's first industrial dyeing machine that uses supercritical carbon dioxide to dye textile materials. Over the past decade or so, research groups in the United States, Korea and Taiwan have invested more than a billion American dollars to introduce environmental technologies to preventing water pollution caused by the dyeing, printing and finishing industry. Advantages of supercritical carbon dioxide technology are that it is waterless, does not use chemicals and does not need to dry the material, results in production twice that of conventional dyeing and saves on energy costs (Joshi et al., 2006). Supercritical carbon dioxide dyeing eliminates the drying process of fabric after dyeing. The result is savings of energy and power during the dyeing of textile materials.

There is pressure for the industry to implement sustainable processes to avoid using environmentally damaging organic solvents. Supercritical fluids (carbon dioxide) have the solvating properties of a wide range of organic solvents (Edward et al., 2009). Carbon dioxide can act as supercritical fluids and work as solvent for dye particles, thereby partially or completely eliminating waste water treatment and the problem of discharge pollution. The use of supercritical fluids as a medium to dye synthetic fabrics eliminates the use of ozone-depleting solvents such as chlorofluorocarbons. Fernandez Cid and van Spronsen (2006) reported that dyeing duration is reduced up to 3 h in the case of cotton fabric dyed with the supercritical method.

van der Kraan et al. (2007) explained the dyeing of polyester, nylon, silk and wool fabric with disperse reactive dyes using the supercritical carbon dioxide method. They reported that maximum colour yield was obtained in all fabric samples with vinylsulphone dye at the saturation point of both the supercritical fluids and the textile materials with water. Deep shades were achieved in polyester, silk, wool and nylon fibres with good fixation between 70% and 92% during a treatment time of 2 h. Water in the supercritical method is used either to swell fibres or to increase the reactivity of the dye-fibre system. van der kraan et al. (2007) concluded that disperse dyes (vinylsulphone/dichlorotriazine group) were suitable for dyeing polyester, nylon, silk and wool and its blends. Moreover, dye absorption in polyester material is independent of water and silk and nylon depends on the addition of water during dyeing. The percentage of fixation of disperse dye containing dichlorotriazine and vinylsulphone groups on silk and wool at the saturation point was 88%, 79%, 85% and 90%, respectively. They reported that water must be added to the supercritical carbon dioxide dyeing of animal fibre (silk and wool) and nylon with disperse dyes.

Bach et al. (2002) reported the advantages of supercritical dyeing. The group achieved high diffusivity owing to the low surface tension of supercritical carbon dioxide and the short duration needed for dyeing compared with that in water. Furthermore, the residual dye and carbon dioxide can be separated and recycled after dyeing. Moreover, there is no need for heat energy to dry the material after dyeing; normally 3800 kJ energy is needed to evaporate 1 kg water in textile material. McHugh and Krukonis (1994) stated that among supercritical fluids, carbon dioxide is the most widely used owing to its low cost and nontoxic and nonflammable characteristics.

9.4.3 Dyeing of cellulosic fibres

Veugelers et al. (2001) obtained superior colour yield in natural fibres with reactive dye in the presence of water. McDonald (1995) explained the possible mechanism of the positive effect of water in the supercritical process by breaking intermolecular hydrogen bonds in natural fibres, thereby allowing the fibres to swell. This facilitates the dye to penetrate easily into the fibrous structure. A dyeing process for cotton fabric has been developed by Ozcan et al. (1998a,b) in which the fabric was chemically modified by benzoyl chloride and dyed using disperse dyes in a supercritical carbon dioxide method. Good colour yield and wash-fastness were obtained in cotton fabric compared with polyester fabric. Gebert et al. (1994) suggested that the SCO_2 dyeing method was improved in terms of colour yield and colour fastness in natural fibres.

Broadbent et al. (1993) reported that modified cotton fibres have increased substantivity of disperse dyes. Beltrame et al. (1998a) obtained increased dye uptake and high fixation of disperse dye, and satisfactory wash- and light-fastness properties on cotton fabric pretreated with polyethylene glycol and dyed using benzamide crystals with the supercritical method. Studies by Chavan and Jain (1988), Ragheb et al. (1996), Reinhardt et al. (1992) and Kreituss et al. (1983) in transfer printing evidenced that pretreated cotton material using polyethylene glycol and polypropylene glycol had increased dye affinity for disperse dyes. Rossbach et al. (2000), Knittel et al. (1994) and Ozcan et al. (1998a,b) reported that cellulosic fabric modified using cyclohexyl carbodiimide and benzoyl chloride were more hydrophobic and had increased interaction between the disperse dye and cellulosic material. Moreover, this method needs pretreatment of cotton fibre with chemicals to achieve good colour fastness. Furthermore, an additional plant is necessary to maintain high pressure and to handle organic co-solvent to be explosion-proof and also a separate recovery plant is needed for solvent and carbon dioxide recovery.

9.4.4 Dyeing of wool and silk fibres

Nonpolar reactive dyes such as disperse-reactive ones are not suitable for dying silk and wool fibres using the supercritical carbon dioxide method, as confirmed by Schmidt et al. (2002, 2003); they found no colouration of silk fabric with vinylsulphone dye and dichlorotriazine dye with the supercritical carbon dioxide dyeing method. Giehl et al. (1999) and Guzel and Akgerman (2000) observed a high percentage of colour yield on wool fabric with disperse/vat dyes and achieved good wash-fastness on wool fabric while treating it with metal cations in supercritical carbon dioxide dyeing.

9.4.5 Dyeing of synthetic fibres

van der Kraan et al. (2007) reported that carbon dioxide at the critical state causing swelling of fibres leading to increased dye uptake and uniform dyeing of polyester material. The solubility of disperse dye is based on the density of supercritical carbon dioxide. Polyester, nylon, silk, cotton and wool fabrics were dyed successfully with this technique. Liao et al. (2000) stated that polyester and nylon fabric may be dyed with disperse dye with the $SciCo_2$ dyeing technique. Saus et al. (1992) developed a new process for dyeing polyester material using supercritical carbon dioxide as a dye transfer medium. Clifford and Bartle (1995) achieved better quality of dye fixation in polyester material with disperse dye with the supercritical carbon dioxide technique. Moreover, dye solubility may easily be controlled by altering the temperature and pressure during dyeing process.

Supercritical carbon dioxide has low density and viscosity, which can diffuse into the fibre more rapidly; as a result, a short duration is needed for dyeing polyester material. Liao et al. (2000) optimized the process conditions for the supercritical carbon dioxide dyeing of nylon 6,6 at 120°C and 315 bar pressure. Saus et al. (1993a,b) reported the use of supercritical carbon dioxide as a dyeing medium to dye synthetic fibres and fabrics to avoid water pollution. The conventional method of dyeing polyester material requires a large amount of dispersing agent and surfactants to achieve a high diffusion rate, uniform dyeing and high evenness (Knoop and Schollmeyer, 1986; Richter, 1983; Schollmeyer et al., 1985).

Maeda and Knitou (2001) reported that polyester/cotton blended fabric can be dyed in one bath using reactive disperse dyes in the supercritical carbon dioxide method in the range of 353–393 K and 10–20 MPa. The research group achieved good colour strength and wash-fastness properties at 393 K and 20 MPa, which were better than for fabrics dyed by thermosol dyeing. Furthermore, the polyester/cotton fabric has to be pretreated with 1% sodium carbonate and 10% *N*-methyl-2-pyrro-lidinone for 1 h at room temperature for swelling to achieve efficient dyeing. It acts as a good swelling agent for cotton fabric and alkaline conditions are maintained to fix the reactive disperse dyes on cotton fabric. Hydrocarbons, chlorofluorocarbons and other organic solvents can be replaced by carbon dioxide (Desimone et al., 1992; Goel and Beckman, 1994a,b; Guzel and Akgerman, 1999; Johnston, 1994; McFann et al., 1994; McHugh and Kurukonics, 1988; Mishima et al., 1999; Montero et al., 2003; Young et al., 1999). Supercritical carbon dioxide is used as a transport medium to dye polyolefin materials (Bach et al., 1998a,b) and wool fabric with mordant dye (Guzel and Akgerman, 2000). Maeda (2003) and Montero et al. (2003) set up pilot plant systems to dye polyester material using supercritical carbon dioxide dyeing and discussed its techno-economic feasibility.

9.4.6 Structural changes in textile fibres

Beltrame et al. (1998b) reported that polyester material may undergo important morphological changes such as a decrease in glass temperature in supercritical carbon dioxide dyeing. Reactive disperse dyes have been used in the supercritical dyeing process (Schmidt et al., 2003; Liao et al., 2000). A dyeing method was developed by Maeda et al. (2002) to dye cotton fibres with fluorotriazinyl reactive disperse dyes at 120°C in a supercritical method. A high temperature of 160°C is needed to dye natural fibres in the supercritical method to obtain high colour yield where fibre degradation occurs.

Initially polyester fibre was dyed in the supercritical carbon dioxide dyeing method on a laboratory scale (Poulakis et al., 1991; Saus et al., 1992), and earlier on a pilot plant scale (Bach et al., 1998c; Bach et al., 1999). Cotton, wool, silk, nylon and polyester fibre can be dyed in supercritical carbon dioxide dyeing method at a high temperature, up to 160°C, which is needed for dye fixation, good colour yield and fastness. The treatment temperature may cause physical and chemical damage to fibre polymers in the supercritical method. At a high temperature, polymer chains may be cleaved, substances can melt or crystallinity may increase (Berndt and Bossmann, 1976; Cleve et al., 1998; Drew and Jordan, 1998; Knittel et al., 1994). At 280 bar pressure in the supercritical carbon dioxide dyeing method, Cleve et al. (1998) observed a decrease in the melting point of 13–14°C, shrinkage of 6% and no change in the tensile strength of polyester material. Carbonic acid formed in the presence of water or from the moisture content of the fibre may also cause fibre damage. This acid cleaves the chemical bonds in monomeric units of nylon 6,6 whereas polyester is resistant to saponification by weak acids (Bobeth, 1993). Furthermore, cotton and viscose fibres can be depolymerized under acidic conditions (Rogowin, 1982). Schmidt et al. (2002) observed fibre damage in cotton, viscose, polyester, nylon 6,6, silk and wool in supercritical carbon dioxide dyeing at 100°C, 120°C, 140°C and 160°C for 1 and 4 h at a pressure of 300 bar. The following damage was also observed in various fibres: shrinkage in nylon 6,6, a decrease in the degree of polymerization in cotton and viscose after only 4 h at 120°C and damage to wool and silk after 4 h at 160°C. Dyeing conditions such as temperature, treatment duration and carbon dioxide seem to be responsible for the damaging effects.

9.4.6.1 Applications and their challenges in textiles

Plasma technology can be used in the chemical processing of textiles for different applications such as the removal of PVA size material from cellulosic fibres and to introduce antifelting to wool fabric by increasing the hydrophilic groups and reducing the directional frictional effect in wool fibre (Kan et al., 2010; Millard, 1975). Plasma treatment can be used to improve wettability, capillary action and surface area, to enhance dyeability and to create reactive groups on both natural and synthetic fibres. In finishing, this technology can be used to introduce various types of functionality. The cost of implementing this technology is high, but it offers great benefits in terms of production rate, lowered production costs, better results and freedom from environmental pollution (Shah and Shah, 2013).

The ultrasound-assisted dyeing technique can be applied to dye natural, animal and synthetic fibres. The treatment increases the dye exhaustion rate, improves colour strength, reduces processing time and gives better colour-fastness properties. This technique promises significant benefits to the textile industry through improvements in two vital areas: the cost of manufacturing and solutions to environmental impact problems. The manufacturing cost would be reduced as a result of the speed of processing and improved quality. Because there is a reduction in the need for energy, water, and processing chemicals, significant environmental benefits could be realized.

Advantages of using the supercritical carbon dioxide dyeing method to dye natural, animal and synthetic fibres/fabrics are the complete elimination of water pollution, savings in heat energy in drying textile materials, savings in the use of auxiliary chemicals to either exhaust or fix the dyes on materials, a high degree of exhaustion and fixation, a short dyeing duration, no after-treatments and the reuse of residual dyestuff materials (Ozcan et al., 1998a,b; Saus et al., 1993a,b; van der Kraan et al., 2007; Edward et al., 2009). Demerits of this technology are its high investment cost and the need for high pressure and temperature for dyeing (Miah et al., 2013).

9.5 Conclusion

Interest in ecologically friendly wet processing techniques has increased because of increased consciousness regarding environmental awareness throughout the world. General technical measures to reduce and check water consumption have to be adopted immediately because they will involve a low cost for implementation. Technologies such as plasma treatment, ultrasonic-assisted dyeing and supercritical carbon dioxide dyeing are promising technologies and have immense scope for further improvement in terms of their ability to process a wide variety of materials at an acceptable cost for processing. Changing the mindset of producers about water conservation has to become an industry priority to bring a positive and accelerated pace of change.

References

Bach, E., Cleve, E., Schollmeyer, E., 1998a. The dyeing of pololefin fibres in supercritical carbon dioxide, part I: thermo-mechanical properties of polyolefin fibres after treatment in CO_2 under dyeing conditions. J. Text. Inst. 89, 647–656.

Bach, E., Cleve, E., Schollmeyer, E., 1998b. The dyeing of pololefin fibres in supercritical carbon dioxide, part II: the influence of dye structure on the dyeing of fabrics and on fastness properties. J. Text. Inst. 89, 657–668.

Bach, E., Cleve, E., Schollmeyer, E., Bork, M., Korner, P., 1998c. Experience with the UHDE CO_2 – dyeing plant on a technical scale Part I: optimization steps of the pilot plant and first dyeing results. Melliand Int. 3, 192–194.

Bach, E., Cleve, E., Schollmeyer, E., Vardag, T., Korner, P., 1999. Experience with the UHDE CO_2 – dyeing plant on a technical scale. Part II: concepts for the development of the pilot plant in respect of a scaling up of the machine. Melliand Int. 2, 165–168.

Bach, E., Cleve, E., Schollmeyer, E., 2002. Past, present and future of supercritical fluid dyeing technology—an overview. Rev. Prog. Color. 32, 88.

Balachandran, S., Rudramoorthy, R., 2008. Efficient water utilization in textile wet processing. IE (I) Journal -TX 89, 26–29.

Basto, C., Tzanov, T., Cavaco-Paulo, A., 2007a. Combined ultrasound-laccase assisted bleaching of cotton. Ultrason. Sonochem. 14, 350–354.
Basto, C., Silva, C.J., Gubitz, G., Cavaco-Paulo, A., 2007b. Stability and decolourization ability of Trametesvillosa laccase in liquid ultrasonic fields. Ultrason. Sonochem. 14, 355–362.
Beltrame, P.L., Castelli, A., Selli, E., Mossa, A., Testa, G., Bonfatti, A.M., Seves, A., 1998a. Dyeing of cotton in supercritical carbon dioxide. Dyes Pigments 39 (4), 335–340.
Beltrame, P.L., Castelli, A., Selli, E., Villani, L., Mossa, A., Seves, A., Testa, G., 1998b. Morphological-changes and dye uptake of poly(ethylene-terephthalate) and 2,5-cellulose diacetate immersed in supercritical carbon-dioxide. Dyes Pigments 39 (1), 35.
Berndt, H.J., Bossmann, A., 1976. Thermal analysis of heat set poly(ethylene terephthalate) fibres. Polymer 17, 241–245.
Beutler, H.J., Gahrs, H.J., Lenhard, U., Lurken, F., 1988. Einfluβ der Losungsmittelfuhrung auf den Hochdrunk-Extraktions-Prozeβ. Chem. Ing. Tech. 60, 773–776.
Bobeth, W., 1993. Textile Faserstoffe Beschaffenheit und Eigenschaften. Springer–Verlag Berlin, Germany.
Broadbent, P.A., Lei, X.P., Lewis, D.M., 1993. Cellulosics: Materials for Selective Separations and After Technologies. Ellis Horwood Ltd., Chichester.
Brogle, H., 1985. CO_2 as a solvent: its properties and applications. Chem. Ind. 385–390.
Bruno Pollet, 2012. Power Ultrasound in Electrochemistry: From Versatile Laboratory Tool to Engineering Solution (Chapter 1). John Wiley & Sons. 1119967864.
Carr, W.W., Beckham, H.W., Michielsen, S., Mock, G.N., McCall, R., 1997. Fundamental Investigations of Ultrasonic Effects in Textile Wet Processing. National Textile Center Research Briefs. http://www.ntcresearch.org/pdf-rpts/anrp98/c95-g13.pdf.
Chavan, R.B., Jain, A.K., 1988. Am. Dyestuff Rep. 77, 40.
Cid, M.V.F., van Spronsen, J., 2006. A significant approach to dye cotton in supercritical carbon dioxide with fluorotriazine reactive dyes. J. Supercrit. Fluids 40, 477–484.
Cleve, E., Bach, E., Schollmeyer, E., 1998. Untersuchungen zum Schmelz und strukturverhalten von polyethylene terephthalatfasern (PETP) in Luftbei 1 bar und uberkritischemCO_2 bis 280 bar. Angew. Makromol. Chem. 256, 39–48.
Clifford, A.A., Bartle, K.D., 1995. Text. Technol. Int. 145–150.
Cowper, C.J., DeRose, A.J., 1988. Supercritical Fluid Chromatography Royal Society of Chemistry, London.
De Giorgi Cadoni, M.R., Maricca, E., Piras, D., 2000. Dyeing polyester fibres with disperse dyes in supercritical CO_2. Dyes Pigments 45 (1), 75–79.
Desai, A., January, 2008. Plasma technology: a review. Indian Text. J. 120, 26–28.
Desimone, J.M., Guan, Z., Elsbernd, C.S., 1992. Sythesis of fluoropolymers in supercritical carbon dioxide. Science 257, 945–947.
Drew, M.J., Jordan, C., 1998. The effect of supercritical CO_2 dyeing conditions on the morphology of polyester fibers. Text. Chem. Color. 30, 13–20.
Edward, R., Qiubai, S., Zhiqiang, Z., Chongmin, Z., Wei, G., 2009. Mini-review: green sustainable processes using supercritical fluid carbon dioxide. J. Environ. Sci. 21, 720–726.
El-Shishtawy, R.M., Kamel, M.M., Hanna, H.L., Ahmed, N.S.E., 2003. Ultrasonic-assisted dyeing: II. Nylon fibre structure and comparative dyeing rate with reactive dyes. Polym. Int. 52, 381–388.
Ferrero, F., Periolatto, M., 2012. Ultrasound for low temperature dyeing of wool with acid dye. Ultrason. Sonochem. 19, 601–606.

Gebert, B., Saus, W., Knittel, D., Buschmann, H.J., 1994. Dyeing natural fibers with disperse dyes in supercritical carbon dioxide. Text. Res. J. 67 (7), 371–374.

Giehl, A., Gießmann, M., Schafer, K., Hocker, H., 1999. Dyeing wool without water—possibilities and limits of supercritical CO_2. In: Wenclawiak, B., Padberg, S. (Eds.), 4th International Symposium Extraction for Sample Preparation–sfe–(X)SE–spme—Book of Abstracts 1999. University GH of Siegen, Siegen, pp. 29–30.

Goel, S.K., Beckman, E.J., 1994a. Generation of microcellular polymeric foams using supercritical carbon dioxide I: effect of pressure and temperature on nucleation. Polym. Eng. Sci. 34, 1137–1147.

Goel, S.K., Beckman, E.J., 1994b. Generation of microcellular polymeric foams using supercritical carbon dioxide II:cell growth and skin formation. Polym. Eng. Sci 34, 1148–1156.

Guzel, B., Akgerman, A., 1999. Solubility of disperse and mordant dyes in supercritical Co_2. J. Chem. Eng. Data 44, 83–85.

Guzel, B., Akgerman, A., 2000. Mordant dyeing of wool by supercritical processing. J. Supercrit. Fluids 18, 247–252.

Hashem, M., Hauser, P., Smith, B., 2003. Reaction efficiency for cellulose cationization using 3-chloro-2-hydroxypropyl trimethyl ammonium chloride. Text. Res. J. 73, 1017–1023.

Hauser, P.J., Tabba, A.H., 2001. Improving the environmental and economic aspects of cotton dyeing using a cationised cotton. Color. Technol. 117, 282–288.

Hirata, M., Ishikawa, T. (Eds.), 1987. The Theory and Practice in Supercritical Technology. NTS, Tokyo.

Hulme, P., March 2015. Modern Technologies Can Help Textile Industry Save Water. http://www.business-standard.com/content/b2b-chemicals/modern-technologies-can-help-textile-industry-save-water-paul-hulme-115032300242_1.html.

Hurren, C., Cooksona, P., Wang, X., 2008. The effects of ultrasonic agitation in laundering on the properties of wool fabrics. Ultrason. Sonochem. 15, 1069–1074.

Johnston, K.P., 1994. Safer solutions for chemists. Nature 368, 187–188.

Joshi, A.S., Malik, T., Parmar, S., 2006. Supercritical carbon dioxide dyeing of polyester. Asian Dyer 51–54.

Kamel, M.M., El-Shishtawy, R.M., Hanna, H.L., Ahmed, N.S.E., 2003. Ultrasonic-assisted dyeing: I. Nylon dyeability with reactive dyes. Polym. Int. 52, 373–380.

Kamel, M.M., El-Shishtawy, R.M., Yussef, B.M., Mashaly, H., 2005. Ultrasonic assisted dyeing. III. Dyeing of wool with lac as a natural dye. Dyes Pigments 65, 103–110.

Kamel, M.M., El-Shishtawy, R.M., Youssef, B.M., Mashaly, H., 2007. Ultrasonic assisted dyeing. IV. Dyeing of cationised cotton with lac natural dye. Dyes Pigments 73, 279–284.

Kamel, M.M., El Zawahry, M.M., Ahmed, N.S.E., Abdelghaffar, F., 2009. Ultrasonic dyeing of cationized cotton fabric with natural dye. Part 1: Cationization of cotton using Solfix E. Ultrason. Sonochem. 16, 243–249.

Kan, C.W., et al., 2001. Development of low temperature plasma technology on wool. In: The 6th Asian Textile Conference Proceedings, August 22–24, Hongkong.

Kan, C.W., et al., 2010. Plasma pretreatment for polymer deposition- improving antifelting properties of wool. IEEE Trans. Plasma Sci. 38 (6), 1505–1511.

Karaboğa, C., Körlü, A.E., Duran, K., Bahtiyari, M.I., 2007. Use of ultrasonic technology in enzymatic pretreatment processes of cotton fabrics. Fibres Text. East. Eur. 15, 97–100.

Khajavi, R., Azari, P., 2007. Effects of ultrasound irradiation on wet wool chlorination treatment. Pak. J. Biol. Sci. 10, 2732–2735.

Khatri, Z., Memon, M.H., Khatri, A., Tanwari, A., 2011. Cold pad-batch dyeing method for cotton fabric dyeing with reactive dyes using ultrasonic energy. Ultrason. Sonochem. 18 (6), 1301e1307.

Knittel, D., Saus, W., Hoger, S., Schollmeyer, E., 1994. Fä rbenausü berkritischem CO_2: Bestimmung von Einflußgrö ßen auf das strukturelleVerhalten von Poly(ethylenterephtalat)-Fasermaterial. Angew. Makromol. Chem. 218, 69–79.

Knoop, K., Schollmeyer, E., 1986. Einfluβ des dispergierhilfsmittels auf die mizellare solubilisierung eines dispersionsfarbstoffes durch nicht-ionogene tenside. Melliand Textilber 67 882–823.

Kreituss, A., Kaimins, I., Ioelovich, M., 1983. J. Appl. Polym. Sci. Symp. 36, 1043.

Lee, K.W., Chung, Y.S., Kim, J.P., 2003. Characteristics of ultrasonic dyeing on poly(ethylene terephthalate). T. Res. J. 73, 751–755.

Lei, N., Gong, D., Ling, X., Shi, Y., 2013. Researches on microwave dyeing cotton fabrics. In: Proceedings of Second International Conference on Textile Engineering and Materials, pp. 343–347.

Lewis, D.M., Lei, S., 1989. Text. Chem. Color. 21, 23–25.

Liao, S.K., Ho, Y.C., Chang, P.S., 2000. Dyeing of nylon 66 with a disperse-reactive dye using supercritical carbon dioxide as the dyeing medium. J. Soc. Dyers Color. 116, 403.

Luo, F.Y., Xu, R.D., Tang, R.C., 2002. Silk textile technology. Overseas 5, 23.

Ma, J., Chen, S., Liu, C., Xu, W., Wang, S., 2008. The influences of ultrasonic on embedding nanoparticles into porous fabric materials. Appl. Acoust. (9), 763–769.

Maeda, S., 2003. Development of dyeing technique using supercritical carbon dioxide. Sanyogijutuzasshi 50, 24–28.

Maeda, S., Knitou, K., 2004. One-bath dyeing of polyester/cotton blends with reactive disperse dyes in supercritical carbon dioxide. Text. Res. J. 74, 989–994.

Maeda, S., Hongyou, S., Kunitou, K., Mishima, K., 2002. Dyeing cellulose fibres with reactive disperse dyes in supercritical carbon dioxide. Text. Res. J. 72, 240–244.

McDonald, R. (Ed.), 1995. Colour Physics for Industry, second ed. Society of Dyers and Colourists, Bradford.

McFann, G.J., Johnston, K.P., Howdle, S.M., 1994. Solubilization of nonionic reverse Micelles in carbon dioxide. Am. Inst. Chem. Eng. J. 40, 543–555.

McHugh, M.A., Krukonis, V.J., 1994. Supercritical Fluid Extraction: Principles and Practice, second ed. Butterworth-Heinemann Press, Boston.

McHugh, M.A., Kurukonics, V., 1988. In: Bikkales, N.M., Overberger, C.G., Menges, G. (Eds.), Encyclopedia of Polymer Science and Engineering, second ed. John Wiley, NY.

Merdan, N., Akalin, M., Kocak, D., Usta, I., 2004. Effects of ultrasonic energy on dyeing of polyamide (microfibre)/Lycra blends. Ultrasonics 42, 165–168.

Miah, L., Ferdous, N., Azad, M.M., 2013. Textiles material dyeing with supercritical carbon dioxide (CO_2) without using water. Chem. Mater. Res. 3 (5), 38–40.

Millard, M.M., 1975. Proceedings of the 5th International Wool Textile Research Conference, Aachen, vol. 11, p. 44.

Mishima, K., Matsuyama, K., Nagatani, M., 1999. Solubilities of poly(ethylene glycol)s in the mixtures of supercritical carbon dioxide and cosolvent. Fluid Phase Equilib. 161, 315–324.

Mistik, S.I., Yükseloğlu, S.M., 2005. Hydrogen peroxide bleaching of cotton in ultrasonic energy. Ultrasonics 43, 811–814.

Moholkar, V.S., Warmoeskerken, M.M.C.G., Ohl, C.D., Prosperetti, A., 2004. Mechanism of mass-transfer enhancement in textiles by ultrasound. Am. Inst. Chem. Eng. J. 50, 58–64.

Montero, G., Hinks, D., Hooker, J., 2003. Reducing problems of cyclic trimer deposits in supercritical carbon dioxide polyester dyeing machinery. J. Supercrit. Fluid 26, 47–54.

Montazer, M., Malek, R.M.A., Rahimi, A., 2007. Salt free reactive dyeing of cationized cotton. Fibers Polym. 8, 608–612.

Nazaria, A., Montazer, M., Moghadam, M.B., Anary-Abbasinejad, M., 2011. Self-cleaning properties of bleached and cationised cotton using nano TiO_2: a statistical approach. Carbohyd. Polym. 83, 1119–1127.

Oner, E., Baser, I., Acar, K., 1995. Use of ultrasonic energy in reactive dyeing of cellulosic fabrics. J. Soc. Dyers Colour. 111 (9), 279–281.

Ozcan, A.S., Clifford, A.A., Bartle, K.D., Lewis, D.M., 1998a. Dyeing of cotton fibres with disperse dyes in supercritical carbon dioxide. Dyes Pigments 36, 103–110.

Ozcan, A.S., Clifford, A.A., Bartle, K.D., Broadbent, P.J., Lewis, D.M., 1998b. Dyeing of modified cotton fibres with disperse dyes from supercritical carbon dioxide. J. Soc. Dyers Colour. 114 (5/6), 169–173.

Padma, S.V., Rakhi, S., Srivastava, J., 2007. Ultrasonic dyeing of cotton fabric with aqueous extract of *Eclipta alba*. Dyes Pigments 72 (1), 33.

Parvinzadeh, M., Memari, N., Shaver, M., Katozian, B., Ahmadi, S., Ziadi, I., 2010. Influence of ultrasonic waves on the processing of cotton with cationic softener. J. Surfactants Deterg. 13 (2), 135–141.

Paulaitis, M.E., Penninger, J.M.L., Gray Jr., R.D., Davidson, P. (Eds.), 1983. Chemical Engineering at Supercritical Fluid Conditions. Ann Aror Science Publishers, Michigan.

Perincek, S., Uzgur, A.E., Duran, K., Dogan, A., Korlu, A.E., Bahtiyari, I.M., 2009. Design parameter investigation of industrial size ultrasound textile treatment bath. Ultrason. Sonochem. 16, 184–189.

Philip Morris Inc, 1971. NY, U.S. Patent 21 43 388.

Poll, H.U., Schladitz, S., 2001. Schreite, Penetration of plasma effects into textile structures. Surf. Coat. Technol. 142–144, 489–493.

Poulakis, K., Spee, M., Schnneider, G.M., Knittle, D., Buschmann, H.J., Schollmeyer, E., 1991. Farbung von Polyester in uberkritischem CO_2. Chemiefasern Textillind. 41 (93), 142–147.

Poulakis, K., Bach, E., Schollmeyer, E., 1997. Peracetic Acid in Textile Finishing.Textilvere. 32, 74.

Radetic, M., Jovancic, P., Puac, N., Petrovic, Z., 2007. J. Phys. Conference Series 71, 012017.

Ragheb, A.A., Hamagy, E., Soltan, M., 1996. Am. Dyestuff Rep. 85, 28.

Reinhardt, R.M., Blanchard, E.J., Graves, E.E., 1992. Am. Dyestuff Rep. 81, 26.

Richter, P., 1983. Synchron aufziehende Dispersionsfarbstoffe fur das HT-Farben von Polesterfasern und Polyester-cellulosefaser-Mschungen. Lenzinger Berichte 54, 53–58.

Rogowin, Z.A., 1982. Chemiefasern. Georg ThiemeVerlag, Stuttgart, Germany.

Rossbach, V., Hudson, A., Fischer, P., Poulakis, K., 02.03.2000. Reaktivfarbstoffe, Herstellungsverfahren und Färbeverfahren. DE 199 37 328 A1, C 09 B 62/00 Germany.

Saus, W., Knittle, D., Schollmeyer, E., 1992. Farben Aus uberkritischem Kohlendioxideine Alternative zur HT-Fabung von Polyester. Textile Praxis 47, 1052–1054.

Saus, W., Hoger, S., Knittel, D., Schollmeyer, E., 1993a. Farben Aus uberkritischem Kohlendioid-Dispersions-farbstoffe und Baumwollgewebe. Textilveredlung 28, 38–40.

Saus, W., Knittle, D., Schollmeyer, E., 1993b. Dyeing of textiles in supercritical carbon dioxide. Text. Res. J. 63, 135–142.

Schmidt, A., Bach, E., Schollmeyer, E., 2002. Use of fiber reactive groups in supercritical carbon dioxide. Melliand Textilberichte 83 (9), 648–650.

Schmidt, A., Bach, E., Schollmeyer, E., 2003. The dyeing of natural fires with reactive disperse dyes in supercritical carbon dioxide. Dyes Pigments 56, 27–35.

Schneider, G.M., Stahl, E., Wilke, G., 1980. Extraction with Supercritical Gases. Verlag Chemie, Weinheim.

Schollmeyer, E., Heidemann, G., Treptau, G., 1985. Zur Egalitat beim diskontinuierlichen Farben von Polyethylenterephthalat-Fasern mit Dispersionsfarbstoffen. Textilveredlung 20, 190–198.
Shah, J.N., Shah, S.R., 2013. Innovative plasma technology in textile processing: a step towards green environment. Res. J. Eng. Sci. 2 (4), 34–39.
Shukla, S.R., Mathur, M.R., 1995. Low-temperature ultrasonic dyeing of silk. J. Soc. Dyers Colour. 111, 342.
Sivakumar, V., Rao, P.G., 2003. Studies on the use of power ultrasound in leather dyeing. Ultrason. Sonochem. 10, 85.
Smith, B., Rucker, J., 1987. Water and textile processing-Part 1. Am. Dyestuff Rep. 15–23.
Sun, D., Guo, Q., Liu, X., 2010. Investigation into dyeing acceleration efficiency of ultrasound energy. Ultrasonics 50, 441–446.
Thakore, K.A., 2011. Ultrasound treatment in exhaust and pad-batch dyeing. AATCC Rev. 11 (4), 66–74.
Thakore, K.A., Smith, C.B., Clapp, G., 1990. Application of ultrasound to textile wet processing. Am. Dyest. Rep. 79, 30–44.
Vajnhandl, S., Le Marechal, A.J., 2005. Ultrasound in textile dyeing and the decolouration/mineralization of textile dyes. Dyes Pigments 65, 89–101.
van der Kraan, M., Fernandez Cid, M.V., Woerlee, G.F., Veugelers, W.J.T., Witcamp, G.J., 2007. Dyeing of natural and synthetic textiles in supercritical carbon dioxide with disperse reactive dyes. J. Supercrit. Fluids 40 (3), 470–476.
Vankar, P.S., Shanker, R., Srivastava, J., 2007. Ultrasonic dyeing of cotton fabric with aqueous extract of Eclipta alba. Dyes Pigments 72, 33–37.
Veugelers, W.J.T., Gooijer, H., Gerritsen, J.W., Woerlee, G.F., 2001. European Patent EP 1 126 072 A2.
Vouters, M., Rumeau, P., Tierce, P., Costes, S., 2004. Ultrasounds: an industrial solution to optimise costs, environmental requests and quality for textile finishing. Ultrason. Sonochem. 11, 33–38.
Wang, W.X., Li, H.H., Dong, S., 2004. Text. Sci. Technol. 4, 52.
Wang, C.X., Liu, Y., Xu, H.L., Ren, Y., Qiu, Y.P., 2008. Influence of atmospheric pressure plasma treatment time on penetration depth of surface modification into fabric. Appl. Surf. Sci. 254 (8), 2499–2505.
Wang, L., Zhaoa, H.F., Lin, J.X., 2010. Studies on the ultrasonic-assisted dyeing of poly(trimethylene terephthalate) fabric. Color. Technol. 126, 243–248.
Water and Chemical Use in the Textile Dyeing and Finishing Industry, 1997. www.wrap.org.uk/sites/files/wrap/GG062.pdf.
Water for People, 2003. Water for Life – UN World Water Development Report (WWDR). http://www.un.org/esa/sustdev/publications/WWDR_english_129556e.pdf.
West, B.L., Kazarian, S.G., Vincent, M.F., Brantley, N.H., Eckert, C.A., 1998. Supercritical fluid dyeing of PMMA films with azo-dyes. J. Appl. Polym. Sci. 69, 911–919.
https://en.wikipedia.org/wiki/Ultrasound.
Young, T.J., Johnston, K.P., Mishima, K., Tanaka, H., 1999. Encapsulation of lysozyme in a biodegradable polymer by precipitation with a vapour-over-liquid anti-solvent. J. Pharm. Sci. 88, 640–650.
Youssef, Y.A., 2000. Direct dyeing of cotton fabric pre-treated with cationising agents. J. Soc. Dyers Colour. 116, 316–322.
Zhu, J.P., Cui, S.L., 2008. Dyestuffs. Colorage 45, 37.
Zhuse, T.P., Jushkevic, G.N., Gekker, J.E., 1958. Lanolinextraktion mit Hilfe von komprimierten Gasen aus Wollfett. Maslob.-Zhir. Prom. 24, 34–37.
Zosel, K., Ger, 1970. Patent 20 05 293 (Studiengesellschaft Kohle GmbH), Afiliation.

Further reading

Allan, G., Fotheringham, A., Weedall, P., 2002. The use of plasma and neural modeling to optimize the application of a repellent coating to disposable surgical garments. AUTEX Res. J. 2 (2), 64–68.

Hoger, W., Knittel, S.D., Schollmeyer, E., 1993a. Färben aus überkritischem Kohlendioxid—Dispersionsfarbstoffe und Baumwollgewebe. Textilveredlung 28, 38–40.

http://www.isasf.net/fileadmin/files/Docs/Versailles/Papers/PMt2.pdf.

Klausen, T., 1992. Diplom-Arbeit RWTH. (Aachen).

Lorimer, J.P., Mason, T.J., Plattes, M., Phull, S.S., 2000. Dye effluent decolourisation using ultrasonically assisted electro-oxidation. Ultrason. Sonochem. 7, 237–242.

Nasadil, P., Benešovský, P., 2008. Plasma in textile treatment. Chem. Listy 102, s1486–s1489.

Schollmeyer, E., Heidemann, G., Treptau, G., 1981. Zur Egalitat beim diskontinuierlichen Farben von polyethylenterephthalat-Fasern mit Dispersionsfarbstoffen. Textilveredlung 16, 147–151.

Supercritical fluids-chemical and engineering principles and applications. In: Squires, T.G., Paulaitis, M.E. (Eds.), 1987. ACS Symposium Series, 1987, vol. 329.

Sustainable chemical technologies for textile production

10

Asim K. Roy Choudhury[1,2]
[1]KPS Institute of Polytechnic, Belmuri, West Bengal, India; [2]Ex-professor, Government College of Engineering and Textile Technology, Serampore, West Bengal, India

10.1 Introduction

Sustainable textiles should be environmentally friendly and should satisfy rational conditions to respect social and environmental quality by preventing pollution or by installing pollution-control technologies.

When considering both the volume generated and the composition of effluent, textile wastewater is considered to be the most polluted of all industrial sectors. With a global textile supply chain that is complex, involving many different stages and often several companies, company owners may therefore be best placed to bring about the required change. As many as 2000 different chemicals are used in textile processing, especially in textile wet processing, many of which are known to be harmful to human (and animal) health.

During production, the chemicals are added at various stages using huge quantities of water, mostly at high to medium temperature. After completion, warm or hot water carrying various types and amounts of chemicals is expelled as wastewater, which, if untreated, may thermally pollute the environment by virtue of the high temperature of the effluent and extreme pH (acidic or alkaline). These effluents are complex mixtures and are contaminated with unexhausted or unfixed dyes, diluents of dyes, various auxiliaries, bleaches, detergents, optical brighteners and many other chemicals including carcinogenic heavy metals used during textile processing.

10.2 Sustainability

Sustainable textiles should be environmentally friendly and should satisfy rational conditions to respect social and environmental quality by preventing pollution or by installing pollution-control technologies. Certification, however, is a voluntary process. Any entity conducting a business for which a standard exists may ask to have its output or services certified. Certification is a procedure through which a third party, the certification body, gives a written assurance that an organizational system, process, person, product or service complies with requirements specified in a standard

Sustainable Fibres and Textiles. http://dx.doi.org/10.1016/B978-0-08-102041-8.00010-X

or benchmark. Certification is awarded for a limited period during which the certification body carries out monitoring. Third-party certification bodies and governments have issued restricted substances lists that link production ecology to human ecology (Oecotextiles, 2012).

The textile industry is the one of the biggest industrial polluter of fresh water (NIEHS, 2012). In some cases, wastewaters are discharged (largely untreated) into groundwater with extreme pH values and temperatures as well as high chemical loading.

10.3 Sustainability issues in current textile production

The life cycles of textiles and clothing are unsustainable because of (Sherburne, 2009):

1. use of toxic chemicals
2. high water consumption
3. high energy consumption
4. huge waste generation
5. heavy transportation
6. excessive packing materials

10.3.1 Use of toxic chemicals

It is difficult to ensure that all of the chemicals used in the textile industry are eco-friendly. The main difficulty is that too many dyes are used in the dyeing and printing of textiles; the Colour Index International lists 27,000 individual products under 13,000 Colour Index generic names (SDC, 2016). Volatile chemicals pose particular problems because they evaporate into the air or are absorbed into foods or through the skin. Some chemicals are carcinogenic or may cause harm to children even before birth, whereas others may trigger allergic reactions in some people.

Lists of restricted substances are constantly changing as more information from scientists and health professionals becomes available, leading to an enhanced understanding of chemicals and their effect on human health and the environment. The substances listed in the Restricted Substances List (RSL) are based in large part on global legislation regulating chemicals use in the manufacturing of apparel products. The European Union (EU) has developed the 'Regulation Concerning the Registration, Evaluation, Authorization and Restriction of Chemicals', aimed at ensuring a high level of protection of human health and the environment from the risks that can be posed by chemicals (AAFA).

Other countries which have developed or are developing similar lists of restricted substances are China, Canada and South Korea. In the United States, several states, including California, Washington and Maine, have adopted laws regulating chemicals in consumer products. These regulatory requirements are incorporated into the RSL.

The RSL released by the American Apparel and Footwear Association includes only those materials, chemicals and substances that are restricted or banned in finished home textile, apparel and footwear products because of a regulation or law. The list includes the names of selected arylamines, certain disperse dyes, solvents, pesticides, asbestos, certain fluorinated compounds, greenhouse gases (GHGs), dioxins and furans, flame retardants, metals, organotin compounds, phthalates and miscellaneous other chemicals.

Nonylphenol ethoxylate (NPE) is used in textile manufacture as a cleaning, dyeing and rinsing agent. NPE degrades in the environment into substances including nonylphenol, which accumulates in the bodies of fish and disrupts their hormones, harming fertility, growth and sexual development. Use of NPE in textile manufacture in Europe was banned a long time ago but the substance is still released into the aquatic environment through imported textiles being washed. EU member states have agreed to ban hormone-disrupting NPE widely found in clothing because it poses an 'unacceptable risk' to the environment (The Guardian, 2015).

The various approaches being undertaken by the industry include avoiding using hazardous chemicals, reducing the extent of use of such chemicals or totally substituting the process by using safe chemicals. Because of driving factors such as the increased demand for eco-friendly processing and intensified control on polluting technologies, the use of biotechnology is increasing day by day (Hasan et al., 2015). Various enzymes are being used to substitute a number of hazardous chemicals in the textile industry. The global market for industrial enzymes increased from US $1.5 billion in 2000 to US $2.25 billion in 2007 (Roy Choudhury, 2014a).

10.3.2 High water consumption

After agriculture, the textile and related industries are considered to be the second highest consumer and polluter of clean water (Oecotextiles, 2012). The textile services sector is an essential adjunct to the textile industry and is needed to manufacture, finish, market and distribute the products (Nagaraj, 2012). Water is used in many steps of the textile dyeing process both to convey the chemicals used during the step and to wash them out before beginning the next step. In a traditional dyeing and finishing operation, for example, one ton of fabric could result in the pollution of up to 200 tons of water by a suite of harmful chemicals and in the process consume large amounts of energy for steam and hot water (Moore and Wentz, 2009). With the industry now centred in countries with still-developing environmental regulatory systems, such as China, India, Bangladesh and Vietnam, textile manufacturing continues to have a huge environmental footprint. Some commonly observed routes of water waste (Oecotextiles, 2012) are:

- excessive use of water in washing operations;
- poor housekeeping measures such as broken or missing valves;
- unattended leaks through pipes and hoses;
- instances when cooling waters are left running when machinery is shut down;
- choice of inefficient washing equipment;
- excessively long washing cycles; and
- use of fresh water at all points of water use.

The reuse of wastewater can present important savings, namely in the reduction of water, energy and chemical consumption. The recycling of waste water is effected in process baths and rinsing waters, before water is taken for treatment to remove remaining chemicals and other effluents generated. Steam condensate and cooling water are easily recoverable because they are clean. Their thermal energy can quickly pay back the investment.

With the continuous growth of global production and consumption of textile-related products and services, there are both opportunities and challenges for the textile industry.

10.3.3 High energy consumption

The textile industry is a major energy-consuming industry with low efficiency in energy use. About 23% of energy is consumed in weaving, 34% in spinning, 38% in chemical processing and another 5% for miscellaneous purposes. Thermal energy dominates in chemical processing while electrical power dominates the energy consumption pattern in spinning and weaving. Thermal energy in textile mills is mainly consumed to heat water and dry textile materials (Rupp, 2008).

The textile industry is one of the largest sources of GHGs, not least because of its enormous size. In 2008, the annual global production of textiles was estimated at 60 billion kilograms of fabric with associated (estimated) energy and water needs of 1074 billion KWh of electricity (or 132 million tons of coal) and between 6 and 9 trillion litres of water (Oecotextiles, 2011).

A large quantity of nonrenewable energy sources is eventually consumed in the form of electricity, not so much in the process of textile production (15%–20%) but mostly in subsequent laundering processes during consumer use (75%–80%) (Sherburne, 2009). It is reported (Ray and Reddy, 2008) that the total thermal energy required per metre of cloth is 18.8–23 MJ and the electrical energy required per metre of cloth is 0.45–0.55 kWh. The energy used is mostly consumed by heating process water or for use in laundering and drying materials after laundering. Whereas data on energy use for the textile industry are readily available, complications arise in estimating the associated CO_2 emissions that arise from the sources (coal, electricity, natural gas or other sources) from which the energy is produced because the textile industry is a fragmented and heterogeneous sector dominated by small and medium enterprises.

Energy is one of the main cost factors in the textile industry. Especially in times of high energy price volatility, improving energy efficiency should be a primary concern for textile plants, and various energy-efficiency opportunities exist in every textile plant, many of which are cost-effective but not implemented because of limited information or high initial cost. For example, the use of electricity is associated with built-in inefficiencies compared with the direct use of thermal energy, and the use of steam is less efficient than direct-fired gas heating in a mill. The share of total manufacturing energy consumed by the textile industry in a particular country depends on the structure of the manufacturing sector in that country. For instance, the textile industry accounts for about 4% of the final energy use in manufacturing in China (LBNL, 2007), whereas this share is less than 2% in the United States (U.S. DOE, 2010).

Electricity is the main energy consumed in the textile industry; it is used for driving machinery, cooling, temperature control, lighting and office equipment, whereas fuel oil, liquefied petroleum gas, coal and city gas are widely used to generate steam. Efficiencies have been achieved; between 1990 and 2005, the carbon emission intensity in the textile industry decreased for grey cloth, jute goods and polyester chips by 1.90%, 2.07% and 0.72%, respectively. On the other hand, cotton yarn showed the highest increase in emission intensity of 7.37%, which means that cotton yarn continues to be produced inefficiently (ECCJ, 2007). Emission intensity is the average emission rate of a given pollutant from a given source relative to the intensity of a specific activity; for example, grams of carbon dioxide released per megajoule of energy produced, or the ratio of GHG emissions produced to gross domestic product (GDP). Emission intensities are used to derive estimates of air pollutant or GHG emissions based on the amount of fuel combusted, the number of animals in animal husbandry, industrial production levels, distances travelled or similar activity data. Emission intensities may also be used to compare the environmental impact of different fuels or activities. The related terms 'emission factor' and 'carbon intensity' are often used interchangeably, but 'factors' exclude aggregate activities such as GDP, and 'carbon' excludes other pollutants (Wikipedia, 2013).

Parties to the United Nations Framework Convention on Climate Change reached a landmark agreement on 12 December 2015 in Paris, charting a fundamentally new course in the 2-decade-old global climate effort. Culminating in a 4-year negotiating round, the new treaty ends the strict differentiation between developed and developing countries that characterized earlier efforts, replacing it with a common framework that commits all countries to put forward their best efforts and strengthen them in the years ahead. This includes, for the first time, requirements that all parties report regularly on their emissions and implementation efforts and undergo international review (C2ES, 2015).

Spinning consumes the greatest share of electricity (41%), followed by weaving (including weaving preparation) (18%), whereas wet-processing preparation (desizing and bleaching) and finishing together consume the greatest share of thermal energy (35%). A significant amount of thermal energy is also lost during steam generation and distribution (35%), but these percentages vary by plant. Such an analysis of energy-efficiency improvement opportunities in the textile industry points to advantages to be gained from retrofit/process optimization, not just from complete replacement of current machinery with state-of-the-art technology (Hasanbeigi, 2010).

There are various possibilities for using renewable energy in the textile industry. Four examples from a range of possible measures are:

1. installation of wind-powered turbo ventilators on production plant roofs;
2. use of direct solar energy for fibre drying;
3. use of solar energy for water heating in the textile industry; and
4. solar electricity generation

10.3.4 Huge waste generation

Like any other industry, the textile industry generates all categories of industrial wastes: liquids, solids and gases. For greener processes, nonrenewable wastes need to be recycled and renewable wastes need to be composted if recycling is not an option. Various useful materials can be recovered from textile process wastes.

The recovery of chemicals such as sodium hydroxide from mercerization baths is achievable by heating to concentrate the solution; following such a step, 90% of sodium hydroxide can be recovered (Indiamart, 2013). The EVAC vacuum suction system in the textile dyeing process recovers hot alkaline hydrogen peroxide, additives and finishing chemicals (TIFAC (DST), 2012). EVAC is the trade name of a vacuum system introduced in Thailand from the United States. This equipment is installed at the finishing stage to suck excess chemical solution from the fabric and then transfer it to the storage tank for recovery and recycling (UNEP, 1994).

Desizing effluent containing polyvinyl alcohol (PVA) has high chemical oxygen demand (COD) and hence contributes significantly to a textile plant's primary oxygenation treatment of water operation. Because it is biologically inert, it presents a threat to the environment. Unfortunately, no effective and efficient means to treat PVA desize effluent has been implemented in the textile industry. Ultrafiltration (UF) reverse osmosis technology for the recovery and recycling of PVA size is more than 40 years old but it is not widely used because of its many disadvantages. The situation necessitates a new technology for the recovery and recycling of PVA size which can reduce energy and water consumption in an economical and environmentally friendly manner. A new technology that would eliminate the disadvantages of the current ultrafiltration process in the recovery of PVA from desize effluent is vacuum flash evaporation (VFE). The VFE process for recovery and concentration has been used in a variety of other industries but has never been demonstrated for size recovery in the textile industry (Gupta, 2009).

Some waste-water is still being disposed of in an environmentally unfriendly way into the sewage networks where available, or else in cesspools, with no regard to the biochemical oxygen demand (BOD), chemical oxygen demand (COD) and/or heavy metal content of the waste-water. Untreated wastewater generated from textile production and processing can vary greatly depending on the chemicals and treatment processes involved and may include materials with a high BOD and COD, high total suspended solids, oil and grease, sulphides, sulphates, phosphates, chromium, copper and/or the salts of other heavy metals (Ho et al., 2012); of these, the most important are considered to be COD, BOD, pH, fats, oil, nitrogen, phosphorus, sulphates and suspended solids (Tüfekci et al., 2007). Total suspended solids levels are low in raw textile dyeing wastewater compared with many other industries. On the other hand, BOD and COD are relatively high in effluents from sizing operations and wet processing, and therefore are more important pollution prevention targets (Visvanathan et al., 2000). Sulphates and phosphates are toxic at high concentrations. Problems caused by sulphates are most frequently related to their ability to form strong acids which changes the pH, whereas in surface waters phosphates cause eutrophication.

10.3.4.1 Solid waste

Industrial solid wastes from textile production include the following:

- ashes and sludge
- cardboard boxes, bale wrapping film or nonrecyclable soiled fabric

- plastic bags carrying chemical raw material
- nonreusable paper cones and tubes
- waste fabrics, yarns, and fibres from nonrecyclable processing

Unmanaged solid waste is likely to be dumped as landfill.

The textile and footwear industries consume high amounts of water and threaten water quality in China. The latest discharge standard, GB 42872012, serves as a way to enforce China's Environmental Protection Law and encourage the improvement of production methods used in the textile dyeing and finishing industry and of pollution-control technology. The standard covers the discharge limits, requirements for monitoring and controlling water pollutants in the textile dyeing and finishing industry or production facility.

According to the World Bank, 20% of water pollution globally is caused by textile processing. Nongovernmental organizations such as environmental groups say that parts of India and China are among the most polluted.

Major chemical companies as well as startups are responding by developing less harmful textile-processing chemicals. However, such substances are often more expensive than ones in use. Lower-cost reagents and better enforcement by regulators are required, textile chemical executives say, along with a shift in consumer behaviour away from opting for the cheapest products.

The textile industry uses more than 8000 chemicals to make the 400 billion square metres of fabric sold annually around the world. Many are toxic and persist in the environment. They include heavy metal-rich dyes and fixing agents, bleaches, solvents and detergents.

Textile production is a water-intensive business. Producing a pair of jeans requires about 1800 gal of water; a T-shirt takes 700 gal. Treating such large volumes of wastewater is costly, if it is treated at all.

Clothing makers were stung into action in July 2011 by 'Dirty Laundry', a report by the environmental group Greenpeace that claimed they source their textiles from low-cost producers in developing countries while turning a blind eye to the pollution this generates. By November 2011, several major apparel producers and retail organizations had come together to form Zero Discharge of Hazardous Chemicals (ZDHC). ZDHC's members committed to eliminating release of toxic substances into waterways by 2020. Members include Adidas, Gap, H&M, Nike, Puma and the Taiwanese dyestuffs maker Everlight Chemical (Scott, 2015).

Burnt fossil fuels contribute to the emissions of carbon dioxide, a primary contributor to the greenhouse effect. The textile manufacture is also responsible for the following emissions:

- nitrogen oxides and sulphur oxides (from fossil-fuel heated boilers) which create acidity in the natural environment (freshwater lakes, rivers, forests and soils) and lead to the deterioration of metal and building structures. They also contribute to smog formation in urban areas
- solvent escaping into the air from drying ovens used in solvent coating operations
- solvents released from cleaning activities (general facility cleanup and maintenance, print screen cleaning)
- emissions of volatile hydrocarbons which include nonmethane hydrocarbons (NMHC) and oxygenated NMHC (e.g., alcohols, aldehydes and organic acids).

10.3.5 Heavy transportation

Generally textile units requires large areas and a large amount of labour. Hence modern textile units are located in remote places far from big cities. Long-distance transport is required to move the finished products from the factories located in low-paying countries to consumers in developed countries, thus adding to the overall quantity of nonrenewable fuel consumed.

The impact from transport is relatively low except for those arising from the formation of photochemical oxidants (smog), which comes from using trucks, ships and planes. It is estimated that 8% of imported textiles are carried by air freight; the rest (92%) are by ship. Although air freight accounts for only a small share of distribution, its impacts are proportionally much higher (Dodd et al., 2012).

10.3.6 Excessive packing materials

Packaging is the science, art and technology of enclosing or protecting products for distribution, storage, sale and use. Packaging also refers to the process of design, evaluation and production of packages. Packaging can be described as a coordinated system of preparing goods for transport, warehousing, logistics, sale and end use. Packaging contains, protects, preserves, transports, informs and sells (Soroka, 2002). For consumer packaging, the packaging used to present products in stores, materials often used are plastic, paper, metal, aluminium, cotton, hemp and biodegradable materials. Companies implementing eco-friendly actions are reducing their carbon footprint by using more recycled materials, increasingly reusing packaging components for other purposes or products and employing recycled materials (e.g., paper, cotton, jute, hemp, wood), biodegradable materials, natural products grown without the use of pesticides or artificial fertilizers and reusable materials (e.g., cotton bags or hemp). Reducing packaging waste is one of the best ways to minimize environmental impact. EU Directive 94/62/EC specifies a number of requirements relevant for packaging and packaging waste. It also sets specific recycling targets and maximum levels for heavy metals (Tufekci et al., 1998). Sustainable packaging is the development and use of packaging which results in improved sustainability. At the end stage of design it involves increased use of life-cycle inventory and life-cycle assessment (LCA), which consider the material and energy inputs and outputs to the package, the packaged product (contents), the packaging process and the logistics system (Zabaniotou and Kassidi, 2003)

10.4 Sustainable processing

Sustainability of textile processing can be improved by several ways:

1. substitution of unsustainable textile materials and chemicals by greener organic and biodegradable materials
2. elimination or minimization of the use of toxic chemicals in production and packing
3. minimization of the use of water and chemicals and recycling them
4. minimization of the consumption of energy and fuel in production and transport

5. minimization of waste and easy waste disposal
6. strict maintenance of environmental management systems

10.4.1 Cleaner production

As a part of a cleaner approach to approach production, a textile processor has to be lean, efficient and innovative, which can be achieved by the following ways:

Lean
By maintaining good housekeeping, conservation and control.
Efficient
By adopting the 'right-first-time' (RFT) approach, chemical/water/energy/machine audits and optimization/rationalization.
Innovative
Reuse, recovery and recycle initiatives for process change.

Some important areas where chemicals recovery and reuse have proved most effective are (LCPC, 2010):

- reuse of dye solutions from the dye bath
- recovery of sodium hydroxide in mercerizing (by effective evaporation or using membrane technology)
- recovery of size in cotton processing (using technologies such as ultrafiltration)
- recovery of grease in raw wool fibre scouring (by acid cracking, centrifuging or solvent extraction)
- recovery and recycling of PVA size from desize effluent streams, which can be done by UF. However, it is still not widely used in the global textile industry because of certain disadvantages including high energy costs for pumping and high membrane cleaning/replacement costs. Environmentally friendly and economical technology for PVA size recovery and recycling is still needed to reduce energy and water consumption. One example that may prove useful is VFE, which overcomes most of the disadvantages of the current UF process. In this system, water is evaporated under vacuum at low temperature, leaving extracted cotton impurities from the previous desizing process, nonevaporated water and PVA in the concentrate (Gupta and Cook, 2012).

10.5 Greener textile materials

Environmental considerations are becoming vital factors during the selection of consumer goods including textiles all over the world. The basic raw material for textile manufacture is textile fibres which are of two types: natural and man-made (Farrington et al., 2008). The major natural fibres are of vegetable or animal origin. They are not inherently green. They are biodegradable, but more biodegradable polymers can be made by biological means. A major defining difference between biopolymers and other polymers can be found in their structures. Biopolymers often have a well-defined structure. In contrast, most synthetic polymers have much simpler and more random (or stochastic) structures. Synthetic polymer fibres are not biodegradable and hence must preferably be recycled to minimize accumulation.

However because of increased awareness of the polluting nature of textile effluents, social pressures on textile processing units are increasing. Awareness about eco-friendliness in apparel textiles has become one of the important issues in recent times. Owing to the demands of global consumers, research is carried out to establish new eco-friendly technology. Problems related to toxicity and other health hazards have resulted in the replacement of chemical processing by more eco-friendly physical methods (The Textile Institute, 2002). Plasma, biotechnology, ultrasonic treatment, supercritical carbon dioxide and laser technologies, for example, are technologies which are gaining ground because they offer many advantages against wet techniques with fewer or no harmful chemicals in wet processes or waste water (although there may still be some pollution; plasma treatment, for example, can result in some air pollutants from the degradation of fibre surfaces).

Plant-based fibres and eco-friendly chemicals can be used to manufacture green composites. Among natural fibres the bast fibres, extracted from the stems of plants such as jute, kenaf, flax, ramie and hemp, are widely accepted as the best candidates for the reinforcement components of composites owing to their good mechanical properties, with hemp shown to have particular promise (Kolybaba et al., 2003).

10.5.1 Biopolymers

Biodegradable polymers have become of great interest mainly for biomedical applications (Hutmacher, 2000). Biodegradable polymers break down in physiological environments by macromolecular chain scission into smaller fragments, and ultimately into simple stable end products (Bastiol, 1997). Such degradation may be due to aerobic or anaerobic microorganisms, or biologically active processes (e.g., enzyme reactions) or passive hydrolytic cleavage. The past 2 decades of polymer technology have seen a sharp rise in the development and commercial marketing of such materials. It is desirable to make polymers from renewable biological sources and to synthesize and use polymers that will biodegrade after disposal. The use of biopolymers, i.e., plastics made from corn, sugar, starch and other renewable raw materials, has expanded in recent years. A fibre manufactured from soybean has biodegradable, nonallergic and microbiocidal properties. Although not as durable as cotton or hemp fabrics, clothing made from the soy fibre has a soft, elastic handle, in addition to offering softness similar to cashmere. The production of fibre from soybean protein was first investigated in the 1930s by Kajita and Inoue in Japan (Kajita and Inoue, 1940), and by Boyer in the United States (Boyer, 1940). Soy fibre was later commercialized using a wet-spinning process followed by cross-linking with an aldehyde, which is not eco-friendly in the present context. Hydroxyl groups are introduced by co-spinning the extracted soybean protein with polyvinyl alcohol. The Huakang biochemical engineering group in China obtained the patent for the process and now produces soybean fibre in bulk. Reactive dyes are more suitable than acid dyes for exhaust dyeing of soy fibre, giving much better exhaustion and fixation than acid dyes, an effect attributable to the large numbers of hydroxy end-groups in soy fibre and encouraging the formation of a dye–fibre covalent bond (Choi et al., 2005).

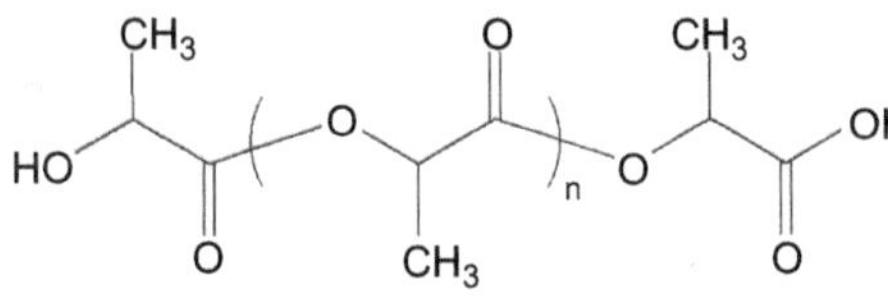

Figure 10.1 Structure of polylactic acid.

10.5.1.1 Polylactic acid fibre

Poly(lactic acid) or polylactide (PLA) (Fig. 10.1) is the only melt-processable fibre from annually renewable natural resources. It is thermoplastic aliphatic polyester derived from renewable resources such as cornstarch (in the United States), tapioca products (roots, chips or starch mostly in Asia) or sugar cane (in the rest of world).

PLA fibres are commercially produced by many fibre companies. As of June 2010, NatureWorks was the primary producer of PLA (bioplastic) in the United States; other companies involved in PLA manufacturing are PURAC Biomaterials (The Netherlands), several Chinese manufacturers (e.g., Xinhui Yuexin Chemical, Fujian Jiayi Chemical and Xinxiang Sunshining Textiles) plus some Taiwanese manufacturers (e.g., Jinsor-Tech Industrial Corporation). Ingeo is the brand name for a range of PLA biopolymers owned by NatureWorks LLC (formerly Cargill Dow LLC), Nebraska, United States (http://www.natureworksllc.com). It was introduced globally in 2003.

It will be interesting to see whether PLA follows a similar course to becoming as widely accepted as Lyocell, for example, with its many technical difficulties in processing (Farrington et al., 2008).

Cargill Dow's technology uses maize starch to produce PLA. The prices of natural products like crops depend on a good growing season; drought creates severe difficulties, as do crop diseases and pests; the process consumes up to 50% less fossil fuel than conventional PLA manufacturing processes using petroleum-based feedstocks, but the price of corn has risen considerably (Lancaster, 2002).

The commercial viability of the polymer also relies on novel processing that can be used to manipulate the molecular weight, crystallinity and polymer chain branching, enabling the production of materials with a wide range of end uses and potential markets. PLA has high strength, good drape and wrinkle and UV light resistance properties.

The route of manufacture of PLA from corn is as follows (Dartee, 2000):

Corn → starch → unrefined dextrose → fermentation → D- and L-lactic acid → monomer production → D-, L- and mesolactides → polymer (PLA) production → polymer modification → fibre, film, plastic, bottle manufacture.

The direct condensation is thus performed in a step-wise fashion in which lactic acid is first oligomerized to PLA oligomers. Thereafter, polycondensation is done in the melt or as a solution, in which short oligomeric units are combined to give a high-molecular weight polymer strand. Water removal by application of a vacuum or by azeotropic distillation is crucial to favour polycondensation over transesterification. The polymerization reaction is shown in Eq. (10.1).

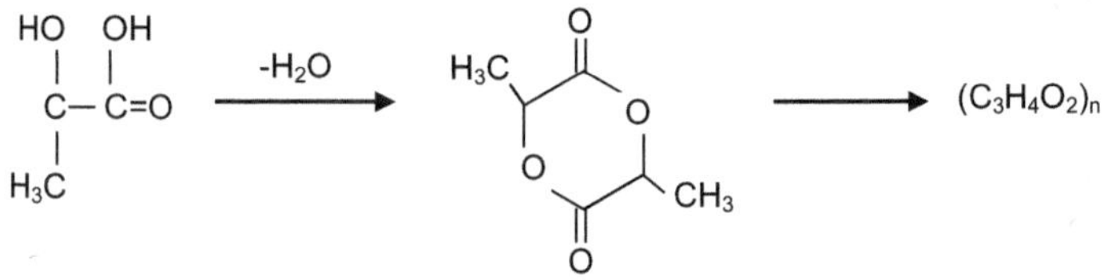

(10.1)

By controlling the D-, L- and mesolactide ratio, together with the molecular weight, the polymer properties can be tailored to meet various product specifications (Reddy and Yang, 2005). They thus have a lower environmental impact than polyester [polyethylene terephthalate (PET)]. The several benefits of cornstarch-derived fibres are expected to make these fibres preferable over the currently available natural and man-made fibres (Farrington et al., 2008).

Compared with PET, PLA is a 'greener' polymer because it is produced from renewable plants and food crops. Whereas scientists estimate that petroleum-based plastic products require thousands of years to decompose, PLA degrades under commercial composting conditions in 75–80 days. For fully biodegradable polymer the percentage of biodegradation of a product made from a single polymer should be a minimum of 60% over a 45-day span. The PLA degradation process produces no toxic wastes and its end products are fully biodegradable (into lactic acid and carbon dioxide) and compostable, because they degrade within 45–60 days. Under normal storage conditions and use, however, the fabrics are durable enough to meet the various market and supply chain requirements (BASF, 2011).

PLA has better elastic recovery, better hydrophilicity, higher limiting oxygen index and lower smoke generation than conventional PET (BASF, 2011). It does not have the bioaccumulation problems associated with the waste derived from conventional PET, which, of course, is derived from finite petroleum resources (Phillips et al., 2004). PLA fibre can be successfully dyed with disperse dyes; the optimum recommended conditions are 30 min at 110°C at pH 4.5–5.0. Its lower refractive index often allows PLA to be dyed to a deeper shade than PET using the same concentration of disperse dye on fibres of similar decitex values. However, the wet-fastness properties of disperse dyes on PLA were found to be lower than those on PET after finishing owing to thermal migration of the disperse dyes (Fiore et al., 2010).

The melting temperature of PLA (130–175°C) is much lower than that of PET 250–255°C; this means that there is a need to use much lower domestic ironing temperatures, particularly on woven constructions. Although the glass transition temperature is between 60°C and 65°C, heat-resistant PLA can withstand temperatures of 110°C (Mitrus et al., 2009).

Hydrolytic degradation of the polymer can occur, particularly under combined aqueous high-temperature and alkaline conditions; the degree of hydrolysis is influenced by time, temperature and pH. This is of particular significance in the dyeing and finishing processes, because it will cause a reduction in the molecular weight of the polymer and therefore the strength of a yarn or fabric, if the appropriate finishing conditions are not observed (Farrington et al., 2008).

PLA can be used for rigid thermoforms, films, labels and bottles, but not for hot-fill containers or gaseous drinks such as beer or sodas. PLA can also be used for noncarbonated beverages such as water, juices and milk, as well as for edible oil products. It provides a flavour and aroma barrier comparable to that provided by PET and readily accepts coatings, inks and adhesives (Gupta et al., 2007).

The medical applications of PLA arise from its biocompatibility. The fibres may be fabricated into various forms and may be used for implants and other surgical applications such as sutures. Tissue engineering is a domain in which PLA is being used and is found to be one of the most favourable matrix materials (Martins et al., 2008).

10.5.1.2 Polyhydroxyalkanoates

The most active current area of research in natural polymers involves the bacterial polyesters, polyhydroxyalkanoates (PHAs) with poly-(R)-3-hydroxybutyrate as the first homologue (Fig. 10.2). These polymers were found to be produced by fluorescent pseudomonads and other bacteria which generate and store them as reserves of carbon and energy. Their production has been achieved in transgenic (genetically engineered) plants. Advantages include production from fully renewable resources, fast and complete biodegradability and excellent strength and stiffness. However, serious drawbacks of these polymers include high thermal degradability and brittleness with consequent difficulty in processing, and their high price (Chodák, 2009).

10.5.1.3 Polycaprolactone

Polycaprolactone (Fig. 10.3) is a synthetic polymer prepared by the ring-opening polymerization of caprolactone. The mechanical properties are similar to those of the polyolefins. Similar to the PHAs it is fully biodegradable but it degrades at a lower rate compared with PHAs. Owing to the lower melting temperature of about 60°C, the polymer is mainly used in polymer blends or as a matrix for biodegradable composites.

10.5.2 Greener fibres

Textile fibres are either natural or manufactured. Natural fibres such as cotton, flax, jute, silk and wool fibres are obtained from nature. Manufactured textile fibres are of

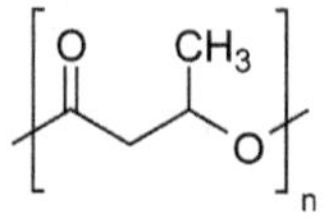

Figure 10.2 Structure of poly-(R)-3-hydroxybutyrate.

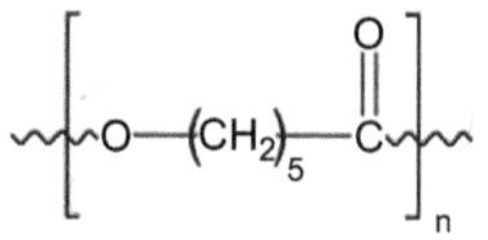

Figure 10.3 Structure of polycaprolactone.

two types: regenerated (such as viscose rayon, Lyocell and modal fibres) or synthetic (for example, polyester, nylon, acrylic, modacrylic and polyolefin fibres).

Natural fibres have the following benefits over the manufactured fibres:

- a smaller carbon footprint in the production of the fibre
- ability to be degraded by microorganisms and to be composted (improving soil structure), whereby the fixed carbon dioxide in the fibre is released and the cycle is closed
- sequesters carbon: the process through which carbon dioxide is absorbed by plants from the atmosphere through photosynthesis and stored as carbon in the biomass (leaves, stems, branches, roots, etc.) and soils. Jute, for example, absorbs 2.4 tons of carbon per ton of dry fibre.

Traditional synthetic fibres do not decompose readily, and in landfills they release heavy metals and other additives into soil and groundwater. Recycling requires costly separation, whereas incineration produces pollutants. For example, in the case of high-density polyethylene, 3 tons of carbon dioxide emissions is produced for every ton of material burnt (FAO, 2009). Left in the environment, synthetic fibres contribute, for example, to the estimated 640,000 tons of abandoned fishing nets in the world's oceans.

Substituting organic fibres for conventionally naturally grown fibres is better in many respects. For example, organic fibre uses less energy in its production, emitting fewer GHGs and supporting organic farming (which has myriad environmental, social and health benefits). A study published by Innovations Agronomiques (2009) found that 43% less GHGs is emitted per unit area under organic agriculture than under conventional agriculture (Aubert et al., 2009). A study done by Dr David Pimentel of Cornell University found that organic farming systems used just 63% of the energy required by conventional farming systems, largely because of the massive amounts of energy needed to synthesize nitrogen-based fertilizers for use in conventional farming. Furthermore, it was found in controlled long-term trials that organic farming adds between 100 and 400 kg of carbon per hectare to the soil each year, compared with nonorganic farming. When this stored carbon is included in carbon footprint calculations, it reduces the total measured GHG even further (UNCTAD/WTO and FIBL, 2007).

The carbon footprint of any fabric is assessed by 'embodied energy' in the fabric: that is, all of the energy used in each step of the process needed to create the fabric. The embodied energy in any fabric is estimated by adding the energy required in two separate fabric production steps:

- the energy needed to make the fibres used in the yarn (the energy required to produce various fibres varies widely)
- the energy used to weave those yarns into fabric, which is then added to that used in yarn production. The amount of energy and conversion process to weave that yarn into a textile is consistent. Whether the yarn is composed of wool, cotton, nylon or polyester, the thermal energy required per metre of cloth is 4500–5500 Kcal and the electrical energy required per metre of cloth is 0.45–0.55 kWh.

For natural fibres one must look at field preparation, planting and field operations such as mechanized irrigation, weed control, pest control and fertilizers (which may be manure or synthetic chemicals), harvesting and yields. Synthetic fertilizer use is a major contributor to the high cost of conventional agriculture.

For synthetic fibres, a crucial fact is that the fibres are made from fossil fuels. High amounts of energy are used to extract the oil from the ground as well as produce the polymers.

A study (Oecotextiles, 2011) conducted by the Stockholm Environment Institute on behalf of the Bio-Regional Development Group concluded that the energy used (and therefore the mass of carbon dioxide emitted) to create 1 ton of spun fibre is much higher for synthetics (9.20 kg) than for hemp (4.05 kg) or cotton (5.90 kg). Other synthetics have even more of an impact: acrylic is 30% more energy intensive in its production than polyester. Not only is the quantity of GHG emissions of concern with synthetic fibres, so, too, are the kinds of gases produced during the production of synthetic fibres. Nylon, for example, creates emissions of N_2O, which is 300 times more damaging than carbon dioxide, and because of its long lifetime in the environment (120 years), it can reach the upper atmosphere and deplete the layer of stratospheric ozone, an important filter of UV radiation. The study also showed that the energy required (expressed as kilograms of CO_2 emitted per ton of spun fibre) for organic cotton grown in the United States was 2.35 kg, compared with 5.90 kg for conventional cotton, whereas the energy required for organic cotton in India was 3.80 kg. Other work conducted for the New Zealand Merino Wool Association (Barber and Pellow, 2006) showed that a much lower total energy is required for the production of natural fibres than for synthetics.

10.5.2.1 Hemp fibre

By far, the crop with the most potential for eco-friendly textile use is hemp. The ecological footprint of hemp is considerably smaller than that of most other plants considered for their fibres. Hemp plants grow quickly and densely, which makes it difficult for weeds to take hold, eliminating the need for herbicides and artificial fertilizers. It requires no irrigation because it thrives on the amount of water in the average rainfall, and it is highly pest resistant. Hemp has naturally long fibres, which makes it suitable for spinning with a minimum of processing. Those fibres are also long-lasting. Hemp fabrics come in a variety of weights and textures.

10.5.3 Traditional cotton

Analysts in the 71st plenary meeting of the International Cotton Advisory Committee reported that polyester dominates the global fibre market and cotton's share continues to decline. It is expected that polyester fibre quality will improve and that production capacity will increase further. The secretariat of the plenary meeting reported that world cotton production was expected to decrease by 6% to 26 million metric ton in 2012–13 compared with 2011, whereas the global market for polyester was expected to reach 39.3 metric tons by 2015 (ICAC, 2012). The low-cost and easy-care characteristics of polyester are expected to continue to attract consumers in various end-use sectors including apparel and furnishings.

For a given weight of cotton harvested, a farmer uses one-third of that weight in chemical fertilizer (Sayitgreen.com, 2010). Cotton plants are highly susceptible to pests, especially in humid areas (Grose, 2009). Clay (2004) reported that whereas cotton production is restricted to 2.4% of the cultivatable land globally, an estimated 25% of insecticide and 11% of global pesticide production is consumed in cotton cultivation. A report (Blécourt et al., 2010) claims that in 2008 the global insecticides share used on cotton had declined from 19% (in 2000) to 15.7%. In the same year, cotton's pesticide consumption claimed to be 6.8% of global use. Further details are available in this report. This 'thirsty' crop also requires 7000–29,000L of water to produce 1kg of cotton fibre (ISIS, 2007).

Historically, cotton was planted at low densities and rotated with other crops to ensure the optimum health of the soil. Pest cycles were taken into consideration before planting and harvesting. Significant amounts of pesticides began to be applied from the mid-20th century, not least because of the advent of dichlorodiphenyltrichloroethane and other neurotoxins that were considered to be cheaper ways of controlling pests than strategic crop management and the efforts of agricultural labourers.

However, there are increasing concerns that the pesticides used in 'conventional' (as against 'organic') cotton farming increasingly threaten people, wildlife and the environment; as insects gradually become resistant to pesticides, ever-increasing amounts need to be applied to be effective, resulting in ecological damage and crop failures.

10.5.4 Organic cotton

Organic cotton production is more environmentally friendly and better for the health of the community and for the local economy than GM cotton, according to a study by the Centre for Sustainable Agriculture in Andhra Pradesh, and Bt cotton was found to be more prone to pests and diseases (UCSAREP, 2012).

Organic cotton is cotton from plants that are not genetically modified and are certified to be grown without the use of any synthetic agricultural chemicals, such as fertilizers, pesticides or defoliants, produced according to the internationally recognized organic farming standards of the EU regulation 834/2007, the US National Organic Program, the Indian National Programme for Organic Production and the Japanese Agricultural Standard. Organic cotton production does not simply mean replacing synthetic fertilizers and pesticides with organic types; it is a systemic approach which aims to establish a diverse and balanced farming ecosystem ideally including all types of crops and farm activities. Farms typically need to complete a 2-year conversion period to change their production system from conventional to organic. An essential element of organic cotton production is the careful selection of varieties adapted to local conditions in terms of climate, soil and robustness to pests and diseases. Soil fertility management and crop nutrition are based on crop diversification and organic inputs such as compost, mulch and manures. Pest management focuses on pest prevention and the stimulation of a balanced agro-ecosystem through crop rotation, mixed cultivation, trap crops, and the use of natural pesticides when pest infestation rises above the economic threshold. The beneficiaries of organic cotton are farmers, traders/retailers and consumers.

Benefits for farmers are (OTA, 2010):

1. a balanced ecosystem and enhanced health;
2. improved economic situation and food security;
3. improved access to markets; and
4. training and education.

Benefit for traders and retailers are:

1. participation in a dynamic market;
2. traceability, risk management and quality management;
3. contribution to ecological and social sustainability; and
4. credibility and a good image.

Benefits for consumers are:

1. buying a healthy product;
2. traceability, clear standards and labels;
3. environmental friendliness; and
4. a positive impact on producers' livelihoods.

According to the fourth annual Organic Exchange Farm and Fibre Report produced in 2009, organic cotton production grew by an impressive 20% over 2007–08 to 175,113 metric tons (802,599 bales) grown on 625,000 acres (253,000 hectares). Organic cotton in 2010 represented 0.76% of global cotton production. Organic cotton was grown in 22 countries worldwide with the top 10 producer countries led by India, and in order of rank were Turkey, Syria, Tanzania, China, the United States, Uganda, Peru, Egypt and Burkina Faso. Approximately 220,000 farmers grew the organic fibre (Ferrigno, 2012).

Major trends in the 2010–11 were (Ferrigno, 2012):

- Global organic cotton production in 2010–11 dropped by 37% to 151,079 tons.
- India, Syria, China, Turkey, and the United States were the top five producers in 2010–11.
- Production in India fell by 48% from 195,412 tons to 102,452 tons owing to stringent regulatory control by the Agricultural and Processed Food Products Export Development Authority with the intention of improving integrity and chances for capturing the market in the future.
- Twelve countries increased organic cotton production (most significantly Benin, Brazil, Mali, Nicaragua, Kyrgyzstan and Tajikistan).

In the United States it is required by the law that any producer wanting to label and sell a product as 'organic' must meet the standards established by the Organic Food Production Act of 1990, enforced by The State Organic Program (SOP); This act specifies the procedures and regulations for production and handling of organic crops (Coleman, 2012). The Global Organic Textile Standard (GOTS) was developed in 2006 through the collaboration of standard setters for trade in organic products with the aim of defining requirements that are recognized worldwide and that embrace all steps, from harvesting of the raw materials through environmentally and socially responsible manufacturing to labelling, to provide assurance of the organic status of textiles to the consumer (GOTS, 2013).

The preparatory processes required before dyeing and printing are similar for organic cotton and conventional cotton processing. However, some chemicals such as substances with high adsorbable organic halogen (AOX) values, blueing agents, chelating agents, chlorine compounds and formaldehyde are prohibited for use on organic textiles. All dyestuffs should conform to Ecological and Toxicological Association of Dyes and Organic Pigments Manufacturers (1997) restrictions regarding residual heavy metals and banned aromatic amines. The first choice for dyeing organic fabrics, where, applicable, could be plant-based natural vegetable dyes; however, they have never been subjected to rigorous eco-toxicological testing, and their commercial availability is limited. In the La Rochelle Natural Dyes Conference (ISEND, 2011), for example, it was stated that one of the components in madder was a carcinogen. The best choice could be low-impact dyes, e.g., fibre-reactive dyestuffs made from petrochemicals. Permitted synthetic and nonsynthetic chemicals are listed in eco-labels such as the GOTS (www.global–standard.org). The use of synthetic flame retardants and many functional finishes is prohibited. Mechanical finishing techniques must be explored instead of chemical finishes wherever possible (Wakelyn and Chaudhry, 2009).

Most of the top apparel brands and retailers in the world such as Nike, Levi, Walmart, Patagonia, Timberland, Orvis, Adidas, Marks & Spencer, Roots, Cotton Ginny and Target have already introduced organic cotton into their product range, and they are expecting manifold increases in the demand for organic textiles in coming years, given demand particularly in the health conscious high-end society (Hanu, 2010).

Even so, some companies deemphasize its use because organic cotton is too thirsty a crop for most areas where it is grown and it is difficult to ensure that the organic cotton comes from fields that enjoy natural rainfall sufficient to irrigate the crops and does not cause desertification such as in Kazakhstan and Uzbekistan around the Aral Sea. On a positive front, organic cotton is primarily produced in areas where pest problems are manageable.

10.5.5 Organic linen

The flax plant (*Linum usitatissimum*) is one of the oldest fibre crops in the world and has been used to produce linen for over 5000 years. High in oil (40% on average), linseed is pressed into industrial oils (linseed oil) and food-grade oils and because of its high omega-3 essential fatty acid content, it has been gaining in popularity as a source of healthy fats.

Organic linen refers to linen made from flax fibres grown without the use of toxic pesticides or chemical fertilizers. Although there are products on the market claiming to be 'organic linen' or 'eco-friendly linen', and some of these products may be made from flax fibres, many are made from other fibres (www.natural-environment.com, 2008).

Like all crop farming, flax cultivation has environmental impacts which can be greatly reduced if a certified organic method of crop production is used. Compared with other crops, flax performs poorly in soils with low fertility and can require significant use of fertilizers, but by using crop rotation and multiseeding methods, as well as biological pest control and the use of green manure and compost, organic flax farming can produce the seeds without reduced environmental impact. Crop rotation is essential not only from an organic certification standpoint but also for maintaining soil quality (www.extension.iastate.edu, 2012).

10.5.6 Organic wool

Organic certification standards vary among countries. In some countries, the standards are set and overseen by the government, whereas in others the standards are set by a non-profit-making organization or by private companies. Requirements for certification of organic wool by the Organic Trade Association in North America are (OTA, 2011):

- Livestock feed and fodder used from the last third of gestation must be certified organic.
- The use of synthetic hormones and genetic engineering is prohibited.
- The use of synthetic pesticides (internal, external and on pastures) is prohibited.
- Producers must encourage livestock health through good cultural and management practices.

10.5.7 Organic silk

Organic silk must be obtained not only by feeding silkworms with the leaves from mulberry bushes that have been grown organically, but no 'cruelty' must have been employed (i.e., not by the conventional production method of placing the cocoons containing the live silkworms into boiling water). Thus, in the so-called 'peace' or 'vegetarian' varieties, the silkworms are allowed to develop and emerge as moths. As a consequence, the silk is obtained in the form of short staple instead of continuous filament, yielding a fabric with a different appearance and handle but with a warmer handle. In conventional silk manufacture, the worms are fed mulberry leaves which they convert into body mass, and hence the proteins sericin and fibroin which are used to spin the cocoon. If the feed contains harmful substances the highly sensitive pupae react adversely; either they die or they do not develop properly and are liable to many different kinds of diseases leading to both quantitative and qualitative losses, yielding lower standard cocoons which are smaller, spotted, thin-walled and yellowish in colour instead of pearl white (ALKENA, 2012). For organic silk, silkworms are given feed which is free from harmful substances.

10.5.8 Greener regenerated cellulose fibres

Regenerated cellulose filaments and fibres such as viscose rayon, modal and cupro fibres are manufactured by dissolving cellulose in carbon disulphide in the case of viscose and modal fibres and cuprammonium hydroxide solvent for cupro fibres, followed by precipitation in filament form in a suitable solution (usually mineral acid). Both solvents cause environmental problems; hence, attempts have been made to develop an alternate process. Lyocell fibres are produced by regenerating cellulose in an organic solvent, *N*-methylmorpholine-*N*-oxide hydrate, which is nontoxic and biodegradable and is almost completely recycled (Albrecht et al., 1997). The life cycle of a Lyocell fibre has minimal environmental impact and is significantly more sustainable than oil-derived synthetic fibres (e.g., polyester, nylon, acrylic) or natural fibres such as cotton; it uses less land, irrigation, pesticides or fertilizers to cultivate the eucalyptus or beech trees from which Lyocell fibres are made if the forests are sustainably managed than is the case for cotton (White et al., 2005). Lyocell fibres are available as Tencel (Lenzing, Austria), Lyocell (Lenzing, Austria) and NewCell (Akzo-Nobel, Germany); 98% of Lyocell fibre production is attributable to Lenzing.

Because of their very high degree of crystallinity compared with other cellulosic fibres, unmodified Lyocell type fibres can be easily fibrillated by any mechanical action on the woven fabric surface (e.g., during winch- or jet-dyeing in rope form or during mechanical washing processes). The fibrillated Lyocell always displays a lower visual colour yield than its nonfibrillated counterpart because of greater light scattering (Goswami et al., 2007). Fibrillation is generally uneven; hence Lenzing enzyme-treats the fabric to remove surface fuzz, then subjects it to a mechanical tumble to achieve a more regular effect. Fibrillation is useful for the production of the popular suede or peach-skin surface effect, which can be accomplished by the dyeing method itself, instead of by separate enzyme treatment and mechanical abrasion. As a consequence, Lyocell fibre has become popular for certain styles: for example, in the production of fashion jeans with a pleasing softer handle compared with those made from cotton.

10.5.9 Greener synthetic fibres

A large quantity of adipic acid [$HOOC(CH_2)_4COOH$] is used in the production of nylon 6.6, polyurethanes, lubricants and plasticizers. Conventionally it is produced from benzene (a known carcinogen), as shown in Eq. (10.2). Benzene is converted to cyclohexane and then into cyclohexanone and cyclohexanol, which are then converted into adipic acid.

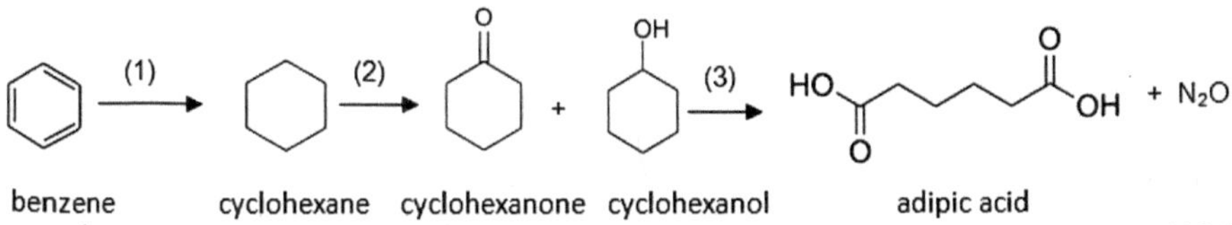

(10.2)

The conditions for various reaction steps are as follows:

1. reaction in the presence of $Ni\text{-}Al_2O_3$ under 370–800 psi pressure.
2. oxidized in the presence of Co catalyst.
3. reaction with HNO_3 in the presence of Cu and NH_4VO_3.

However, natural glucose can be converted into adipic acid by an enzyme (*Escherichia coli*) discovered in genetically modified bacteria, as shown in Eq. (10.3). Steps 1 and 2 are treatment with enzyme (*E. coli*) and Step 3 is reduction with H_2 in the presence of platinum catalyst at 50 psi.Polyurethane polymers are presently produced from toxic diisocyanates, but a series of polyurethanes based on *bis*-carbamate diols can be synthesized using the *Candida antarctica* lipase B as catalyst (McCabe, 2004).

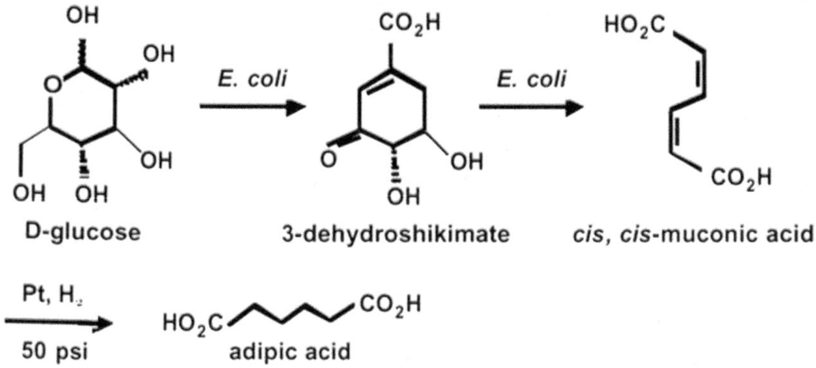

(10.3)

10.6 Greener preparatory processes

Textile materials possess a variety of impurities. Some are natural or acquired by contamination (e.g., soil) and others may be added purposefully for better textile manufacturing. All such impurities have to be removed before dyeing or printing processes by preparatory processes (or pretreatments), which may vary from fibre to fibre. They may be broadly classified into two groups:

1. cleaning processes such as singeing (removal of projecting fibres), desizing (removal of starch and other sizes applied to warp yarns) and scouring (removal of natural, acquired or applied impurities)
2. bleaching or destruction of colouring matters

The most important application of green chemistry in textile preparatory processes is the use of enzymes.

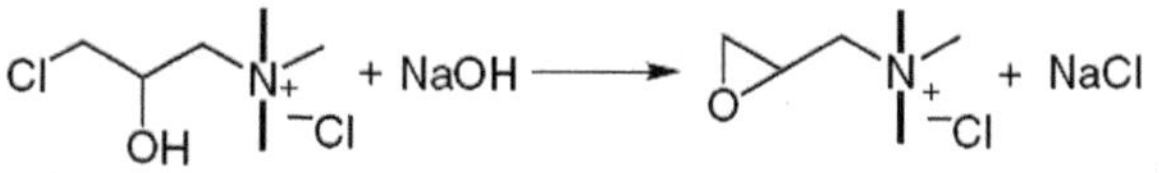

(10.4)

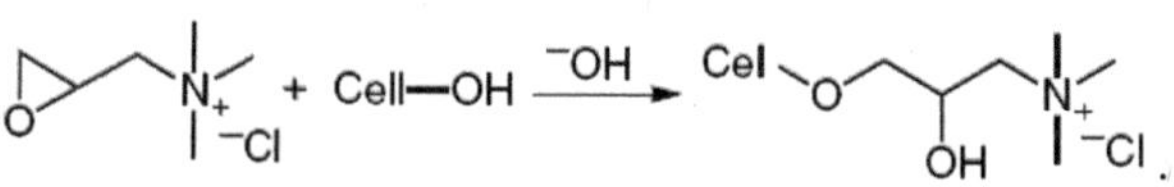

(10.5)

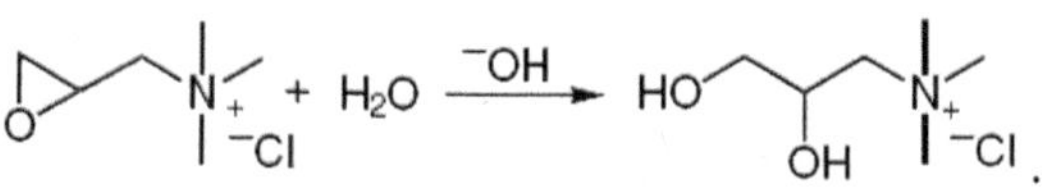

(10.6)

10.7 Enzyme processing

Bioprocessing is the application of living organisms and their components to industrial products and processes, mainly based on enzymes, and it offers the potential for highly effective industrial processes that require less energy and less water and cause fewer effluent problems. Enzymes qualify as green chemicals because they are biodegradable and are produced with renewable resources; enzymatic processes often work at lower temperatures, at normal atmospheric pressures and at more favourable (even neutral) pH, and use less water and energy (Thiry, 2011).

Enzymes are biological catalysts. The term 'enzyme' was first used by Kühne in 1878 and is derived from the Greek word *enzumé* meaning *in (en) yeast* (*zumé*). Chemically enzymes are proteins of high molecular weight: for example, that of enzyme α-amylase of *Bacillus subtilis* is about 100,000. They consist of complex three-dimensional proteins composed of polypeptide chains. They range from individual proteins with a relative molecular mass (RMM) of around 13,000 catalysing a single reaction, to multienzyme complexes of RMM of several millions catalysing several distinct reactions. In addition, some enzymes require some specific small nonprotein molecules, known as cofactors, to function as catalysts. Enzymes are relatively fragile substances and are susceptible to denaturing: that is, degradation due to temperature, ionizing radiation, light, acids, alkalis and biological factors, and thereby become inactive.

Commercially enzymes are obtained from three primary sources: animal tissue, plants and microbes. Natural enzymes are not readily available in sufficient quantities for industrial use. They are largely manufactured by fermentation, the technique has been well-known for more than 3000 years. Microorganisms producing enzymes important for the textile industry are listed in Table 10.1 (Shaikh, 2010).

10.7.1 Advantages of enzymes

The use of enzymes has following benefits (Cavaco and Gübitz, 2003):

- The rate of a reaction is accelerated by lowering the activation energy of reaction.
- The enzyme remains intact at the end of reaction by acting as catalyst.

Table 10.1 Applicability of enzymes at various textile processing steps

Process	Enzyme
Desizing	Amylase, lipase
Scouring	Pectinase, cellulase, cutinase
Bleaching	Oxidoreductase, xylanase
Dyeing	Oxidoreductase
Finishing	Cellulase, oxidoreductase, lipase
Composting (textile waste)	Laccase, cellulase, protease, nylonase, polyesterase

- There is remarkable chemical precision owing to far greater reaction specificity even in the absence of functional group protection. Enzyme engineering can further improve their stability and specific activity.
- Processes are easy to control because the enzyme activity depends on operating conditions.
- Enzymes are biodegradable and eco-friendly.
- They are applicable under milder conditions of temperature (below 100°C), pressure (atmospheric) and pH (around neutral). However, high-temperature-stable enzymes are also available for faster reactions.
- Lower energy input leading to lower cost and lower emission of GHGs to the environment.
- Enzymatic reactions are several orders of magnitude faster.
- By-products are rarely generated.
- Fewer wastes are generated; therefore disposal problems are minimized.
- They are safe to handle and noncorrosive.
- Reduced pollution loads result owing to the easy biodegradability of enzymes.
- They are safely dischargeable after use, sometimes after inactivation by changing pH or temperature.
- They reduce global warming, save acidification, and enrich nutrients (Nielsen et al., 2009).

10.7.2 Demerits and challenges of enzymes

The problems of enzyme application are (Leitat, 2013):

- enzyme recovery and reuse
- scale-up in industrial environment
- higher process time
- may not always be economically feasible

10.7.3 Safety measures

During enzymatic processing the following safety measures should be strictly adopted:

- Live steam should not be used.
- Chemicals should be added in prediluted form.
- Compatible ionic surfactants should be used.
- Nonionic wetting agents with appropriate cloud points are preferred (Saravanan et al., 2008)
- Temperature, pH, and heavy metal contamination should be closely monitored.
- The enzymes should be carefully stored under prescribed conditions.

10.7.4 Specific actions of enzymes

Enzymes work mostly on renewable materials. Fruit, cereals, milk, fats, cotton, leather and wood are some typical candidates for enzymatic conversion in industry. However, their use is not limited to natural materials. It has been demonstrated that enzymes are able to modify the surfaces of synthetic textile materials such as polyester and polyamide.

Enzymes are specific in action. Nobel laureate organic chemist Fischer E (1894) suggested the simplistic 'lock and key' model. Both the enzyme and the substrate possess specific complementary geometric shapes that fit exactly into one another.

The high substrate specificity of enzymes results from the individual architecture of the active site in which only certain molecules can 'stereo-fit in'. The enzymes have a true activity centre in the form of fissures, holes, pockets, cavities or hollows. The active site is a perfect fit for a specific substrate, and once the substrate binds to the enzyme no further modification is necessary.

Enzymes are principally classified and named according to the chemical reaction they catalyse, because this is the specific property that distinguishes one enzyme from another. In other words, they are named by the substrate on which they act. An enzyme name is assigned not to a single enzyme protein, but to a group of proteins with the same catalytic property, even if they are obtained from different sources.

Specific enzymes, alone or as mixture, are used in various textile-processing steps as shown in Table 10.1 (Holmes, 1998).

The enzymes may be applied in various stages of textile processing, including desizing, scouring, bleaching, dyeing and finishing, and also later when the product is being disposed of by composting (Preša and Tavčer, 2009).

10.7.5 Enzymatic desizing

Warp yarns are coated with sizes to increase their tensional and frictional strengths. Sizes mostly consist of starch, modified starch or organic polymers such as PVA, and polyacrylates. Conventionally, starch desizing is done by hydrolysis with water, by mineral acids or by oxidation with sodium bromite or peroxy compounds. The alternative eco-friendly method is to treat it with amylase enzymes in the presence or absence of lipase (to remove lipids). Amylase enzymes are obtained mainly from malt, bacterial or sometimes pancreatic sources.

Amylases were the first and most successful enzymes used in the textile industry for desizing. They were first used in the early 1900s when malt extracts from barley containing the active enzymes were applied to greige fabrics. In 1917 bacterial enzymes were isolated and ultimately bacteria became the source for industrial production.

Amylases are of two types: dextrinogenic or α-amylases (EC 3.2.1.1), and saccharogenic or β-amylases (EC 3.2.1.2). They are similar in that they hydrolyse glucosidic linkages in the starch molecules, but the point at which the reaction occurs is different (Peters, 1967). The α-amylase is capable of acting at random locations along the starch chain to break down long-chain carbohydrates, ultimately yielding maltose from amylase, or maltose and glucose from amylopectin. Because it can act anywhere on the substrate, α-amylase tends to be faster-acting than β-amylase and is therefore used for textile desizing. A regular amylase may be applied at a pH 5.5–7.0 and at a temperature of 25–55°C. Amylases for use at medium temperature can be used in the range of 50–95°C, whereas high-temperature amylase can be used successfully above 95°C and also in the pad-steam process. The end products of the enzymatic desizing process are various types of sugars and dextrins which are nontoxic. However, they have a negative impact on the BOD of the wastewater.

10.7.6 Enzymatic scouring

Scouring is the process of removing unwanted oils, fats, waxes, dirt and soluble impurities adhering to fibres with the aim of making textile materials uniformly absorbent. Conventionally, cotton scouring is done by prolonged boiling (at atmospheric or high-pressure) with aqueous sodium hydroxide and other chemicals, or by shorter pad-steam methods.

Scouring is an alkaline cleaning process for raw fibres (mostly natural) before the colouration process. Bioscouring has a number of potential advantages over traditional scouring. It is performed at neutral pH, which reduces total water consumption. The treated yarn/fabrics retain their strength properties, the weight loss is reduced or limited compared with processing in traditional ways, and it increases cotton fibre softness. The temperature of bioscouring is much lower compared with classical scouring; the optimal temperature is from 40 to 60°C. Bioscouring can create the desired hydrophilicity in the cotton fabric but at the same time leave sufficient wax materials on the fabric surface to lessen the required addition of softeners for consumer satisfaction.

An extremely powerful alkaline pectinase has been isolated. The major benefit of this enzyme in biopreparation is that the enzyme does not destroy the cellulose of the cotton fibre. The enzyme is a pectate lyase, and as such rapidly catalyses hydrolysis of salts of polygalacturonic acids (pectins) in the primary wall matrix. Pectinase, as the name suggests, hydrolyses pectin present in cotton as a noncellulosic impurity. The best kinds of pectinase are those that can function under slightly alkaline conditions even in the presence of chelating agents. Such enzymes are called 'alkaline pectinases'. Most conventional pectinases are usually inactive under these commercially useful conditions; their optimum activity lies in the slightly acidic region (Sundar et al., 2007).

Cellulases are especially suited for scouring of cotton fabrics. Some pectinase enzyme preparations contain cellulase. Those impurities are then removed by subsequent washing. However, the combined actions of both types give greater weight loss and strength loss compared with the action of pectinase or lipase alone.

Cotton fibres, or their blends with other fibres, can be treated with aqueous solutions containing protopectinases for 18 h at 40°C to give scoured yarn with good tensile strength retention. Pectinases and cellulases are effective compared with the proteases and lipases. The change in water absorbency of cotton is rapidly catalysed by pectinases, cellulases or their mixtures. Pectinases can destroy the cuticle structure by digesting the inner layer of pectins in the cotton cuticle. Cellulases can destroy the cuticle structure by digesting the primary wall cellulose immediately under the cotton cuticle. Cellulases break the linkage from the cellulose side and pectinases break the linkage from the cuticle side. The result of the synergism is a more effective scouring in both the speed and evenness of the treatment. The destruction of the cuticle during enzymatic scouring of cotton was revealed by scanning electron microscopy photographs (Li and Hardin, 1998).

The combined enzyme system for simultaneous desizing and scouring may contain amylase, lipase and pectinase enzymes to achieve the necessary fabric properties without the use of harsh chemicals (Karmakar, 1998).

BioPrep 3000L (Novo Nordisk, Denmark) is an alkaline pectinase free from cellulases (Lange, 2000). It works optimally at pH 7–9.5 and at temperatures up to 60°C in exhaust systems and at somewhat higher temperature in pad systems. The BOD and COD of the enzymatic scouring process are 20%–45% compared with alkaline scouring (100%). The total dissolved solid of the enzymatic scouring process is 20%–50% compared with alkaline scouring (100%). The handle is soft in enzymatic scouring compared with the harsh feel in the alkaline scouring process.

Enzymatic scouring makes it possible to scour fabric effectively without negatively affecting the fabric or the environment. It also minimizes health risks inasmuch as operators are not exposed to aggressive chemicals (Pawar et al., 2002).

Despite all the research on bioscouring, it has yet to be applied on an industrial scale. There is a need for pectinases with higher activity and stability at high temperatures and alkaline conditions.

10.7.7 Enzymatic bleaching

Glucose oxidase enzyme (EC 1.1.3.4) was used for bleaching when the whiteness index was improved with lower strength loss (Shaikh, 2010). In the presence of molecular oxygen, glucose is oxidized by the enzyme glucose oxidase to gluconic acid and hydrogen peroxide. D-Gluconic acid acts as a sequestering agent during bleaching. Amyloglucosidases, pectinases and glucose oxidases are compatible concerning their active pH and temperature range, and were selected.

A combination of two or all three preparation steps with minimal amounts of treatment baths and rinse water showed compatible results in whiteness, absorbency, dyeability and tensile properties of the treated fabrics. Studies were done of biobleaching of wool under both oxidative and reductive conditions. The studies showed that hydrogen peroxide bleaching in the presence of protease preparation, Bactosol SI (Clariant), considerably improved whiteness and hydrophilicity.

10.7.8 Peroxide killer

Catalase, more correctly hydroperoxidases, or peroxidase, is an oxidoreductive class of enzyme produced by a variety of different microorganisms including bacteria and fungi (Mueller et al., 1997). Catalases catalyse the degradation of H_2O_2 to H_2O and O_2 and have optima at moderate temperatures (20–50°C) and neutral pH. In addition to the protein part of the molecule, catalase enzyme contains a nonprotein part, which is a derivative of heme and includes metal iron. The catalase enzyme (Jensen, 2000) attacks only hydrogen peroxide and nothing else. The reaction is as follows:

$$H_2O_2 + \text{catalase} \rightarrow 2H_2O + O_2 + \text{catalase} \tag{10.7}$$

10.7.9 Laccases

Laccases are a group of oxidative enzymes whose exploitation as biocatalysts in organic synthesis has been neglected in the past, probably because they were not readily commercially available. They are glycoproteins, which have been reported as present in higher plants and in virtually every fungus that has been examined for them (Barreca et al., 2003). Laccases are oxidoreductases (EC 1.10.3.2, *p*-diphenol: dioxygen oxidoreductase) belonging to the so-called blue-copper family of oxidases (Riva, 2006), the multinuclear copper-containing oxidases. They catalyse the mono-electronic oxidation of substrates at the expense of molecular oxygen.

The search for new, efficient and environmentally benign processes for the textile and pulp and paper industries has increased interest in these essentially 'green' catalysts, which work with air and produce water as the only by-product. Their uses span the textile, pulp and paper industries, from food applications to bioremediation processes. Laccases also have uses in organic synthesis, where their typical substrates are phenols and amines, and the reaction products are dimers and oligomers derived from the coupling of reactive radical intermediates (Bourbonnais et al., 1997).

At present, the main technological applications of laccases are in the textile dyeing or printing industries, in processes related to decolourization of dyes (Li et al., 1999), and in the pulp and paper industries for the delignification of woody fibres, particularly during the bleaching process (Galante and Formantici, 2003). In most of these applications, laccases are used together with a chemical mediator.

A laccase-mediator system process has also been introduced in the textile industry, especially for indigo extraction on denim garments in industrial laundries; the dye chromophore is split to yield two molecules of the colourless isatin. The system is currently composed of a granular formulation of recombinant laccase from *Mycelyopthora thermophila*, which is applied with the mediator, the methyl ester of syringic acid (Fig. 10.4), in a series of 15- to 20-min treatments (Marzorati et al., 2005).

10.8 Greener bleaching methods

Conventionally bleaching is carried out using sodium hypochlorite, sodium chlorite or hydrogen peroxide. However, because sodium hypochlorite and sodium chlorite can generate AOX they are being substituted with the nonpolluting oxidant, hydrogen peroxide (H_2O_2).

Manganese complexes with 1,4,7-trimethyl-1,4,7-triazacyclononane and related ligand systems may act as highly effective catalysts for the bleaching of stains by hydrogen peroxide at low temperatures. These complexes also catalyse the epoxidation of alkenes and the oxidation of phenolic substrates by hydrogen peroxide. Results demonstrate the considerable potential of these systems for clean and efficient low-temperature bleaching (Hage et al., 1994).

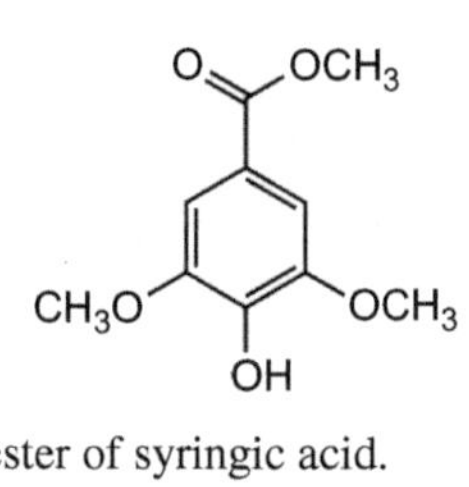

Figure 10.4 Structure of methyl ester of syringic acid.

Cellulosic fabrics can be effectively photobleached using various excimer lasers (KrF, XeCl and XeF), a low-pressure mercury lamp, and a black-light fluorescent lamp in the presence of sodium percarbonate ($2Na_2CO_3 \cdot 3H_2O_2$) or mixtures of sodium carbonate and hydrogen peroxide aqueous solutions at room temperature. The efficiency of photochemical bleaching was found to be comparable to that of commercial thermal chemical bleaching treatments (Ouchi et al., 2003). The experiment was also carried out in the presence of several reducing agents; the best bleaching efficiency was achieved with sodium borohydride ($NaBH_4$) (Ouchi et al., 2004).

Huntsman Textile Effects and Genencor announced the results of an LCA on Gentle Power Bleach, an enzymatic bleaching system introduced into the textile pretreatment market in 2009; potential savings could equate to 10 trillion litres of fresh water and 30 million metric ton of CO_2 emissions per year (Genencor and Huntsman, 2010).

10.9 Colouring matters

For ages, the dyeing of textiles was based on natural products. William Perkin discovered the first synthetic dye, called mauveine, in 1856. Spectacular changes were initiated by the relatively new science of organic chemistry. Within 50 years of Perkin's discovery synthetic dyes accounted for over 90% of the dyes used. Two other remarkable developments achieved by Scottish Dyes Ltd. (later part of ICI) were the manufacture of copper phthalocyanine dyes in 1927 and the development of fibre-reactive chlorotriazinyl dyes in 1954 by ICI Dyestuffs Division. With the introduction of hydrophobic fibres, namely polyester and acrylic (polyacrylonitrile), new classes of dyes were developed, but these were merely modifications of dyes that were already available. In fact, a high percentage of the current organic chemical industry (such as the pharmaceuticals and agrochemicals) is an offshoot of the dyestuff industry. Perkin is rightly acclaimed the father of the dyestuff industry, so the dyestuff industry itself could be called the mother of the modern organic chemical industry (Roy Choudhury, 2006).

Health and safety were absent from the list of priorities in the early decades of the synthetic dyestuffs industry. Practical experience in the primitive working conditions of the time made workers aware of more obvious dangers, such as corrosive acids, flammable solvents and potentially explosive nitro compounds.

10.9.1 Azo dyes

Azo (—N=N—) groups, such as those incorporated in benzenoid compounds, and which form the basis of most synthetic dyes, are not found in naturally occurring products. Approximately 70% of all dyes (belonging to various dye classes) used in the textile industry are azo dyes (Carliell et al., 1995). Under reductive conditions some azo dyes may produce carcinogenic amines (Lacasse and Baumann, 2004). During the synthesis of dyes, aromatic amines are treated with nitrous acid to form diazonium compounds, which are then reacted with naphthols to form either azo dyes (water-soluble) or pigment (water-insoluble) (Roy Choudhury, 2006). For many years, *ortho*-toluidine had been a foundation of the dye industry; it was used to make magenta and several other important colours. It is an aromatic amine similar in structure to benzidine, α-naphthylamine, β-naphthylamine (BNA), and other chemicals used in the dye and rubber industries that had been widely recognized to cause bladder cancer in humans. Acknowledging that it could not be manufactured safely, Switzerland banned BNA production in 1938 and Great Britain followed in 1952. In the absence of regulation in the United States, DuPont did not stop producing this carcinogen until 1955 and Allied Chemical Corporation continued to manufacture BNA-containing chemicals through 1960 (Michaels, 1995).

Following the introduction of the German Consumer Goods Ordinance in 1994, the use of certain azo dyes was restricted in consumer goods. Several other EU member states introduced similar but different regulations. In the interests of transparency and the maintenance of the single market, the European Parliament accepted the 19th amendment of the Council Directive 76/769/EEC relating to restrictions on the marketing and use of certain dangerous substances and preparations (azo colourants) (EC, 2002). A small number of aromatic amines are classified as being carcinogenic or potentially carcinogenic to humans; hence, to protect human health, those azo dyes which can break down under reductive conditions to release any of a group of defined aromatic amines were prohibited from being used in consumer goods considered to have regular skin contact.

Under reductive conditions, using sodium dithionite azo groups may be cleaved to form two amines, as shown in Eq. (10.8).

$$A - N = N - B \xrightarrow[\text{pH 6}]{Na_2S_2O_4} A - NH_2 + B - NH_2 \tag{10.8}$$

It has been estimated that less than 4% of known azo dye structures would release the corresponding amines; only a few azo dyes can release these amines upon reductive cleavage (Půntener and Page, 2004). The regulation, EU Directive 2002/61/EC, identified 22 aromatic amines as harmful and prohibited their use. These amines are listed elsewhere (Table 9 of Roy Choudhury, 2013). For Oeko-Tex Standard 100, two more arylamine substances are added to the list of harmful amines, which are also listed in that table (Serial Nos. 23 and 24).

Fig. 10.5 shows the chemical structures of four banned amines listed in Table 9 mentioned earlier, which are definitely carcinogenic. The four amines are (a) 4-aminodiphenyl; (b) benzidine; (c) 4-chloro-*o*-toluidine; and (d) 2-naphthylamine.

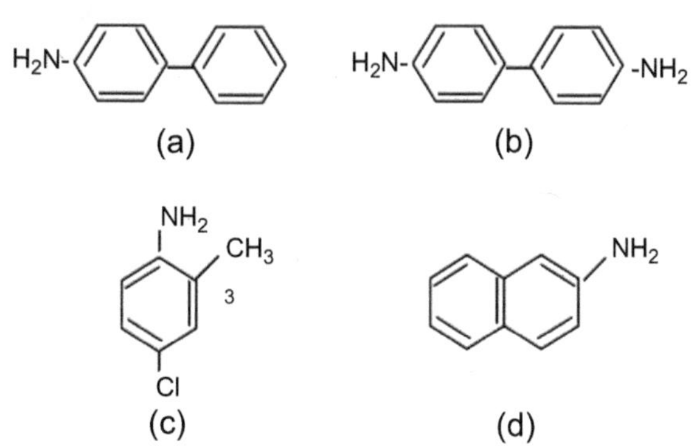

Figure 10.5 Structures of four carcinogenic aromatic amines.

Cleavable arylamines, carcinogens, allergens and other banned colourants should not be detectable in the finished articles or in their dyed parts. Further information is available on the Oeko-tex website (Oeko-tex, 2013).

The EU Directive was published in September 2002 and all EU member states had until 11 September 2003 to enact appropriate legal restrictions in their countries (EC, 2002). There may still be some dyes and pigments available on the market which are not officially registered owing to confidentiality from dyestuff manufacturers. The list of the 187 banned azo dyes (106 direct dyes, 30 acid dyes, 11 basic dyes, 11 azoic colours, 27 disperse dyes and two other dyes) out of about 2000 is available elsewhere (EC, 2002); such lists are difficult to keep up to date because of new chemical developments and sector changes, and for that reason it is recommended to check the internet links periodically (Oeko-tex, 2013), where the update list of prohibited dyes is available.

10.9.2 Metal complex dyes

Metal complex dyes are used for dyeing wool and nylon in dark shades. The metal, typically chromium (Cr III) or cobalt (Co III), acts as a mordant. The metal in metal-complex dyes is usually linked to some organic chromogen and the links to the fibre are between the $-SO_3^-$ or $-COO^-$ groups on the dye molecules and the $-NH_3^+$ groups in the fibre. The colour fastness of such dyes is usually good, in particular against light and washing. Owing to the known ecological and toxicological issues surrounding the use of chromium and cobalt, research has been carried out into their substitution with more environmentally friendly metals such as iron and aluminium. A group of scientists at North Carolina State University are continuing to build on previous success in developing iron-complexed azo and formazan dyes, which exhibit lower toxicity than the corresponding Cr and Co analogues. As environmentally friendly alternatives to Cr(III) and Co(III) metal-complex acid dyes, a range of 1:2 iron complex formazan dyes was synthesized which produce violet, blue, black and brown shades on wool and polyamide with good wet and light fastness (Carliell et al., 1995)

10.9.3 Reactive dyes

Reactive dyes differ fundamentally from other dye classes in that they chemically react with the textile fibre-forming strong covalent bonds between the dye molecules and the fibre-forming polymer. Metal-free reactive dyes are popular in the textile dyeing and printing trade. They are comparatively safe and not many adverse effects have been reported (Hannemann and Runser, 1999). However, the most obvious deficiency in the conventional reactive dyestuffs lies in the fact that dye fixation on the fibre is no more than 70% even in printing applications. The remainder of the dyestuff undergoes hydrolysis instead of reaction with the fibre, increasing the pollution load in the effluent.

Reactive dyestuffs have been developed which provide much higher degrees of fixation than conventional reactive dyes (Morris et al., 2008). These high-fixation (above 90%) reactive dyestuffs contain multiple (mostly two) reactive groups attached to each dye molecule. The dye may contain two similar reactive groups, i.e., homofunctional dyes, e.g., Procion HE dyes (DyStar) or two different reactive groups, i.e., heterofunctional dyes, e.g., Procion ME (DyStar).

In 2000 Dystar was awarded the UK Green Chemistry Award for its bifunctional, high-fixation, reactive dyes. When reactive dyes were first introduced in 1956 by ICI Dyestuffs Division, the dichlorotriazine Procion MX dyes (now withdrawn from the range) were also homobifunctional but these dyes did not have sufficiently high substantivity to give good exhaustion (e.g., on average only 40% at 30:1 liquor-to-goods ratio) before the alkali-fixation stage during batch-wise dyeing, which limited the efficiency of dye fixation. The dyes did, of course, have the desirable property (from a green chemistry point of view) of being applied at only 40°C.

Some synthetic dyes such as reactive dyes are difficult to remove from effluent because of their high water solubility; thus they are not captured and are held by the activated sludge in wastewater treatment plants, and they also have poor biodegradability (Cooper, 1995). A few easily degradable synthetic dye structures are being developed (Moozyckine and Davies, 2002). Because of increasing awareness and concern by the global community about the discharge of synthetic dyes into the environment and their persistence, much attention has been focused on the remediation of these pollutants. Among current pollution-control technologies, biodegradation of synthetic dyes by different microbes is emerging as an effective and promising approach to the development of an economic, effective, bio-friendly and environmentally benign process. The bioremediation potentials of many microbes for xenobiotics including synthetic dyes have been demonstrated; it is hoped that bioremediation by various microbes will prove a green solution to the problem of environmental soil and water pollution in the future (Ali, 2010).

10.9.4 Natural dyes

Natural dyes can be placed into three categories on the basis of origin: from plants (such as indigo), from animals (such as cochineal) and from minerals (such as ochre). The major types of natural dyes and their origin have been tabulated by Hill (1997).

Before the advent of synthetic dyes, natural dyes were widely used, often together with mordants such as alum, to dye natural fibres including wool, linen, cotton and silk, but their use declined after the discovery of synthetic dyes. However, interest in natural dyes has been revived owing to increasing demands on manufacturers to produce more environmentally friendly alternatives to petrochemical-derived dyes. One main issue associated with the use of natural dyes in the colouring of textiles is their poor to moderate light-fastness, and despite their long tradition, not all natural dyes are especially environmentally friendly. Some natural dyes have no or little affinity for the textile materials and they require heavy-metal salts as mordants for fixation and colour-fastness. Natural dyes may be sustainable but they need water and land to produce and there is insufficient dye yield per acre of plant material to sustain industrial-scale production.

Natural dyes can be classified into two groups, depending on their method of application: substantive and nonsubstantive. Substantive dyes require no pretreatment to the fabric (e.g., indigo, orchil and turmeric) and there are three types: direct (for cotton, e.g., turmeric, safflower; acid (for silk and wool, e.g., saffron, lac) or basic (for silk and wool, e.g., berberine). Nonsubstantive dyes, by contrast, can dye only already mordanted material or work with the addition of a mordant to the dye bath (e.g., logwood, madder/alizarin, cochineal or fustic)

Natural mordant dyes are either monogenetic or polygenetic; monogenetic dyes produce only one colour irrespective of mordant, whereas polygenetic dyes produce different colours according to the mordant employed (e.g., logwood, alizarin, fustic and cochineal).

Based on their chemical nature the dyes are as follows:

1. diarylol methane (turmeric)
2. carotenoid (saffron, annatto)
3. alkaloids (barberry)
4. quinonoid (dolu, henna)
5. flavonoid (French marigold, sandalwood)
6. benzoquinone (safflower)
7. anthraquinone (alroot, moyam)

The fixing agents recommended are alum, stannous chloride, tartaric acid, ferrous sulphate and Turkey Red oil. Tannins are used as natural mordants in the natural dyeing process because they can enhance the affinity of different dyes. Tannins are naturally occurring phenolic compounds, and those with *o*-dihydroxy groups form metal chelates after pretreatment or posttreatment of the substrate with metal salts. Tannin–metal salt combinations, rather than the tannins themselves, can be considered the mordant for natural dyes. Some important raw materials for tannins are oak bark and wood (*Quercus infectoria*), pomegranate rind (*Punica granatum*) and cutch (*Acacia catechu*) (Roy Choudhury, 2006).

With the exception of indigo and logwood, most natural dyes have poor light stability. A nickel sulphonate derivative of a phenyl ester (the nickel complex of 1-benz oyloxynaphthalene-3-sulphonic acid) may be applied as an effective stabilizer against the fading of natural dyes; such nickel complexes were deemed safe (with regard to the environment and human health) after evaluation of antimicrobial activity; hence they could prove useful in practice for improving the light-fastness of natural dyed fabrics (Oda, 2012).

10.10 Textile chemicals

In addition to dyes, three other categories of chemicals are widely used in the manufacture of the vast majority of products from the textile industry: auxiliaries, surfactants and stain removers. It is also necessary to consider textile finishes and lubricants. Three studies focussing on the EU, United States and China highlight the problems that may arise from discharges into the environment and the large number of solutions that still need to be sought.

The Oslo and Paris (OSPAR) Convention for the Protection of the Marine Environment of the North-East Atlantic (the 'Convention') was open for signature at the Ministerial Meeting of the Oslo and Paris Commissions in Paris on 22 September 1992. It was adopted together with a Final Declaration and an Action Plan. The OSPAR Convention entered into force on 25 March 1998. OSPAR publishes such general assessments in the form of Quality Status Reports (QSRs) of the Northeast Atlantic and its subregions. QSRs are published periodically as major benchmark assessments resulting from the joint efforts of the contracting parties.

Some important findings mentioned in OSPAR Quality Status Report 2010 (OSPAR, 2013) are:

1. Surface water temperatures have risen since 1999. Water in the North Sea has warmed 1–2°C since 1985.
2. Warm-water plankton and some fish species are shifting northward.
3. Ocean acidification could have significant impacts on calcareous organisms.
4. The extent of winter Arctic sea ice has fallen 2.5% per decade since 1979, reducing ice-associated habitat.
5. The sea level is rising, making low-lying coasts vulnerable to flooding and coastal erosion.
6. Nutrient discharges to eutrophication problem areas are substantially reduced compared with 1985 for phosphorus (up to 85% lower), but less so for nitrogen (up to 50% lower).
7. Nitrogen inputs from the atmosphere remain high and air emissions from shipping are increasing.
8. A third of OSPAR's 26 priority (groups of) chemicals are expected to have been phased out by 2020. More effort is needed to move towards the cessation target for releases of the remaining priority chemicals.
9. Levels of cadmium, mercury, lead, polycyclic aromatic hydrocarbons and polychlorinated biphenyls are unacceptable in fish, shellfish and sediment in many coastal areas.
10. Persistent organic pollutants such as brominated flame retardants are widespread and accumulating in marine life, partly as a result of atmospheric transport.
11. Discharges of radionuclides have fallen [a radionuclide or radioactive isotope is an atom that has excess nuclear energy, making it unstable (excess energy that can either create and emit, from the nucleus, new radiation (gamma radiation) or a new particle (α particle or β particle) or transfer this excess energy to one of its electrons, causing it to be ejected from nuclear installations], and radiation doses to humans and marine life from this pollution are low in all OSPAR Regions (β-activity down by 38% since 1995).

10.11 Greener dyeing processes

Dyeing is a process of uniform colouration of textile materials using dyes or pigments in an aqueous medium. The process demands the use of a large number of chemicals including acids or alkalis. The dyeing process is generally conducted at high temperature for a prolonged time.

To become environmentally responsible, dyers need to adopt the well-established '3R' principles of pollution prevention: i.e., reduce, reuse and recycle. The most effective pollution prevention practice in textile wet processing is RFT dyeing. Corrective measures such as shading additions, or stripping and redyeing processes, consume additional dyes and chemicals, water and energy (Holmes, 2012).

There are two basic approaches to pollution control:

- effluent treatment or end-of-pipe solutions
- waste minimization or source-reduction solutions

The first approach has no financial payback whereas the waste minimization approach not only reduces environmental impact but also minimizes the cost of production and has the potential to make dyeing processes greener in several ways:

1. process optimization aimed at reducing process time and energy consumption, for example, by the use of combined scour/dye processes
2. reduced consumption of water (ultralow liquor ratio dyeing: 3:1 to 4:1 for polyester and 6:1 for cotton), electrical power and steam consumption
3. promoting high exhaustion, e.g., through the use of polyfunctional fibre-reactive dyestuffs which can give exhaustion in excess of 95%, thus decreasing pollution
4. continuous preparatory and dyeing methods instead of batch-wise methods minimizing consumption of energy, water and chemical to less than half
5. semicontinuous cold pad-batch dyeing processes with reactive dyes instead of batch-wise or pad-steam methods (Kehry and Strohle, 2006)
6. chemical-free water-free denim processing using laser technology to burn away quickly the surface of the dyed denim fabric or a pair of jeans on a mannequin to replicate an authentic worn look. Conventionally, this is done with stones, oxidizing agents or enzymes by aqueous methods (Tarhan and Sariisik, 2009)
7. sustainable digital printing and heat transfer printing which require far less water and produce far less waste than traditional printing methods (Fashion-incubator, 2010)
8. cold transfer printing process (Cooltrans) of reactive dyes at room temperature on pretreated cotton, viscose, linen and silk, thereby saving water and heat (Thiry, 2010)
9. supercritical carbon dioxide dyeing or waterless dyeing (Miah et al., 2013)

10.11.1 Pad-batch process

Cold pad-batch processing (for both pretreatment and dyeing) is a superb means of reducing resource use in textile mills, and saving money (Eksoy, 2012). In this method, alkali/hydrogen peroxide (in the case of pretreatment) or dyes are applied to the fabric using a padder, and the fabric is then stored to allow complete reaction between the fabric and chemicals before rinsing. Compared with exhaustion methods, as much as 50% of the water and electricity and 38% of the steam used in pretreatment can be decreased with this method, with an overall cost reduction of about 50% (CCME, 2012). The results for dyeing are similarly impressive, particularly for energy savings, in large part because cold pad-batch dyeing does not require drying or the use of steam. Both dye-penetration and fixation rates are high (showing 15%–25% improvement), which also reduces dye consumption and wastewater colour problems. However, cold pad-batch technology is currently limited to woven cotton fabric and works best with heavyweight fabrics and dark colours.

10.11.2 Sulphur dyeing

Sulphur dyes are economical and exhibit good colour strength and good colour-fastness on cellulosic materials. Sulphur dyes constitute the largest class of dyes in terms of quantity consumed, with an estimated worldwide consumption of more than 80,000 ton per year, even though a different picture emerges if the calculations are made on the basis of price (Shore, 1991). C. I. Sulphur Black one alone represents a substantial portion (20%–25%) of the dyestuff market for cotton, which is the highest consumption of any single textile dye in the world (Blackburn, 2006).

All sulphur dye molecules contain disulphide linkages and are applied by reducing them with the addition of a suitable reducing agent; to yield a green process, the reducing agent must be sustainable. Sulphur dyes themselves have a relatively low detrimental environmental impact because the dyes are free from heavy metals and AOX (Khatri and Brohi, 2009), but significant environmental problems occur with the sulphur dyeing process owing to the use of sulphides as reducing agents (90% of all sulphur dyes are reduced using sodium sulphide). Discharge of sulphides is permissible only in small amounts (usually the legal allowance is 2 ppm). The problems are:

- danger to life from liberated hydrogen sulphide
- corrosion to sewerage systems
- damage to treatment works
- high pH
- aquatic life downstream significantly affected
- damage to the DNA of tadpoles
- the micropolluting nature of sulphur dyes
- over time the substance can build high concentrations

Sulphur dyes are available in their insoluble oxidized (pigment) form. Addition of reducing agent (usually sodium sulphide) cleaves a proportion of the disulphide linkages to form the partially soluble 'leuco' form. Further addition of reducing agent and increases in redox potential cause a reduction of the remaining disulphide linkages and quinoneimine groups. After exhaustion of the dye onto the fibre, the reduced, adsorbed dye is reformed in situ within the fibre by air or chemical oxidation.

A number of alternative reducing agents to replace the sodium sulphide have been suggested, but they are less effective as reducing agents than sodium sulphide, reacting more slowly with the dyes and requiring higher vatting temperatures and increased concentrations of sodium hydroxide to achieve a sufficient degree of dyestuff reduction (Broadbent, 2001). Some examples are:

Thiourea dioxide
- Dyeing is comparable, but environmental effects are unclear.
- Cost is significantly higher than sodium sulphide.

Indirect cathodic reduction processes
- They successfully reduce sulphur dyes.
- Some reducing agent is required to prevent premature reoxidation of the dye.
- Dyeing is comparable.

- Electrolysis is an appreciably more expensive technology than use of the chemical reducing agent.

Glucose and caustic soda

- Above 90°C, the mixture has sufficient reducing potential.
- This mild reducing agent may be used to reduce sulphur dyes.
- The dyed material shows lower colour strength and fastness.

No fundamental work has been done to determine the optimum conditions for the reducing sugar and caustic soda system.

Attempts have been made to replace the sodium sulphide used in the Sulphur-dyeing process with a variety of environmentally friendly reducing sugars. The redox potential of various hexose and pentose monosaccharides and reducing disaccharides was recorded and compared. It was observed that reducing sugars gave comparable, and in many cases superior, colour strength and wash-fastness results with respect to the commercial sulphide-based reducing agents, which still account for the vast majority of Sulphur-dyeing processes and that pose significant environmental concern. Employment of reducing sugars in sulphur dyeing could provide a sustainable, nontoxic, biodegradable, cost-effective alternative to sodium polysulphide and sodium hydrogen sulphide. The sugars examined as possible reducing agents by Blackburn and Harvey (2004) were D-arabinose, D(−)-fructose, D(+)-galactose, α-D-glucose, β-D-lactose and D-maltose.

Comparison of the redox potential of reducing sugars against the colour strength of the dyeing obtained demonstrated that there was an optimum redox potential of around −650 mV for maximum colour strength achieved. The same redox potential also conferred the lowest colour loss upon washing. These observations were attributed to reduction of the polymeric dye molecules to an optimum size for fibre affinity and diffusion into the fibre which would also confer maximum wash-fastness upon oxidation.

10.11.3 Reactive dyeing

Not many environmental problems have been reported against reactive dyes. However, a number of problems are being encountered during application of reactive dyeing:

1. high electrolyte concentration
2. poor dye exhaustion
3. excessive water use

10.11.3.1 High electrolyte concentration

For optimum fastness, the exhaustion values of reactive dyes are set to a moderate range. As a result, large quantities of sodium chloride or sodium sulphate are required for maximum exhaustion of reactive dyes by exhaustion methods. Anionic reactive dyes, which are rendered water-soluble by virtue of their sulphonic acid groups ($-SO_3^-$), are repelled by the negative charge carried by the $-COO^-$ groups created on the cellulose fibre surface by oxidative bleaching treatments. Electrolytes such as NaCl, Na_2SO_4 and $(NH_4)_2SO_4$ provide cations that neutralize the charge on the fibre, thereby enhancing dye exhaustion. A high level of salt in the effluent water causes several problems, such as:

- difficulty in removal from effluent
- contaminated soil that becomes too alkaline to support crops
- high salinity that kills aquatic life
- high volumes of water required in the wash-off of salt and hydrolysed reactive dye

High electrolyte concentrations in dye-bath discharges are undesirable, because increased salinity in rivers upsets the delicate balance of aquatic flora and fauna. Efforts have been made to produce reactive dyes demanding less salt during application. The most notable of these is the Cibacron LS range (CIBA, 1996); Hoechst also introduced the Remazol EF range (Taylor, 2000). The other way to minimize salt requirement for reactive dyeing is to precationize cellulosic fabric so that it can be dyed either without or with only a minimal amount of salt (Roy Choudhury, 2006). There are also low-salt reactive dyes which are trifunctional and applied at 60°C (Huntsman), and Bezema has a similar range applied at 40°C. The pad application methods avoid the use of sodium chloride or sodium sulphate completely (Cho et al., 2007).

10.11.3.2 Poor dye exhaustion

A significant quantity (10–40%) of residual reactive dye remains in the exhaust dye bath owing to incomplete exhaustion and/or fixation; hence the effluent water is highly coloured and the pollution load is too high and aesthetically unpleasant, blocking sun light and causing overpopulation of algae, reducing the level of dissolved oxygen and suffocating flora and fauna in watercourses.

Dystar's Sera Fast C-RD and their Sera Wash process enable hydrolysed reactive dyestuffs to be washed off at 60°C with much lower water consumption than normal (Dystar, 2013).

10.11.3.3 Excessive use of water

A large volume of water is consumed for multistage washing (up to 10 separate rinses) after dyeing with reactive dyes to remove unfixed dye. Because these washes include hot washes and washes at boil, a large amount of energy is also required during washing. Therefore, the cost is very high, contributing about 50% of total dyeing cost.

One technology that is continuing to be developed is the use of cationizing agents to form cationized cotton before dyeing (Blackburn, 2006; Roy Choudhury, 2014b); treatment of cellulose with cationic, nucleophilic polymers enables reactive dyeing at neutral pH without electrolyte addition. Dyeing cationized cotton results in greater use of dye and higher colour values. In addition, the strong dye–fibre interactions resulting from cationizing cotton allow dyeing to be accomplished with no added electrolytes and minimal rinsing and after-washing.

A typical cationization process involves the reaction of 3-chloro-2-hydroxypropyl-trimethylammonium chloride (CHPTAC) with an alkali to form epoxypropyl-trimethyl-ammonium chloride (EPTAC), which then reacts with the cellulose to form cationic cotton (Dow Chemical). The cationic sites on the modified cotton strongly attract anionic dye molecules, leading to performance and process improvements. Chemical modification of cotton fabric via cationization with CHPTAC was studied (Hashem et al., 2003). The reaction was carried out in a strong alkaline medium (usually sodium hydroxide) either by the exhaustion method at high temperature or by the cold pad-batch method. The quaternary ammonium group, $-N(CH_3)_3^{\ +}$, has a persistent positive charge and can thus lead to the formation of ionic bonds (salt linkages) with negatively charged anionic groups, such as those found in a wide array of anionic dye classes or carboxy compounds. The former imparts the no-salt dyeing properties on the cotton fabric. In addition, the presence of cationized groups in the cellulose imparts antimicrobial properties (Soares et al., 2005).

Reaction of CHPTAC with cotton cellulose has been the focus of a number of studies (Soares et al., 2005). The suggested reaction scheme features are:

1. The chlorohydrin form of the reagent is converted to the epoxy intermediate (2,3-epoxypropyltrimethyl-ammonium chloride) (Eq. 10.9).
2. The epoxy compound reacts with the cellulose itself (Eq. 10.10)
3. The epoxy compound is converted via hydrolysis to no-reactive 2,3-dihydroxy derivatives (Eq. 10.11)

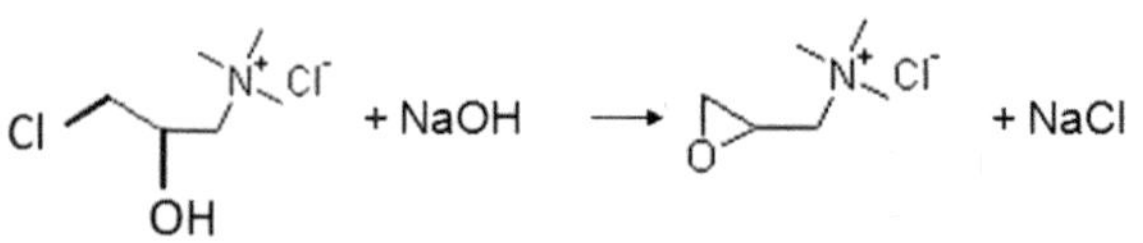

(10.9)

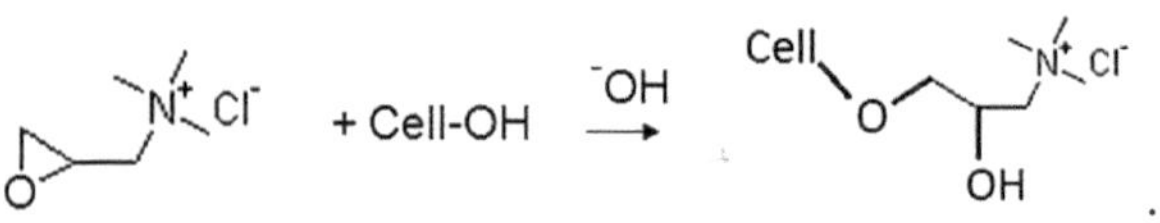

(10.10)

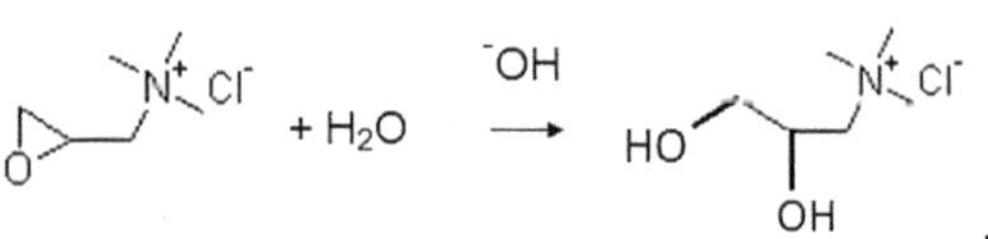

(10.11)

Considering these reaction schemes, the magnitude of cationization of cotton fabric would rely on the sodium hydroxide concentration, reaction temperature, reaction time, solvent used and liquor-to-goods ratio, as well as the method of application.

Studies (Hashem, 2006) have been conducted on combining cationization with cotton pretreatment processes. The results obtained showed that in the exhaustion method, the highest efficiency of scouring and cationization could be attained when the desized cotton fabric was treated with an aqueous bath containing sodium hydroxide (5 g/L) with CHPTAC (10 g/L) using a liquor-to-goods ratio of 20:1 at 80°C for 90 min. In the cold pad-batch technique, maximum scouring and cationization efficiency were observed at CHPTAC and sodium hydroxide concentrations equal to 100 and 49.3 g/L, respectively. The results further indicated that the cold pad-batch technique gave a higher fixation percentage (50%) than the exhaustion technique (25%), whereas scouring efficiency was comparable in the two cases. Another study showed that the cationic pretreatments often decrease colour-fastness to light (Kannan et al., 2006), but elsewhere it was reported that the light-fastness of some of the reactive dyes on cationized fabric samples was higher (Montazer et al., 2007). It may be said that the variation of light-fastness on cationization of the cotton substrate is dye specific.

Two cationic polymers were studied by Blackburn and Burkinshaw (2003):

1. A copolymer of diallyldimethylammonium chloride and 3-aminoprop-1-ene (PT1).
2. A copolymer of 4-vinylpyridine quaternized with 1-amino-2-chloroethane (PT2).

Both (PT1 and PT2) pretreatment polymers are highly substantive to cellulosic fibre. Because of their relatively low pKa values, the anionic carboxylic acid groups in the substrate are ionized at the pH values of the application (pH 6–7). Ion–ion interactions occur between cationic groups in the agent and the anionic carboxylic acid groups in the substrate. The other forces of attraction present are H-bonding and Van der Waals forces.

The ionic interaction between the positively charged nitrogen of PT1 and the hydroxyl group of cellulose is shown in Fig. 10.6. In the case of PT2, the interactions may be:

1. ion–dipole interactions between cellulose hydroxyl groups and pyridinium residues of PT2
2. Yoshida H-bonding between cellulose hydroxyl groups and pyridine residues in PT2 (Blackburn, 2006).

For cationized cotton, little or no salt is required during dyeing with reactive dyes. The nucleophiles in the polymer are more reactive than the cellulosic hydroxyl groups in the fibre. Consequently:

1. application may be done at neutral pH,
2. hydrolysis of dyes is minimized,
3. colour fixation is maximized,
4. there is less colour in effluent,
5. there is a reduced wash-off requirement,
6. it shows a significant reduction in operation time, and
7. there is a significant reduction in water consumption.

Aspects of physical, chemical, human safety, environmental safety and safe handling, storage, and disposal of CHPTAC have been examined and reviewed by Farrell (2012). In general terms, CHPTAC is a high-volume chemical (in excess of 1 million pounds produced per year in the United States) that is only slightly toxic to freshwater organisms

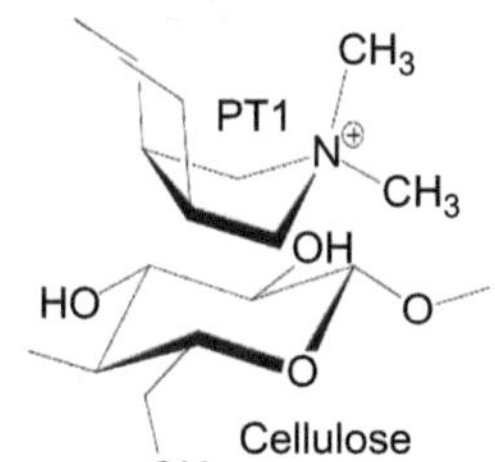

Figure 10.6 Cellulose-PT1 complex.

and has low levels of oral and dermal toxicity for humans. CHPTAC is not a skin or eye sensitizer. With a very low vapour pressure, inhalation risks are not expected. CHPTAC should be stored in sealed, lined, recommended containers. Owing to the unique transitioning chemical structures of CHPTAC to EPTAC and/or 2,2-dihydroxypropyltrimethylammonium chloride (DIOL) from CHPTAC, it is important to understand and specify test conditions (pH, temperature, purity and CHPTAC purity in test) for toxicity, mutagenicity, etc. On the other hand, EPTAC, one of the conversion products of CHPTAC, is a potent skin sensitizer and Carcinogen Class 2. Care and proper PPE should be exercised and used in the direct handling and/or processing of both CHPTAC and EPTAC. The half-life, degradability and toxicity to the environment all depend on the pH and temperature of the surroundings (water or soil) and the corresponding formation of EPTAC and/or DIOL.

10.11.4 Indigo dyeing

Indigo is a dark blue crystalline powder that melts at 390–392°C. It is insoluble in water, alcohol or ether but soluble in chloroform, nitrobenzene or concentrated sulphuric acid.

Indigo is among the oldest dyes to be used for textile dyeing and printing. Many Asian countries such as India, China and Japan have used indigo as a dye for centuries. The dye was also known to ancient civilizations in Egypt, Greece, Rome, Britain, Peru and Africa. Indigo is a subtropical shrub 1–2 m tall that possesses the woad pigment in the leaves. Leaves are crushed and then soaked in water for fermentation (removal of the sugar). World consumption of indigo in the 1800s was large, so in 1866 a German chemist named Adolph von Baeyer began his studies of the pigment and eventually elucidated its chemical structure so that it could be synthesized commercially. At the end of the 19th century, Germany was able to produce synthetic indigotine that was cheaper than the natural dyestuff (Indigo, 2016). Today, indigo retains its position as one of the world's most important industrial chemicals. The annual world production of indigo dye is about 80,000 tons (Ghaly et al., 2014)

Indigo is a vat dye (CI Vat Blue 1), which in its solid blue pigment form is practically insoluble in water but is made soluble by reducing with sodium hydroxide and sodium hydrosulphite.

Textile researchers in Tunisia claim to have developed an unconventional, eco-friendly exhaust dyeing process for precationized indigo denim fabrics that uses glucose dissolved in water as a reducing agent in the presence of alkali and an optimal temperature range (Ecotextile News, 2013).

It is only recently that investigation into the types of bacteria involved in indigo fermentation vats has been conducted. The best-characterized bacteria responsible for indigo reduction were first isolated from a woad vat prepared essentially as described in medieval Europe. In principle, and with modern microbiology techniques, bacteria-driven reduction of indigo and other vat dyes would offer a sustainable, eco-friendly alternative to current chemical methods of vat dye reduction. Indigo particles are at least 50 times the diameter of the bacterial cells and the way in which bacteria interact with the indigo to effect its reduction is as yet unknown. In terms of future developments, it is foreseen that the bacterial system will remain a small-scale process compared with the large-scale chemical and electrochemical systems for delivering prereduced indigo (Blackburn et al., 2009).

10.11.5 Polyester dyeing

Polyester fabric is usually dyed in pressurized dyeing machines (jet-dyeing machines) at 130°C. Economy can be achieved by reducing the dyeing time using selected optimized designs for the dyeing machines/processes or by using low-molecular weight disperse dyes (to avoid uneven dyeing during the shorter dyeing period) (Baldwinson, 1985); the use of compatible dyes that are adsorbed at the same rate is also important in dye selection.

To reduce dyeing times for polyester, dyeing should be started at the highest possible temperature and the rate of temperature rise should be the highest possible without hampering levelness. The critical temperature test has been developed in which visual estimations are made of the temperature at which 50% exhaustion of a specified amount of dye is obtained within 20 min. According to Clariant, the critical temperature zone for high-temperature dyeing is the temperature zone in which the last 80% of the dye is exhausted into polyester fibre (Clariant (India), 1983). Level dyeing can be obtained if the heating rate is controlled through this temperature zone.

In conventional jet-dyeing machines, the maximum heating rate is 2°C per min, whereas in rapid jet dyeing, the rate normally is 6°C per min, and in advanced machines such as the Hisaka and Callebaut de Blicquy Jet-Dyeing machines, 8–10°C per min.

Below the critical temperature, the temperature can be increased at a higher rate. In rapid jet-dyeing, the rate may be 6°C/min followed by about 2°C/min in the critical temperature zone. In conventional machines, the maximum heating rate may be 2°C/min, which is to be further reduced to 1–1.5°C/min in the critical zone.

To reduce the dyeing time still further, quick-fixable rapid-dyeing disperse dyes may be selected so that satisfactory diffusion and fixation may be achieved within 10–20 min in the rapid-dyeing machine compared with the much higher time (up to 50 min) by conventional dyes in conventional jet machines. The total dyeing cycle may be reduced to 50–70 min compared with180 min in conventional process (Sherrill et al., 1981).

10.11.6 Mass colouration

Mass colouration of polymers is a colouration method in which colourants are added to the polymer solution or melt before spinning into fibre, mostly used to colour polypropylene. It can usually be carried out in a controlled manner with minimal pollution and it yields shades with excellent light and wet fastness. However, it is a much less flexible process commercially and the colour range produced needs to be carefully chosen to avoid large stockholding, but it is viable for certain large volume shades such as blacks. Mass colouration was used to a limited extent with polyester and acrylics and is still largely the only method of producing fast colours on polypropylene fibres which are difficult to dye via conventional water-based dyeing methods. It has been demonstrated that it is possible to colour PLA fibres permanently during manufacture by using a catalyst containing a suitable chromophore for the polymerization process itself. The so-called DyeCat process (Macrae et al., 2011) enables even deep shades to be attained with good fastness properties, but again such a process will permanently be of interest permanently if much larger quantities of PLA fibres are produced.

10.11.7 Right-first-time dyeing

Acceptable colour or shade of the dyed material at the first attempt is called 'RFT dyeing', 'no addition dyeing' or 'blind dyeing'. Elimination of the inspection stage makes a significant savings, but the following objectives must be achieved in the dyed material. The successful practical dyer must achieve the following objectives (Park and Shore, 2009):

- The material must match the target colour (usually being colour constant or isometric under different illuminants);
- The colour must be applied uniformly and the processed material must have satisfactory colour fastness and physical properties;
- The cost of production must be within the allowed budget; and
- The material must be delivered on time.

The RFT dyeing concept was introduced in 1970 and became a desired feature of textile dyeing. This concept meant that at each dyeing, the target shade was achieved at the first attempt, and hence did not requiring redyeing. Originally it was termed 'no addition' dyeing if the colour was acceptable at the assessment stage, or 'blind dyeing' if colour assessment was carried out after the batch had been removed from the dyeing machine. However, the successful evolution of the concept depended on work carried out over many years by a relatively small number of organizations. Many application research and development projects were carried out mainly by the laboratory, pilot-plant and bulk-scale equipment of the major users and manufacturers of dyes and equipment, as well as by universities (Park and Shore, 2009).

Since the end of the 20th century, with increased competition, dye houses have been asked to meet more exacting requirements while also being under pressure to reduce cost. To stay competitive and be in business, they were required to exercise tighter quality control and seek ways to optimize dyeing necessitating better understanding of dyes, chemicals and substrates and their compatibilities, and the parameters that influence the rate and extent of dye uptake by the substrate (Park and Shore, 2007).

Continual refinement of textile machine design and its control equipment, together with dye selection and reliable recipe prediction systems in the laboratory all have contributed to enabling the achievement of a high proportion of RFT dyeing. Whether visual or instrumental methods are used, a standard operating procedure is required and this must include factors such as preparation, handling and measurement of samples; conditioning of samples for illuminant, humidity and temperature is essential. A number of factors which must be monitored or controlled to achieve RFT processing in the dyeing process have been identified (Park and Shore, 2004). Most of these factors, called assignable variables, are controlled directly by standard operating procedure.

A relatively few, known as random variables, such as water quality and substrate dyeability, require routine and regular testing. These variables can prevent dye houses from achieving a high level of RFT production in blind-dyeing techniques. The RFT philosophy is intended to provide higher quality with a quick response mode of operation and just-in-time concept, meaning the right product in the right place at the right time. RFT production can be obtained in manually operated dye houses.

Marks & Spencer (2011) identified key performance indicators (KPI's) which are essential to assess the capability and capacity of a mill to deliver to the Marks & Spencer standards for quality, safety and the environment. One of the most important KPI's for a dye house is the percentage of RFT dyeing, the number of batches produced that comply with the customer's standards without the need for any reworking, additions or reprocessing. Marks & Spencer uses RFT to determine whether a mill has the capability to supply standard products. Factories with low RFT performance figures will have:

- poor control of their internal processes
- higher than average costs of production
- unacceptable levels of water, energy and chemical use
- poor delivery records
- poor quality control

There are two stages in RFT: laboratory-to-bulk and bulk-to-bulk. The success of achieving the customer's standards for shade and fastness for the first bulk batch is indicated by the lab-to-bulk RFT performance. Success in achieving the customer's standards for subsequent bulk batches is indicated by the bulk-to-bulk RFT performance. The most experienced world-class dye houses can regularly achieve RFT for lab-to-bulk and bulk-to-bulk in excess of 90%.

The KPIs for sustainability that a mill will need to measure include water use, energy use, chemical use, production volumes per day, week and month and machine use.

10.11.8 Dyeing in ionic liquids

Nonaqueous systems can either reduce or completely eliminate the amount of water used in textile processing. Reducing the amount of water in a textile process provides environmental benefits as well as cost savings. Not only are water and energy conserved in a nonaqueous process, operating costs may be lower than in conventional processes. Research has shown that ionic liquids have the potential to be used in place of water in some textile processes (Anderson, 2010).

Ionic liquids (ILs) are liquids composed entirely of ions; they have low vapour pressure and thus exhibit little emission by evaporation at ambient temperatures (Scammells et al., 2005). Thus, molten sodium chloride is an IL whereas a solution of sodium chloride in water (a molecular solvent) is an ionic solution (Wenzel et al., 1997). Regardless of whether the solvent is polar or nonpolar, the common solvents (e.g., water, ethanol and benzene) are generally molecular liquids. ILs are composed of ions and when they are used as solvents, they behave differently from molecular liquids. Most ILs consists of nitrogen-containing organic cations and inorganic anions.

The most commonly studied systems contain phosphonium, imidazolinium or tricaprylmethyl ammonium cations, with varying hetero-atom functionality. The range of possible ILs is infinite; the number of ILs synthesized by 2009 was in excess of 1000 different compounds.

A possible alternative to using water as a solvent in the dyeing process of textiles is to use ionic liquids. The term 'room temperature ILs' has been assigned to organic salts that are liquid at ambient conditions. ILs have unique dissolving properties in the presence of organic compounds; therefore they can replace conventional solvents in many applications.

ILs have excellent stability over a wide temperature range and are neither volatile nor readily flammable, which gives them advantages when used in a production environment. Because of their nonvolatility, they are considered to have a low impact on the environment and human health. Another advantage to using ILs is that they have negligible vapour pressure at room temperature; therefore, they will not volatilize, which makes them recoverable for reuse with zero emissions of volatile organic compounds.

10.11.9 Dyeing in supercritical carbon dioxide

A supercritical fluid, a substance above its supercritical temperature and pressure (which for supercritical CO_2 ($scCO_2$) are 31.1°C and 73.8 bar, respectively, as shown in Fig. 10.7 (Cid et al., 2005), has the properties partway between a gas and a liquid. When a gas is compressed it becomes increasingly dense, and finally it is nearly as dense as a liquid, expanding to fill its container like a gas but with a density similar to that of a liquid. The practical application of supercritical fluids as dyeing media

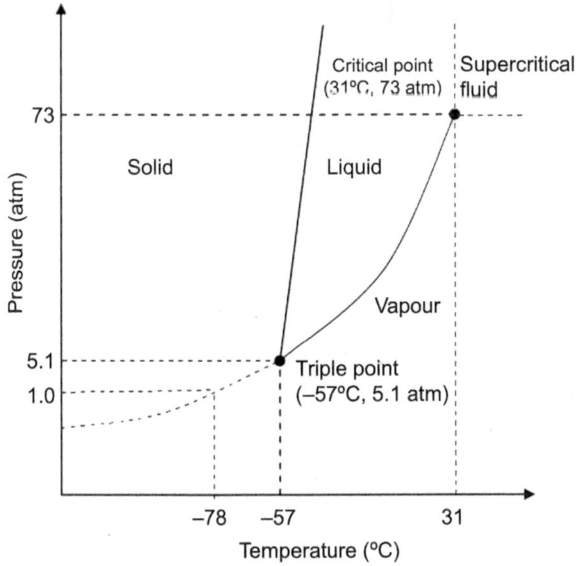

Figure 10.7 Temperature-pressure diagram of supercritical carbon dioxide.

received considerable attention from researchers in the last decade of the 20th century, as illustrated by the review by Bach and coworkers (Bach et al., 2002). Carbon dioxide has been the most widely used because of its convenient critical point, cheapness, chemical stability, nonflammability, stability in radioactive applications and nontoxicity. Owing to small viscosity of supercritical carbon dioxide and to high diffusion coefficient of dyestuff molecules in this condition, disperse dyestuff molecules easily penetrate into fibres, and as a result supercritical carbon dioxide dyeing provides a better dyestuff transportation compared with dyeing in water. The CO_2 used is a waste product of combustion, fermentation and ammonia synthesis, so it does not have to be specially produced for dyeing.

To allow solubilization of low vapour pressure compounds such as solid dyes, both the density and the temperature of the CO_2 need to be sufficiently high, on the order of 600 kg/m^3 and 120°C, respectively (Poliakoff and Licence, 2007). To obtain this set of conditions, the pressure needs to be around 300 bar, so that both the temperature and pressure are far above the critical point (31°C and 74 bar).

The main advantages (Poliakoff and Licence, 2007) of using $scCO_2$ instead of water in a dyeing process are:

- The easy separation of the CO_2 and the unused dye remaining after the dyeing process. Depressurization leads to precipitation of the excess dye and gives clean, gaseous CO_2, so that both compounds can be recycled and no waste is generated.
- After the dyeing process, the textile does not need any energy-intensive drying step as in the case of aqueous dyeing.
- Because $scCO_2$ is a nonpolar solvent, no dispersing agent is needed when nonpolar dyes are used, an additional advantage of $scCO_2$ in the disperse dyeing of polyester. This means that simpler dye formulations can be used than in aqueous polyester dyeing, in which the dispersing agent constitutes around 50% of the whole dye powder.
- One of the reasons for the superior economy of the supercritical process over the aqueous process is the higher rate of dyeing.

A further advantage of the use of supercritical carbon dioxide is its growing acceptance in processes other than textile dyeing:

- It is becoming an important commercial and industrial solvent owing to its role in chemical extraction in addition to its low toxicity and environmental impact. The relatively low temperature of the process and the stability of carbon dioxide also allow most compounds to be extracted with little damage or denaturing.
- The solubility of materials in supercritical carbon dioxide can be varied merely by changing the pressure of the gas. This opens up separation and purification methods that are not possible with conventional solvents. For example, a modest pressure of carbon dioxide can cause some dissolved compounds to separate out without the need for distillation, which could greatly simplify industrial-scale processes.
- $scCO_2$ is also revolutionizing the chromatographic purification of many active pharmaceutical ingredients because the high rates of diffusion of compounds in carbon dioxide make separation more effective than with conventional solvents.
- $scCO_2$ is nontoxic (except, of course, in a sealed environment); for example, it has been used for many years to decaffeinate coffee. It need not contribute to climate change, because the carbon dioxide used is a by-product from other processes, but there is a downside: compressing carbon dioxide to a supercritical form is energy-intensive (Poliakoff and Licence, 2007).

The applicability of the technology is currently limited by three factors:

- First, there is not enough knowledge on reactive and nonreactive dyeing in $scCO_2$.
- Second, because the process operates at conditions of typically 120°C and 300 bar, high-pressure equipment is needed, which requires high investment costs that could stand in the way of industrial implementation.
- Third, the batch size for dyeing is relatively small (maximum 200 kg) at present.

The supercritical dyeing process was investigated experimentally for both reactive and nonreactive dyeing and new equipment was designed to minimize the equipment and process costs (Kraan, 2005). Silk, wool, nylon and polyester were dyed successfully in $scCO_2$ with two nonpolar dyes containing vinylsulphone or dichlorotriazine reactive groups. It was found that the presence of a small amount of water increased the colouration of wool and nylon significantly and that silk only reacted with the dyes when water was added. Another interesting development was in the reactive dyeing of cotton used the supercritical carbon dioxide method to apply unsulphonated difluorotriazine dyes under mildly acidic conditions, to attain excellent fastness properties and 100% fixation (Cid et al., 2005). Polypropylene fibres are extremely difficult to dye by conventional methods because of their high crystallinity and nonpolar aliphatic structure, which does not contain reactive sites for dyestuffs in the molecule. The alternative mass-colouration method commits the textile manufacturer to particular colours at the earliest stage in production, so there is considerable interest in new approaches. One such method is dyeing with $scCO_2$ using dyes with long alkyl substituents. Such hydrophobic dyestuffs can be more readily dissolved in $scCO_2$ compared with aqueous dyeing systems, so the effect on the dyeability of polypropylene of the chain length of the alkyl groups attached to the colour-forming part of the dyestuff molecule was investigated using 1,4-bis(alkylamino)anthraquinone as the base dyestuff. For good colour-fastness and good colour strength to coexist, the optimum alkyl chain substituent for 1,4-bis(alkylamino)anthraquinone was found to be 8–12 carbon atoms in length (Michaels, 1995). Success in supercritical dyeing of polypropylene could have significant beneficial effects. Because of its low density (0.91 g/cm^3) compared with that of polyester (1.38 g/cm^3), as well as good elasticity and comparatively low cost, the greater bulk per unit weight of polypropylene leads to widespread use in applications that traditionally consume large quantities of fibre, such as carpets and nonwoven fabrics. Additional properties supportive to these and other applications include low moisture absorption, excellent chemical resistance to acids and alkalis, high abrasion resistance and resistance to insects and pests (Balasubramanian, 2012), so it is an important fibre in nonwoven processing and dominates in many nonwoven markets (Hegde et al., 2004).

A commercial-scale 1000-L supercritical dyeing machine was designed for treating 300 kg polyester, recycling all dye and 96% of the CO_2 (Kraan, 2005). An economic analysis showed that although the purchase cost for a supercritical dyeing machine is higher (€500,000) than for an aqueous machine (€100,000), the operating cost is lower (€0.35 instead of €0.99 per kg of polyester), because of the higher rate of dyeing and the simpler dye formulations that can be used in $scCO_2$, delivering a

50% lower process cost in total for the supercritical process compared with traditional methods.

DyeCoo Textile Systems B.V., based in the Netherlands, is believed to be the first company to apply the $scCO_2$ process successfully to the commercial dyeing of polyester fabric. It installed $scCO_2$ polyester dyeing machines in the Tong Siang plant in Thailand to dye fabric marketed under the name DryDye (Yeh Group) for Adidas and Nike (Forbes.com, 2013). Research is under way to apply the technology to other natural and synthetic fabrics.

Two industry leaders, Huntsman Textile Effects and DyeCoo Textile Systems, joined forces in October 2012 to develop and grow $scCO_2$ textile processing technology. The joint collaboration is set to create more sustainable products which will benefit the industry as a whole. By using carbon dioxide as the application medium, DyeCoo's innovative technology completely eliminates the use of water in the textile dyeing process. Huntsman Textile Effects is working with DyeCoo to develop and deliver innovative dye and chemical products to support the waterless dyeing process and obtain the high level of colour-fastness and performance that consumers demand. DyeCoo uses $scCO_2$ gas rather than water to infuse fabric with colour. Special temperature-controlled pressure chambers force the carbon dioxide to act as a fluid similar to water (the supercritical fluid CO_2), which causes the polymer fibre to swell, allowing the dispersed dye to diffuse easily within the polymer, penetrating the fibres and carrying the dyes into the fabric and causing perfect dyeing. Research is under way to apply the technology to other natural and synthetic fabrics (Huntsman, 2016).

For natural fibres the diffusion of $scCO_2$ is hampered by its inability to break the hydrogen bonds present in many natural fibres, including cotton, wool and silk. A further problem is that reactive dyes, direct dyes and acid dyes which are suitable for the dyeing of natural fibres are insoluble in $scCO_2$. Further studies might be conducted on methods to modify natural fibres or to develop a new fixation mechanism to improve the dyeing efficiency of natural fibres, because they are not easily dyed under supercritical conditions (Nalan, 2016).

A review (Banchero, 2012) shows that the supercritical dyeing of PET has achieved the correct commercial requirements. The dyeing of polyamide fibres such as nylon 6 and nylon 6,6 can also be successfully conducted in $scCO_2$, but the process is less simple than that of PET because it requires the use of specific reactive dyes and the addition of small amounts of water. Aramid, polypropylene and PLA textiles require additional studies to improve the fastness performance and mechanical resistance of the dyed product. Future investigations could involve the use of reactive dyes and swelling solvents to improve fastness, or heat-set textiles to improve the mechanical strength. Obtaining a good colouration efficiency of natural textiles, such as cotton, wool and silk, has been the main goal of the past decade. Cotton, in particular, is the most important natural fibre, but its colouration in $scCO_2$ is the most difficult task. It is clear that the only way to dye cotton is through ad hoc synthesized reactive disperse dyes and the concomitant presence of swelling and cosolvent auxiliaries.

10.12 Future trends

Most chemicals are used during wet processing, such as dyeing, washing, printing and fabric finishing. According to surveys, as much as 200 tons of water is consumed for every ton of textiles produced. Most of the chemicals used in textile production are non-hazardous, but a relatively small proportion of these chemicals are potentially hazardous.

Sustainable production and consumption can be achieved only if industrialists, retailers and consumers take ecological factors into account in every decision-making process. There has been an increasing awareness that many products could be produced under better conditions with greater respect for the environment: for example, by using less energy, attaining better yields, creating less pollution of the water or air and generating less waste and fewer (or no) unwanted by-products. There is an international set of standards under the International Standards Organization 14,000 series concerning the assessment and improvement of environmental management within industry (Boiral, 2007).

The challenges for sustainable textile processing are various:

1. To avoid using petroleum as the basic feedstock, the chemical industries need to find a renewable source of carbon; the only such renewable source of carbon is biomass derived from plants. However, biomass has a different chemical composition from oil and it contains more oxygen. Therefore, new reaction pathways will have to be developed to turn it into useful compounds efficiently. Approaches using enzymes show promise. Other problems with biomass as a resource for fuel or chemical feedstock include the seasonal nature of crops, the variability of biomass composition and possible conflicts arising from land use for biomass crops rather than crops for food supply.
2. Reactions that have minimum environmental impact, achieved, for example, by replacing conventional toxic metal-based catalysts with ecologically acceptable organic catalysts or enzymes, may be described as eco-friendly reactions.
3. Industrial processes should show maximum efficiency with reactions generating maximum yields to produce minimum waste.
4. Products with reduced toxicity and increased biodegradability should be substituted for environmentally harmful chemicals.
5. Flammable, toxic and volatile solvents with the potential to pollute the atmosphere need to be substituted by more eco-friendly and less hazardous solvents or by the use of solvent-free chemical syntheses.
6. Perhaps the greatest challenge facing the 'green chemist' is the eventual elimination of all environmentally harmful chemical products. In other words, when designing compounds for a particular application, synthetic chemists need to ensure low toxicity and rapid biodegradability while retaining the desired effect, but chemists are as yet unable to predict both the chemical and biological properties of compounds from their chemical structures.

As globalization of the wet-processing part of the textile industry continues, it is clear that there continues to be scope for it to continue to 'green' its processes and the chemistry involved. In almost every case of pollution, the fundamental problem for the textile industry is found to be low process inefficiency and a poor understanding of the life cycle of textile chemicals. The application of the principles of green chemistry and other aspects of clean technology will increasingly lead to more environmentally compatible manufacturing systems.

References

AAFA (American Apparel & Footwear Association). Restricted Substances List. https://www.wewear.org/industry-resources/restricted-substances-list/.

Albrecht, W., Reintjes, M., Wulfhorst, B., 1997. Lyocell fibers (Alternative regenerated cellulose fibers). Chem. Fibres Int. 47, 298.

Ali, H., 2010. Biodegradation of synthetic dyes–a review. Water Air Soil Pollut. 213 (1/4), 251–273.

ALKENA, 2012. Sichuan ALKENA Textiles Co., Sichuan, China. Available at: http://www.alkena.com/.

Anderson, K., 2010. Ionic Liquids: An Environmentally Friendly Alternative, 12/02/2010. www.techexchange.com.

Aubert, C., et al., 2009. Organic farming and climate change: major conclusions of the Clermont–Ferrand seminar. In: Agriculture biologique et changement climatique : principales conclusions du colloque de Clermont-Ferrand (2008).

Bach, E., Cleve, E., Schollmeyer, E., 2002. Past, present and future of supercritical fluid dyeing technology – an overview. Rev. Prog. Color. 32, 88.

Balasubramanian, N., 2012. Polyester vs. Polypropylene. ww.fibre2fashion.com.

Baldwinson, T.M., June 1985. The compatibility of disperse dyes for the rapid dyeing of polyester. Rev. Prog. Color. Relat. Top. 15 (1), 6–14.

Banchero, M., 2012. Supercritical fluid dyeing of synthetic and natural textiles – a review. Color. Technol. 129, 2–17.

Barber, A., Pellow, G., 2006. LCA: New Zealand merino wool total energy use. In: 5th Australian Life Cycle Assessment Society (ALCAS) Conference, Melbourne, 22–24 November.

Barreca, A.M., et al., 2003. Laccase-mediated oxidation of lignin model for improved delignification procedures. J. Mol. Catal. B: Enzym. 26, 105–110.

BASF, 2011. BASF Safety Data Sheet, Biodegradable Plastic, Product: ECOVIO* F FILM C2331. Revised on 11.9.11 Version: 2.0. Available at: http://www.bioplastics.basf.com.

Bastiol, C., 1997. Biodegradable materials. In: Brody, A.L., Marsh, K.S. (Eds.), Wiley Encyclopedia of Packaging Technology, second ed. Wiley, New York, p. 77.

Blackburn, R.S., 2006. New Green Chemical Techniques in Textile Coloration Processes, WUN Presentation (Power Point). Centre for Technical Textiles, University of Leeds, UK.

Blackburn, R.S., Burkinshaw, S.M., 2003. Treatment of cellulose with cationic, nucleophilic polymers to enable reactive dyeing at neutral pH without electrolyte addition. J. Appl. Polym. Sci. 89, 1026–1031.

Blackburn, R.S., Harvey, A., 2004. Green chemistry methods in sulfur dyeing: application of various reducing d-sugars and analysis of the importance of optimum redox potential. Environ. Sci. Technol. 38 (14), 4034–4039.

Blackburn, R.S., Bechtold, T., John, P., 2009. The development of indigo reduction methods and pre-reduced indigo products. Color. Technol. 125, 193–207. http://dx.doi.org/10.1111/j.1478-4408.2009.00197.x.

Blécourt, M.de, Lahr, J., Brink, P.J.van den, 2010. Pesticide Use in Cotton in Australia, Brazil, India, Turkey and USA. Alterra, Wageningen. http://www.icac.org/seep/documents/reports/2010_alterra_report.pdf.

Boiral, O., 2007. Corporate greening through ISO 14001: a rational myth? Organ. Sci. 18 (1), 127.

Bourbonnais, R., Paice, M.G., Freiermuth, B., Bodie, E., Borneman, S., 1997. Reactivities of various mediators and laccases with kraft pulp and lignin model compounds. Appl. Environ. Microbiol. 63 (12), 4627–4632.

Boyer, R.A., 1940. "Soybean protein fibers," experimental production. Ind. Eng. Chem. 32, 1549.

Broadbent, A.D., 2001. Basic Principles of Textile Coloration. Society of Dyers and Colourists, Bradford, England, p. 383.

C2ES, December 2015. Outcomes of the U.N. Climate Change Conference in Paris. The Center for Climate and Energy Solutions (C2ES). www.c2es.org.

Carliell, C.M., Barclay, S.J., Naidoo, N., Buckley, C.A., Mulholl, D.A., Senior, E., 1995. Microbial decolourisation of a reactive azo dye under anaerobic conditions. Water SA 21, 61–69.

Cavaco-Paulo, A., Gübitz, G.M. (Eds.), 2003. Textile Processing With Enzymes. Woodhead, Cambridge.

CCME (Canadian Council of Ministers of the Environment), 2012. What Is Smog? http://www.ccme.ca/assets/pdf/pn_1257_e.pdf.

Cho, H.J., Lewis, D.M., Jia, B.H., 2007. Improved reactive dyeing of wool with novel trifunctional reactive dyes, 03. Rev. Prog. Color. Relat. Top. 123 (2), 86–95. http://dx.doi.org/10.1111/j.1478-4408.2007.0006.

Chodák, I., 2009. Sustainable synthetic fibres: the case of poly(hydroxyalkanoates) (PHA) and other fibres. In: Blackburn, R.S. (Ed.), Sustainable Textiles: Life Cycle and Environmental Impact. Woodhead, Cambridge, UK, p. 88.

Choi, J.H., Kang, M.J., Yoon, C., 2005. Dyeing properties of soya fibre with reactive and acid dyes. Color. Technol. 121, 81.

CIBA, April 1996. Low salt Cibacron dyes for cotton. Colourage 43 (4), 73.

Cid, M.V.F., van Sprosen, J., van der Kraan, M., Veuglers, W.J.T., Woerlee, G.F., Witkamp, G.J., 2005. Excellent dye fixation on cotton dyed in supercritical carbon dioxide usingfluorotriazine reactive dyes. Green Chem. 7, 609–616.

Clariant (India), 1983. Dyeing with disperse dyes. Colour Chron, (July/September).

Clay, J., 2004. Cotton (Chapter 12). In: World Agriculture and Environment. World Wildlife Fund, Island Press, Washington, DC, USA, pp. 283–304.

Coleman, P., November 2012. Guide for Organic Crop Producers. National Center for Appropriate Technology (NCAT). https://www.ams.usda.gov.

Cooper, P., 1995. Colour in Dyehouse Effluent. Society of Dyers and Colourists, Bradford, UK.

Dartee, M., 2000. NatureWorksTM PLA: a sustainable performance fibre. In: Green-Tech 2000 Conference. Utrecht, The Netherlands.

Dodd, N., et al., 2012. Revision of the European ecolabel and green public procurement criteria for textile products. In: 1st AHWG Meeting for the Revision of the Ecolabel Criteria for Textile Products.

Dystar, 2013. Sera Eco Wash Resource Saving Wash-off Process for Reactive Dyeings in Exhaust Process. http://www.dystar.com/sera-eco-wash.cfm.

EC, 2002. Directive 2002/61/EC of the European parliament. Off. J. Eur. Communities L 243/15, 9/11/2002.

ECCJ (Energy Conservation Centre, Japan), 2007. Energy Saving Measures & Audit of Dyeing & Finishing Processes in Textile Factories. Available at: http://www.aseanenergy.org/download/projects/promeec/2007-2008/industry/.

Ecotextile News, 2013. Improvements in Denim Dyeing. Published online on Wednesday, 3 April 2013. http://www.ecotextile.com/.

Eksoy, 2012. Cold Pad Batch Dyeing. http://www.eksoy.com.tr.

ETAD, 1997. Handling Dyes Safely. (Washington, DC).

FAO, 2009. Why Natural Fibers. Available at: http://www.naturalfibres2009.org/en/iynf/sustainable.html.

Farrell, M.J., 2012. Sustainable Cotton Dyeing (Ph.D. thesis). North Carolina State University, Raleigh, North Carolina. European Union Risk Assessment Report, (3-Chloro-2-Hydroxypropyl)Trimethylammoniumchloride. 2008.

Farrington, D.W., Lunt, J., Davies, S., Blackburn, R.S., 2008. Poly(lactic acid) fibers. In: Deopura, B.L., Alagirusamy, R., Joshi, M., Gupta, B. (Eds.), Polyesters and Polyamides. Woodhead, Cambridge, UK. ISBN: 1 84569 298 5.

Fashion-incubator, 2010. Introduction to Digital Fabric Printing. http://fashion-incubator.com Posted on June 24.

Ferrigno, S., 2012. Sustainable Cottons. Available at: http://icac.org/.

Fiore, G.L., Jing, F., Young Jr., V.G., Cramer, C.J., Hillmyer, M.A., 2010. High T_g aliphatic polyesters by the polymerization of spirolactide derivatives. Polymer Chem. 1, 870–877. http://dx.doi.org/10.1039/c0py00029a.

Fischer, E., 1894. Einfluss der configuration auf die Wirkung der Enzyme. Ber. Dtsch. Chem. Ges. 27 (3), 2985–2993. http://dx.doi.org/10.1002/cber.18940270364.

Forbes.com, 2013. http://www.forbes.com/sites/davidferris/2012/08/30/nikeadidas-want-to-color-your-shirt-with-no-water/.

Galante, Y.M., Formantici, C., 2003. Enzyme applications in detergency and in manufacturing industries. Curr. Org. Chem. 7, 1399–1411.

Genencor, Huntsman, 2010. Genencor and Huntsman Announce LCA Results. Available at: http://primagreen.dupont.com/sustainability/lca_results/.

Ghaly, A.E., Ananthashankar, R., Alhattab, M., Ramakrishnan, V.V., 2014. Production, characterization and treatment of textile effluents: a critical review. J. Chem. Eng. Process Technol. 5, 1.

Goswami, P., Blackburn, R.S., Taylor, J., Westland, S., White, P., 2007. Dyeing behaviour of lyocell fabric: effect of fibrillation. Color. Technol. 123, 387. http://dx.doi.org/10.1111/j.1478-4408.2007.00113x.

GOTS (Global Organic TextileStandard), 2013. Available at: http://www.global-standard.org/.

Grose, L., 2009. In: Blackburn, R.S. (Ed.), Sustainable Textiles: Life Cycle and Environmental Impact, Chapter 3. Woodhead, Cambridge, UK, pp. 33–60.

Gupta, K.K., May 2009. Polyvinyl Alcohol Size Recovery and Reuse via Vacuum Flash Evaporation (Ph.D thesis). Gorgia Institute of Technology.

Gupta, K., Cook Fred, L., May-June 2012. Recovery and reuse of Polyvinyl alcohol from desize effluent streams via vacuum flash evaporation. AATCC Rev. (issue), 47–56. www.aatcc.org.

Gupta, B., Revagade, N., Hilborn, J., 2007. Poly(lactic acid) fiber: an overview. Prog. Polym. Sci. 32, 455.

Hage, R., et al., 1994. Efficient manganese catalysts for low-temperature bleaching. Nature 369, 637.

Hannemann, K., Runser, P., 1999. Metal-free reactive dyes meet strict Oeko-Tex standard. Colourage 46 (10), 33.

Hanu, 2010. Organic Textiles. Available at: http://dyeingworld1.blogspot.com/.

Hasan, M., Nabi, F., Mahmud, R., 2015. Benefits of enzymatic process in textile wet processing. Int. J. Fiber Text. Res. 5 (2), 16–19.

Hasanbeigi, A., September 2010. Energy-Efficiency Improvement Opportunities for the Textile Industry. Ernest Orlando Lawrence Berkeley National Laboratory, p. 10. LBNL-3970E www.energystar.gov/.

Hashem, M.M., 2006. Development of a one-stage process for pretreatment and cationisation of cotton fabric. Color. Technol. 122, 135–144.

Hashem, M., Smith, B., Hauser, P., 2003. Reaction efficiency for cellulose cationization using 3-chloro-2- hydroxypropyl trimethyl ammonium chloride. Text. Res. J. 73, 1017.

Hegde, R.R., Dahiya, A., Kamath, M.G., 2004. Olefin Fiber. http://web.utk.edu/ (Updated: April).

Hill, D.J., 1997. Is there a future for natural dyes? Rev. Prog. Color. 27, 18–25.

Ho, Y.C., Show, K.Y., Guo, X.X., Norli, I., Alkarkhi Abbas, F.M., Morad, N., 2012. Industrial discharge and their effect to the environment. In: Show, K.Y. (Ed.), Industrial Waste. Available from: http://www.intechopen.com/books/industrial-waste/industrial-emissions-and-theireffect-on-the-environment-. ISBN: 978-953-51-0253-3.

Holmes, I., 1998. Recent advances in chemical processing. Colourage 45 (annual), 41–56.
Holmes, I., February 2012. Right-first-time dyeing: the sustainable approach. ATA J. Asia Tex. Apparel (Web Version).
Huntsman, 2016. Waterless Dyeing for the Textile Industry. http://www.huntsman.com/corporate/a/Innovation/.
Hutmacher, D.W., 2000. Scaffolds in tissue engineering bone and cartilage. Biomaterials 21, 2529.
ICAC, 2012. Polyester Dominance to Hit Cotton Industry. Press release, Last Updated: 12 October 2012 http://www.commodityonline.com/news.
Indiamart, 2013. http://www.indiamart.com/unitop-projects-services/caustic-recovery-plant.html.
Indigo, 2016. http://www.primaryinfo.com/indigo.htm.
ISEND (International Symposium and Exhibition on Natural Dyes), 25–30 April 2011. La Rochelle, France. Available at: http://www.isend2011.com/.
ISIS, 2007. The Institute of Science in Society, Picking Cotton Carefully. ISIS Report 15/03/07 http://www.i-sis.org.uk/Picking_Cotton_Carefully.php.
Jensen Niels, P., 2000. Catalase enzyme. Text. Chem. Color. Am. Dyest. Rep. 32 (5), 23–24.
Kajita, T. and Inoue, R., 1940. US2198538, US2237832, US2192194 (Showa Sangyo, Japan).
Kannan, M.S.S., Gobalakrishnan, M., Kumaravel, S., Nithyanandan, R., Rajashankar, K., Vadicherala, T., 2006. Influence of cationization of cotton on reactive dyeing. J. Text. Appar., Technol. Manag. 5 (2), 1–16.
Karmakar, S.R., 1998. Application of biotechnology in the pre-treatment process of textiles. Colour. Annu. 45, 75–86.
Kehry, S., Strohle, J., July 2006. Cold pad batch dyeing and wash of knitwear. Int. Dyer 191 (6), 26–29.
Khatri, Z., Brohi, K.M., July 28, 2009. Environmental Friendly Textiles–A Road to Sustainability. http://susthub.blogspot.in.
Kolybaba, M., Tabil, L.G., Panigrahi, S., Crerar, W.J., Powell, T., Wang, B., 2003. Biodegradable polymers: past, present, and future. In: Paper Number: RRV03–0007, an ASAE Meeting Presentation. Available at: http://www.biodeg.net/.
Kraan, M.van der, 2005. Ph. D thesis at Technische Universiteit, Delft and J. Supercrit. Fluids. Process and Equipment Development for Textile Dyeing in Supercritical Carbon Dioxide, vol. 34, p. 99.
Lacasse, K., BaumannW, 2004. Textile Chemicals: Environmental Data and Facts. Springer, Berlin; New York.
Lancaster, M., 2002. Principles of sustainable and green chemistry. In: Clark, J., Macquarrie, D. (Eds.), Handbook of Green Chemistry and Technology. Blackwell Science, Oxford, UK.
Lange, N.K., 2000. Biopreparation in action. Int. Dyer 185 (2), 18–21.
LBNL (Lawrence Berkeley National Laboratory), 2007. China Energy Databook Version 7.0. China Energy Group. Available at: http://china.lbl.gov/research/china-energy-databook.
LCPC (Lebanese Cleaner Production Center), 2010. Cleaner Production Guide for Textile Industries. Beirut. www.lebanese-cpc.net/CP_Textile.pdf.
Leitat, 2013. Biotechnology and Microbiology in Textile Sector. Leitat, Terrassa, Bercelona, Spain. www.t-pot.eu/docs/Workshops/LEITAT_6_Biotechnology.pdf.
Li, Y., Hardin, I.R., 1998. Enzymatic scouring of cotton—surfactants, agitation, and selection of enzymes. Text. Chem. Color. 30 (9), 23–29.
Li, K., et al., 1999. Comparison of fungal laccases and redox mediators in oxidation of a non-phenolic lignin model compound. Appl. Environ. Microbiol. 65, 2654–2660.
Macrae, R.O., Pask, C.M., Burdsall, L.K., Blackburn, R.S., Rayner, C.M., McGowan, P.C., January 3, 2011. The combined synthesis and coloration of poly(lactic acid). 1/2011 Angew. Chem. Int. Ed. 50(1), 324. http://dx.doi.org/10.1002/anie.201006804.

Marks & Spencer, May 2011. Environmental & Chemical Policy for Textile Processing. Version 1.0. http://corporate.marksandspencer.com/documents/.

Martins, A., Reis, R.L., Neves, N.M., 2008. Electrospinning: processing technique for tissue engineering scaffolding. Int. Mater. Rev. 53 (5), 257.

Marzorati, M., et al., 2005. Selective laccase-mediated oxidation of sugar derivatives. Green Chem. 7, 310–315.

McCabe, R.W., Taylor, A., 2004. Synthesis of novel polyurethane polyesters using the enzyme *Candida antarctica* lipase B. Green Chem. 6, 151. http://dx.doi.org/10.1039/b400372c.

Miah, L., Ferdous, N., Azad, M.M., 2013. Textiles material dyeing with supercritical carbon dioxide (CO2) without using water. Chem. Mater. Res. 3 (5). www.iiste.org. ISSN: 2224-3224 (Print) ISSN: 2225-0956 (Online).

Michaels, D., July–September 1995. When science isn't enough: Wilhelm Hueper, Robert A. M. Case, and the limits of scientific evidence in preventing occupational bladder cancer. Int. J. Occup. Environ. Health 1 (3), 278–288.

Mitrus, M., Wojtowicz, A., Moscicki, L., 2009. Biodegradable polymers and their practical utility. In: Janssen, L.P.B.M., Moscicki, L. (Eds.), Biodegradable Polymers and Their Practical Utility: Thermoplastic Starch. WILEY-VCH Verlag GmbH & Co. KGaA, Weinheim. ISBN: 978-3-527-32528-3.

Montazer, M., Malek, R.M.A., Rahimi, A., December 2007. Salt free reactive dyeing of cationized cotton. Fibres Polym. 8 (6), 608–612.

Moore, S.B., Wentz, M., 2009. Chapter 10. In: Blackburn, R.S. (Ed.), Sustainable Textiles: Life Cycle and Environmental Impact. Woodhead, Cambridge, UK, pp. 214–229.

Moozyckine, A.U., Davies, D.M., 2002. Green S as a prototype for an environmentally-degradable dye: the concept of a 'green dye' in future Green Chemistry. Green Chem. 4, 452–458.

Morris, K.F., Lewis, D.M., Broadbent, P.J., June 2008. Design and application of a multifunctional reactive dye capable of high fixation efficiency on cellulose. Color. Technol. 124 (3), 186–194.

Mueller, S., Ruedel, H.D., Stemmel, W., 1997. Determination of catalase activity at physiological hydrogen peroxide concentrations. Anal. Biochem. 245, 50–60.

Nagaraj, A.R., 2012. Energy Management in Textile Industry. (e-newsletter) published on Tuesday, May 29. http://www.fibre2fashion.com/.

Nalan, D., 2016. Waterless Dyeing [Review]. Annals of the University of Oradea Fascicle of Textiles, Leatherwork, Rumania. http://textile.webhost.uoradea.ro/.

NIEHS, 2012. National Institute of Environmental Health Sciences. http://www.niehs.nih.gov/.

Nielsen, P.H., Kuilderd, H., Zhou, W., Lu, X., 2009. Enzyme biotechnology for sustainable textiles. In: Sustainable Textiles. Woodhead, Cambridge, pp. 113–138.

Oda, H., 2012. Improving light fastness of natural dye: photostabilisation of gardenia blue. Color. Technol. 128 (1), 68–73. http://dx.doi.org/10.1111/j.1478-4408.2011.00324.x.

Oecotextiles, January 19, 2011. https://oecotextiles.wordpress.com/2011/01/19/estimating-the-carbon-footprint-of-a-fabric/.

Oecotextiles, 2012. Textile Industry Poses Environmental Hazards. http://www.oecotextiles.com/PDF/textile_industry_hazards.pdf.

Oeko-tex, 2013. Limit Values and Fastness. https://www.oeko-tex.com/en/manufacturers/test_criteria/limit_values/limit_values.html.

OSPAR, 2013. http://qsr2010.ospar.org/en/media/content_pdf/ch00/Keyfindings_EN.pdf.

OTA (Organic Trade Association), 2010. Organic Cotton Facts. Available at: http://www.ota.com/organic/mt/organic_cotton.html.

OTA (Organic Trade Association), 2011. Organic Wool Fact Sheet. http://www.ota.com/organic/woolfactsheet.html.

Ouchi, A., Sakai, H., Oishi, T., Hayashi, T., Ando, W., Ito, J., 2003. Oxidative total chlorine free photochemical bleaching of cellulosic fabrics. Green Chem. 5, 516–523. http://dx.doi.org/10.1039/b304680j.

Ouchi, A., Obata, T., Oishi, T., Sakai, H., Hayashi, T., Ando, W., Ito, J., 2004. Reductive total chlorine free photochemical bleaching of cellulosic fabrics, an energy conserving process. Green Chem. 6, 198–205. http://dx.doi.org/10.1039/b315580c.

Park, J., Shore, J., 2004. Practical Dyeing. Society of Dyers and Colourists, Bradford, UK.

Park, J., Shore, J., 2007. Significance of dye research and development for practical dyers. Color. Technol. 123, 209–216 (J).

Park, J., Shore, J., June 2009. Evolution of right-first-time dyeing production. Color. Technol. 125 (3), 133–140. http://dx.doi.org/10.1111/j.1478-4408.2009.00194.x.

Pawar, S.B., Shah, H.D., Andhorika, G.R., 2002. Man-Made Text. India 45 (4), 133.

Peters, R.H., 1967. Textile Chemistry, vol. II. Elsevier, Amsterdam, Netherlands.

Phillips, D., Suesat, J., Taylor, J.A., Wilding, M., Farrington, D., Bone, J., Dervan, S., 2004. Color. Technol. 120 (5), 260.

Poliakoff, M., Licence, P., 2007. Green chemistry. Nature 450 (6), 810–812. http://dx.doi.org/10.1038/450810a.

Preša, P., Tavčer, P.F., January 2009. Low water and energy saving process for cotton pretreatment. Text. Res. J. 79 (1), 76–88, 13 pp.

Pűntener, A., Page, C., 2004. European Ban on Certain Azo Dyes. Available at: http://www.tfl.com (Updated on 5.1.04).

Ray, B.K., Reddy, B.S., June 2, 2008. Understanding Industrial Energy Use: Physical Energy Intensity Changes in Indian Manufacturing Sector. Indira Gandhi Institute of Development Research, Mumbai. http://www.igidr.ac.in/pdf/publication/.

Reddy, N., Yang, Y., 2005. Properties and potential applications of natural cellulose fibers from cornhusks. Green Chem. 7, 190–195.

Riva, S., 2006. Laccases: blue enzymes for green chemistry. Trends Biotechnol. 24 (5), 219.

Roy Choudhury, A.K., 2006. Textile Preparation and Dyeing, second ed. Oxford & IBH, New Delhi, and Science Publishers, New Hampshire, USA. The second edition published by The Society of Dyers and Colourists Education Charity, India Charity, India in 2010. www.sdc.org.in.

Roy Choudhury, A.K., 2013. Green chemistry and the textile industry. Textile Prog. 45 (1), 3–143. http://dx.doi.org/10.1080/00405167.2013.807601.

Roy Choudhury, A.K., 2014a. Sustainable textile wet processing: applications of enzymes, roadmap to sustainable textiles and clothing. In: Muthu, S.S. (Ed.), Textile Science and Clothing Technology Series. Springer Science+Business Media, Singapore, pp. 203–238. http://dx.doi.org/10.1007/978-981-287-065-0_7.

Roy Choudhury, A.K., 2014b. Coloration of cationized cellulosic fibers—a review. AATCC J. Res. (U. S. A.) 1 (3), 11–19 May/June 2014.

Rupp, J., November/December issue 2008. Ecology and Economy in Textile Finishing. Textile World.

Saravanan, D., Nalankiili, G., Ramachandran, T., 2008. Enzyme Inactivation in Textile Processing. Man-Made Text India, India, pp. 44–47.

Sayitgreen.com, 2010. Your Grandkids Will Thank You. http://www.sayitgreen.com/organic/fairtrade.html. Retrieved 7 December 2010.

Scammells, P.J., Scott, J.L., Singer, R.D., 2005. Ionic liquids: the neglected Issues. Aust. J. Chem. 58, 155–169.

Scott, A., 2015. Cutting Out Textile Pollution, Cleaning up one of the world's dirtiest industries will require new technology and more. Chem. Eng. News 93 (41), 18–19.

SDC, 2016. Colour Index International. The Society of Dyers and Colourists, UK. http://www.colour-index.com/.

Shaikh, M.A., 2010. Enzymes: a revaluation in textile processing. Pak. Text. J. 48–51.

Sherburne, V., 2009. In: Blackburn, R.S. (Ed.), Sustainable Textiles: Life Cycle and Environmental Impact, Chapter 1. Woodhead, Cambridge, UK, pp. 3–32.

Sherrill, W.T., Hipp, J.M., McCall, M.T., Maslen, S.H., 1981. Rapid dyeing concepts for polyester fabrics. In: Book of Papers, National Technical Conference of AATCC, p. 179.

Shore, J., 1991. Developments in direct dyes for cellulosic materials. Rev. Prog. Color. 21, 23–42.

Soares, G., Silva, R., Vieira, R., Santos, J., Gomes, J.R., 2005. How to Improve Dyeing and Antimicrobial Properties of Cotton Materials. Universidade do Minho, Portugal. Available at: http://www.repositorium.sdum.uminho.pt/.../4/IFATCC%2BFinal%2Btext.pdf.

Soroka, 2002. Fundamentals of Packaging Technology. Institute of Packaging Professionals. ISBN: 1-930268-25-4.

Sundar, P.S., Bhatoye, S.K., Karthikeyan, N., Prabhu, K.H., 2007. Enzyme applications in textiles. Indian Text. J. 117 (8), 25–31.

Tarhan, M., Sariisik, M., 2009. A comparison among performance characteristics of various denim fading processes. Text. Res. J. 79 (4), 301–309. http://dx.doi.org/10.1177/0040517508090889.

Taylor, J.A., 2000. Recent developments in reactive dyes. Rev. Prog. Color. 30, 93–107.

The Guardian, Tuesday 21, July 2015. EU Countries Agree Textile Chemical Ban. http://www.theguardian.com/environment/2015/jul/21/.

The Textile Institute, 2002. Textile Terms and Definitions, eleventh ed., Manchester, UK, p. 407.

Thiry, M.R., 2010. Color it greener. AATCC Rev. 10 (3), 32–39.

Thiry Maria, C., 2011. Moving in a greener direction. AATCC Rev. 33. Available at: http://www.aatcc.org.

TIFAC (DST), 2012. Recovery from Textile Industry Waste. Code No: TMS121 Technology Roadmaps, Govt. of India. http://www.tifac.org.in/index.php?option=com_content&view=article&id=675&Itemid=205.

Tufekci, N., San, H.A., Aydın, S., Ucar, S., Barlas, H., February 1998. Wastewater treatment problems in the operation of woven and knit fabric industry. Turk. J. Fish. Aquat. Sci. 7, 795–802.

Tüfekci, N., Sivri, N., Toroz, I., 2007. Pollutants of textile industry wastewater and assessment of its discharge limits by water quality standards. Turk. J. Fish. Aquat. Sci. 7, 97–103.

UCSAREP (University of California Sustainable Agriculture Research and Education Program), 2012. Available at: http://www.sarep.ucdavis.edu.

UNCTAD/WTO (International Trade Centre) and FIBL (Research Institute of Organic Agriculture), 2007. Organic Farming and Climate Change. ITC, Geneva.

UNEP, 1994. Cleaner Production in the Asia Pacific Economic Cooperation Region. United Nations Environment Programme Industry and Environment (UNEP), Paris, France.

U.S. DOE (United States Department of Energy), 2010. Manufacturing Energy Consumption Survey (MECS)-2006. Available at: http://www.eia.doe.gov/emeu/mecs/mecs2006/2006tables.html.

Visvanathan, C., Kumar, S., Han, S., 2000. Cleaner production in textile sector. In: Asian Scenario at National Workshop on "Sustainable Industrial Development through Cleaner Production" Held at Colombo, Sri Lanka on 12–13 November.

Wakelyn, P.J., Chaudhry, M.R., 2009. Organic cotton: production practices and post-harvest considerations. In: Blackburn, R.S. (Ed.), Sustainable Textiles: Life Cycle and Environmental Impact. Woodhead, Cambridge, UK, p. 231.

Wenzel, H., Hauschild, M., Alting, L., 1997. Environmental Assessment of Products. Methodology, Tools and Case Studies in Product Development, vol. 1. Chapman and Hall, London, UK.

White, P., Hayhurst, M., Taylor, J., Slater, A., 2005. In: Blackburn, R.S. (Ed.), Biodegradable and Sustainable Fibres. Woodhead, Cambridge, UK, p. 157.

Wikipedia, 2013. http://en.wikipedia.org/wiki/Emission_intensity.

www.extension.iastate.edu, 2012. Organic Flax Production in Iowa. Available at: www.extension.iastate.edu/Publications/PM2058.pdf.

www.natural-environment.com, 2008. What Is Organic Linen? Available at: http://www.natural-environment.com/blog/2008/01/24/what-is-organic-linen/.

Zabaniotou, A., Kassidi, E., August 2003. Life cycle assessment applied to egg packaging made from polystyrene and recycled paper. J. Cleaner Prod. 11 (5), 549–559. http://dx.doi.org/10.1016/S0959-6526(02)00076-8.

Further reading

Shore, J., 2002. Colorants and Auxiliaries, vol. 1, second ed. Society of Dyers and Colourists, Bradford, UK.

Sustainable wastewater treatments in textile sector

11

Ponnusamy Senthil Kumar, Anbalagan Saravanan
SSN College of Engineering, Chennai, India

11.1 Introduction

Water pollution is a major environmental threat throughout the world. Water is an important source for all living beings in the ecosystem. Effluent from the textile industry is the foremost contribution to water pollution. Textile effluents are rich in chemical essences and a variety of dyes (Liu et al., 2011). In this context, dye is considered to be one of the most troublesome components in textile wastewater treatment because of its complex chemical structure. Chemical additives in textile effluent consist of a high concentration of organic and inorganic compounds (Bes-Piá et al., 2002). To minimize water scarcity and water contamination, different conventional methods have been employed in wastewater (domestic and industrial) treatment.

11.1.1 Types of textiles industries and their operations

In India, the textile industry is one of the most important types of manufacturing in the economy. Different types of textiles are created, such as textile fibres, including natural fibres such as cotton, jute, silk and wool; synthetic/man-made fibres such as polyester, viscose, nylon, acrylic and polypropylene and multiple blends of such fibres and filament yarns such as partially oriented yarn (Holkar et al., 2016). The type of yarn used is dictated by the end product that is manufactured. In the textile industry, fibre is prepared in a series of steps. Some of the unit operation steps in the wet processing of textile fabrics are shown in Fig. 11.1.

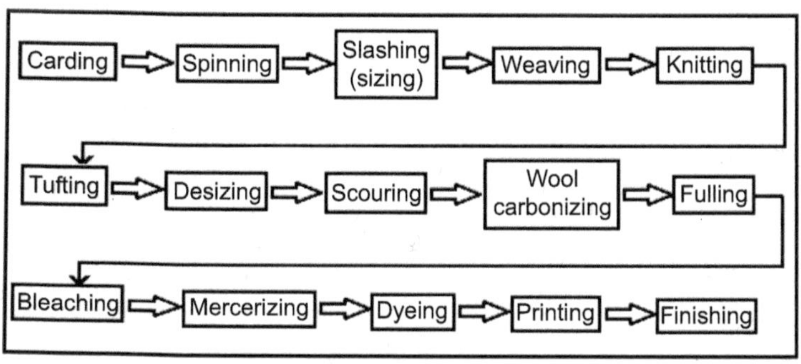

Figure 11.1 Flow diagram of different unit operation steps in fabric.

Sustainable Fibres and Textiles. http://dx.doi.org/10.1016/B978-0-08-102041-8.00011-1

Table 11.1 **Standards for effluent from textile industry**

S. No	Parameter	Standards
1	pH	5.5–9.0
2	Total suspended solids	100 mg/L
3	Biochemical oxygen demand	30 mg/L
4	Chemical oxygen demand	250 mg/L
5	Total residual chlorine	1 mg/L
6	Oil and grease	10 mg/L
7	Total chromium	2 mg/L
8	Sulphide	2 mg/L
9	Phenolic compounds	1 mg/L

GPM, gallons per minute; *MGD*, millions of gallons per day.

11.1.2 Textile effluent: permissible limits

Because of concerns about environmental pollution, the textile industry must develop on-site or in-plant treatment plants to treat effluents before their discharge into water. Based on effluent treatment, the Department of National Environmental Protection from the Central Pollution Control Board determined permissible limits for effluents for the textile industry (Paul et al., 2012). Standards for effluent from the textile industry are listed in Table 11.1. In developed countries, state and federal agencies in the United States have fixed lower effluent colour limits (<200 units according to the American Dye Manufacturers Institute) (McCurdy et al., 1992). Similarly, in Egypt, environmental protection agencies promoted the prevention of transferal of effluent disposal problems from one place to another.

11.1.3 Discharge of textile wastewater into water streams

Textile industries consume a large amount of chemicals and water for manufacturing. After the product is manufactured, untreated effluent from the textile industries is directly discharged into water streams. Dye constituents in the effluent are highly variable because of the different types of clothes, including different structural varieties of dyes such as acidic, basic, metal complex, azo, diazo and reactive dyes. These effluents are typically highly coloured, usually alkaline, and contain higher amounts of biochemical oxygen demand (BOD) and chemical oxygen demand (COD) and a variety of toxic contaminants such as suspended solids, dyes, acids, bases, salts, surfactants, chlorinated compounds, oxidizing and reducing agents (Mahmoued, 2010; Badani et al., 2005).

11.1.3.1 Colour

Dye is the major constituent in textile colouring effluent. Different types of dyes are used in the textile industry, based on the characteristics of the fibre. During the dyeing process, most of the dye is exhausted in the fabric material; the unfixed dye is discharged into the wastewater and creates a coloured solution (Santos and Boaventura, 2015). Generally, textile dyes are classified into three categories: (1) anionic, (2) cationic and (3) nonionic. Major anionic dyes are direct, acid and reactive. Cationic dyes are azo basic, anthraquinone and reactive dyes, and major nonionic dyes are dispersing. Among these, cationic dyes are considered to be carcinogenic; they are made from benzidine and other aromatic compounds. The most problematic dyes are the reactive and acid ones that are brightly coloured and water soluble; they cannot be removed using conventional treatment methods. Similarly, organic dyes in effluents have high photolytic stability and higher chemical compounds. Discharge of dyes in the water streams consists of a low concentration of dyes, but it is highly visible because of the staining properties. Removal of colour from the effluent is an important environmental problem for the following reasons:

1. The effluent is toxic, carcinogenic and mutagenic, which leads to several health effects.
2. Generally, dyes have a complex aromatic structure.
3. They have low biodegradability, which affects sunlight penetration and photosynthesis.
4. They are recalcitrant in nature.

11.1.3.2 Dissolves solids

Textile effluent is heavily contaminated with pollutants such as dyes, dissolved solids, suspended solids and toxic metals. The main factor to be considered in textile effluent is total dissolved solids (TDS). Because of the use of common salt and Glauber salt, the level of TDS increases in textile wastewater. The direct discharge of textile effluents may increase the level of TDS in ground water and surface water (Islam and Guha, 2013). The presence of TDS (high or low concentration) in water may affect the osmotic balance, causing swelling or dehydration in aquatic organisms and a change in taste. The quality of irrigation of water depends on the salt content. When the salt content increases, it contaminates the water and makes it unsuitable for domestic, industrial and agricultural use.

11.1.3.3 Suspended solids

Suspended solids seem to be a major pollutant in textile effluents. Suspended solids can form by combining with an oily scum such as oil, grease, clay, silt and gritty materials. The concentration of suspended solids in wastewater has greatly affected the aquatic environment, interfering with the oxygen transfer mechanism in the air–water interface. In an aquatic system, suspended solids are capable of choking the breathing organ of fish, leading to slowed growth and disease resistance in the fish (Ghaly et al., 2014). Moreover, suspended solids do not allow light penetration into the aquatic environment, which reduces the ability of algae to produce food and oxygen. In addition, suspended solids absorb heat from sunlight, which increases the water temperature and simultaneously decreases the level of dissolved oxygen.

11.1.3.4 Toxic metals

Heavy metals are natural components; they cannot be destroyed or degraded. The substantial concentration of heavy metals in textile effluent is a worldwide environmental problem. These effluents are directly discharged into the surface water, ultimately contaminating both surface and ground water by leaching. Generally, a small amount of heavy metals such as copper, cadmium, chromium, mercury, zinc and manganese is essential to maintain the metabolism of the human body (Sarker et al., 2015). At higher concentrations, heavy metals can cause toxic health effects to humans and other aquatic organisms; they can enter the human body through drinking water, food and air. The Bureau of Indian Standards and the US Environmental Protection Agency have determined the maximum permissible limit of metals in drinking water (BIS, 1992; US Environmental Protection Agency (USEPA), 1990).

11.1.4 Toxic threats

In the textile industry, various types of dyes are used to manufacture several textile products. The textile industry also employs several chemicals during the production process. Textile effluents consist of several synthetic dyes and toxic chemicals containing acids, sulphur, alkalis, naphthol, nitrates, hydrogen peroxide, surfactant-dispersing agents and toxic heavy metals such as Cu, Cr, Cd, Zn, Ni, As and Pb, which are directly discharged into water streams. The textile effluents which flow in sewers clog the pipes, resulting in a change in the quality of drinking water and making it unfit for human consumption. This textile effluent–contaminated water is not biodegradable and carcinogenic and poses a major threat to human health and the environment (Kant, 2012). The wastewater can cause haemorrhage, dermatitis, ulceration of skin, nausea, risk of bladder cancer, vomiting gastritis, lacrimation, skin irritation and dermatitis in humans; it may also be toxic to autotrophic organisms.

11.1.5 Conventional methods

Different treatment methodologies have been employed to treat textile wastewater, such as adsorption, ion exchange, membrane filtration [reverse osmosis (RO), ultrafiltration (UF) and nanofiltration (NF)], ozonation, evaporation (multiple-effect evaporation, mechanical vapour compression and direct contact evaporation) electrochemical oxidation, flocculation, phytoremediation, photochemical and crystallization. Among these different methods, the selection of a suitable methodology for effective treatment depends entirely on several categories; the lists are shown in Fig. 11.2. Technologies performed for textile wastewater treatment are mainly divided into three categories: (1) separation and concentration processes, (2) decomposition and degradation processes and (3) exchange processes (Jegatheesan et al., 2016).

11.2 Wastewater treatment in textile industry

11.2.1 Process of wastewater treatment

Textile industry effluent consists of dyes, chemicals and sizing materials. These effluents which are directly discharged into water sources contaminate water streams

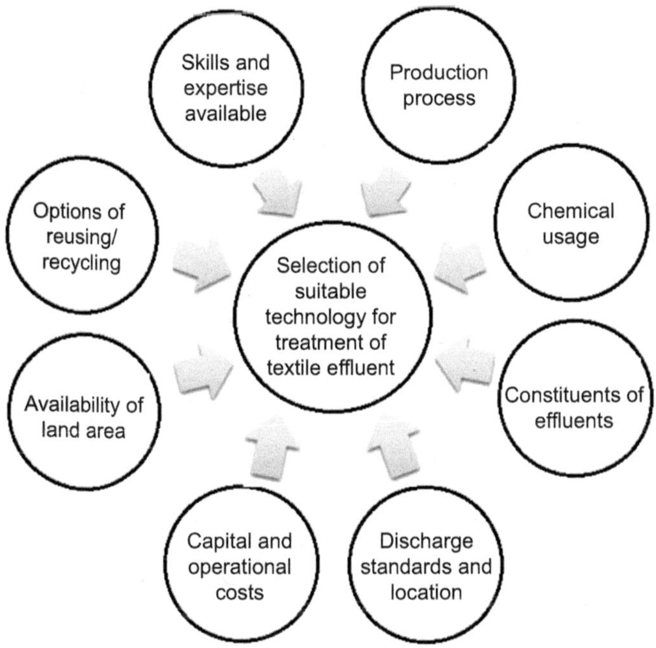

Figure 11.2 Selection criteria of suitable methodology for effective treatment of textile effluent.

and cause environmental problems and health issues for humans. Therefore, the effluent must be treated before discharge from the industry. Generally, there are three mechanisms for wastewater treatment: (1) physical, (2) chemical and (3) biological (Kumar and Bhat, 2012). These three mechanisms are performed together in a treatment plant, as shown in Fig. 11.3. Wastewater treatment plants are classified into three groups:

1. primary treatment
2. secondary treatment
3. tertiary treatment

11.2.1.1 Primary treatment

In the treatment of textile effluent, the first step is to remove floating and settleable materials, i.e., suspended solids, organic matter, excessive quantities of oil and grease and gritty materials (physical and chemical). In primary treatment, textile effluent undergoes screening for coarse suspended particles such as yarn, lint, pieces of fabric, fibres and rags, using bars and fine screens. The screening process is carried out by sedimentation, which uses gravity to settle suspended particles. In addition, effluent from the textile industries undergoes mixing and equalization to ensure that effluent possesses uniform characteristics in terms of the pollution load, pH and temperature and that the acidic content of the effluent is reduced. The pH

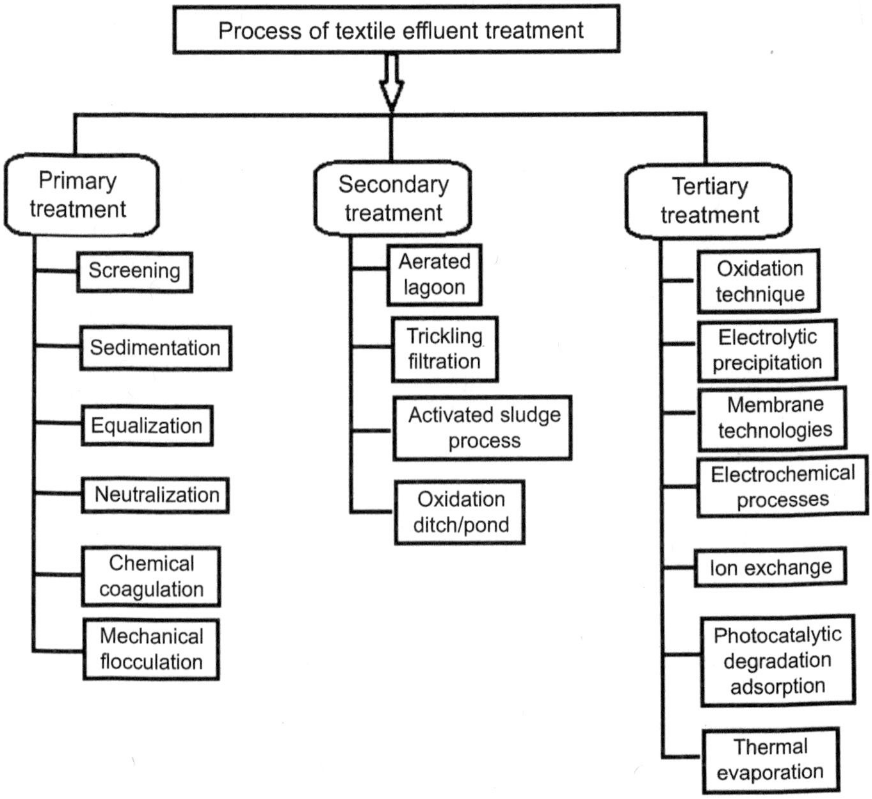

Figure 11.3 Process of textile effluent treatment.

of the effluent is considered to be an important parameter in the treatment process; it must be in the ultimate range of 5.0–9.0. The pH value of the textile effluent is altered by adding chemicals such as sulphuric acid and boiler flue gas (Babu et al., 2007). By performing chemical treatment, a substantial reduction in BOD and COD is found in the effluent. Coarse screening is performed through two types: (1) fine screening (1.5–6 mm) and (2) very fine screening (0.2–1.5 mm). Then the screened effluents are allowed to settle for the removal of suspended particles. The main disadvantage of the sedimentation process is that it is not capable of removing colloidal particles from the effluent because the colloidal particles carry some charge on their surface. By adding chemicals such as ferric sulphate, ferric chloride, lime, alum and ferrous sulphate, the surface property of the colloidal particles changes and can easily cluster together, followed by settling as sludge (Chen et al., 2005). Mechanical flocculation is another physical process which helps mix the effluent using paddles. By mixing, the small particles clump together, are converted into heavier particles and settle. The settled particles are collected as sludge (Tripathy and De, 2006). The main disadvantage in both coagulation and flocculation is disposal of the sludge.

11.2.1.2 Secondary treatment

Different technologies have been employed in secondary treatment, such as aerated lagoon, trickling filtration, an activated sludge process and oxidation. The biological mechanism has a significant role in secondary treatment; it is mainly associated with reducing the colour, BOD, COD, oil, grease and phenol contents in the effluent. Dissolved organic matter is generally oxidized by aerobic bacteria, which uses the organic matter as an energy source and nutrients (Rasheed et al., 2014). Nitrogenous and dissolved organic matter can be converted into ammonia, methane, CO_2 and water.

Aerated lagoon is one of the most commonly used biological treatments. It consists of a large tank. Effluent from primary treatment undergoes aeration for about 2–6 days in aeration tanks and is completely mixed. The wastewater is treated either on a flow-through basis or with solids recycling. BOD, total suspended solids and phosphorus removal can be achieved in aerated lagoons (Chavan, 2001). The main disadvantage of this technique is that it is expensive, it requires a large space and there is a risk of bacterial contamination.

A trickling filter is another common method for removing organic matter from effluent. It mostly operates under aerobic conditions. Effluent from primary treatment is sprayed onto the filter medium. The medium usually consists of stones, polyvinyl-chloride, synthetic resins, coal and gravel (Etter et al., 2011). The surface of the filter medium coated with microorganism oxidizes the organic matter in the effluent into CO_2 and water. Main disadvantages are vector and odour problems, high clogging and the need for additional treatment. The trickling filter method is cost-effective.

The activated sludge process is the most effective method for treating textile dyeing effluent. Activated sludge has strong decomposition characteristics and adsorbs toxic organic compounds (Yasui et al., 1996). Activated sludge is mainly composed of many microorganisms. The main advantage is that there is no sludge disposal problem.

Oxidation is the most commonly used chemical decolouration process. Generally, in textile effluent, colour removal from several dyes and toxic organic compounds is complicated. In the oxidation process, different oxidizing agents such as free hydroxyl radicals, ozone and hydrogen peroxide are used effectively. These oxidizing agents are used with catalysts which oxidize the dye and toxic organic compounds (Kestioglu et al., 2005). The main disadvantages of this process are flocculation, the generation of sludge and the transfer of impurities from wastewater to sludge.

11.2.1.3 Tertiary treatment

Tertiary treatment reduces organic colour compounds and removes specific contaminants. The most commonly used methods of tertiary treatment are electrolytic precipitation, membrane technology, electrochemical processes, ion exchange, photocatalytic degradation, adsorption and thermal evaporation (Amar et al., 2009).

11.2.2 Advance methods for treatment

Some advanced treatment processes may require the textile effluent to be treated, which enables the treated wastewater to be reused in the manufacturing process.

The recovery and reuse of water from textile effluent streams totally depends on the proper selection and application of advanced treatment methods.

11.3 Pollution loads for various textile materials

11.3.1 Textile printing and dyeing process

The primary sources of wastewater in the textile industry process are the rinse water and dye bath. In the manufacturing process, more than 50% of water is consumed in dyeing and finishing processes, which are responsible for 80% of wastewater generated. In the textile process, around 8000 chemical products are used which are associated with the dyeing process. These dyes include different varieties of dyes such as acidic, basic, reactive, azo, diazo and metal-complex dyes. Effluent streams from the textile dyeing process are highly toxic and coloured, and may contain suspended solids, and organic and inorganic compounds.

11.3.2 Textile materials and their classification

In the textile industrial process, processes that use water are desizing, scouring, bleaching, mercerizing, dyeing, washing and neutralizing. Their pollutants in wastewater are shown in Table 11.2. Different types of materials are used and processed in textile operations, such as cotton, wool, nylon, acrylic, polyester, viscose and acetate. Textile operations vary greatly in their water consumption. Wool and felted fabric processes are more water-intensive than other processing subcategories such as wovens, knits, stock and carpets (Correia et al., 1994).

11.3.3 Textile fibre

Textile fibres are classified into two categories: (1) natural and (2) man-made.

Natural fibres are greatly elongated substances produced by plants and animals that can be spun into filaments, thread or rope. Woven, knitted, matted or bonded, they form fabrics that are essential to society. Natural fibre is divided into two groups: (1) animal and (2) vegetable fibres.

Man-made fibre is divided into two groups: (1) organic polymer and (2) inorganic polymer.

11.3.3.1 Animal fibre

Animal fibres are generally made up of proteins. They include wool, hair and secretions such as silk.

Wool

Wool is a fibre that has been traditionally used in the textile industry; it is commonly obtained from sheep. Wool fabric is soft to the touch and provides warmth, so it is the preferred for winter apparel. Wool has other features such as elasticity and good drape.

Table 11.2 Textile industry operation process and its pollutants in wastewater

S. No	Unit process	Possible pollutants in wastewater	Nature of wastewater
1	Desizing	Starch, glucose, carboxymethyl cellulose, polyvinyl alcohol, resins, fats and waxes	High BOD (35%–50% of total); carboxymethyl cellulose and polyvinyl alcohol do not exert high BOD
2	Bleaching	Hypochlorite, chlorine, caustic soda, hydrogen peroxide, acids, etc.	Alkaline, constitutes approximately 5% of BOD
3	Mercerizing	Caustic soda	Strongly alkaline, low BOD (<1% of total)
4	Dyeing	Various dyes, mordants and reducing agents such as sulphides, hydrosulphide, acetic acid and soap	Strongly coloured, fairly high BOD (6% of total)
5	Printing	Colours, starch, gums, oil, china clay, mordants, acids and metallic salts	Highly coloured, fairly high BOD and oily appearance
6	Finishing	Traces of starch, tallow, common salt, Glauber salt, etc.	Slightly alkaline, low BOD

BOD, biochemical oxygen demand.

Moreover, it can easily be dyed in different colours, which makes it suitable for use in fashionable winter apparel.

Speciality hair fibres

In the textile industry, the most commonly used hair fibres are goat and the camel family fibres. These fibres have desirable properties such as a fine diameter and natural lustre, and the ability to impart pleasing and handle to fabrics. Special hair fibres obtained from the goat family include mohair from the Angora goat, and cashmere. Fibres obtained from animals of the camel family include camel hair, mainly from the Bactrian camel, and guanaco, llama, alpaca and vicuna fibres, all of which are from members of the genus *Llama*.

Silk

Silk is a natural fibre used in the textile industry. It is obtained from silkworms. The most popular kind of silk is obtained from the mulberry silkworm. Silk obtained from other varieties of silkworms are called wild silk. China, India, Nepal and Europe have been traditional producers of good-quality silk on a large scale. Silk fibre has a unique sheen. It is smooth to the touch and is strong. These qualities make silk a favourite fabric for saris and dress materials. Apart from this, silk is also used in nightwear, bed linen and underwear as well as home furnishings.

11.3.3.2 Vegetable fibre

Cellulose is the main component of vegetable fibres, which can be effectively used to manufacture paper and cloth. Examples include cotton, jute, flax, ramie, sisal and hemp. Vegetable fibres can be divided into the following groups: seeds, leaves, skins, fruit and stalk fibres.

Cotton fibres

Cotton fibre is obtained from the cotton plant. It is a traditional fibre used in the textile industry. It is one of the most preferable fibres because cloth made from it is durable and has good drape. Moreover, it is moisture-absorbent and smooth to the touch. One quality of cotton fabric is that it takes time to dry. It also creases easily and requires regular ironing.

Linen or flax fibre

Linen fabric is obtained from the flax plant. It is a fibre that has been used in the textile industry for ages. The properties of linen are quite similar to those of cotton. Like cotton, linen is also highly moisture-absorbent and durable. It easily creases and requires ironing. However, it is stiffer than cotton. Linen is usually used to manufacture summer clothes and home linens.

Jute fibre

Jute is a natural fibre that has been used in the textile industry for centuries. It is obtained from the jute plant and is popularly known as golden fibre because of the golden sheen it possesses. Because of its high strength, it is perfect for use in packaging material. Jute is sometimes blended with other fabrics or even used individually to produce apparel.

11.3.3.3 Organic polymer

Organic polymers have an important role in the textile industry. The solid parts of all plants are made up of polymers including cellulose, lignin and various resins. Cellulose is a polysaccharide, a polymer composed of sugar molecules. Lignin consists of a complicated three-dimensional network of polymers. Organic polymers can be classified into polyamide, polyester, polyacrylonitrile and polypropylene fibre.

11.3.3.4 Inorganic polymer

Many inorganic polymers are found in nature, including diamond and graphite. Both are composed of carbon. In diamond, carbon atoms are linked in a three-dimensional network that gives the material its hardness. In graphite, which is used as a lubricant and in pencil 'leads', carbon atoms link in planes that can slide across one another. Inorganic polymer is categorized into two parts, such as in glass and metal fibre.

11.4 Laws for wastewater textile pollutants in various countries

11.4.1 Textile wastewater disposal standards

The textile industry is one of the largest and oldest industrial sectors throughout the world. It has undergone modernization incorporating the latest developments in textile technology. Wet processes in the textile industry require large volumes of water of a high degree of purity and the discharge of equally large volumes of liquid waste which are complex and highly variable in regard to both their quantity and characteristics. The wastewaters generally discharge into the water streams with or without treatment, or on land. The ultimate disposal of the waste along with sewage may be into streams or the sea or on agricultural land. Wastewater from smaller textile processing units are small in volume but they carry a significant pollution load and should be discharged only after treatment.

11.4.1.1 Textile industry standards for water pollutants in China

China's water pollution has been on the rise as a result of rapid urbanization and industrialization. In this context, the textile industry has had a major role in water pollution. The textile industry consumes high amounts of water and threatens the quality of water in China. The latest discharge standard serves as a way to enforce China's environmental protection law; it encourages the improvement of production methods used in the textile dyeing and finishing industry and of pollution control technologies. Emission standards for different indicators in textile industry standards for water pollutants in China are shown in Table 11.3.

Table 11.3 Emission standards for different indicators in textile industry standards for water pollutants in China

S. No	Parameters	Limit value	
		Direct discharge	Indirect discharge
1	pH	6.0–9.0	6.0–9.0
2	Chemical oxygen demand	80 mg/L	200 mg/L
3	Biochemical oxygen demand	20 mg/L	50 mg/L
4	Total suspended solids	50 mg/L	100 mg/L
5	Colour	50	80
6	Ammonia nitrogen	10 mg/L	20 mg/L
7	Chlorine dioxide	0.5 mg/L	0.5 mg/L
8	Adsorbable organically bound halogens	12 mg/L	12 mg/L
9	Sulphide	0.5 mg/L	0.5 mg/L
10	N total	15 mg/L	30 mg/L
11	P total	0.5 mg/L	1.5 mg/L

Table 11.4 **Effluent discharge limits for textile effluent in Germany**

S. No	Parameters	Limit value at point of discharge
1	Chemical oxygen demand	200 mg/L
2	Biochemical oxygen demand	20 mg/L
3	Total nitrogen	20 mg/L
4	Total phosphorus	3.0 mg/L
5	Total hydrocarbons	10 mg/L
6	Adsorbable organically bound halogens	0.5 mg/L
7	Nitrite	1.0 mg/L
8	NH_3-N	10 mg/L

11.4.1.2 Textile industry standards for water pollutants in Germany

The textile industry is one of the most important sectors in Germany. Textiles and clothing are the second largest consumer goods market after the food and beverage industry. The emission of organic and inorganic water pollutants from the textile industry causes several toxic health problems for humans as well as water pollution. Therefore, stringent environmental regulations have been enacted. Emission standards for different indicators in the textile industry for water pollutants in Germany are shown in Table 11.4.

11.4.1.3 Textile industry standards for water pollutants in the United States

In the United States, textile manufacturing is part of the manufacturing sector. This sector consists of establishments engaged in spinning natural and man-made fibres into yarn and thread. These are then woven or knitted into fabric, and the fabric is dyed and finished. In the United States, the Federal Environmental Protection Agency has set certain limits on the disposal of effluents into the environment; these are listed in Table 11.5.

11.4.1.4 Textile industry standards for water pollutants in India

The textile industry is one of the largest segments of the Indian economy, based on its contribution to industrial production, employment and experts. Textile manufacturing effluents are discharged directly into water streams. As a result, the water is unfit for drinking and domestic purposes; this also causes many health hazards to humans. Therefore, the regulatory authorities of the Central Pollution Control Board at the central level and the State Pollution Control Boards at the state level have fixed stringent regulations limiting textile industrial effluents. These are listed in Table 11.6.

Table 11.5 Federal Environmental Protection Agency limits of effluents discharged from textile mills in United States

S. No	Parameter	Limit value
1	pH	6.0–9.0
2	Biochemical oxygen demand	50 mg/L
3	Chemical oxygen demand	80 mg/L
4	Total dissolved solids	2000 mg/L
5	Total suspended solids	30 mg/L
6	Sulphide	200 μg/L
7	Free chlorine	1000 μg/L
8	Nitrate	20,000 μg/L
9	Ammonia	0.2 μg/L
10	Phosphate	5.0 μg/L

Table 11.6 Discharge standard limits for textile effluent in India

S. No	Parameters	Limit value
1	pH	6.0–8.5
2	Biochemical oxygen demand	30 mg/L
3	Chemical oxygen demand	250 mg/L
4	Total suspended solids	100 mg/L
5	Total dissolved solids	5034 mg/L
6	Total residual chlorine	1 mg/L
7	Sulphide	100 mg/L
8	Chloride	100 mg/L
9	Colour	20 Hazen units
10	Iron/manganese	0.25 mg/L

11.5 Sustainable chemicals

11.5.1 Reason for choosing sustainable chemicals

The main toxic constituents in textile effluents are organic and inorganic chemical compounds such as acids, alkalis, hydrogen peroxide, sodium hypochlorite, polyamine derivatives, propanol, ethoxylates, polyglycol ether, surfactants and dispersing agents. Some chemicals are so compact and toxic that they are not biodegradable (Arputharaj et al., 2015). More than 50% of chemicals are directly discharged from the processes, depending on the ratios of fixation in the textile production processes. The softening agents (residual chemicals) are most responsible for the highest concentration of COD content of textile effluents. They cause several toxic and harmful health effects to humans and ecosystems. Therefore, the primary focus for preventing pollution should be on textile processing chemicals. In the recent decades, the potential replacement of toxic chemicals by other pollution-free sustainable chemicals such as desizing agents, dyes and auxiliaries have been attempted to prevent pollution.

11.5.2 Results of using sustainable chemicals

After the replacement of toxic chemicals by the substituted chemicals, it was found that the biodegradability of those chemicals increased by 26% and the sulphide concentration of the textile effluent was reduced by 76%. By changing the chemical consumption and implementing chemical substitution, the COD concentration of wastewater was reduced 10%–25% and iron was removed from the fabric's surface. The most important advantage was that the operational cost of waste treatment was reduced 49%.

11.5.3 Conditions for the selection of sustainable chemicals

The replacement of toxic chemicals using substantial pollution-free chemicals may vary substantially among mills because of:

1. environmental conditions
2. process conditions
3. products
4. raw materials

11.6 Sustainable wastewater treatment methods

11.6.1 Treatment methodologies and their classifications

Textile effluent consists of a high concentration of organic compounds, BOD and COD, and is fully loaded with salt and colour. Effluent is discharged directly into rivers and other water streams, which makes the river water unusable for drinking purposes and other domestic uses. Therefore, several research studies reported that treatment textile effluent is mandatory. In textile effluent treatment, many methodologies have been employed to find an economic and effective way to treat wastewater. These methodologies are usually classified as:

1. physiochemical
2. biochemical
3. combined treatment process

Industrial textile wastewater treatment is an assortment of physical, chemical and biological processes. In biological wastewater treatment, the efficiency of removal is fully based on the ratio between the organic load and biomass present in the treatment tank. Biological treatment is mainly classified into two groups: (1) aerobic and (2) anaerobic. The application of microorganisms in biological treatment is an attractive method and has considerable advantages. Biological treatment is affected by several parameters such as the pH, temperature, level of aeration and redox potential of the system.

Different types of biodegradable organic matter such as polyvinyl alcohol slurry and new additives are added into the water environment, resulting in an increase in the concentration of BOD, COD and other organic complexes. Therefore, the combined physiochemical and biological process has a better role in removing new synthetic dyes and organic complexes from effluents.

In the different kinds of treatment techniques, some of the methods have been used effectively.

11.6.1.1 Adsorption

Adsorption is a surface phenomenon used to remove soluble organic pollutants that are present in solution using a suitable interface. The most toxic chemicals (phenols, pesticides and cyanides) and dyes (methylene blue, malachite green and Congo red) can also be removed using an adsorption process (Santos et al., 2007). Different parameters have an influence on adsorption, such as the adsorbent surface area, particle size, contact time, adsorbate–adsorbent concentration, pH and temperature.

In adsorption, the most commonly used adsorbent for effluent treatment is activated carbon, which is prepared from carbonaceous materials such as sawdust, coconut husks, cellulosic materials, palm kernel husks, peanut skins, peanut hulls, maple, hazelnut shells, coffee and tea waste, palm fruit bunches, modified sugar beet pulp and maize leaves. Activated carbon is effective in adsorbing metal ions, dyes, cations and mordants.

11.6.1.2 Coagulation

In several developing countries, coagulation is a widely used method for textile wastewater treatment. Different copolymers such as pentaethylene, hexamine and ethylediene dichloride and different coagulant agents such as ferric chloride, ferrous sulphate, aluminum sulphate, ferric sulphate, copper sulphate and calcium chloride are used to form flocs with contaminants. Then the clustered impurities are separated by sedimentation and filtration. Decolouration of textile wastewater can be effectively performed by coagulation.

11.6.1.3 Membrane separation

Membrane separation is a technology often used to treat textile dyeing effluent. In this process, the membrane filter is used for filtration. It consists of micropores and separates organic substances from the effluent using membrane-selective permeability. This methodology is classified into three categories: (1) UF, (2) NF and (3) RO.

Ultrafiltration (UF) is a promising methodology for separation. UF works at low transmembrane pressure to remove dissolved substances. UF membranes may be made from polyelectrolyte complexes, cellulose acetate and inert polymers. These membranes are capable of handling high flux and are free from microbes. Pore size ranges between 0.001 and 0.02 μm.

NF is an advanced membrane technology that is effective in removing heavy metals. NF membranes have a nonporous thin skin layer which provides high permeability. NF possesses the property of intermediate UF and RO. NF processes consume less energy and result in the high removal of pollutants.

RO technology has many applications in the treatment of wastewater. It gives a clear permeate and leaves an RO concentrate. This RO concentrate will have a high level of organic concentration. Because RO membranes have fine pores, they are prone to fouling by suspended particles. To prevent this fouling, the feed water must undergo a pretreatment process to control the turbidity level.

11.6.1.4 Flotation

In flotation, suspended organic compounds attach to the air bubbles that are generated. These air bubbles are less dense than water and make the cluster of suspended particles rise to the surface of the feed water. Many types of flotation process, such as dissolved air flotation, precipitate flotation and ion flotation, are available for the treatment of textile effluent. However, flotation has several disadvantages such as higher energy consumption and a high operational cost, and it involves intense mechanical operations. Hence it is unsuitable for areas where maintenance work is neglected.

11.6.1.5 Ozonation

Among the different effluent treatment techniques, ozonation has some special advantages for textile dyeing treatment. Commercially available oxidizers such as sodium hypochlorite can break down organic molecules, detergents and phenols into smaller compounds by ozonation. In the ozonation process, ozone reacts with organic compounds in two different ways: (1) by a direct molecular reaction, and (2) by an indirect free radical-type reaction. In addition, ozone can easily break double bonds present in dye molecules. Also, it can destroy the foaming properties of residual surfactants. The main disadvantage is the release of carcinogenic aromatic amines and toxic molecules.

11.6.1.6 Ion exchange

Ion exchange is widely used to remove inorganic salts and organic anionic components. Ion exchange materials are capable of exchanging soluble cations and ions with electrolyte solutions and transferring them into sludge. Ion exchange resin can be classified on the basis of functional groups as anion exchange resin, cationic exchange resin and chelating exchange resin. Ion exchange resins have specific metal and organic compound uptake capacity and are used during ion exchange. The resins may be either synthetic or natural. The main disadvantage is the transfer of impurities from the effluent stream to sludge that need to be disposed of because of added regeneration chemicals.

11.6.1.7 Evaporation

In the evaporation process, concentrated liquid is allowed to pass through the multiple evaporator system. At the same time, steam is applied to the evaporator. During evaporation, the liquid is recirculated until it attains the desired concentration of salt or the specific density. The vapor and steam are collected in the condenser. The temperature in the evaporator may vary based on the length of the tubes. The evaporation process is classified into three groups:

1. multiple effect evaporator
2. mechanical vapour compression
3. direct contact evaporation

11.6.1.8 Crystallization

Crystallization is used effectively for solid–liquid separation. This method refers to the formation of solid crystals from a homogeneous solution. This process can be classified into two categories: (1) single stage and (2) multistage. In the crystallization technique, some crystallizers can be used to remove chemicals such as calcium sulphate, sodium sulphate, sodium chloride and calcium chloride.

11.6.2 Comparison studies

All of these techniques have been found to be capable of removing toxic substances from textile effluent wastewater. Comparative studies were carried out to find a suitable technology to treat different types of effluent water. Generally, performance comparisons of different treatment processes are based on:

1. operational issues: This includes maintenance requirements and process failures experienced during pilot testing.
2. water quality: This can be produced from each demineralization process.
3. projected costs: This is determined for the construction and operation of full-scale facilities required to produce 50 million gallons per day of recycled water with a final TDS concentration of either 350 or 500 mg/L.

11.6.3 Advantages and disadvantages

Several technologies have been developed and used to remove toxic organic and inorganic compounds, pollutants and other impurities from textile effluent water. Each technology has its own advantages and disadvantages, as shown in Table 11.7.

11.6.4 Advance wastewater treatment process

Different methods were used in advanced waste treatment to satisfy the following criteria, including the removal of: (1) suspended solids, (2) BOD, (3) plant nutrients, (4) dissolved solids and (5) toxic substances. These methods can be employed at any stage of the total treatment process to industrial waterways and may also be used for the complete removal of pollutants after secondary treatment (Upadhye and Joshi, 2012).

Table 11.7 Advantages and disadvantages of different methods for textile effluent treatment

S. No	Conventional methods	Advantages	Disadvantages
1	Adsorption	High efficiency of metal ion removal, fast kinetics, extensive range of target pollutants	High cost of activated carbon, performance depends on adsorbent
2	Coagulation	Dewatering characteristics	More sludge settling, expensive reagents, chemical consumption
3	Membrane separation	Metal dye selective, small space requirements	Membrane fouling, capital cost, maintenance, limited flow rates, less efficient
4	Flotation	Low retention time	High maintenance and operational cost, initial capital cost
5	Ozonation	Constant pressure performance, no disposal of concentrates or solids	Capital incentive, complex chemistry must be tailored to specific application
6	Ion exchange	Selective removal of metal ion, lower number of metal ions removed	Expensive (synthetic resins), large-scale application, lower number of metal ions removed, high maintenance and operational costs
7	Evaporation	Simple and cheap to construct, easy to use, clean and maintain	Water may dry up, products result in foaming,
8	Crystallization	Products are relatively pure, crystal size can be controlled	Need for large amounts of solvents

11.7 Case studies

11.7.1 Innovation potential

Innovations will have an important role in achieving sustainable and secure water resources for the future.

1. There has been little innovation in the water industry over the past few years.
2. New innovative technologies have been developed and used based on categories such as effective products, technology, services, design processes and financial resources.
3. Innovative ideas have been implemented in various ways in domestically, globally and in the water sector and supply chain.
4. The United Kingdom government has removed barriers and partnered innovation leadership groups: (1) TSB and (2) the UK Water Research and Innovation Framework (UKWRIF).
5. There is a need to ensure an effective approach to wastewater management that contributes to improved water quality and reduced flood risk.

6. It is important to create a modern regulatory system that protects customers and minimizes regulatory burdens.
7. Incentivizing water conservation can be done through changing attitudes and behaviours.
8. The contribution of water to economic growth should be maximized.
9. UKWRIF–Vision: By 2030, the United Kingdom will be a critical contributor to providing integrated solutions for water security and sustainability.
10. UKWRIF–approaches: Innovations need to be focus on the following factors:
    a. technological (high or low)
    b. partnership/collaboration
    c. research ideas
    d. approaching steps
11. M-Water (technological innovation): Combines smart metering and mobile banking technology: waterpoint data transmitter
12. Netherlands Water Partnership (NWP): Focuses on developing sectors and exports such as the North American Energy Reliability Corporation, WS KE Programme, BWP, TSB, Science and Innovation Network, etc.
13. NWP: Supports Dutch international profile in water research and innovation.

11.8 Challenges

11.8.1 Selection of advanced wastewater treatment for process water requirements

In the recent decades, the advance treatment process seems to be effective in removing pollutants, contaminants and salts from textile effluents. For effective removal, the advance treatment process needs to meet stringent environmental or regulatory requirements. The selection process can be divided into two groups: (1) quantifiable and (2) nonquantifiable.

'Quantifiable' criteria (technology comparison) are normally based on:

1. capital costs
2. energy use
3. availability and reliability
4. greenhouse gas emissions

Results from quantifiable analyses are generally further sectioned into 'nonquantifiable' criteria according to their:

1. reliability and maintainability of technology
2. flexibility of technology
3. future technology developments

11.8.2 Benefits of advanced wastewater treatment

Application of the advanced treatment process to the effluent treatment has shown substantial benefits:

1. In membrane filtration, the quality of permeate is reasonably good for reuse in textile operations.

2. In RO, permeate water consists of a low concentration of total dissolved oxygen (196 mg/L).
3. Permeate water has negligible hardness. This avoids the use of chemicals for water softening and leads to low maintenance costs, less breakdown and more output.
4. In an evaporation system, condensate water has lower concentrations of TDS.
5. In an NF system, the final permeate water consists of saltwater (brine); other major contaminants are rejected from the textile effluent. The brine water is further reused in the dyeing process, which requires less additional salt.
6. In crystallization, salts are recovered from the effluents. This method can also reduce problems related to the disposal of a higher concentration of TDS in textile effluent streams.
7. Finally, installation of advanced treatment methods for textile wastewater treatment is helpful for environmental pollution problems. The reuse of treated effluent is an alternative that conserve natural resources and increases sustainable development.

11.8.3 Limitations of advanced wastewater treatment

The advanced treatment process for the effective removal of organic and inorganic compounds from textile effluents has major limitations:

1. In RO, different filters are used, which need:
   a. to meet feed effluent requirements
   b. regular monitoring at each stage to maintain the desired effluent quality
2. In the adsorption process, activated carbon intermittently needs to be replaced or regenerated
3. In the evaporation process, evaporators are used to reduce the volume of effluent. It requires regular cleaning to remove the deposition of unwanted contaminants on the tube walls.
4. In the membrane separation process, some membrane units deliver treated effluents with a higher concentration of COD (81 mg/L) and higher colour content (20 Hazen units).
5. RO is a method for water treatment that requires an additional treatment to dispose rejects from membrane units. The membrane units consist of a high load of pollutants which is difficult to treat.

11.9 Future trends

11.9.1 Implementation of suitable advanced wastewater treatment in textile industries

The implementation of a suitable advanced water treatment process in the textile industry is driven by three primary factors:

1. discovery of rarer contaminants
2. promulgation of new water quality standards
3. construction, design and production costs

However, the past 20 years have seen a dramatic change in the water industry's approach to water treatment, in which water utilities have started to seriously consider alternative treatment technologies to the traditional filtration/chlorination treatment approach. For a new technology to be considered it must have advantages over traditional treatment processes: (1) lower capital and operational maintenance costs, (2) higher efficiency, (3) easier operation, (4) better effluent water quality and (5) lower

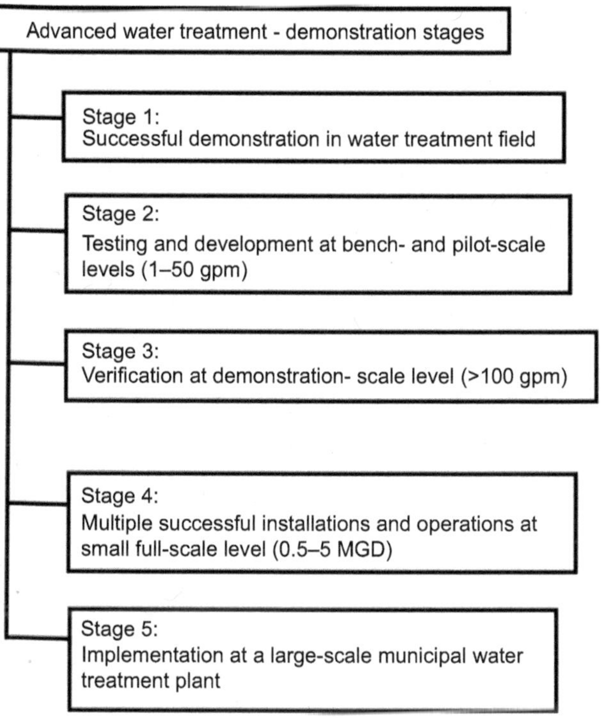

Figure 11.4 Demonstration stages of advanced water treatment.

waste production. Nevertheless, for a water treatment technology to be accepted and implemented on a large municipal scale, it must be demonstrated in stages. A typical sequence of these stages is summarized in Fig. 11.4.

11.10 Conclusion

The textile industry is one of the oldest and largest industrial segments worldwide. It is responsible for environmental water pollution unless the water is properly treated before its disposal. The textile industry has a substantial impact on the aquatic system in two ways: (1) it consumes a lot of water (fresh water), and (2) it directly discharges effluents into water streams.

Therefore, based on environmental safety, effluent treatment methodologies are mostly expected and accepted. Unfortunately, no particular treatment methodology is appropriate for or universally adaptable to all kinds of textile effluents. Consequently, a combination of different treatment methods is the best solution and one that can be adopted and is capable of treating all kinds of textile effluent. The combination treatment process should consist of physical, chemical and biological methods which will effectively treat wastewater based on the pollution load. Based on the combined treatment system, analysis results indicated that membrane filtrations have been in operation. During operation, a multiple-effect evaporation system has been in use to minimizing the volume of effluent. In addition, a crystallizer installed with an evaporator is used to recover the Glauber salt.

Similarly, other treatment methodologies such as adsorption using activated carbon and ozonation are employed to make effluent suitable for use in membrane filtration. Membrane filtration such as RO can be used for water purification to produce colourless treated effluent with dissolved solids as low as 196 mg/L and zero hardness. In this process, the main disadvantage is the disposal problem of membrane rejects which contain high contaminants. The reject stream of RO can be disposed of through solar evaporation ponds or other evaporation systems.

In recent decades, the reuse of textile effluents has increased attention towards reducing the scarcity of water resources and increasing stringent regulatory requirements for disposing of textile effluent. Therefore, suitable treatment methodologies (combined treatment plants) can be used and made fit for recycling and the reuse of treated effluent from the textile industry.

Sources of further information and advice

Generally, textile industry effluent can be treated in two ways: with or without the segregation of effluent streams. In textile wastewater treatment, the recovery of water can be achieved without segregation, but this treatment is not feasible for recovering salts. With segregation, both the salt and water can easily be recovered from the effluent streams. When subjecting the effluent streams to the advanced water treatment techniques, the level of toxic pollutants is reduced and there is possible recovery of water and salts.

Textile effluent streams can be classified into two types based on pollutants: (1) highly polluting effluent streams and (2) low-polluting effluent streams. Normally, highly polluting effluent streams consist of spent dye bath of low volume (10% of the total effluent discharge), which needs to be segregated and treated separately. Low-polluting effluent streams consist of a huge volume of wash water (90% of total effluent discharge), which can be treated by primary, secondary or tertiary treatment to meet stringent regulatory requirements and reuse in the industry for appropriate operations. For reuse, the treated water is further purified using RO.

References

Amar, N.B., Kechaou, N., Palmeri, J., Deratani, A., Sghaier, A., 2009. Comparison of tertiary treatment by nanofiltration and reverse osmosis for water reuse in denim textile industry. J. Hazard. Mater. 170, 111–117.

Arputharaj, A., Raja, A.S.M., Saxena, S., 2015. Developments in sustainable chemical processing of textiles. Environ. Footpr. Ecodes. Prod. Process. 217–252. http://dx.doi.org/10.1007/978-981-10-0111-6_9.

Babu, R.B., Parande, A.K., Raghu, S., Kumar, P.T., 2007. Textile technology-cotton textile processing: waste generation and effluent treatment. J. Cotton Sci. 11, 141–153.

Badani, Z., Ait-Amar, H., Si-Salah, A., Brik, M., Fuchs, W., 2005. Treatment of textile waste water by membrane bioreactor and reuse. Desalination 185, 411–417. http://dx.doi.org/10.1016/j.desal.2005.03.088.

Bes-Piá, A., Mendoza-Roca, J., Alcaina-Miranda, M., Iborra-Clar, A., Iborra-Clar, M., 2002. Reuse of wastewater of the textile industry after its treatment with a combination of physico-chemical treatment and membrane technologies. Desalination 149, 169–174.

BIS, 1992. Methods of Sampling and Test (Physical and Chemical) for Water and Waste Water. IS No. 3025.

Chavan, R.B., 2001. Indian textile industry – environmental issues. Indian J. Fibre Text. Res. 26, 11–21.

Chen, X., Shen, Z., Zhu, X., Fan, Y., Wang, W., 2005. Advanced treatment of textile wastewater for reuse using electrochemical oxidation and membrane filtration. Water SA 31, 127–132.

Correia, V.M., Stephenson, T., Judd, S.J., 1994. Characterization of textile wastewaters – a review. Environ. Technol. 15, 917–929.

Etter, B., Tiley, E., Khadka, R., Udert, K.M., 2011. Low-cost struvite production using source-separated urine in Nepal. Water Res. 45, 852–862.

Ghaly, A.E., Ananthashankar, R., Alhattab, M., Ramakrishnan, V.V., 2014. Production, characterization and treatment of textile effluents: a critical review. J. Chem. Eng. Process. Technol. 5, 182. http://dx.doi.org/10.4172/2157-7048.1000182.

Holkar, C.R., Jadhav, A.J., Pinjari, D.V., Mahamuni, N.M., Pandit, A.B., 2016. A critical review on textile wastewater treatments: possible approaches. J. Environ. Manage. 182, 351–366. http://dx.doi.org/10.1016/j.jenvman.2016.07.090.

Islam, A., Guha, A.K., 2013. Removal of pH, TDS and color from textile effluent by using coagulants and aquatic/non aquatic plants as adsorbents. Resour. Environ. 3, 101–114. http://dx.doi.org/10.5923/j.re.20130305.01.

Jegatheesan, V., Pramanik, B.K., Chen, J., Navaratna, D., Chang, C.-Y., Shu, L., 2016. Treatment of textile wastewater with membrane bioreactor: a critical review. Bioresour. Technol. 204, 202–212, http://dx.doi.org/10.1016/j.biortech.2016.01.006.

Kant, R., 2012. Textile dyeing industry an environmental hazard. Nat. Sci. 4, 22–26.

Kestioglu, K., Yonar, T., Azbar, N., 2005. Feasibility of physico-chemical treatment and advanced oxidation processes (AOPs) as a means of pretreatment of olive mill effluent (OME). Process. Biochem. 40, 2409–2416.

Kumar, G.N.P., Bhat, S.K., 2012. Decolourization of azo dye Red 3BN by bacteria. Int. Res. J. Biol. Sci. 1, 46–52.

Liu, M., Lu, Z., Chen, Z., Yu, S., Gao, C., 2011. Comparison of reverse osmosis and nanofiltration membranes in the treatment of biologically treated textile effluent for water reuse. Desalination 281, 372–378. http://dx.doi.org/10.1016/j.desal.2011.08.023.

Mahmoued, E.K., 2010. Cement kiln dust and coal filters treatment of textile industrial effluents. Desalination 255, 175–178. http://dx.doi.org/10.1016/j.desal.2009.12.025.

McCurdy, M., Boardman, W., Michelson, G., Woodby, M., 1992. Chemical reduction and oxidation combined with biodegradation for the treatment of a textile dye. In: 46th Oroc. Purdue Industrial Waste Conf. Lewis Publishers, MI, pp. 229–234.

Paul, S.A., Chavan, S.K., Khambe, S.D., 2012. Studies on characterization of textile industrial waste water in solapur city. Int. J. Chem. Sci. 10, 635–642.

Rasheed, S., Kavitha, A., Hayavadana, J., 2014. A critical review on effluent treatment for textiles. Trans. Eng. Sci. 2, 18–21.

Santos, S.C.R., Boaventura, R.A.R., 2015. Treatment of a simulated textile wastewater in a sequencing batch reactor (SBR) with addition of a low-cost adsorbent. J. Hazard. Mater. 291, 74–82. http://dx.doi.org/10.1016/j.jhazmat.2015.02.074.

Santos, A.B., Cervantes, J.F., Lier, J.B., 2007. Review paper on current technologies for decolourisation of textile wastewaters: perspectives for anaerobic biotechnology. Bioresour. Technol. 98, 2369–2385.

Sarker, B.C., Baten, A., Haque, E.U., Das, A.K., Hossain, A., Hasan, Z., 2015. Heavy metals concentration in textile and garments industries wastewater of bhaluka industrial area, Mymensingh, Bangladesh. Curr. World Environ. 10, 61–66. http://dx.doi.org/10.12944/CWE.10.1.07.

Tripathy, T., De, B.R., 2006. Flocculation: a new way to treat the wastewater. J. Phys. Sci. 10, 93–127.

Upadhye, V.B., Joshi, S.S., 2012. Advances in wastewater treatment – a review. Int. J. Adv. Chem. Sci. Appl. 3, 264–268.

US Environmental Protection Agency (USEPA), 1990. Environmental Pollution Control Alternatives: Drinking Water Treatment for Small Communities. USEPA, Washington, DC.

Yasui, H., Nakamura, K., Sakuma, S., Iwasaki, M., Sakai, Y., 1996. A full scale operation of a novel activated carbon process without excess sludge production. Water Sci. Technol. 34, 395–404.

Sustainable chemical management and zero discharges

12

K. Amutha
Bharathiar University, Coimbatore, India

12.1 Introduction

Chemicals have turned out to be an indispensable part of human life on earth. The chemical industry occupies an important role in the economic development of a country. The United Nations (UN) declared 2011 to be the 'International Year of Chemistry', thereby laying emphasis on meeting future challenges. With rising concerns about sustainability, the chemical industry is striving hard to meet demands. Sustainability requires systematic solutions. The chemical industry serves as the raw material supplier for many other industries such as food, textiles, cosmetics, agriculture, electronics, automotive and construction.

The textile industry is a significant cluster among manufacturing industries. It consumes large quantities of chemical substances in the form of dyes, pigments, mordants, finishing agents and other auxiliaries for various processes such as sizing, bleaching, scouring, desizing, mercerizing, dyeing, printing and finishing. Colour is an integral part of human life and fashion. In the past, natural dyes and pigments were used to colour textiles. The first synthetic dye was developed in 1856 and it led to the origin of the dyestuff industry in Europe. With the advent of synthetic dyes and changes in fashion trends, more and more chemicals are being used along the supply chain. Some of these chemicals are potentially harmful to both humans and the environment. Effluent discharge in water bodies is a major problem resulting from the textile wet processing sector. An investigation by Greenpeace International in 2014 found that children's clothing and footwear, which is manufactured and sold in many countries across the globe, contained a wide range of hazardous chemicals that are above specified limits of detection. Certain chemicals are categorized as priority substances, and the use of these chemicals is prohibited by many European countries. The use of these priority chemicals should be at the zero level because there is *no safe level* for these hazardous chemicals. It could be achieved only by the commitment of both the government and the manufacturer (textile industry). This chapter discusses the Restricted Substances List (RSL), Manufacturing Restricted Substances List (MRSL) and management of these chemicals and zero discharge of hazardous chemicals (ZDHC) and related issues and solutions.

12.2 Sustainability issues in the current scenario

The textiles and clothing supply chain features sustainability issues throughout the chain (Fig. 12.1).

Sustainable Fibres and Textiles. http://dx.doi.org/10.1016/B978-0-08-102041-8.00012-3

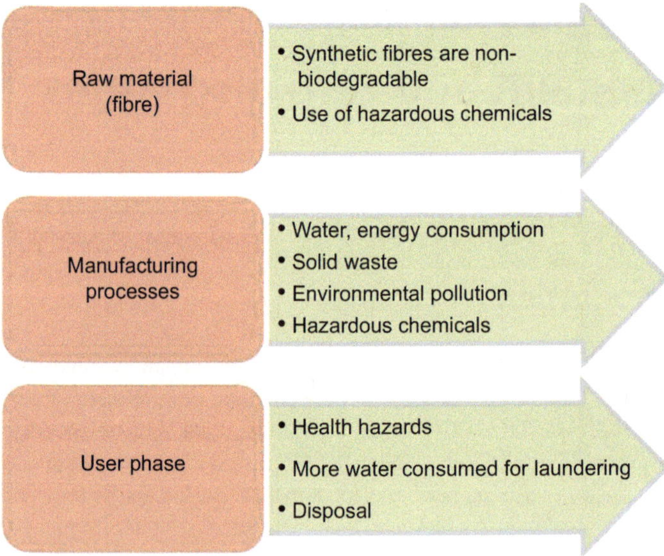

Figure 12.1 Sustainability issues of textiles.

Major threats to sustainability faced by the textile dyeing and finishing sector are:

- the use of large quantities of water
- the use of an enormous quantity of chemicals
- the hazardous nature of effluent
- direct discharge of effluent or wastewater into waterways

Water, the 'universal solvent', is the most vital requirement for the entire globe. It is used for various domestic and industrial applications. However, the water resource is becoming scarce. The UN declared 2013 to be the International Year of Water Cooperation and released facts and figures about water resources around the world. The UN estimated that by 2025, 1800 million people globally will experience water scarcity and two-thirds of the world population may experience stress. The conditions are poorer in developing countries than in developed ones. Approximately 783 million people lack access to clean water; in China alone 320 million people have no access to clean drinking water. Up to 40% of surface water is considered polluted. Underground water is also contaminated, sometimes with carcinogenic chemicals.

The consumption of water for wet processing of textiles depends on the fibre type, the method employed, the equipment used and the nature of the dye. The wastewater or effluent is discarded in rivers, lakes, seas and seepage in ground water streams. A large quantity of water becomes polluted with various dyes, chemicals and textile auxiliaries. An increase in the salinity of water affects agriculture.

The World Bank has estimated that 17%–20% of the world's industrial water pollution comes from the dyeing and treatment of textiles. According to KEMI, a Swedish Chemicals Agency, approximately 10% of 2400 identified textile-related substances are considered to be of potential risk to human health. Health effects include skin allergy or contact dermatitis, carcinogenicity, mutagenicity, reproductive toxicity and respiratory allergy.

Effluent contains high suspended solids, mineral oils, nonbiodegradable or low biodegradable surfactants, halogenated organics and other compounds including phenols. Effluent streams from dyeing processes are typically hot and coloured and contain significant concentrations of heavy metals. When let into waterways, these chemicals break down and release hormone-disrupting and cancer-causing chemicals.

A study by Greenpeace ('Toxic Threads') revealed that Indonesia, one of the world's leading exporters of clothing, faces severe water pollution: Effluent let into water bodies is highly alkaline, with a pH of 14, which could burn skin when it comes into direct contact. Many more facts from various parts of the globe reveal the hazardous nature of pollutants which are mixed into water bodies such as lakes, rivers and seas. This is a threat to both human and aquatic life. The most degraded system owing to water pollution is the biodiversity of the freshwater ecosystem.

To a greater extent, synthetic or man-made fibres are replacing natural fibres either completely or partially. For example, traditionally woollen blankets were used during winter, but polyester fleece has replaced it. Also, hazardous chemicals are employed in the manufacture of synthetic fibres, such as the use of antimony to produce polyethylene terephthalate (polyester fibre). In the same way, the production of blended yarn is increasing annually. Recycling or disposal of clothing made from blended yarn is tedious because one fibre may be biodegradable (e.g., cotton) and the other may be nonbiodegradable (e.g., polyester, polypropylene).

12.3 Detox campaign

The Detox campaign was launched in July 2011 by Greenpeace, a nongovernmental organization. It serves as a means to reduce environmental impacts resulting from the use of hazardous chemicals in the manufacture of textiles and apparel. It encourages the world's leading brands to eliminate hazardous chemicals in products and the supply chain by 2020.

Fig. 12.2 shows the Detox logo.

Major retailers such as H&M, Puma, Burberry, Adidas, Mango, Primark, Disney and Nike have committed to the campaign. To achieve its goal, Greenpeace has released many reports since its launch which revealed the presence of hazardous chemicals in

Figure 12.2 Detox logo.

Figure 12.3 Detox captions for Levi's.

branded clothes and in the public waterways where effluent in discharged. Greenpeace is pursuing legislative changes within the textile industries in several Asian countries and the European Union (EU) to protect rivers and the communities and ecosystems they support. The focuses of the Detox campaign are:

- ZDHC: complete elimination of hazardous chemicals in water, air and soil by 2020
- prevention and precaution: precautionary action headed for the elimination of hazardous chemicals
- right to know: Brands and their supply chains need to be transparent through unrestricted disclosure of information about the hazardous chemicals used and discharged during textile and apparel manufacturing.

Greenpeace also insists that governments, especially those of countries that are major manufacturers and suppliers of textiles and clothing, establish chemical management policies towards ZDHC. Greenpeace has initiated and catalysed a change in the apparel supply chain, which is urgently needed. Events run by Detox are sophisticated, which makes them different from other campaigns by Greenpeace.

Values-Based Segmentation is a strategy adopted by Greenpeace to motivate the supporters of Detox. It is a model with the objective of understanding people's choices.

Detox has been designed in such a way as to break through the fashion industry. The materials and strategies used for campaigning are innovative and media-friendly so that they reach far and wide (Figs 12.3 and 12.4). *Fashion Victim* is one such theme in which professional models staged a catwalk and were made to look sick. The Detox campaign set up the world's largest coordinated striptease outside brand stores as part of its offline mobilization efforts.

In November 2012, activists organized a mannequin revolt in 80 cities across the globe. The storyline was that Zara's mannequins were sick of wearing hazardous chemical-filled clothes every day and were quitting their jobs to join the Detox campaign. The Detox team works towards their objectives while Greenpeace plays the role of mentor. It encourages people to get more involved, realize the value of their actions and inspire others. Campaigning is about learning, developing and executing; the most important aspect is to have fun.

Figure 12.4 Detox logo at H&M retail outlet.

Ize Smith, a Detox campaigner at Greenpeace, said, 'From budget to luxury, we have a right to demand that our clothes are free from hazardous chemicals, and fashion brands have a responsibility to do something about it.'

The Detox logo, which has been widely used as part of the campaign, features on everything from T-shirts to videos and temporary tattoos handed by the thousands at music festivals and shopping malls, includes the Chinese symbol for water ('X'), which helps to cement both the connection with water and the importance of China into all communications.

As a part of its Detox campaign, Greenpeace tested the football kit (football shirts, boots and other accessories) used in the World Cup 2014 held in Brazil between 12th June and 13th July 2014. For the study, 33 products were bought from different countries around the world. The products represented three major sportswear brands: Adidas, Nike and Puma. Analysis of these products revealed the presence of hazardous chemicals such as PFC (perfluorinated and polyfluorinated chemicals) in both ionic and volatile forms, nonylphenols, nonylphenol ethoxylates, phthalates, dimethylformamide, antimony and organotin compounds. The study substantiated that hazardous chemicals have been used along the textile supply chain despite government and industry regulations and corporate responsibility programmes. Hence it is implicit that current regulations are failing to protect human health and the environment. The Detox campaign urges the Detox leaders to join the Detox revolution and respond immediately to inadequacies in policy and practice.

12.4 Zero discharge of hazardous chemicals and similar industrial initiatives

ZDHC: ZDHC is a programme inaugurated in 2011 with a commitment to ZDHC in the apparel supply chain by 2020. The beginning was a joint initiative by six textile brands; later many others joined the stream. ZDHC aims to implement the challenging Joint Roadmap for a cleaner environment and safer conditions for people. Participants of ZDHC include the Adidas group, C&A, G-Star Raw, H&M, Jack Wolfskin, Levi Strauss & Co., Li Ning, Nike Inc., PUMA and many others. The chief goal is to eliminate the use of priority chemicals and substitute them with safer chemicals. The mission of the ZDHC programme is set with high ambition and provides a new standard for the environmental performance of the apparel and footwear industry across the world. The programme acts as a channel that guides and trains the textile industry towards zero discharge. Areas of focus are:

1. MRSL and conformity guidance
2. wastewater quality
3. audit protocol
4. research
5. data and disclosure
6. training

The programme adopts a holistic system change that catalyses a constructive modification in the discharge of hazardous chemicals across the product life cycle. The first Joint Roadmap, released in November 2011, identified the areas in which ZDHC members could collaborate to conduct research and take corrective action to achieve the goal of zero discharge. Since 2012, ZDHC has published annual report featuring research and activities. The programme identified projects in seven main categories in the first year of work:

1. benchmarking and phase-out
2. water repellency
3. chemical identification and hazard assessment
4. 'green' chemistry
5. audit protocol
6. joint training
7. disclosure

12.4.1 Hazard and risk

Chemicals which pose only a small hazard but to which there is frequent or excessive exposure may pose as much risk as chemicals which have a high degree of hazard but to which only limited exposure occurs.

Risk = hazard × exposure
Exposure = dose × time
Hazard = potential of a substance to cause harm

As a group, the textile industry will need to identify and prioritize chemical substances of concern and substitute their use with safer alternatives. Once the chemical inventory has been screened for risk, the chemicals will be organized under the following groups:

1. *black list*: Chemicals found in this list are of high concern. It is good to avoid or phase out the use of these chemicals.
2. *manage list*: Chemicals in this list can be used but it is better to look for safer chemicals and/or processes.
3. *positive list*: This is a list of preferred, safer chemicals.

The ZDHC adopts a transparent screening process to promote safer chemicals. Common tools are being implemented. In some cases, ZDHC has aligned with industry associations such as the European Outdoor Group (EOG), the Outdoor Industry Association (OIA) and the Sustainable Apparel Coalition (SAC).

SAC: The SAC is a nonprofit alliance meant for sustainable production in the apparel, footwear and home textile industries. It provides a single approach to measure sustainability. It developed a sustainability measurement tool called the Higg Index. It is the first public rollout from the motley crew of manufacturers, retailers, non-governmental organizations, and academic experts which includes such recognizable names as Adidas, Esprit, Gap, H&M, Levi Strauss, Nike, Marks & Spencer, Patagonia, Timberland, Target, Wal-Mart, the National Resource Defence Council, and the US Environmental Protection Agency.

EOG: EOG is an association founded in 2003 by 19 of the world's largest outdoor companies; it represents the common interests of the European outdoor industry. EOG is an active member of the Federation of the European Sporting Goods Industry. It has launched the new Responsible Wool Standard, in association with Textile Exchange, for the outdoor industry. SAC is also a partner of EOG.

OIA: OIA was founded in 1989 by a group of 14 outdoor industry leaders. It constitutes a Sustainability Working Group to implement environmental and social best practices in the supply chain of the outdoor industry.

Registration, Evaluation, Authorization and Restriction of Chemicals (REACH): REACH is a regulation for the chemical industry framed by the EU; it was put into force on 1 June 2007 with the aim of improving the protection of human health and the environment from threats that may be caused by chemicals. It is applicable to all chemical substances used for both industry and domestic purposes. REACH is applicable for a wide range of industries and sectors including the textile industry. It encourages the use of alternative methods for the hazard assessment of chemical substances.

REACH uses 'restrictions' as a tool and therefore limits or bans the manufacture or use of certain chemicals. The most hazardous chemical substances are categorized as Substances of Very High Concern; and these chemicals have to be substituted with safer alternatives.

The European Chemicals Agency (ECHA) releases the data about the candidate list of substances in articles; these data are updated twice a year.

Table 12.1 lists out the restricted substances in textiles which is an excerpt from the Official Journal of the European Union.

12.5 11 Flagship chemicals and zero discharge of hazardous chemicals audits

Greenpeace identified 11 priority groups of hazardous chemicals in the textile supply chain and insisted that industry participants completely eliminate the use of these chemicals. The ZDHC aimed to phase out the use of these chemicals by the top 50% of wet processors by 2015; the outcome has to be evaluated. Table 12.2 lists the 11 priority chemical groups, their application areas in the textile supply chain, the hazards caused by these chemicals and safer alternative substances.

Table 12.1 **Restricted substances in textiles**

Substance	Designation	Condition of restriction
Tris(2,3 dibromopropyl) phosphate	CAS no. 126-72-7	Should not be used in textile articles such as garments, undergarments and linen intended to come into contact with skin
Tris(aziridinyl) phosphinoxide	CAS no. 545-55-1 EC no. 208-892-5	
Polybromobiphenyls; polybrominated biphenyls	CAS no. 59536-65-1	
Mercury compounds		Should not be used in impregnation of heavy-duty industrial textiles and yarn intended for their manufacture
Cadmium	CAS no. 7440-43-9 EC no. 231-152-8 and its compounds	Should not be used in impregnated, coated, covered or laminated textile fabrics
Nickel	CAS no. 7440-02-0 EC no. 231-111-4 and its compounds	Should not be used in articles intended to come into direct and prolonged contact with skin: rivet buttons, tighteners, rivets, zippers and metal marks used in garments

CAS, Chemical Abstracts Service.
Official Journal of the European Union, 26.6.2009 (excerpt).

12.5.1 Zero discharge of hazardous chemicals audit

The ZDHC Audit Protocol provides brands with the capability of initiating and self-assessing the safe handling of chemicals with objectives at the foundational, progressive and inspirational levels. In 2013, the ZDHC Audits and Assessment team developed a comprehensive environmental audit tool, Version 1.0, with an emphasis on chemicals management to ensure the performance and improvement of the environment all the way through the supply chain. The audit was designed with the hopes of scaling adoption and becoming the apparel and footwear industry's one consistent standard environmental audit protocol. Version 1.0 is composed of six areas of concentration: environmental management systems, permits, emissions, health and safety, resources and hazards. The audit tool was designed with more than 300 questions covering all of these areas.

To evaluate the efficacy of the ZDHC audit tool, pilot audits were conducted in 23 dye house and finishing facilities and two washing facilities spread across eight countries: Bangladesh, China, El Salvador, India, Taiwan, Turkey, South Korea and the United States. The pilot audit provided a good opportunity to improve the tool. Feedback was obtained from the auditors, the facilities that were pilot audited and the key stake holders: SAC, OIA, the Leather Working Group and the Global Social Compliance Programme.

Table 12.2 11 priority chemical groups

S. No	Chemical	Use in textile industry	Nature of hazard	Alternative
1	Alkylphenol ethoxylates/ alkylphenols	Softeners, wetting and scouring agents, preparation of dyes and pigments, silk degumming, etc.	Toxic to aquatic life; accumulates in body tissue; causes feminization of fish	C9–11 alcohols, ethoxylated (6EO), C12–15 alcohols, ethoxylated (9EO), oxirane, methyl-, polymer with oxirane, mono(2-ethylhexyl ether)
2	Brominated and chlorinated flame retardants	Fireproof and flame-retardant finish of children's sleep-wear, home furnishings, upholstery, etc.	Interference in hormone systems associated with growth and sexual development	Aluminium hydroxide, magnesium hydroxide, ammonium compounds, trichloropropyl phosphate
3	Chlorinated solvents	Used to dissolve chemical substances and clean fabrics	Depletes ozone; affects liver, kidney and central nervous system	Nonchlorinated solvents such as acetone, mineral spirits, *n*-methylpyrrolidone and terpenes; supercritical fluids; plasma and laser technologies
4	Chlorobenzenes	Solvents and biocides in manufacture of dyes; chemical intermediaries	Affects liver, thyroid and central nervous system; hexachlorobenzene is most toxic of this group, a hormone disruptor	Aromatic esters and substituted phenols; benzyl benzoate, fluorous ether F-626, benzotrifluoride, fluorous dimethylformamide and perfluorohexanes (FC-72)
5	Chlorophenols	Biocides, preservative and mould control agent	Highly toxic to humans and aquatic species; affects many organs in body	Zinc-2-pyridinthiol-*N*-oxide–2,2′-dihydroxy-5,5′-dichlorodiphenylmethane-ester
6	Heavy metals	Dyes and pigments used to colour textiles	Accumulates in body over time and is highly toxic; lead and mercury damage nervous system, cadmium damages kidney; chromium VI is highly toxic; pollutes surface water	Heavy metal-free dyes and pigments; suppliers such as DyStar and Greener Shades offer metal-free dyestuff; natural dyes are good alternatives

Continued

Table 12.2 **Continued**

S. No	Chemical	Use in textile industry	Nature of hazard	Alternative
7	Organotin compounds (e.g., tributyltin)	Biocides and antifungal agents in socks, shoes and sports clothes as odour control agents	Persists in environment, accumulates in body and affects immune and reproductive systems.	Calcium-zinc stabilizers; titanate catalysts for polyester production; bismuth, titanate, titanium and zirconium catalysts for polyurethane production
8	Perfluorinated chemicals	Waterproof and stain-proof textiles	Persists in environment, accumulates in body and affects liver; disrupts growth and reproductive hormones	Siloxanes and silicone polymers; stearamidomethyl pyridine chloride
9	Phthalates (*ortho*-phthalates)	Artificial leather, rubber; polyvinyl chloride and plastisol prints and some dyes for textiles	Reprotoxic in mammals; toxic to reproduction	Tributyl *O*-acetylcitrate, (2-ethylhexyl) adipate, cyclohexane-1,2-dicarboxylic acid diisononyl ester, trioctyl trimellitate, dioctyl terephthalate
10	Short-chain chlorinated paraffin	Flame retardants and finishing agents for textiles and leather	Highly toxic to aquatic organisms; accumulation in living organisms	Alkylphosphates and sulfonated fatty-acid esters; aluminium hydroxide and phosphate-containing compounds
11	Toxic azo dyes	Dyeing of textiles	Carcinogenicity caused by aromatic amines released from some azo dyes	Natural dyes and eco-friendly pigments

Table 12.3 Scoring criteria of audit

Score	Meeting requirements	Risk level
4	Fully established, documented and well-implemented	Minimum
3	Well-established, documented and adequately implemented	Low
2	Established and documented; gaps in implementation; lapse in control	Moderate
1	Nonconformance: not formally established, documented; inadequate implementation	High
0	Critical nonconformance: not established, documented or implemented	Potentially critical
N/A	Not applicable	

This feedback was taken cautiously and incorporated into Version 2.0 of the audit protocol tool. The assessment and audit team worked in collaboration with Sustainable Textile Services, the Hohenstein Institute and OEKO-TEX for reasons of transparency. ZDHC published its Audit Protocol Report on 21 July 2014.

In Version 2.0, the Assessments and Audits team changed the scoring criteria from 0 to 4 for all questions except text responses. Score breakdowns are shown in Table 12.3.

12.6 Restricted substances list and manufacturing restricted substances list

The RSL is unique for any product and in most cases is prepared by the manufacturer. It provides information on the regulations and laws which either ban or restrict the presence of certain chemicals and substances in finished products. In the case of the textile and apparels sector, a general RSL is prepared by the following organizations:

1. American Apparel and Footwear Association
2. REACH (EU)

Leading textile manufacturers around the world have released their own RSL. Levi Strauss & Co. implemented an RSL in 2013; the latest version was released in October 2015. NIKE, Inc. published RSL Version 1.0 in January 2016.

There are many benefits to RSLs:

- protect consumer health and the environment
- enhance brand image through sustainability
- manage chemicals all the way through the supply chain and product life cycle
- protect revenues by reducing the costs of recalls
- comply with various substance-restricting laws

The ZDHC MRSL is a list of all banned chemical substances at the manufacturing facility. For the textile dyeing and finishing sector, chemicals in the MRSL include dyes, solvents, cleaners, adhesives, colourants, detergents, coating and finishing substances, and other auxiliaries. The MRSL consists of hazardous substances potentially used while manufacturing that are discharged into the environment as a result of the wet processing of textiles. These substances may be especially harmful to workers and/or the environment if not handled properly. Processes include raw material production, maintenance, wet processing, wastewater treatment, sanitation and pest control. To protect the health of workers and minimize environmental impact, restricted chemicals are not allowed to be used intentionally in production. Chemical formulations containing restricted substances at excessive limits are not compliant with the ZDHC MRSL.

The MRSL serves as a guide for brands, the supply chain and the industry as a whole, to achieve the target of zero discharge. At the same time, MRSL cannot be considered a replacement for legal restrictions on hazardous substances. In some cases, the threshold limit values of restricted chemical substances in MRSL are significantly higher than the limits in RSL. This is because the manufacturing process employs chemicals of higher concentration whereas the final product contains the same chemicals at lower concentration and some unfixed chemicals remain in the wastewater.

The first version of ZDHC MRSL was released in June 2014. Revised Version 1.1 was released in 2015. Because a large number of chemicals are used to process textiles, the identification of particular chemical substance is essential for the preparation of MRSL, RSL, etc. For this purpose, the Chemical Abstracts Service number or Colour Index number is used. This number is unique to any chemical substance and helps in easy identification.

The ZDHC MRSL includes chemical substances from the 11 priority chemical groups by the Joint Roadmap and other chemical substances recommended by the experts of the Technical Advisory Committee of ZDHC. The MRSL has to be consistent with industry standards and hence is revised through collaboration with other technical experts. Sometimes, independent third-party technical experts and industry associations are involved in the peer-review process. The MRSL is composed of two chapters:

Chapter 1: MRSL for Textiles and Synthetic Leather Processing
Chapter 2: MRSL for Natural Leather Processing

For the convenience of manufacturing facilities, MRSL is categorized into two groups:

Group A: Raw Material and Finished Product Supplier Guidance
Group A substances are banned from intentional use in facilities that process raw materials and manufacture finished products.
Group B: Chemical Supplier Formulation Limit
Group B substances are restricted to concentration limits in chemical formulations that are commercially available from chemical suppliers. These limits ban intentional use while allowing for reasonable expected manufacturing impurities that should be consistently achievable by responsible chemical manufacturers.

12.7 Substance of very high concern

SVHC: Substance of very high concern. Generally, an SVHC is a substance meeting one or more of the following criteria:

- Class 1 or 2 carcinogen, mutagen, or toxic for reproduction
- Substance which is PBT (persistent, bioaccumulative and toxic) or very persistent, very bioaccumulative and toxic) in accordance with Annex III of REACH
- Other substances for which there is evidence of an equivalent degree of concern (e.g., endocrine disruptors)

There are over 1000 substances that potentially fulfil these criteria. However, in terms of REACH compliance, SVHCs refer specifically to those substances that meet the hazard categories mentioned and that have been placed by the Commission on the Candidate List of substances considered for authorization. The Candidate List of SVHCs is Annex XIV of the REACH Regulation. In the case of the textile supply chain, the substance may be present at any stage beginning from fibre production to the spinning of yarn, the fabrication of fabric, processing and finishing, and finally storage.

The process of preparing the Candidate List of SVHCs is composed of the following steps:

1. SVHC is identified in articles.
2. Proposals are made for inclusion of a substance on the list of SVHCs by either the European Commission or one of the member states of the EU.
3. The substance is notified to the Registry of Intentions to give companies advance warning of the possible inclusion of the new substance on the Candidate List.
4. Public consultation: A dossier is prepared by the ECHA and submitted for consultation for 45 days.
5. Based on the draft recommendations, the substance may be included in the Candidate List.

Once a substance is added to the SVHC Candidate List, the EU REACH Regulation imposes immediate obligations on manufacturers and importers to declare the substances if they are present. Article 33(1) of the REACH Regulation states that 'manufacturers and importers of articles (products) are required to notify their customers of the presence of SVHC in their products exceeding 0.1% by weight and provide instructions on safe use of the product'. REACH SVHC Compliance and Disclosure requirements:

- determine whether products contain any SVHCs in excess of 0.1% weight by product
- disclose the presence of SVHCs in products within 45 days upon request from consumers or customers
- ensure environmental compliance and safety of products

The recent Candidate List of SVHCs, as of 12 January, 2017, includes 173 substances in total. More substances will be added in the future and the list is regularly updated. Expansion of the list from time to time adds more challenges for the manufacturers of articles. Hence, manufacturers need to adapt according to changes in the Candidate List. Table 12.4 provides a list of SVHCs that may be found in textile articles.

Table 12.4 **Probable SVHC found in textiles**

S. No.	Substance name	CAS No.	Point of use	Reason	Date of inclusion
1	Dibutyl phthalate (DBP)	84-74-2	Plasticizers for textiles made from/with PVC (up to 30 % in the plastic material): - Rubber boots and rainwear - PVC coated technical textiles - Shower curtains - Back coating of carpets for the bathroom - Prints on T-Shirts (but often PU, not PVC)	R	28/10/2008
2	Benzyl butyl phthalate (BBP)	85-68-7		R	28/10/2008
3	Bis (2-ethyl(hexyl)phthalate (DEHP)	117-81-7		EQC/R	28/10/2008
4	Hexabromocyclododecane (HBCDD)	25637-99-4; 3194-55-6; 134237-50-6; 134237-51-7; 134237-52-8	Flame retardant for natural and synthetic fibres: used for theatre curtains, furniture, sunscreen, back coating of carpets	PBT	28/10/2008
5	Alkanes, C10-13, chloro	85535-84-8	Short chain chlorinated paraffins : - Flame retardant - Additive for the leather treatment (renders smoothness to leather): leather clothing and belts	PBT / vPvB	28/10/2008
6	Bis (tributyltin)oxide (TBTO)	56-35-9	Biocide for textiles against fungal attack (outdoor) - Biocide against bad smell after sweating (used in sportswear) - traces in plastics: silicon, polyurethane, PVC (normally don't exceed 0,1%!)	PBT	28/10/2008
7	Sodium dichromate	7789-12-0; 10588-01-9	Additive for a special practice of wool dying to improve the fastness of the coloration; traces of chrome VI as residues in the textiles possible (normally don't exceed 0,1%!)	C / M / R	28/10/2008

12.8 Zero liquid discharge

Zero liquid discharge (ZLD) is a strategic wastewater management system that ensures that there will be no discharge of industrial wastewater into the environment. It is achieved by treating wastewater through recycling and then recovery and reuse for industrial purpose. Hence ZLD is a cycle of closed loop with no discharge. Although ZLD is a costly process, it paves the way for economic benefits by recovering salts and other chemical compounds. Factors that motivate the ZLD system are:

1. water scarcity
2. water economics
3. environmental regulations

Different methods can be employed to achieve ZLD:

1. thermal process (evaporation)
2. reverse osmosis: RO (membrane technology)
3. electrodialysis
4. forward osmosis
5. membrane distillation

A systematic ZLD is made up of the following components:

- pretreatment (physicochemical and Biological)
- RO (membrane processes)
- evaporator and crystallizer (thermal processes)

In addition to ZLD, the recovery and reuse of salt reduces the amount of sludge. This achieves ZWD. The adoption of ZLD requires extensive research and a pilot study because the wastewater generated is not the same for different processes. Above all, it must be financially viable. The type of industry that can afford it may have an effluent treatment plant to achieve ZLD; and those that cannot afford it may become a member of a common effluent treatment plant (CETP). CETP is a project that has been implemented successfully in Tirupur, the leading hosiery hub of India.

There are benefits of ZLD:

1. Wastewater discharge is avoided by recycling.
2. Recovery of water and salt aids in reducing the cost of operation of ZLD.
3. It promotes the sustainability of the industry and the environment at large.
4. There is less use of water by the textile industry, which means water is available for other purposes such as irrigation (agriculture) and domestic utilities.
5. ZLD helps recover the environment.
6. The sludge can be effectively used by the cement industry.

A study by Saravanan et al. (2015) used cadmium oxide-modified zinc oxide nanocomposites for the degradation of textile effluent by photocatalytic and electrochemical activity. The nanocomposite serves as the most efficient catalyst for effluent treatment at a low cost and high efficiency.

Muhammad Bilal et al. (2016) employed chitosan-immobilized manganese peroxide for the degradation and detoxification of textile-dyeing effluents. It showed tremendous bioremediation of textile effluents with degradation up to 90% within 5 h.

Sharista Raghunath et al. (2016) synthesized a novel, cost-effective proline polymer composite that removes selective dye material, especially anionic dyes, from effluent by adsorption. The proline polymer is reported to be a cheaper alternative to commonly used adsorbents such as alumina, silica and activated carbon.

Insa Friha et al. (2015) developed a pilot membrane bioreactor (MBR) system for textile wastewater treatment and operated the MBR continuously for 7 months; they achieved high efficiencies despite variations and complications in the wastewater. A significant reduction in cytotoxicity was observed.

Research by Rashid Mahmood et al. (2015) revealed a method to treat textile effluent and use it for irrigation purposes. The researchers designed a bioreactor, treated textile effluent and then used the treated water for irrigation and carried out a bioremediation study. Results showed that the bioreactor was successful in reducing the toxicity of the effluent.

Ouhammou et al. (2016) used two natural flocculants, a polymer based on polyacrylamide and juice extracted from cactus cladodes, to treat of textile-dyeing effluent by coagulation and flocculation. A purification efficiency above 85% was observed.

12.9 Industry trends: sustainable chemical management and zero discharges

Leading manufacturers and retail brands of textiles and clothing across the globe have realized the importance of sustainable production and started working towards zero discharge in association with leaders in sustainability such as Greenpeace International, SAC and the Roadmap to Zero: ZDHC Programme. These include C&A, H&M, PUMA, Levi Strauss, Esprit, Adidas, Burberry, Primark and Mango. India is one of the countries to implement ZLD successfully in the textile industry.

The DyStar Group offers colouring substances, chemicals and services to the textile industry. It is a leader in the global market with wide-ranging products and services. Through 'econfidence', DyStar monitors about 500 regulated or restricted substances. Through Sustainable Textile Solutions, chemical management and RSL training were offered to about 700 customers. Since the establishment of DyStar in 1994, about 300,000 samples were tested for eco-parameters. A total of 5000 products by DyStar are preregistered through REACH. In addition, 1700 DyStar products comply with ZDHC MRSL 1.1. DyStar offers a state-of-the-art, patented Indigo Vat 40% solution claimed to be the 'cleanest indigo' on the market. This prereduced indigo dye reduces the consumption of sodium hydrosulphite by 70%.

Archroma, a leader in specialty chemicals, published its first formulation list of 'ZDHC MRSL–compliant' colourants and chemicals for textile and apparel in July 2014. H&M Chemical Restriction MRSL: Version 1.0 was released in January 2016, and Version 2.0 in May 2016. Burberry came out with a Product RSL and MRSL in January 2016.

12.10 Challenges

The textile supply chain is long and complicated; hence the implementation of sustainable practices along the entire chain is challenging. As per the guidelines of ZDHC, many brands are working to achieve zero discharge by 2020. However, it is really hard to achieve in practice. Is zero discharge really 'zero'? This is a question for which the answer is sought through scientific study. The textile supply chain lacks transparency. Cost is another important criterion that decides the choice of materials and methods. Safer materials and methods are indeed dearer than existing ones. Research and innovation become mandatory to keep in pace with the demands of brands, governments and consumers.

The textile industry leaves one of the largest footprints on the global environment. It includes the carbon footprint and water footprint, the use of hazardous chemical substances, wastewater quality, the disposal of used clothing and recycling. The carbon footprint represents all greenhouse gas (GHG) emissions throughout the life cycle of a product. Patagonia (United States) reports that the CO_2 footprint of a normal T-shirt is eight times the weight of the shirt. Studies estimate that the net carbon footprint of the average T-shirt is 6 kg, which is about 20 times the weight of the product itself. Our customers and suppliers face increasing pressure from the public to behave more sustainably.

The water crisis is increasing around the world. Water pollution resulting from the dyeing and processing of textiles intensifies this. According to the report of the World Bank, 20% of freshwater pollution is caused by the textile wet processing industry. There is a huge difference between water and wastewater in terms of ecology and economy. Water or fresh water is a resource used for various purposes on earth whereas wastewater is a polluted entity that poses a threat to humans and the environment.

Although ZLD reduces water pollution and recovers wastewater efficiently, the process emits a considerable amount of GHG, increasing the carbon footprint. Also, it consumes large amounts of energy and requires more chemicals. The nature of effluent varies widely depending on the materials used and the process employed; hence adoption of ZLD is unique for each process. The process generates large quantities of solid waste, which leads to problems such as odour, negative impacts on wildlife, risk of leakage and leaching of chemicals into the groundwater when disposed into landfill. Unrecovered common salt in solid waste continues to accrete daily and by the end of a year it results in thousands of tons of solid waste. Management of this solid waste will be yet another problem. The ZLD system is sophisticated, and hence it requires skilled labour so that operation and maintenance are efficient. Implementation of ZLD is challenging, especially for units operating on a small or medium scale. Use of mild steel pipes leads to corrosion and leakage; high-density polyethylene pipes are preferable, which adds to the cost of ZLD. A large space is required for effluent treatment. The operation and maintenance of the plant require additional cost and skilled labour. Ways to generate and save energy have to be explored.

12.11 Future trends

The future requirements of people and the environment as a whole are a toxic-free life with the elimination or substitution of hazardous chemicals. This demands a cradle-to-grave approach along the textile supply chain. Solutions to current challenges are:

1. the adoption of appropriate technology
2. an integrated design approach
3. maintenance of transparency along the supply chain
4. proper recovery and reuse of water, salts and other chemical compounds
5. amalgamation of various dimensions of sustainability
6. adoption of cradle-to-grave approach

12.11.1 Fashion revolution

The fashion revolution came into existence after the collapse of the Rana Plaza Complex in Dhaka, Bangladesh, on 24 April 2013. In 2016, more than 92 countries across the globe took part in Fashion Revolution Week. The objective was to make fashion safe and clean in a beautiful manner. Equal importance was given to creativity, quality, the environment and people. In 2017, Fashion Revolution Week was designated for April 24–30.

12.11.2 The Foundation for Strategic Environmental Research Future Fashion Programme

The Foundation for Strategic Environmental Research (Mistra) Future Fashion Programme is a research programme that aims for closed-loop fashion in clothing. It was started in 2011 for an initial period of 5 years and then was extended to 2019 in a second phase. It is funded by Mistra and coordinated by the SP Technical Research Institute of Sweden. Mistra developed designer tools for sustainable fashion, published the 'Green Textiles Guide', and conducted a life-cycle assessment of Swedish fashion products and many more activities.

12.12 Conclusion

The apparel supply chain is so long and complicated, implementation of complete sustainable practices has to be preceded by careful research and pilot programmes. It may not be achievable in a short span. Worldwide, many associations, coalitions and organizations can lead sustainable development. Environmental degradation can be prevented only through a committed paradigm shift by the entire world. Adequate consumer awareness regarding products and processes may oblige manufacturers to pay attention to sustainability. Industry, government and organizations (that develop standards) need to unite to establish standards and regulations. Our ancestors left us a beautiful and safe earth in which to live; it is our duty to promise future generations a safe world and toxic-free life.

Sources of further information and advice

1. ZDHC Programme: Annual Reports: 2015, 2014, 2013, 2012; Audit Protocol Report, 2014; MRSL Version 1.1, 2014; Joint Roadmap Update, 2015 (www.roadmaptozero.com).
2. Reports by Greenpeace International: Toxic Threads, Dirty Laundry, Dirty Laundry: Reloaded, Dirty Laundry 2: Hung Out to Dry, A Red Card for Sportswear Brands, A Little Story About a Fashionable Lie, A Little Story About the Monsters in Your Closet.
3. PUMA Sustainability Handbook–Chemical Management, Version 5, 2016.
4. Case Study of a Zero Liquid Discharge Facility in Textile Dyeing Effluents at Tirupur, I. Sajid Hussain (2014), Head O&M (Water Reuse) Tamil Nadu Water Investment Company Limited. www.twic.co.in
5. BURBERRY Product Restricted Substances List, January 2016.
6. C&A Global Sustainability Report, 2015.
7. Manufacturing Restricted Substances List, ZDHC Programme, June 2014.
8. G-STAR: ZDHC Progress Report (2015), April 2016.
9. DyStar's Sustainability Performance Report, 2015.
10. Chemsec's Textile Guide.
11. Candidate List Substances in articles by REACH (ECHA).
12. Roadmap for SVHC identification and implementation of REACH risk management measures: Annual Report, April 4, 2016

References

Bilala, M., Asgherb, M., Iqbalc, M., Hua, H., Zhang, X., 2016. Chitosan beads immobilized manganese peroxidase catalytic potential for detoxification and decolorization of textile effluent. Int. J. Biol. Macromol. 89, 181–189.

Friha, I., Bradai, M., Johnson, D., Hilal, N., Loukil, S., Amor, F.B., Feki, F., Han, J., Isoda, H., Sayadi, S., 2015. Treatment of textile wastewater by submerged membrane bioreactor: in vitro bioassays for the assessment of stress response elicited by raw and reclaimed wastewater. J. Environ. Manage. 160, 184–192.

Hussain, S., 2014. Indian Case Study on ZLD – the Tirupur Textile Cluster Experience.

Mahmood, R., Sharif, F., Hayyat, M.U., Ali, S., Siddiq, Z., Rizwan, S.T., 2015. Biotreatment of textile effluent for irrigation purpose using Indigenous Bacterial Consortium (BMP1/SDSC/01): a step towards reducing fresh water scarcity. Environ. Ecol. Res. 3 (1), 9–14.

Ouhammou, M., Lahnine, L., Mghazli, S., Bouchdoug, N.H.M., Jaouad, A., Hejjaj, A., Mandi, L., Mahrouz, M., 2016. Green chemistry for treatment of liquid discharges from a dyeing industry: bio coagulation and flocculation. Mor. J. Chem. 4 (3), 684–694.

Raghunath, S., Ananda, K., Gengan, R.M., Nayunigari, M.K., Maity, A., 2016. Sorption isotherms, kinetic and optimization process of amino acid proline based polymer nanocomposite for the removal of selected textile dyes from industrial wastewater. J. Photochem. Photobiol. B 165, 189–201.

Saravanan, R., Gracia, F., Khan, M.M., Poornima, V., Gupta, V.K., Narayanan, V., Stephen, A., 2015. ZnO/CdO nanocomposites for textile effluent degradation and electrochemical detection. J. Mol. Liq. 209, 374–380.

Further reading

Das, D., 2014. Approach to ZLD – Special Reference to Textile and Papers Sector.

https://data.gov.in/catalog/production-textile-products-india.

http://www.greenpeace.org/international/en/campaigns/detox/fashion/about/eleven-flagship-hazardous-chemicals/.

http://www.greenpeace.org/international/en/campaigns/detox/water/detox/intro/.

http://levistrauss.com/wp-content/uploads/2016/01/RSSP-June-2015.pdf.

http://levistrauss.com/wp-content/uploads/2015/11/RSL-2015.pdf.

http://materialimpacts.c-and-a.com.

Mendoza-Castillo, D.I., Rojas-Mayorga, C.K., García-Martínez, I.P., Pérez-Cruz, M.A., Hernández-Montoya, V., Bonilla-Petriciolet, A., Montes-Morán, M.A., 2015. Removal of heavy metals and arsenic from aqueous solution using textile wastes from denim industry. Int. J. Environ. Sci. Technol. 12, 1657–1668.

http://www.roadmaptozero.com/fileadmin/pdf/2014_Annual_Report.pdf.

http://www.roadmaptozero.com/fileadmin/pdf/MRSL_v1_1.pdf.

http://www.unwater.org/water-cooperation-2013/water-cooperation/facts-and-figures/en/.

https://www.thereachcentre.com/site/content_svhc_information_page.php.

https://www.ec.europa.eu/DocsRoom/documents/1935/attachments/1/translations/en/.../pdf.

http://mistrafuturefashion.com.

http://www.intertek.com/reach/svhc/.

Sustainable production and application of natural fibre-based nonwoven

13

Sanjoy Debnath
ICAR–National Institute of Research on Jute and Allied Fibre Technology, Kolkata, India

13.1 Introduction: nonwovens and their types, consumption and world share of nonwovens and natural fibre-based nonwovens

Nonwovens are a vast area when it comes to their application. From paper to textiles, all are under nonwovens. Nonwovens are also known a type of fabric for which no traditional methods of spinning or weaving or their preparatory are involved. It is a process developed almost a century ago consisting of forming fibre into fabric in its shortest method. As far as the manufacturing process is concerned, nonwovens can be categorized as mechanical bonding, chemical bonding and thermal bonding. However, some advanced methods of bonding exist, such as spun bonding and hydro-entanglement. The spun-bonded method of nonwoven preparation is most popular in the case of synthetic fibres. Out the different types of bonding, a large volume of nonwovens are produced globally using the mechanical bonding method for sustainable development. Needle-punching and stitch bonding are most popular types of mechanically bonded nonwovens. However, of these two types of mechanically bonded nonwovens, needle-punched are the most widely used.

Many of us know that the United Nation designated 2009 as the International Year of Natural Fibres, as declared by the Food and Agriculture Organization (Jan, 2016). The transition towards a bio-based economy and sustainable development as a consequence of the Kyoto Protocols on greenhouse gas reduction and production of neutral CO_2 offers high perspectives for the natural fibre markets. Changing to a bio-based economy requires the substitution of common raw materials that are currently largely produced from fossil (petrochemical) or mineral resources with by-products from renewable (plant- and animal-based) resources. Development of a sustainable global economy, which permits improved purchasing power and living standards without exhausting resources for future generations, requires a fundamental change in attitude. Upon ecological grounds, products should thus be preferred that are based on photosynthetic CO_2 fixation. The benefit of those sustainable resources is that they can be regrown within the foreseeable future, without negative side-effects on global biodiversity. Therefore, competitive products based on renewable resources need to be developed that are of high quality, show excellent technical

Sustainable Fibres and Textiles. http://dx.doi.org/10.1016/B978-0-08-102041-8.00013-5

performance and harm the environment less than do current products based on petrochemical materials. Natural fibres can be defined as bio-based fibres or fibres of vegetable and animal origin. This definition includes all natural cellulosic fibres (cotton, jute, sisal, coir, flax, hemp, abaca, ramie, etc.) and protein-based fibres such as wool and silk. Excluded are mineral fibres such as asbestos that occur naturally but are not bio-based. Asbestos-containing products are not considered sustainable owing to their well-known health risk; this has resulted in the prohibition of their use in many countries. On the other hand, man-made cellulose fibres (e.g., viscose-rayon and cellulose acetate) are produced through chemical procedures from pulped wood or other sources (e.g., cotton and bamboo). Similarly, regenerated (soybean) protein, polymer fibre [bio-polyester, poly(hydroxyalkanoate)] and poly(lactic acid) (PLA)] and chitosan fibre are examples of semisynthetic products that are based on renewable resources (Jan, 2016).

The global nonwoven market reached $33.1 billion in 2013 and was projected to grow 7.5% per year to $47 billion by 2018. Nonwoven media used in filtration applications made up about 14% of the total nonwovens market or a value of nearly $3.7 billion in 2013. In terms of applications, nonwovens for transportation are a category that includes off-road vehicles used in mining, agriculture and construction, as well as aircraft; it is considered the largest market. Media used in these applications include materials for both air and liquid filtration. Nonwovens and technical textiles are predominantly synthetic polymer-based products, owing to inherent advantages of strength and versatility of such fibres. Polyester and polyolefin account for almost 50% of the total raw material consumption. Improvements in textiles which are nonwoven technology are leading to significant cost savings in some cases and enhanced performance and durability in a wide array of applications. Only 3% natural fibre-based nonwovens are consumed globally from natural fibres. However this figure is expected to increase in the near future (Memon, 2016). In a survey, Memon (2016) described the global production of nonwovens in which China was the largest producer, followed by Japan, India and Korea until 2013.

Synthetic fibre nonwovens have a major role in the area of nonwovens when it comes to their applications. Mostly synthetic nonwoven are used for different industrial applications, such as geotextiles, agrotextiles, insulation material, home textiles, packaging material, battery separators, lightweight carry bags, etc. With consciousness about environmentally friendly processes, natural fibre-based nonwovens are increasingly become popular globally. Among different natural fibres, jute is still the cheapest and most commercially available natural fibre used for industrial applications. Hence, this chapter deals with the design, application and evaluation of jute and similar fibre-based nonwoven fabrics. As far as jute-based material is concerned, including nonwovens, Debnath et al. (2009) has done much work focusing on sustainable rural development. The main advantage of this natural fibre is that it is biodegradable and mostly does not affect the natural balance of the environment and flora or fauna. In general, the demand for natural fibre-based nonwovens is gradually increasing in many domestic as well as industrial areas. More applications and awareness of the benefits of natural fibres for society may help the nonwoven fibre industry become sustainable.

13.2 Design, development and application of natural fibre-based nonwovens: current status of natural fibre-based nonwovens and an evaluation of their different properties for specific application, product diversification and added value

The current chapter discusses the use specifically of lignocellulosic fibre (jute, hemp, flax, sisal, pineapple, ramie, etc.) based nonwovens and their sustainable development and applications. Jute is a bast fibre grown mainly in Bangladesh, India, Nepal and China. In the middle of the 18th century, the East India Company (which was then British) introduced jute processing to India and Bangladesh. At first, jute processing was done through flax machinery, which was found to be unsuitable. The industry then developed jute processing machinery because the import of flax fibre from Russia and Europe became costly. Jute processing machinery was invented in Dundee, Scotland and the machinery was imported to Asian jute-growing countries to begin economical processing. In 1856, the industry began in Bengal, India by the East India Company. In those days, jute fibre was imported from these Asian countries and products were developed in Europe (Dundee, Scotland). Jute nonwoven products have developed and diversified from their conventional jute sacks as packaging material and carpet backing, (Debnath, 2017). Jute fibre is comparatively coarse in nature and has wide variation in fineness apart from its mesh-like structure. Its high moisture retention also makes its suitable for specific applications. However, these properties make it more popular for needle-punched nonwovens rather than other types of nonwovens that have thermal-bonded and adhesive-bonded nonwoven structures. Apart from these properties, jute fibre is one of the cheapest natural fibres available commercially in large amounts.

As far as the properties of jute fibre are concerned, it is one of the strongest fibres among natural fibres that are commercially available. It has a mesh structure and long staple with considerable high tensile strength and low extensibility, which makes it suitable for certain applications. Its lustre determines its quality: The more it shines, the better the quality is, and accordingly the price of the end product varies. It also has a certain amount of heat and fire resistance added to its biodegradable features.

Almost all natural fibres including jute fibre have good insulating and antistatic properties, as well as low thermal conductivity and moderate moisture retention. These fibre materials also provide acoustic insulating properties and are manufactured with no skin irritation, as long vegetable fibres like jute, can be blended with most of the synthetic and natural fibres. Jute fibres also accepts different classes of dyes (natural, basic, vat, sulphur, reactive and pigment dyes).

Lignocellulosic natural fibres can also be blended well with wool to develop of value-added industrial nonwoven material because of its good insulation properties. By treating jute with alkali (caustic soda), the crimp, softness, pliability and appearance are improved, which aids in its ability to be spun with wool. Liquid ammonia has a similar effect on jute, as well as the added characteristic of improving flame resistance when treated it with flame-proofing agents.

Among the lignocellulosic natural fibres, jute has a long history of use in sacking, carpets and wrapping fabrics (cotton bale, woolen bale, etc.) and as packaging material in the manufacturing industry. However the major breakthrough came when the automobile, pulp and paper, and furniture and bedding industries started to use jute and its allied fibres with nonwoven and composite technology to manufacture nonwovens, technical textiles, composite production of sheet moulding compound, resin transfer moulding, vacuum-pressing techniques and injection. This chapter emphasizes the design and development of jute needle-punched nonwovens and their characterization for specific applications. Different important applications of jute and jute-based needle-punched nonwovens are also covered.

13.2.1 Design and development of jute and jute-like natural fibre-based needle-punched nonwovens

As far as the design of jute-based needle-punched nonwovens is concerned, they have an important role in the end product. Based on the product's properties, durability and applicability, there are variations in the fabric's properties. The fabric's properties depend on various factors such as the fabric's weight, the needling density, the depth of needle penetration, needle gauge, blending of jute with other fibres, the type of jute fibre used, pretreatment of jute fibre, etc. Of these various factors, the most important parameters are the fabric weight, needling density, the depth of needle penetration, and the blending of jute with other fibres. Much research is therefore concentrated on these areas. Furthermore, for the design and development of jute-based needle-punched nonwovens, two main types of designs are used: (1) statistical and (2) artificial neural network (ANN) designs. The following discussions describe these aspects.

13.2.1.1 Statistical design

Researchers have used many types of statistical design to design and predict different properties of jute-based needle-punched fabrics. Of the many statistical designs, second-order polynomial relationships is an empirical equation of second-order polynomial by Box and Behnken (1960) which was derived to predict mechanical properties. Debnath et al. (2000a) predicted properties such as tenacity and initial modulus, and physical properties such as air permeability (Debnath et al., 2000b) from results obtained from samples produced using the Box and Behnken factorial design:

$$Y = \beta_0 + \beta_1 X_1 + \beta_2 X_2 + \beta_3 X_3 + \beta_{11} X_{12} + \beta_{22} X_{22} + \beta_{33} X_{32} + \beta_{12} X_1 X_2 + \beta_{13} X_1 X_3 + \beta_{23} X_2 X_3$$

where Y = the predicted fabric property (tenacity or initial modulus or air permeability); X_1 = the fabric weight; X_2 = the needling density; X_3 = the percentage of polypropylene; β_0 is the constant; and β_i is the coefficient of the variable X_i. The predicted values of fabric properties were then compared with the actual values and error (E) was calculated:

$$E\% = \left(\frac{A - P}{A}\right) \times 100$$

where E is the error (in percentage), A is the actual experimental values, and P is the predicted values from models.

13.2.1.2 Artificial neural networks modeling

The physiology of neurons in a biological neural system such as the human nervous system was the fundamental idea behind developing ANNs. This ANN is basically complex computational modelling and generally is used when the relation between the input and output parameters is nonlinear in nature. This computational model was trained to capture the nonlinear relationship between input and output variables with scientific and mathematical basis. Currently, the most commonly used model is a layered feed-forward neural network with multilayer perceptions and back-propagation learning algorithms (Vangheluwe et al., 1993; Rajamanickam et al., 1997).

ANNs are computing systems composed of a number of highly interconnected layers of simple neuron-like processing elements which process information by their dynamic response to external inputs. Information passes through the complete network by linear connection with linear or nonlinear transformations. The weights are then determined by training the neural nets. Once the ANN is trained, it is used to predict new sets of inputs. A multilayer feed-forward neural network architecture (Fig. 13.1) is used to predict the tenacity, initial modulus, air permeability, initial thickness, percent compression, thickness loss and compression resilience properties of fabrics (Debnath et al., 2000a,b; Debnath and Madhusoothanan, 2008; Debnath, 2011, 2014b). The circles in Fig. 13.1 represent neurons arranged in five layers as one input, one output and three hidden layers. There are three neurons in the input layer and three hidden layers, with each layer consisting of three neurons and one neuron in the output layer. HL-1, HL-2 and HL-3 are the first, second and third hidden layers, respectively, whereas i and j are two different neurons in two different layers. The neuron (i) in one layer is connected to the neuron (j) in next layer with weights (W_{ij}), as presented in Fig. 13.1.

The data are then normalizing with their respective values scaled down between 0 and 1. The ANN is trained with known experimental sets of input–output data pairs.

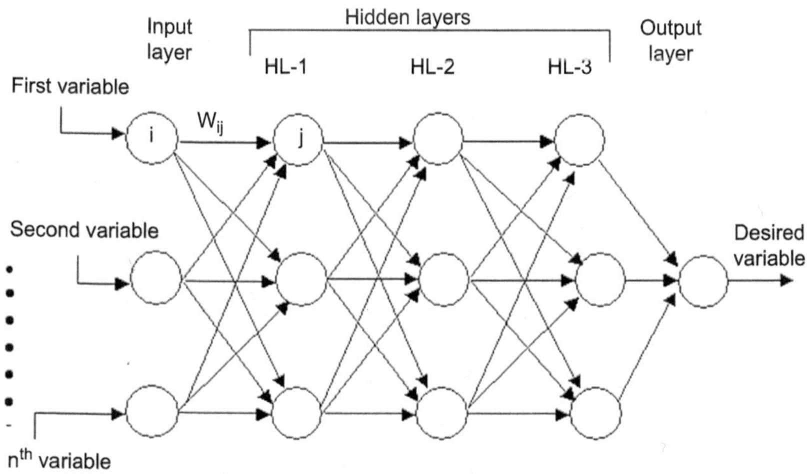

Figure 13.1 Neural architecture of the fabric's property (Debnath, 2016a).

13.2.2 Evaluation of different properties of nonwovens for specific end uses for product diversification and added value

Unlike other textile materials, natural fibre-based needle-punched nonwovens, including jute needle-punched ones, require specific tests to be evaluated to match their suitability for specific applications. This is because the jute fibre has unique features which others do not have, such as high moisture retention, coarseness, biodegradability and high surface roughness. The subsequent discussion involves some important tests that are carried out to determine jute's suitability in some of those areas of application.

13.2.2.1 Measurement of tensile property

Tensile is one of the most important properties to be evaluated for any textile material. To evaluate this property, all samples are subjected to a tensile strip test on a Universal Tensile Tester following standard test method American Society for Testing and Materials (ASTM) D1682-64 under standard atmospheric conditions [$65\% \pm 2\%$ relative humidity (RH) and $27°C \pm 2°C$]. For needle-punched nonwovens, the sample specimen width is 25.4 mm and the gauge length is set at 75 mm. The strain rate has to be set in such a way that failure of the sample occurs at 20 ± 3 s, depending on the type of fabric sample. Maximum load (in terms of Newton units) and breaking extension in percentage have to be recorded. From the value of the breaking load, tenacity values are calculated by the following formula (Debnath et al., 2006):

$$T = \frac{BL}{SW \times FW} \times 100$$

where T = fabric tenacity (cN/tex); BL = breaking load (N); SW = specimen width (mm); and FW = fabric weight (g/m^2).

The initial modulus and secant modulus are also calculated from the respective stress–strain curves.

13.2.2.2 Measurement of air permeability

Jute needle-punched nonwoven fabrics are porous in nature (they have voids) and hence they are permeable. To study air permeability, a test is carried out. Evaluation of air permeability nonwoven fabric samples can be done using any reliable air permeability tester (e.g., the Shirley Air Permeability Tester). Results are expressed as the units of volume of air in cubic centimetres passed per second through 1 cm^2 fabric at a pressure difference of 10 mm or 1 cm head of water. In some cases, however, the range of flow meter available in the instrument may not achieve the high flow rate required for this specific pressure difference. In such a case, superimposed layers of fabric are tested one at a time, and the flow rate reading for a single layer is calculated by multiplying the result by the number of layers. The air permeability value is calculated by dividing the flow rate reading in cubic centimetres per second at 1 cm pressure head of water by the test area, which in this instrument is $5.07\,cm^2$ ($1\,in^2$). The sectional air permeability (SAP) value is used to compare the permeability of different fabric samples. The SAP value of all samples is determined from the following formula (Debnath and Madhusoothanan, 2010b; Debnath et al., 2006; Debnath, 2016a):

$$SAP = AP \times T_0$$

where *SAP* is the sectional air permeability in $cm^3/s/cm$; *AP*, the air permeability in $cm^3/cm^2/s$; and T_0, the mean thickness in metres at 1.55 kPa pressure of the fabric sample. It is recommended that this test be carried out at 65% ± 2% RH and 27°C ± 2°C after the sample is conditioned for 24 h. For each sample, 10 tests are performed.

13.2.2.3 Measurement of bending property

Stiffness of samples of the jute needle-punched nonwoven fabric used for technical textiles is investigated by examining different bending parameter, e.g., flexural rigidity, bending modulus at 65% ± 2% RH and 27°C ± 2°C after the sample is conditioned for 24 h. The flexural rigidity of the fabric samples is determined by measuring the bending length following the standard cantilever principle using the stiffness tester. The average value is calculated from 10 specimens in both the warp and weft directions of the woven fabrics. Similarly as for nonwoven samples, both machine and cross-directions of the fabric specimens are considered. The specimen size of each sample is 150 mm × 25 mm for more flexible fabrics. The bending length is measured directly from the instrument. Flexural rigidity is calculated using the following formula (Debnath et al., 2006; Debnath, 2016a):

$$FR = \frac{WC^3}{10^4} \times 9.81$$

where *FR* = flexural rigidity in Newtons (Newton/cm); *W* = fabric area density (g/m^2); and *C* = bending length (cm).

For bending modulus, the thickness of each sample was measured under the pressure of $1\,lb/in^2$ using a thickness tester. The bending modulus of the samples was determined from the following formula:

$$BM = \frac{12G}{g^3}$$

where *BM* = the bending modulus in Newton/cm^2; and *G* = *g* = the thickness (cm) of the sample at $1\,lb/in^2$.

In the case of a thick and rigid fabric, this method of measuring the bending length is not valid. In such a case, a loop method is suggested (Debnath et al., 1994).

13.2.2.4 Design measurement of water imbibition property

To measure water retention characteristics, samples are cut into equal size of 4×4cm and weight under the following natural atmospheric conditions: for example, a dry bulb temperature of 87°F, wet bulb temperature of 81°F and RH of 77%.

The samples are soaked in distilled water for about 48h. Later, all samples are hung in free air for about 40min to allow excess water absorbed by the samples to drip out. These samples then kept on blotting paper for 5min to absorb excess water present on the surface of the samples. Then the weight of the wet samples is taken to find the absorbency of water using the following formula (Debnath et al., 2006, Debnath, 2014a):

$$WA = \frac{W_{wet} - W_{dry}}{W_{dry}} \times 100$$

where WA is water absorbency (percentage), W_{wet} is the weight of the fabric under wet conditions and W_{dry} is the weight of the fabric under dry conditions.

The release of moisture with respect to time is determined for all samples they achieve a constant weight.

13.2.2.5 Compression

One recent interesting area of investigation into mechanical properties is compression as far as geotechnical applications are concerned. It is an important property for evaluating the performance of needle-punched nonwoven jute fabric for some specific application such as for a carpet or geotextile. For this test category, some major properties such as initial thickness, compression, loss of thickness and compression resilience can calculated from compression and decompression curves. To measure these properties, a thickness tester is essential. The pressure foot area is 5.067 cm^2 (diameter = ϕ2.54cm). The dial gauge with a minimal count of 0.01mm and maximum displacement of 10.5mm is attached to the thickness tester. The compression is studied under a pressure range between 1.55 and 51.89kPa. This can also be measured using an advanced electronic/digital thickness gauge.

The initial thickness of the needle-punched fabrics is observed under 1.55kPa pressure. Corresponding thickness values are observed from the dial gauge for each corresponding load of 1.962N. A delay of 30s is given between the previous load and the next one applied. Similarly, a 30-s delay is allowed during the decompression cycle at each individual load of 1.962N. This compression and the recovery thickness values for corresponding pressure values are used to plot the compression–recovery curves.

From the compression decompression curve, the percent compression, percent thickness loss and percent compression resilience are estimated using the following three relationships (Debnath and Madhusoothanan, 2007):

$$C = \frac{T_0 - T_1}{T_0} \times 100$$

$$TL = \frac{T_0 - T_2}{T_0} \times 100$$

$$CR = \frac{W_c'}{W_c} \times 100$$

where C is the compression (percentage), TL is the thickness loss (percentage), CR is the compression resilience (percentage), T_0 is the initial thickness, T_1 is the thickness at maximum pressure, T_2 is the recovered thickness, W_c is the work done during compression and w_c' is the work done during the recovery process.

An average of 10 readings from different places for each sample needs to be considered. The coefficient of variation should be within 6%; if not, more readings are necessary.

All of these tests must be carried out under standard atmospheric conditions of 65% ± 2% RH and 20°C ± 2°C. The fabrics are conditioned for 24 h under these atmospheric conditions before testing.

Compression under wet condition

For some specific applications such as geotextiles, jute and jute-based needle-punched nonwoven fabrics require special conditioning to test fabrics under wet conditions. To study the compression behaviour of needle-punched nonwoven fabrics under wet conditions (Debnath and Madhusoothanan, 2012a,b, 2013), samples have to be cut into pieces 25 × 25 cm and soaked in distilled water for 24 h before the experiment is conducted. After 24 h of soaking in distilled water, the samples are passed through a padding mangle with uniform pressure exerted upon them by applying a constant load on either side of the top roller. The pressure between the padding mangle and the speed are maintained constant. Under this condition, it can be assumed that the water pickup of the sample remains constant. The sample is then immediately tested for compression. The initial thickness, compression, loss of thickness and compression resilience are calculated from the compression and decompression curves. The compression, loss of thickness and compression resilience can be calculated from the relationship mentioned in Section 13.2.2.5.

13.2.2.6 Compression creep

There are some instances in which jute-based needle-punched nonwoven fabrics are used for floor mats and carpets. In such an application, compression creep is an essential parameter to be evaluated for compression deformation over a time. An instrument has been suggested, as shown in Fig. 13.2, to measure the compression creep of needle-punched non-woven fabrics (Debnath and Madhusoothanan, 2012a,b).

The instrument is capable of measuring compression creep four samples at a time. Fabric samples are mounted under a known compression load with a dial gauge (D) to measure the thickness under the compression. The pressure foot (C) diameter is chosen to be 25.4 mm (1 in) for all samples. The smallest count of the dial gauge (D) is 0.01 mm. The nonwoven sample is placed between the pressure foot (C) and the anvil (I). The initial

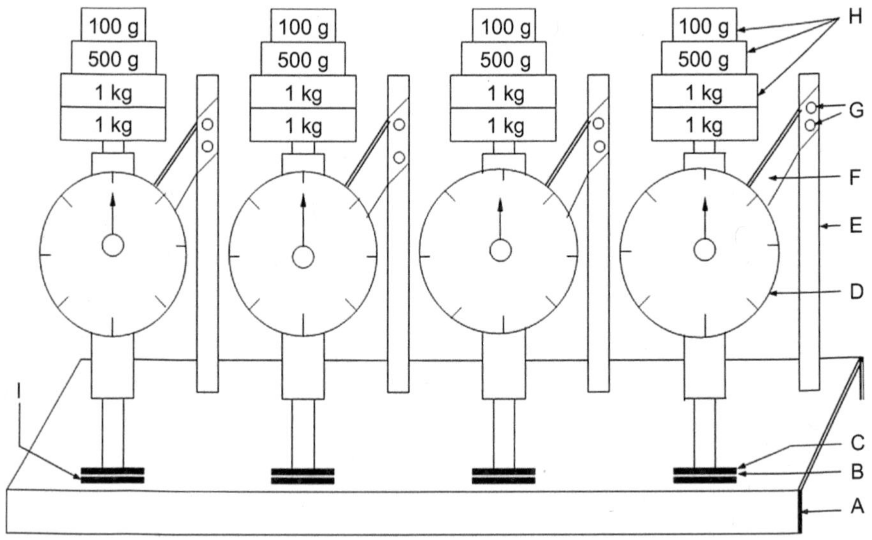

Figure 13.2 Instrument for measurement of compression creep (Debnath, 2016a).

thickness can be noted without adding an additional load. The instant deformation (T_0) is noted from the dial gauge after applying the compression load (H) of 25.51 N (2.6 kg) at the top of the weight pan (J) of dial gauge (D). The selection of this pressure foot diameter of 25.4 mm and compression load of 2.6 kg are meant to simulate the area of the foot of a normal chair and the weight of a normal empty chair, respectively, placed on a carpet surface through each leg of the chair. To study compression creep, each sample is allowed to remain under this compression load for 168 h (7 days). This compression creep time was selected because it was found from a previous study that all the fabric compression creep reaches equilibrium within 168 h. The compression creep (percentage) is evaluated from thickness deformation values, which are measured using the following relation:

$$CR = \frac{T_0 - T_t}{T_0} \times 100$$

where CR is the compression creep (percentage), T_0 is the instant thickness after applying the maximum load and T_t is the thickness after 7 days.

An average of 10 readings from different places is considered for each sample. The coefficient of variation should be less than 6% in all cases. The fabrics required conditioning for 72 h in this condition before carried out in the experiment.

13.2.2.7 Thermal resistance

The thermal resistance of a textile material is usually defined as the ratio of the temperature difference between two faces of the fabric material to the rate of the flow of heat per unit area normal to the faces. It is analogous to electrical resistance in the case of current flow through an electrical conductor. Disc method, an application of Lee's disc apparatus to textiles, was employed to evaluate the thermal resistance of polyester needle-punched nonwoven fabric samples. In this method, the material under test is held between two metal discs, of which one has known thermal resistance. Under steady conditions, the temperature drop across the metal disc with known thermal resistance and across the material under test is measured; from the values obtained, the thermal resistance of the specimen is determined by the following techniques.

Let TR_k and TR_s be the thermal resistance of the known disc and the sample under test, respectively. Let t_1 be the temperature registered by the lower surface of known disc, t_2 be the temperature registered by the lower surface of the sample under test and t_3 be the upper surface of the sample under test. Assuming a constant rate of flow of heat at steady state, TR_s is computed from the following formula (in degrees $\text{K m}^2/\text{W}$):

$$\frac{t_1 - t_2}{TR_k} = \frac{t_2 - t_3}{TR_s}, \quad \text{or } TR_s = TR_k \times \frac{t_2 - t_3}{t_1 - t_2}$$

In this experiment, a guarded two-plate thermal resistance instrument is used to measure the thermal resistance of jute-based needle-punched fabrics (Debnath and Madhusoothanan, 2010b; 2011a,b; Debnath, 2016b) (Fig. 13.3). The thermal resistant instrument is based on a microprocessor and provides automatic results of thermal resistance value in 'tog'. The area of the test specimen used is $706.85\,\text{cm}^2$ (diameter of 30 cm). The test is nondestructive and the process of preparing the sample is free from human error. Thermal insulation of each fabric sample is measured randomly at five different places under a pressure of 0.3352 kPa. An average of five readings was considered and the coefficient of variation of readings was <2%.

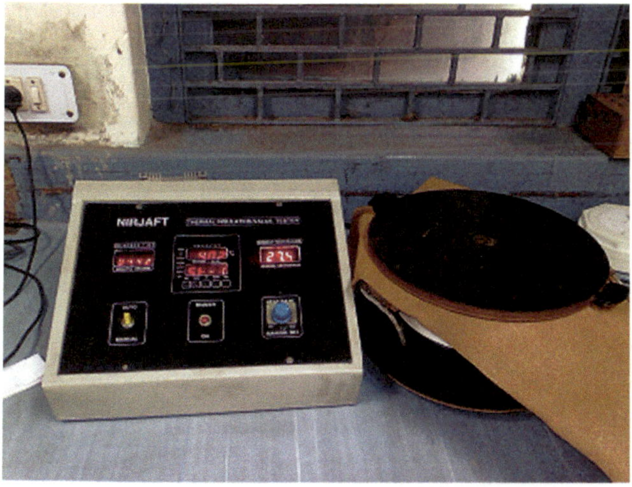

Figure 13.3 Instrument for measuring the thermal resistance of fabrics (Debnath, 2016a).

The specific thermal resistance (STR_s) value is use to compare the thermal resistance of different fabric samples. STR_s values of all samples are determined using the following equation:

$$STR_S = \frac{TR_S}{T_0} \times 100$$

where STR_s is the STR_s (in Km^2/W); TR_s, the thermal resistance value of fabric (in Km^2/W); and T_0, the mean thickness in metres at 1.55 kPa pressure of the fabric sample.

13.2.2.8 Evaluation of water absorbency

The fabric samples are cut into equal size of 4×4 cm and placed under standard atmospheric conditions of 65%±2% RH and 20°C±2°C. The fabric is conditioned for 24 h under these atmospheric conditions and the dry weight (W_D) is measured. To study water absorbency, samples are dipped in distilled water for 24 h to ensure uniform soaking of water, and then wet samples are hung in free air for about 30 min to allow excess water absorbed by the samples to drip out. The weight (W_W) of wet samples is then measured. Water absorbency (Debnath and Madhusoothanan, 2010a) is calculated using the following relationship:

$$WA\,(\%) = \frac{(W_W - W_D)}{W_D} \times 100$$

The average of 10 results is considered.

13.2.2.9 Evaluation of soil moisture control

Soil moisture can be controlled efficiently by jute agrotextile. This is essential to maintain soil moisture for a longer duration, and thus the need for water during irrigation can be minimized. A suitable jute agrotextile can be used as a mulching material. To find the efficacy of the performance of a jute-based mulching material to control soil moisture, two simulated methods are recommended: noncontact and contact. The importance of these two different methods are that during application of the agrotextile the fabrics are not likely to touch the contours of the soil surface uniformly along their length or width. Following are two methods recommend for measuring soil moisture control using jute needle-punched nonwoven fabrics (Debnath et al., 2010).

Noncontact method

In some places the mulching needle-punched non-woven agrotextile will not come into contact with the soil surface because of the contours of the soil. To study the moisture of soil on those noncontact areas of the fabric, this method is considered. In this measurement, six 100-mL beakers are used containing 25 mL water in each beaker. The mouths of five of these beakers are tightly covered with the fabric samples; the rest are kept as controls with no fabric covers. The beakers that contain water and the samples are initially weighed; and their weights are taken repeatedly at certain time intervals until they reached a constant weight. This experiment is conducted under normal atmospheric conditions. At each time interval, the percent evaporation loss is calculated from the weight difference of the individual beaker assembly at that point in time with respect to its initial weight. The results are plotted as a time versus cumulative evaporation loss to compare the performance of various agrotextile fabrics of different fabric weights.

Contact method

In this test, six similarly sized Petri dishes are used containing 90g standard alluvial soil with saturated moisture in each dish. The five Petri dishes are covered with fabric samples in such a way that the fabric has direct contact with the moist soil. The control dish is not covered with sample fabric. To determine the properties of the moisture barrier, the weight of each dish with the soil and sample is initially measured. These weights are taken again at certain time intervals until all dishes reach a constant weight. The cumulative evaporation loss is calculated in a similar method as in the noncontact process. The results are plotted as time versus cumulative evaporation loss to compare the performance of various geotextile fabrics of different fabric weights. This experiment is replicated five times and the average of five readings for each sample is calculated.

13.2.2.10 Evaluation of soil temperature control

The application of jute agrotextile mulch has an influence on controlling the soil temperature. This factor depends on various aspects of the design of jute needle-punched nonwoven fabric and the parameters selected during the manufacture of the fabrics. In this test, jute needle-punched nonwoven and commercial plastic mulch (50 g/m^2 black polyethylene sheet and 20 μm thick) materials are compared with respect to control soil (with no mulch). A plot is made to determine soil temperature behaviour with respect to time against control (Debnath et al., 2010).

Three clay pots of the same size and shape are used and the same quantity of alluvial soil is filled in each pot. The same quantity of water is poured into each pot until saturation is achieved and excess water pours out. Under these conditions, the surface water is allowed to evaporate for 24h so that the same ambient conditions are attained in all pots. Needle-punched jute nonwoven and commercial black polyethylene sheets (50 g/m^2 microns) are laid on the top of the soil on each pot. Utmost care is to be taken no part of the soil surface is left exposed or uncovered by the agrotextiles or experimental materials. Before the mulching materials are covered, the soil temperature is measured in each pot. The temperature is measured approximately 1 in below the soil surface at five different spots and averaged to determine the mean temperature. Soil temperatures are determined identically as previously at a fixed interval of 2h. This study needs to continue for 4 days and the temperature should be recorded for all three pots including the control; it can be represented graphically.

13.2.2.11 Measurement of sound insulation

Natural fibre-based nonwoven materials are used for different sound insulation purposes; hence the evaluation of sound insulation is essential. An unwanted noise as a part of sound is irritating and causes mental and psychological problems for humans. Jute needle-punched nonwoven textile materials have good properties related to controlling sound if they are properly designed for this purpose. A simple design for an instrument to measure the sound insulation of textile nonwoven materials was been explained by Sengupta (2010). It consists of a sound insulating box made of 12-mm-thick rigid plastic (acrylic sheet) and has a removable top lid. Inside one vertical wall of this box, a sound source and a decibel meter (S) are fixed. In another movable vertical wall (to adjust the distance between sound source and receiver), a decibel meter (R) is fixed coaxially opposite to the sound generator to measure the sound intensity. Between these 2-dB meters, a sliding arrangement is there (to adjust the distance between the sound source and the fabric) to fix the fabric sample vertically. An electrical panel is used to control the sound intensity. Fig. 13.4 is a schematic diagram of the sound insulation tester as described (Debnath, 2016a).

By operating the control panel, a particular decibel sound is created. The source decibel and receipt decibel are measured by 2-dB meters S and R, respectively, without and with fabric samples. The sound reduction for which the fabric is responsible, in which sound insulation is expressed as the difference between the decibel reduction with sample and decibel reduction without the sample, is:

$$db_F = (dB_S - dB_R)_{WS} - (dB_S - dB_R)_{WOS}$$

where db_F is the sound reduction caused by the fabric; dB_S is the sound intensity at the source; dB_R is the sound intensity at the receiver; WS is with the sample; and WOS is without the sample.

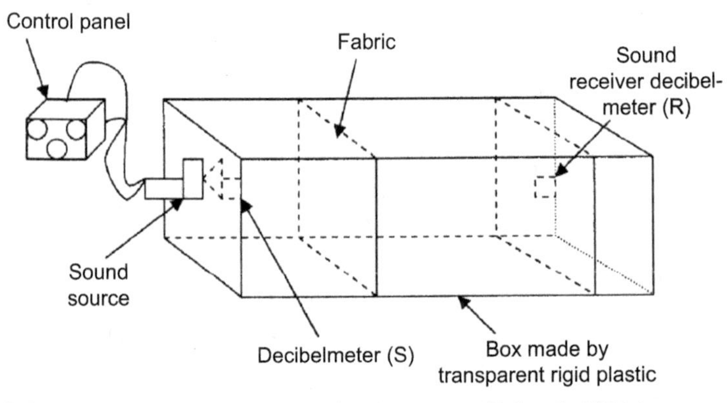

Figure 13.4 Schematic diagram of a sound insulator tester (Debnath, 2016a).

13.2.2.12 Measurement of electrical resistance

After electricity was discovered, textile materials began to be used extensively in as electrical insulation. Textile materials having good electrical insulation properties provided they are dry. There are many such applications where textile materials (fibres, yarn and fabric) are used for electrical insulation. To understand these electrical resistance properties of these textiles, industrial nonwoven materials are subjected to tests. For these measurements, the electrical insulation of the textile materials is used at medium voltage ranges (within 415 V). To measure the electrical insulation, a circuit is necessary which measures the current–voltage (V-I) characteristics of textile material, as shown in Fig. 13.5 (Sengupta and Sengupta, 2013).

This circuit consists of a sample holder (S), an autotransformer (B), a rectifier unit (D), a ammeter (A), a voltmeter (V) and 10 MΩ discrete resistance (R). Two bulldog clips are used between which the sample is placed (Fig. 13.6) with gauge lengths of 2.54, 5.08 and 10.16 cm. The sample is in series with a known resistance (10 MΩ) and is connected to a DC power supply. The voltage varies from 80 to 220 V in five steps, and corresponding current through the sample is measured after 10 s to calculate resistance (Sengupta and Debnath, 2014). The measurement is done at 27°C and 65% RH. Five sets of voltage–current readings are taken to plot V-I characteristic curves. The slope (voltage–current) or resistance for each V-I characteristic is determined. The specific resistance is calculated normalizing the resistance by thickness [multiplying or dividing resistance (mega ohm) by thickness (cm)].

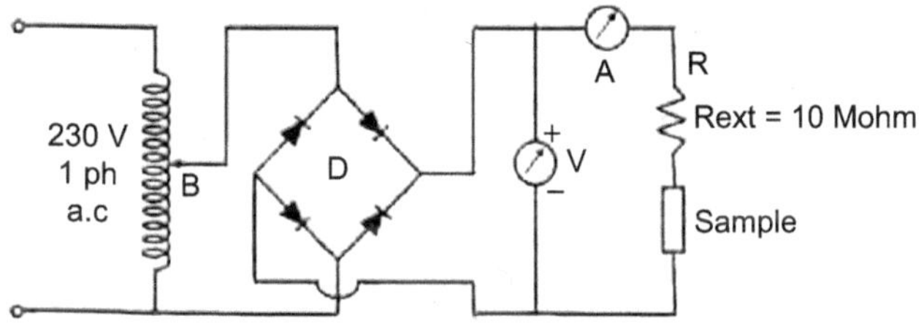

Figure 13.5 Circuit for electrical resistance of nonwoven (Sengupta and Sengupta, 2013).

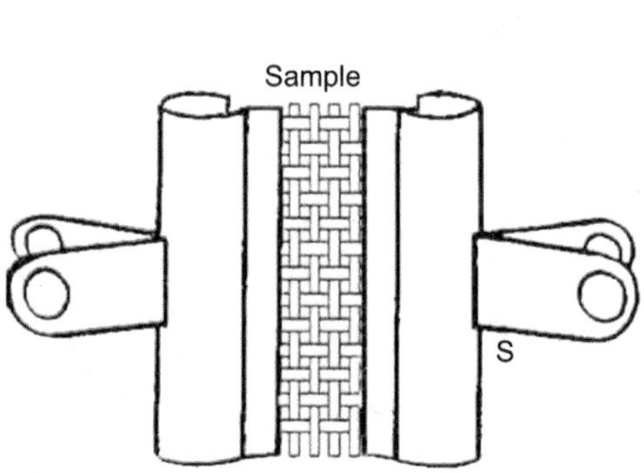

Figure 13.6 Sample holder.

13.3 Sustainability of lignocellulosic fibre-based nonwovens in future technical textiles

13.3.1 *Jute-based nonwovens for sustainable technical applications*

Debnath (2017) commented on the sustainability of the jute industries in the context of India and other jute-growing countries. Roy and Ray (2009) optimized different fabric properties, such as the bursting strength, compressibility, bending modulus, and air permeability from jute needle-punched nonwovens, which are of great importance in different fields of application. They found that the bursting strength of fabric increases with the increase in fabric weight, punch density, and depth of needle penetration, but beyond some optimum punch density and depth of needle penetration, the bursting strength of fabric shows a declining trend. The changes in fabric thickness loss and energy loss follow a similar trend during thickness compression and recovery. The compressibility of fabric in terms of thickness loss or energy loss is initially reduced with an increase in the fabric's weight, punch density, and depth of penetration, and after attaining a minimum value it starts to increase if the increase in these variables is continued further. The bending modulus of fabric increases with the increase in the fabric's weight, whereas after attaining a maximum value the fabric-bending modulus starts to decline with an increase in the punch density and depth of needle penetration. With the increase in fabric weight, punch density, and depth of penetration, the air permeability of the fabric is first reduced owing to better consolidation of the fibres in the fabric. Then it shows an upward trend owing to channels or near-hole formation in the fabric. From the manufacturers' point of view of the fabric, the most important conclusion they drew about the equation was that a particular value of the response (bursting strength, air permeability, etc.) can easily be achieved from different sets of values of the same variables, such as fabric weight and punch density. The manufacturer can select any of these sets of values of variables according to their choice. As far as jute nonwovens are concerned, Debnath (2016a) thoroughly described the design, development, characterization and applications of jute-based needle-punched nonwovens regarding their sustainability. Because jute is an important component for lignocellulosic material productivity, there is much discussion in the literature about sustainable thermal insulating material for technical textiles (Debnath, 2013). Debnath (2014a) established that jute-based nonwovens can be a suitable alternative in sustainable agriculture to control the soil moisture through mulch, control of soil temperature, soil loss/erosion control, cannel lining material, etc. Jute nonwoven agrotextiles will be an integral part of soil after crop harvesting, which leads to sustainable agricultural practices. Nag et al. (2010) studied the application of jute nonwovens in agrotextile, especially in horticultural crops. They concluded from the study that jute nonwovens can manage the soil moisture efficiently and are suitable for the cultivation of horticultural crops in red lateritic soil. Lupu et al. (2013) studied the effect of needling density and needle gauge on the functional characteristics (fabric weight, fabric thickness and thermal conductivity) of nonwovens used as agrotextiles for the frost protection of crops in a temperate region. They determined the optimal processing parameters that affected performance characteristics using a central composite experimental design for a second-order model. Results showed that an increase in needling density influenced the fabric weight more prominently than did an increase in needle gauge. For the same needle gauge, an increase in needling density also increased the thermal conductivity. At the same time, needling density and needle gauge had opposite effects on the fabric thickness. A detailed study on the drapeability of woven and nonwoven structures jute fabric were compared with polypropylene woven, polypropylene nonwoven and polyester nonwoven structured fabrics by Ghosh et al. (2015). The test results concluded that nonwoven materials as well as those woven from jute had good potential in geotechnical engineering as well as agrotextile management, where technical textiles were gradually improving in efficacy. As far as sustainability is concerned, jute-based nonwovens have a huge market apart from their biodegradable aspects. The current jute industry will be benefited by diversifying conventional products such as sacking and Hessian, and the earth can be saved.

13.3.2 Sisal-based nonwovens towards sustainable green composites

Over the past few decades, there has been a lot of research on the use of natural fibres such as jute, silk and sisal as important fillers or reinforcing material in plastics because of their low cost, low density, flexibility, rough surface and reduced wear from processing machinery (Raj et al., 1989). The application of natural fibres in composite making can reveal a new direction for the material sciences as far as sustainability is concerned if it is considered in the broader spectrum. The composite consists of two or more components: one is fibrous material widely known as reinforcing material and another is matrix, also called resin. They are combined to provide structural or functional properties which are not present in any individual component used. In most cases the fibrous nonwoven web in the form of parallel-laid/cross-laid/air-laid/random-laid is used. Sometimes for easy handling during composite preparation, needle-punched/thermal-bonded/adhesive-bonded fibrous materials are used as reinforcing material. Considering these factors, the composites are made from natural fibres as reinforcing material and natural resin/gum/adhesive are used, known as green composite, which is purely biodegradable. Nair et al. (1996) studied the tensile properties of polystyrene reinforced with short sisal fibre and benzoylated sisal fibre for composite making. They examined the influence of fibre length, fibre content, fibre orientation, and the chemical modification of the fibre (benzoylation of the fibre) on the tensile properties of the composites. The benzoylation of the fibre improves adhesion of the fibre to the polystyrene matrix. The benzoylated fibre was analysed by infrared spectroscopy and other experiments; it indicated better compatibility between benzoylated sisal fibre and polystyrene. Benzoylation of the sisal fibre was found to enhance the tensile properties of the resulting composite. The tensile properties of unidirectionally aligned composites showed a gradual increase with fibre content and leveling off beyond 20% fibre loading. The properties were almost independent of fibre length, although the ultimate tensile strength showed marginal improvement at a fibre length of 10 mm. The thermal properties of the composites were analysed by differential scanning calorimetry. Scanning electron microscopy was used to investigate the fibre surface, fibre pullout, and fibre–matrix interface. Finally they developed theoretical models to fit the experimental mechanical data of sisal fibre-based composite. Considering the high strength and durability of sisal fibre, over early 2000 years, development of biodegradable composites based on treated and untreated sisal fibre have been demonstrated by Iannace et al. (2001). Their study disclosed the effect of processing conditions (temperature, speed of rotation and time of mixing) and alkaline treatment on the dimensions of sisal fibre. They found that the length and diameter of the initial fibres were reduced during the mixing process and this effect was correlated with the magnitude of the shear stress developed in the mixer. They also concluded that an increase in the speed of rotation during the mixing process and/or a reduction in temperature produced fibres of smaller dimension but with a higher aspect ratio, *l/d* (length/diameter), of sisal fibres. Furthermore, alkaline-treated fibres increased the kinetics associated with the reduction in the fibre'2s dimensions. They also developed a semiempirical model to predict the size of the fibres versus the time of mixing, which was accepted by the experimental values (Iannace et al., 2001). A sisal fibre web nonwoven has been tried widely in the composite development for different applications. The nonwoven fabric made out of sisal has tremendous potential in development of thermal insulation material, as evidenced by research conducted by Dorivalda and George (2009). They analysed the effects of superficial treatment on the thermal performance of needle-punched nonwovens made of sisal fibre used as thermal insulation material through a series of experiments. They used heavy nonwoven fabric weighing between 1300 and 1400 g/m^2 and further treated it with 2% w/w sodium hydroxide to improve the insulation of the nonwovens. External electrical energy was used to heat the samples to 220°C at different durations in the experiment, to study the thermal stability of the sisal-based nonwovens. From the thermogravimetric analysis, they verified that the temperature limit was 200°C for the untreated sisal nonwovens and 220°C for the NaOH-treated fabrics. It is clear that the heat resistance of any material depends on the exposure time, a fact demonstrated in their study, in which fibres under both conditions began a carbonization process after being submitted to 112°C for 30 h. Both treated and untreated samples showed satisfactory thermal performance when applied to the insulation of tubes heated with a thermal power of 22.5, 40.0 and 62.5 W. Hence, it was believed that better results would be obtained if superficial fibre treatment were performed before the nonwoven mat manufacturing process. Visual analysis showed practically no coloration change on the mat surface (treated and untreated) in contact with the heated pipe surface at a mean temperature of 77°C after 30 h. However, at a mean temperature of 112°C, the surface colour of the untreated nonwoven changed from yellowish-white to light brown, a process attributed to the onset of carbonization of the fibres in contact with the heated tube surface. This alteration was less intense in the treated nonwoven. When the tube surface was maintained at a mean temperature of 155°C, fibre carbonization occurred in both cases, although a little less intensely in the treated nonwoven. Indu and Senthilkumar (2016) developed a nonwoven from coir and sisal-blended material and found that the nonwoven fabric made from 30% sisal and 70% coir performed at higher absorbency. This was because of the high moisture content of the coir, 12.3%, compared with the moisture content of sisal, which is 10.339%. The ultimate sustainability in the composite industry is with green/biocomposites, in which both the resin and reinforcement are biodegradable. Industry based on this biocomposite will be benefited and the biodegradability of the end product will not be problematic for the society.

13.3.3 Flax, kenaf, hemp and ramie fibre-based nonwoven mats for industrial application

Within the category of long vegetable bast fibre, an extensive review was made by Pandey (2007) on developments in ramie fibre and its potential industrial uses, especially in the area of nonwoven textiles apart from traditional textile applications. Researchers also investigated the other similar long vegetable fibre-based composites. Hemp fibre is a strong natural fibre similar to flax, kenaf and ramie; it possesses good fibre quality and physical properties of strength, durability and absorbency. To explore the possibilities in this line, Chen et al. (2007), studied the physical properties of a hemp–polypropylene composite and compared them with those of a kenaf–polypropylene composite, a ramie–polypropylene composite and a bagasse–polypropylene composite. The hemp–polypropylene nonwoven was produced by carding and needle-punching techniques and was thermo-bonded to form a hemp composite. They conducted a preliminary study on the ability of the hemp fibre composite to resist termite attack. The hemp composite featured better thermal property; the kenaf composite had better mechanical properties; the ramie composite indicated better acoustical properties; and the bagasse composite exhibited better wet properties. Neither the hemp nor the bagasse fibre composite was capable of resisting termite attack. Maity et al. (2014) made an extensive review of flax fibre-based nonwovens. Their review described different technologies used to manufacture nonwoven fabrics from flax fibres and flax waste. Carding, air-laying and wet-laid techniques could be used for web formation. Needle-punching, hydro-entangling, stitch bonding and thermal bonding can be used for web bonding. These processes are described in brief with the results for flax fibre. Furthermore, different physical, mechanical and functional properties of flax nonwovens reported by various researchers have been described. Finally, potential areas of applications for flax nonwovens, such as filtration, medical textiles, automotive textiles, composites and geotextiles were highlighted. In the area of civil engineering, Rawal and Anandjiwala (2007) studied needle-punched nonwoven geotextile structures made from flax and polyester fibres, considering their wide applicability in geotextiles used in civil and geotechnical engineering. In this study, they compared the properties of needle-punched nonwoven geotextiles produced from polyester and flax fibres. The properties of nonwoven geotextiles investigated were density, pore size and permeability. They found that large inherent variations in flax fibre length and fineness could result in a loss of tensile strength and cause large variations in the smallest detected pore diameter. Nevertheless, flax fibre-based geotextiles have great potential in various civil engineering applications because they are less anisotropic and more compact and have an 'open structure'. Furthermore, the influence of process parameters such as feed rate, stroke frequency and depth of needle penetration on the properties of the geotextiles was observed (Rawal and Anandjiwala, 2007). Xue et al. (2011) suggested in their study that the resin flow characteristics of natural fibre mats made from flax fibre must be taken into consideration in the design and optimization of liquid composite moulding processes. In their study they studied two common types of nonwoven mat structures produced by the parallel-laying process and the cross-laying process. Flax fibre orientation distribution characteristics of these mats were determined using a fast Fourier transform image analysis algorithm. Flax fibre mats (nominal weight, 750 g/m^2) produced by the parallel-laying method had a prominent principal direction of fibre orientation, whereas mats produced by the cross-laying method had a nearly random fibre orientation. The parallel-laid mats had a higher overall compressibility than the cross-laid mats did because of the tendency to nest between layers of parallel fibres, leading to a lower porosity under the same vacuum level, and thus lower overall permeability. On the contrary, the cross-laid nonwoven mats confirmed nearly isotropic in-plane permeability, whereas the parallel-laid mats had much higher permeability in the principal fibre direction than in the orthogonal direction. A study by Fortea-Verdejo et al. (2016) examined flax fibre nonwovens typically in a web which are mechanically bonded (via entanglement and interlocking of fibres) or thermally bonded (by melting of polymer fibres). In that study they showed that bacterial cellulose (BC) can be used as an effective binder to produce rigid and robust natural fibre nonwovens without the need for polymer binders. They further used various types of nanocellulose, including nanofibrillated cellulose (NFC), BC and pulp fibres, to manufacture flax nonwovens. Two preform manufacturing processes were investigated: single-step filtration and layer-by-layer filtration. Both BC and NFC served as excellent binders for loose flax fibres owing to their high surface area, whereas pulp fibres were a poor binder for flax fibres. This was attributed to the low surface area of pulp compared with BC and NFC, which led to a lower contact area between the flax fibres and pulp. Moreover, the larger fibre diameter of the flax fibre pulp resulted in poorer packing efficiency and therefore a higher porosity of 67% compared with preforms made with BC or NFC as a binder, which had a porosity of 60%. The manufactured preforms possessed excellent tensile and flexural properties. The researchers concluded that the layer-by-layer filtration process resulted in flax nonwovens that produced better tensile and flexural properties of composite, owing to the better distribution of the fibres in the form of a nanocellulose network throughout the preform. Although lot of work has been carried out in this area, more awareness is needed to involve the industry in making these natural fibre-based nonwoven materials sustainable.

13.3.4 Regenerated resin applied to natural fibre nonwoven for sustainable product

There has been much global research using PLA for different industrial areas in materials science, with the aim of making an end product that is environmentally friendly and sustainable. In this line of research on biocomposites, flax-reinforced PLA was developed using a technique incorporating an air-laying nonwoven process, as demonstrated by Shah et al. (2014). PLA and flax fibres as natural materials were thoroughly mixed and converted into webs in an air-laying process. Nonwoven prepregs were then made from the fibre webs by thermal consolidation. The prepregs were finally converted into composites by compression moulding. This study also investigated the biodegradability and water absorption properties of the composites. The composites from these biodegradable materials were incubated in compost under controlled conditions. The percent weight loss and reduction in mechanical properties of PLA and biocomposites were determined at different time intervals. The study concluded that with increasing flax content, the mechanical properties of the biocomposites decreased more during the soil burial trial. The increasing flax content led to the acceleration of weight loss as a result of preferential degradation of the flax. Furthermore, they confirmed the surface morphology of the biodegraded composites using scanning electron microscopic analysis. Morphological observations indicated severe disruption of the biocomposites' structure between 60 and 120 days of incubation. As far as sound absorption is concerned, Yilmaz et al. (2011) studied the effects of porosity, fibre size and layering sequence on the sound absorption performance of needle-punched nonwovens made from monofibre or multifibre consisting of PLA, polypropylene, glass fibres and hemp fibres. The relationships between the material parameters, i.e., fibre fineness, porosity, areal density, layering sequence and airflow resistivity, with the normal incidence sound absorption coefficient of nonwoven composites consisting of three layers were studied. From this study the air flow resistivity was statistically modeled and was found to increase with decreasing fibre size and nonwoven porosity. The effects of the layering sequence on air flow resistivity and sound absorption showed that when the layer including reinforcement fibres, i.e., hemp or glass fibre, faced the air flow/sound source, air flow resistance and the absorption coefficient were higher than when the layer including reinforcement fibres was farthest from the air flow/sound source. The difference was more pronounced when there was a greater difference between the resistivity values of the constituent layers of the nonwoven composite.

Various types of natural material-based thermoset composites prepared using acrylate-modified soybean oil resins from natural fibres and acrylate-modified soybean oil resins were demonstrated by Akesson et al. (2009). The reinforcing materials used to develop the composite were nonwovens of different types, such as a 450-g/m^2 air-laid flax mat, 460-g/m^2 air-laid hybrid mat with 90% flax and 10% polyester, 940-g/m^2 sheets of dissolving pulp, 400-g/m^2 air-laid hemp mat and 900-g/m^2 coarse hemp mat containing residual stem parts. The structural composites from natural fibres and an acrylated epoxidized resin were produced with a high content of renewable material. The resin was applied to the natural fibre mats by spraying and was cured by heating in a hydraulic press. Results showed that it is possible to produce a composite with high mechanical properties without adding a reactive comonomer such as styrene to the resin. Their study suggested that there was good adhesion between the resin and the matrix. It was possible to use a fibre content as high as 70 wt%. This is advantageous for several reasons. First, a high fibre ratio will enable high mechanical properties. Second, natural fibres are generally low-priced, and hence a high fibre ratio will lower the price of the resulting composite. Finally, because the resin contains some nonrenewable material from the resin synthesis, the higher fibre content is, the higher the amount of renewable material will be in the final composite. These natural fibre composites cannot compete in terms of mechanical properties compared with a conventional glass fibre composite. In particular, the impact strength is poor for natural fibre composites, because the glass fibre composite yielded more than double the value compared with the flax composite. However, these natural fibre composites are not far behind when the flexural properties are compared. Natural fibre composites are known to be sensitive to humidity because of the hydrophilic nature of the fibres. The aging test established this, and therefore the main applications of these composites must be restricted to indoor products.

13.3.5 Uncommon natural fibre-based nonwovens for technical application

In their research paper, Ventura et al. (2015) explored the potential of oil palm frond fibres (OPFf) for technical applications such as composite reinforcement. For this purpose, fibres obtained from frond waste were subjected to various chemical treatments with NaOH and chemically and morphologically characterized. Afterward, composites of PLA matrix and nonwoven mats developed with treated and the untreated OPFf were prepared and the mechanical properties were tested. Alkaline treatment was effective in increasing the wettability of the fibres; it led to composites with good balance between lightness and toughness for potential applications in the packaging or automotive industries.

13.4 Conclusions and suggestions for further research

This study highlighted various aspects of natural fibre nonwovens of lignocellulosic origin, their global market potential, the design and development of nonwovens, different evaluations of their properties, their applications, future research and their sustainability. It is clear from this study that jute, sisal, flax and fibres of plant origin of similar nature have a lot of potential for use in different spheres of industrial application for sustainable technical textiles. In some cases technical textiles made from plants have been found to be inferior; this can be overcome by in-depth research and investigation. However, in the coming era, jute and jute-like natural fibre-based nonwovens can serve many needs as technical textiles and replace their counterparts of synthetic origin. Hence systematic research will open up a new area for nonwovens from natural fibre-based materials. Natural fibres grow naturally and cause no harm to the environment. Their benefits even during degradation may leave the final product sustainable apart from the original benefits during cultivation of the fibre. More research is necessary to strengthen and establish the environmental impacts of natural fibre-based nonwovens after their application, because some residue may be present during their application, such as nonwoven filter media. Overall, natural fibre-based nonwoven industry shall sustain, provided more in-depth research on product development is carried out. Further, refinement of manufacturing technology of the end product and wider application in various spheres of technical textile application is necessary.

References

Akesson, D., Skrifvars, M., Walkenstrom, P., 2009. Preparation of thermoset composites from natural fibres and acrylate modified soybean oil resins. J. Appl. Polym. Sci. 114, 2502–2508.

Box, G.E.P., Behnken, D.W., 1960. Some new three level designs for the study of quantitative variables. Technometrics 2 (4), 455–475.

Chen, Y., Sun, L., Negulescu, I., Wu, Q., Henderson, G., 2007. Comparative study of hemp fiber for nonwoven composites. J. Ind. Hemp 12 (1), 27–45. http://dx.doi.org/10.1300/J237v12n01_04.

Debnath, S., 2011. Chapter 3: Modelling of needle-punched nonwoven fabric properties using artificial neural network. In: Kenji, S. (Ed.), Artificial Neural Networks – Industrial and Control Engineering Applications. InTech publisher, Rijeka, Croatia. ISBN: 978-953-307-220-3, pp. 65–88.

Debnath, S., 2013. Chapter 20: Designing of jute–based thermal insulating materials and their properties. In: Mondal, I.H. (Ed.), Textiles: History, Properties and Performance and Applications. Nova Science Publishers, New York. ISBN: 978-1-63117-274-8, pp. 218–499.

Debnath, S., 2014a. Jute-based sustainable agrotextiles, their properties and case studies. In: Senthilkannan Muthu, S. (Ed.), Roadmap to Sustainable Textiles and Clothing. Springer Publisher, pp. 327–355. http://dx.doi.org/10.1007/978-981-287-065-0. ISBN: 978-981-287-064-3 (Print) 978-981-287-065-0 (Online).

Debnath, S., 2014b. Application of artificial neural network in modelling of jute-based nonwovens. In: Nag, D., Ray, D.P. (Eds.), Jute and Allied Fibres – Processing and Value Addition. New Delhi Publishers, New Delhi, India. ISBN: 978-93-81274-41-5, pp. 189–209 (Print).

Debnath, S., 2016a. Chapter 12: Design, development, characterization, and application of jute-based needle-punched nonwoven. In: Han-Young, J. (Ed.), Non-woven Fabrics. InTech publisher, Rijeka, Croatia. ISBN: 978-953-51-4586-8, pp. 277–295. http://dx.doi.org/10.5772/61705.

Debnath, S., 2016b. Chapter 3: Thermal insulation material based on "Jute". In: Almusaed, A. (Ed.), Insulation Materials in Context of Sustainability. InTech publisher, Rijeka, Croatia. ISBN: 978-953-51-2625-6, pp. 45–56. http://dx.doi.org/10.5772/63223.

Debnath, S., 2017. Chapter 7: Sustainability in jute-based industries. Print ISBN 978-981-10-2638-6, Online ISBN 978-981-10-2639-3. In: Muthu, S.S. (Ed.), Sustainability in Textile Industry, Part of the Series Textile Science and Clothing Technology. Springer Nature Singapore Pte Ltd, Singapore, pp. 139–147. http://dx.doi.org/10.1007/978-981-10-2639-3_7. (Published online on 15th October, 2016).

Debnath, S., Madhusoothanan, M., 2007. Compression behaviour of jute-polypropylene blended needle-punched nonwoven fabrics. Indian J. Fibre Text. Res. 32 (4), 427–433.

Debnath, S., Madhusoothanan, M., 2008. Modeling of compression properties of needle-punched nonwoven fabrics using artificial neural network. Indian J. Fibre Text. Res. 33 (4), 392–399.

Debnath, S., Madhusoothanan, M., 2010a. Water absorbency of jute-polypropylene blended needle-punched nonwoven. J. Ind. Text. 393, 215–231. http://dx.doi.org/10.1177/1528083709347121.

Debnath, S., Madhusoothanan, M., 2010b. Thermal insulation, compression and air permeability of polyester needle-punched nonwoven. Indian J. Fibre Text. Res. 35 (1), 38–44.

Debnath, S., Madhusoothanan, M., 2011a. Thermal resistance and air permeability of jute-polypropylene blended needle-punched nonwoven. Indian J. Fibre Text. Res. 36 (2), 122–131.

Debnath, S., Madhusoothanan, M., 2011b. Thermal resistance behaviour of polyester needle-punched nonwoven. J. Inst. Eng. (India) 91TX (2), 27–33.

Debnath, S., Madhusoothanan, M., 2012a. Studies on compression properties of polyester needle-punched nonwoven fabrics under dry and wet conditions. J. Text. Inst. 41 (4), 292–308. http://dx.doi.org/10.1177/1528083711416394.

Debnath, S., Madhusoothanan, M., 2012a. Compression creep behaviour of polyester needle-punched nonwoven fabrics. J. Text. Inst. 103 (12), 1328–1334. http://dx.doi.org/10.1080/00405000.2012.680696.

Debnath, S., Madhusoothanan, M., 2012b. Compression behaviour of jute-polypropylene blended needle-punched nonwoven under wet condition. J. Text. Inst. 103 (06), 583–594. http://dx.doi.org/10.1080/00405000.2011.592662.

Debnath, S., Madhusoothanan, M., 2012b. Compression creep behaviour of jute-polypropylene blended needle-punched nonwoven. Text. Res. J. 82 (20), 2097–2108. http://dx.doi.org/10.1177/0040517512445336.

Debnath, S., Madhusoothanan, M., 2013. Studies on compression behaviour of polypropylene needle punched nonwoven fabrics under wet condition. Fibres Polym. 14 (5), 854–859.

Debnath, S., Kane, C.D., Kadole, P.V., Patil, C.A., 1994. Needle-punched non-woven blankets from polyesters. Indian Text. J. 105 (3), 72–80.

Debnath, S., Madhusoothanan, M., Srinivasmoorthy, V.R., 2000a. Modelling of tensile properties of needle-punched nonwovens using artificial neural networks. Indian J. Fibre Text. Res. 25 (1), 31–36.

Debnath, S., Madhusoothanan, M., Srinivasmoorthy, V.R., 2000b. Prediction of air permeability of needle-punched nonwoven fabrics using artificial neural network and empirical models. Indian J. Fibre Text. Res. 25 (4), 251–255.

Debnath, S., Nag, D., De, S.S., Ganguly, P.K., Ghosh, S.K., 2006. Studies on mechanical and hydraulic properties of JGT for geo-technical applications. J. Inst. Eng. (India) TX86 (2), 46–49.

Debnath, S., Roy, A.N., Basu, G., Chattopadhyay, S.N., 2009. NIRJAFT's technologies for rural development, Section-V, entitled Natural Fibres and Geotextile Applications. In: Shukla, J.P. (Ed.), New Technologies for Rural Development Having Potential of Commercialisation. Allied Publishers Pvt. Ltd., India. ISBN: 978-81-8424-442-7, pp. 136–142.

Debnath, S., Ganguly, P.K., De, S.S., Nag, D., 2010. Control of soil moisture and temperature by light weight jute fabrics. J. Inst. Eng. (India) 90TX (2), 16–19.

Dorivalda, S.M.N., George, S.M., 2009. Nonwoven sisal fiber as thermal insulator material. J. Nat. Fibers 6 (2), 115–126. http://dx.doi.org/10.1080/15440470802328964.

Ghosh, S.K., Bhattacharyya, R., Dey, C., Mukherjee, B., 2015. An evaluation of drapeability of jute fabric. J. Nat. Fibers 12 (2), 132–140. http://dx.doi.org/10.1080/15440478.2014.903818.

Iannace, S., Ali, R., Nicolais, L., 2001. Effect of processing conditions on dimensions of sisal fibers in thermoplastic biodegradable composites. J. Appl. Polym. Sci. 79, 1084–1091.

Indu, G.K., Senthilkumar, P., 2016. Development of nonwovens from natural fibres for various applications. Int. J. Appl. Eng. Res. 11 (3), 1879–1882.

Jan, E.G., 2016. The Netherlands Environmental benefits of natural fibre production and use. In: Proceedings of the Symposium on Natural Fibres. Van Dam Wageningen University. ftp://ftp.fao.org/docrep/fao/011/i0709e/i0709e03.pdf.

Lupu, I.G., Cramariuc, O., Hogas, H.I., Hristian, L., 2013. Parameters optimization for the production of needle-punched nonwoven agrotextiles. J. Text. Inst. http://dx.doi.org/10.1080/00405000.2013.777581.

Maity, S., Gon, D.P., Paul, P., 2014. A review of flax nonwovens: manufacturing, properties, and applications. J. Nat. Fibers 11 (4), 365–390. http://dx.doi.org/10.1080/15440478.2013.861781.

Memon, N.A., 2016. Nonwovens: Global Demand Is Expected to Reach 9.1 Million Tonnes in 2017. http://www.ptj.com.pk/Web-2014/12-2014/PDF-December-2014/Technical-Textiles-and-Nonwovens-DrNoor.pdf.

Nag, D., Debnath, S., Ganguly, P.K., Ghosh, S.K., 2010. Efficient management of soil moisture by jute geotextile for cultivation of horticultural crops in red lateritic zone. J. Inst. Eng. (India) AG-91 (1), 21–24.

Nair, K.C.M., Diwan, S.M., Thomas, S., 1996. Tensile properties of short sisal fiber reinforced polystyrene composites. J. Appl. Polym. Sci. 60, 1483–1497.

Pandey, S.N., 2007. Ramie fibre: part II. Physical fibre properties. A critical appreciation of recent developments. Text. Prog. 39 (4), 189–268. http://dx.doi.org/10.1080/00405160701706049.

Raj, R.G., Kokta, B.V., Daneault, C., 1989. Effect of chemical treatment of fibers on the mechanical properties of polyethylene-wood fiber composites. J. Adhes. Sci. Technol. 3 (1), 55–64.

Rajamanickam, R., Hansen, S.M., Jayaraman, S., 1997. Analysis of the modelling methodologies for predicting the strength of air-jet spun yarns. Text. Res. J. 67 (1), 39–44.

Rawal, A., Anandjiwala, R., 2007. Comparative study between needlepunched nonwoven geotextile structures made from flax and polyester fibres. Geotext. Geomembr. 25, 61–65.

Roy, A.N., Ray, P., 2009. Optimization of jute needle-punched nonwoven fabric properties: Part II—some mechanical and functional properties. J. Nat. Fibers 6 (4), 303–318. http://dx.doi.org/10.1080/15440470903339514.

Sengupta, S., 2010. Sound reduction by needle-punched nonwoven fabric. Indian J. Fibre .Text. Res. 35 (3), 237–242.

Sengupta, S., Debnath, S., November 17, 2014. A System for Measuring Electrical Behaviour of Textile Material. Indian Patent File no. 1188/KOL/2014.

Sengupta, S., Sengupta, A., 2013. Electrical resistance of jute fabric. Indian J. Fibre Text. Res. 37 (1), 55–59.

Shah, A., Gong, R.H., Akonda, M., 2014. Biodegradability of nonwoven flax fiber reinforced polylactic acid biocomposites. Polym. Compos. 35, 2094–2102. http://dx.doi.org/10.1002/pc.22871.

Vangheluwe, L., Sette, S., Pynckels, F., 1993. Assessment of set marks by means of neural nets. Text. Res. J. 63 (4), 244–246.

Ventura, H., Moro, M., Ardanuy, M., 2015. Characterization and treatments of oil palm frond fibers and its suitability for technical applications. J. Nat. Fibers 12 (1), 84–95. http://dx.doi.org/10.1080/15440478.2014.897670.

Verdejo, M.F., Lee, K.Y., Zimmermann, T., Bismarck, A., 2016. Upgrading flax nonwovens: nanocellulose as binder to produce rigid and robust flax fibre performs. Compos Part A 83, 63–71.

Xue, D., Miao, M., Hu, H., 2011. Permeability anisotropy of flax nonwoven mats in vacuum-assisted resin transfer molding. J. Text. Inst. 102 (7), 612–620. http://dx.doi.org/10.1080/00405000.2010.504566.

Yilmaz, N.D., Lee, P.B., Powell, N.B., Michielsen, S., 2011. Effects of porosity, fiber size, and layering sequence on sound absorption performance of needle-punched nonwovens. J. Appl. Polym. Sci. 121, 3056–3069.

Further reading

Thangadurai, K., Thilagavathi, G., Bhattacharyya, A., 2014. Characterization of needle-punched nonwoven fabrics for industrial air filter application. J. Text. Inst. http://dx.doi.org/10.1080/00405000.2014.895089.

Index

'*Note*: Page numbers followed by "f" indicate figures and "t" indicate tables.'

D

E

S